Adam Clymer's

Edward M. Kennedy

"*Edward M. Kennedy* is a thorough, careful account of a passionately interesting man and politician." —*New York Times Book Review*

"Clymer reconstructs with impressive and sometimes exhausting detail all the major legislative struggles Ted Kennedy has had in his nearly four decades in the Senate." —*Time*

"Though Clymer believes that Kennedy is 'the leading Senator of his time,' his book is not hagiography. He is blunt about Kennedy's errors and shortcomings, but he refrains from moralizing." —*Washington Post*

"In a silly age of buzz, hype, gossip and garbage, Clymer's book supplies an uplifting end to a somewhat shabby decade of American biography. The triumphs, tragedies, successes, flops and misdeeds are all there in profound perspective. Clymer breaks enough new ground to build a small city on."
—Thomas Oliphant, *Boston Globe*

"In this monumental work Adam Clymer gives us not only a rich, thoroughly researched narrative of Edward M. Kennedy's life but a fascinating glimpse into the inner workings of Capitol Hill. Here, in colorful detail, is an absorbing story of the personalities, maneuvers, compromises, ambitions, and principles that combine and collide to shape legislation in our democratic process."
—Doris Kearns Goodwin

"The quality biography of Senator Ted Kennedy that we have been anticipating has finally arrived. . . . This is a political biography in the classic sense, as if it had been written about the parliamentary careers of Disraeli or Gladstone. It is the portrait of a public man constructed around the framework of the issues he had championed and the struggles he had won and lost. . . . The story that the author tells with great skill and understanding."
—Ross K. Baker, *Philadelphia Inquirer*

"Adam Clymer's book is the crowning work of a superb journalist and assiduous student of American politics. *Edward M. Kennedy* is an absorbing, original, judicious, unsentimental, and fastidiously researched life of one of the grand American figures of the late twentieth century." —Michael R. Beschloss

"Clymer lets the reader do the psychological speculation, while he does the reportorial work, finding details that illustrate how the Kennedy legacy has both blessed and cursed its surviving standard-bearer. . . . As someone who started covering Kennedy in 1961, I learned something new on every other page. This is a masterful work worthy of its subject not least because of its impressive display of shoe leather." —Martin F. Nolan, *Washington Monthly*

"The reader yearning for an intelligent, comprehensive history of Kennedy as ace legislator and political survivor must read this book." —*Newsday*

"Dad and I both served with Ted Kennedy in the Senate. I did my rounds in the ring with him for eighteen years. With laserlike keenness, Adam Clymer goes right to the heart of the man—not the myth. A sensitive portrait of a sensitive man. No shrink stuff, no fluff, no guff. Just Ted."
—Sen. Alan K. Simpson

"Clymer . . . gives us a readable and worthy account of a flawed and fascinating politician." —*People*

"Better than any of many previous efforts, Clymer's book fleshes out Edward Kennedy in his Senate years—in the Bork fight, for example, and the few feats of lawmaking that lend themselves to human drama. But essentially this is a senatorial biography of record, well-researched, well-annotated and surely the definitive source for years to come." —*Atlanta Journal-Constitution*

"The best researched and most comprehensive biography yet of Senator Kennedy." —*Baltimore Sun*

"Clymer says Kennedy is one of the most effective senators in American history. By the end of the book, most readers—even devoted Republicans and not-so-liberal Democrats ought to be convinced that Clymer has made his case. . . . This is an old-fashioned, evenhanded, well-organized biography of interest to anybody who cares about post World War II politics." —*Seattle Times*

"The Kennedy we see in Adam Clymer's book, *Edward M. Kennedy: A Biography*, is a more complicated man than any caricature. Kennedy emerges as one of the most important U.S. senators of the twentieth century, a man with an envious record of achievement, as well as personal failings."
—*Providence Journal*

"There are times when his narrative about the nonlegislative Ted shines . . . Clymer shows a flair for the tempo and demands of political biography."
—*Tikkun*

"His research and his balance are impressive . . . Clymer's major goal was to explore Kennedy's public career. And he does so effectively."
—*Deseret News*, Salt Lake City

"Cymer . . . reveals in his book Kennedy's previously unpublished foreign affairs achievements." —Associated Press

"Adam Clymer makes a good and rather startling case that Sen. Edward M. Kennedy is one of the greatest senators of our century."
—Marianne Means, Hearst Newspapers

EDWARD M. KENNEDY

EDWARD M. KENNEDY

A Biography

ADAM CLYMER

Perennial
An Imprint of HarperCollins*Publishers*

For Jane emily

Who volunteered in 1980

A hardcover edition of this book was published in 1999 by William Morrow and Company, Inc.

HarperCollins books may be purchased for educational, business, or sales promotional use. For information please write: Special Markets Department, HarperCollins Publishers Inc., 10 East 53rd Street, New York, NY 10022.

First Perennial edition published 2000.

Designed by Nicola Ferguson

The Library of Congress has catalogued the hardcover edition as follows:

Clymer, Adam.
Edward M. Kennedy : a biography / Adam Clymer.—1st ed.
p. cm.
Includes bibliographical references and index.
ISBN 0-688-14285-0
1. Kennedy, Edward Moore, 1932- . 2. Legislators—United States Biography. 3. United States. Congress. Senate Biography.
I. Title.

E840.8.K35C59 1999 [B] 99-39768
9733.92'092—dc21 CIP

ISBN 0-06-095787-5 (pbk.)

00 01 02 03 04 ❖ / RRD 10 9 8 7 6 5 4 3 2 1

Acknowledgments

The first time I covered Ted Kennedy, he was starting at left end for Harvard against Yale in 1955. The game was played in a blizzard. In the second quarter, with Yale leading, 14–0, he made a diving catch of a deflected pass for a touchdown. Harvard lost anyhow.

Over more than three decades with the *Baltimore Sun* and *The New York Times,* I have covered many of his successes and failures personally—civil rights and education bills that passed and national health insurance bills that did not, nominations that failed like Frank Morrissey's or that succeeded like Stephen Breyer's, and elections he lost, like the run against Jimmy Carter in 1980, or won, like the victory over Mitt Romney in 1994. So occasionally the sourcing in this book will be my own memory.

But while I have used every newspaper morgue and Presidential library that had useful material, the backbone of the research for this book is more than four hundred interviews I have conducted over the last seven years. This book would not exist without the generous help of more than three hundred people. I have named almost all in the notes, though a few Senate aides, used to going nameless, insisted on anonymity. But, named or not, they all provided important facts, anecdotes, and insights.

Twenty-one of those interviews were with Senator Kennedy himself. He gave generously of his time and encouraged friends, relatives, and present and former aides to talk to me. He decided to cooperate without any certainty about what the book would say, other than to treat him as an important figure of his time, a major lawmaker—using the phrase literally. That is a view which has gained currency in recent years; it was hardly conventional when I started in 1992, shortly after Palm Beach and Clarence Thomas.

But from the first interview, on September 10, 1992, he raised one warning flag, saying that he would see, as time went on, how much he would talk about "personal" matters. That turned out, on May 28 of this year, to mean that he would not talk about Chappaquiddick. An aide said Kennedy found it "too painful" for him.

Dozens of friends helped in this work, in dozens of ways. Kathleen Hall Jamieson, dean of the Annenberg School of Communications at the University of Pennsylvania, is one who stands out. She offered a fellowship at the school in 1995, which enabled me to begin writing, and ever since has provided expertise and encouragement. I am also indebted to her associates Deborah Porter and Deb Williams, and two students who served as research associates, Meghin Adams and Paul Waldman.

Three sets of special friends have provided hospitality on trips to Boston and their own special insights over these years. They are Rick and Patty Stearns, Bill and Lynne Kovach, and Nick and Jenny Littlefield.

Librarians at many institutions went far beyond any normal professional obligations to help. They include Barclay Walsh, Monica Borkowski, and Marjorie Goldsborough at *The New York Times;* Will Johnston, Megan Desnoyers, and Stephen Plotkin of the John F. Kennedy Library; Mike Gillette of the Legislative Archive at the National Archives; Lisa Tuite and her staff at the *Boston Globe;* Richard Norton Smith at the Ronald Reagan Library; Regina Greenwell of the Lyndon Baines Johnson Library; Joan Horne of the El Centro (California) Public Library; and Greg Harness, Nancy Kervin, and Kimberly Edwards of the United States Senate Library. Other libraries where I have been greeted and helped with unremitting efficiency and good cheer include those at the Harvard Law School, the *Boston Herald,* the University of Texas at Austin, and the Minnesota State Historical Society, along with the Gerald R. Ford Library and the Bentley Library at the University of Michigan in Ann Arbor. And in Washington, of course, no research facility compares, despite inadequate federal spending, to the Library of Congress, and I have spent hours using its resources, especially the newspaper microfilm and manuscript collections.

Not all documents are in libraries, though, and it is a rare gift to have personal papers generously made available. I was privileged to see the papers belonging to Douglas Brinkley, Jim Cannon, Greg Craig, John Aloysius Farrell, Dun Gifford, Bob Healy, Nick Littlefield, Robert Mutch, Jeanus Parks, and Arthur Schlesinger.

Several friends read all or part of the manuscript. None bears any responsibility for its flaws, but all helped reduce them, and I valued the perspectives of Jill Abramson, Richard Baker, Barbara Bardes, Geraldine Baum, Linda Greenhouse, Jon Margolis, Mike Oreskes, Richard Stearns, and Robin Toner. Other friends whose insights and advice made it better include Caroline Rand Herron, James Carroll, and Tom Oliphant.

My colleagues at the *Times* have helped in too many ways to detail. Especially helpful have been Joe Lelyveld, Howell Raines, Johnny Apple, Andy Rosenthal, Robert Pear, Thom Shanker, Don Van Natta, Lizette Alvarez, Steve Greenhouse, Larry Altman, Stephen Crowley, Paul Hosefros, Justin Lane, Shana Raab, Mike Kagay, Marjorie Connelly, Janet Elder, John Files, Rebecca Knight, Renwick McLean, Melissa McNamara, and Sarah Smucker.

In Senator Kennedy's office itself I knew I could count on the staff to find time for me on even the busiest crowded days, starting with Ranny Cooper and Paul Donovan, who helped at the beginning and moved on but continued to help, and Carey Parker, Melody Miller, and David Nexon, who stayed and continued to help. Jim Manley, the Senator's current press secretary, has never failed to find things and arrange things in the last hectic months, and this project has always leaned heavily on the press aides, especially Pam Hughes, Theresa Bourgeois, Lorrie McHugh, Kathy McKiernan, Lisa Brenner, Matt Ferragutto, and Will Keyser. While I was helped by dozens of "former" Kennedy aides (a description about as real as "ex-Marine"), David Burke's constant availability and thoughtful guidance stand out.

When the manuscript was completed, I suddenly had to become a picture editor, choosing photographs for the book. I managed that only with the help of Jean Shiner and Raina Filip at the *Boston Globe,* Phyllis Collazo and Marilyn Cervino at New York Times Syndication Sales, Don Bowden at AP/Wide World, Kathy-Ann Williams at Corbis-Bettmann, Allyson Whyte at Camera 5, Alan Goodrich at the Kennedy library, Philip Scott at the Johnson library, Josh Tenenbaum at the Reagan library, David Lombard at the CBS Photo Archive, and Kimberly Oster at the Washington Post Writers Group.

Without taking pages to describe what each of them did, I also want to thank Camille Balletto, Christopher Beam, Nancy Bertrand, Chris Black, Bob Boorstin, Tom Brokaw, Paul Claussen, Loretta Crane, Sam Donaldson, Tom Donilon, Tom Dwyer, Neill Edwards, Kathleen Frankovic, Eric Kocher, Josh Lahey, Joe Laurano, Lorne Michaels, Robert November, Bob Pastor, Dan

Rather, Tim Russert, Don Ritchie, Dot Rogers, Margaret Roman, Julie Rovner, Bob Schieffer, Mike Shoemaker, Greg Stevens, Matt Storin, George Watson, and Keith White. Their contributions helped in so many different ways.

Liz Perle, then of Addison-Wesley and later with Morrow, bought the book at a time when other publishers were asking why a Kennedy book would not be about sex. Her confidence was reassuring.

A newspaperman is often suspicious of editors—even if he has been one himself. Henry Ferris not only allayed those fears, as he edited this book, but offered encouragement and experience. He did what editors always hope to do. He made the story better.

David Black, my dynamic agent, was not only a pathfinder in the mysterious ways of the book business but a creative colleague in this enterprise, helping conceive and organize it. He persisted despite my disproving one of his early, cheery assumptions (comparable to his annual hopes for the Knicks), the idea that newspaper reporters are very good at making book deadlines. He is a great friend.

I suppose it is at least theoretically possible that I might have found others who could have helped nearly as much as Henry and David. But Ann Fessenden Clymer, my wife, has been indispensable. From proofing, transcription (she has an edge on me with Boston accents), and computer skills to encouragement and love, I could not have done it without her.

—Adam Clymer

Contents

Part Three

Part One

CHAPTER 1

A Decision for 1984

The early favorite for the Democratic Presidential nomination and ten members of his family gathered on a gray Cape Cod day on the Friday after Thanksgiving in 1982. His chief aide explained how polls and television advertisements and a battle-hardened campaign staff could win not only the nomination but, with difficulty, the Presidency in 1984. In corduroys, jeans, and sweaters, they faced him in a semicircle of old chairs and couches in the living room of President Kennedy's old cottage at Hyannis Port.

Two of the group, Senator Edward M. Kennedy and his brother-in-law Stephen Smith, had been next door at Robert F. Kennedy's house twenty-three years before when John F. Kennedy's decision to run had been settled. They had been the youngest of the sixteen men who were there that day. This meeting looked very different. Not only were they the oldest men present, but two of the Senator's sisters were there, as were one niece, Kathleen Kennedy Townsend, two nephews, Joseph P. Kennedy II and Stephen Smith, Jr., and his own three children, Kara, Ted, and Patrick.

Larry Horowitz, a physician who had taken over as the Senator's administrative assistant the year before, began by distributing two memorandums. One summarized the nation's political mood and described Kennedy's opportunity. The other forecast that the economy, now in the depths of the Ronald Reagan recession with unemployment over 10 percent for the first time since the Great Depression, would stay bad into 1984.

In November 1982, Ronald Reagan's Presidency seemed vulnerable. Democrats did not expect to be out of the White House for another decade. They had just picked up twenty-six House seats, almost wiping away the Republican gains of 1980. With only a year or two of hindsight, it would be clear this was a false Democratic dawn. Reagan's own popularity came back quickly as the recession ended and a boom began in 1983. He was a master at political theater, capitalizing on moments like the ceremonies at Normandy honoring the fortieth anniversary of the D-day landings.

But at the time, the 1984 nomination seemed worth pursuing. Several Democrats were seeking it, lining up supporters early, readying fund-raising machinery for the January 1, 1983, starting line, after which contributions could be matched by federal dollars. George McGovern in 1972 and Jimmy Carter in 1976 had beaten crowded nomination fields by starting early, and Kennedy had lost in 1980 after a late entry. A timely decision was demanded.

Horowitz spoke of a campaign that was ready to go. In June, Kennedy had told a reporter that in 1980 he had made an important mistake that he would not repeat—"worrying about whether to run or not to run instead of what to do after I ran." So for months a shadow campaign had been at work. In early November, four supporters had taken every state party official they could find at a Democratic meeting in New Orleans to fancy restaurants, wooing them, flattering them, and stuffing them. Kennedy's people had worked effectively with aides to his most likely rival, former Vice President Walter F. Mondale, to tilt the nominating rules away from outsiders. Decisions had been made about pollsters, a fund-raising operation, and clearly defined roles for the people at the top. With his own reelection to the Senate from Massachusetts well in hand all year, Kennedy had spent the fall campaigning for any Democrat who wanted him, picking up IOUs in eighteen states while honing a message of economic injustice and nuclear uncertainty.

Horowitz's presentation argued that the campaign was in place, that the race would be tough, no sure thing, but could be won. The staff was quite confident about winning the Democratic nomination over Mondale and the rest of the Democrats. The general election was another matter, they thought, whether the economy stayed weak or not. The Senator had told Horowitz that beating Reagan would be very, very tough, that Reagan had established a singular connection to voters that was independent of what he actually said. So why make an uphill race? Kennedy felt that a major party's nomination is always worth having, because anything can happen in Presidential politics and no general election is a sure thing.

When some of those in the room listened to the Senator prod Horowitz and draw him out, they sensed that he felt a political obligation to run and make the liberal case in a conservative age. But politics was not all there was to it, or else the family would not have been the audience. For perhaps half an hour, Horowitz laid the case out, describing the political pluses and minuses, such as polls showing him very strong in Iowa, very weak in Illinois.

The young people showed little interest in those matters. At first the cousins seemed to speak for Kennedy's own children when they scoffed at the survey data and the calculations generally. Then the Senator's older son, Teddy, interrupted Horowitz. Teddy was twenty-one and used to speaking up for the others when Kara, now twenty-two, or Patrick, fifteen, hung back. He said he really did not care about the numbers, but about "what is best for the family," for Kennedy himself, and for his children. Danger to him was only an undertone. The fear was of how the demands of a campaign would take the Senator away from them and rearrange their own lives.

For more than a year Kennedy had been torn between the expectations of his political followers and the reluctance of his family. He had fired up his supporters with a speech at the party's conference in June 1982 in Philadelphia, sounding the old-time Democratic gospel and proclaiming, "The last thing this country needs is two Republican parties," grinning while Mummers bands in electric green and shocking pink led a completely unauthorized but carefully planned demonstration. In the fall, he had used his easy race for reelection to the Senate to test campaign commercials intended to blunt the "character issue" raised by Mary Jo Kopechne's death at Chappaquiddick. In September, he told an old ally, Tip O'Neill, the Speaker of the House, that he expected to run, though O'Neill urged him not to. In October, he persuaded Senator Harold Hughes of Iowa to abandon Senator Alan Cranston of California, a real long shot, and back him instead. They talked of campaigning for the Iowa caucuses with just the two of them in a car, with no press corps and no Secret Service.

And while Kennedy was insistently coy when asked if he would run, most reporters and most politicians put that down as tactical teasing, not uncertainty.

But it wasn't. The reason he assembled his family, chartering a plane to fly Kathleen Kennedy Townsend and Horowitz up from Washington to join those who had been there for the holiday dinner the day before, was the aftermath of the 1980 run against Jimmy Carter. Not the impact on himself. If anything, the chance to show he could run a better race than the one that found its voice only after it had lost any chance to win was an argument for running. But he came to understand how the race affected his family only after it was finished and they told him about it. For his children, his nieces and nephews, the 1980 campaign had been not only frightening because of the assassination threats against him, but oppressive because they were not spared the hatred he stirred.

Arranging the meeting as he did was not the calculation of a politician driven to seek the Presidency, a Richard Nixon or a Bill Clinton. Since his failed 1980 race, Kennedy had reestablished himself as a power in the Senate. He had feared losing that status after his Presidential defeat in 1980 and the Republican capture of the Senate. But even without the power of a chairmanship, he still held pseudohearings called forums which drew television cameras. With Mark Hatfield, the Oregon Republican, he led a campaign for a mutual nuclear freeze with the Soviet Union. Despite Republican control of the Senate, he was now its unquestioned leader on civil rights legislation, maneuvering the 1982 extension of the Voting Rights Act into law despite Reagan administration opposition. He did it as he had passed laws for nearly two decades, by finding Republicans to work with and sharing the credit, or even letting them have it all. So when he considered whether to run, he did not have to depend on a Presidential candidacy to define himself. He believed he could be a great Senator.

He was not likely to be surprised by what he heard that morning. Three weeks before, he had gone sailing with his children, and after that meeting he told Bob Shrum, his press secretary and speechwriter, to delete a hint about running for President from his Election Night victory statement.

As the meeting went on, Steve Smith and the Senator's sisters, Jean Smith and Patricia Lawford, pointed to both pros and cons. The young people were generally critical and told again about the hateful heckling they had encountered in 1980, how they had been shaken by encounters like one Kara had had with a Catholic priest who introduced her at a candidates' forum by saying her father was unfit to be President. Kathleen said she doubted that the race could really be won and wondered whether the strain would be worth it for just a gallant defeat.

After more than two hours, his nephew Joe told him, "Teddy, what it all boils down to is it's a personal decision. I feel for you. You have to make it." The Senator ended the meeting and walked out onto the porch with Horowitz. Outside he said simply, "That's it."

Kennedy spent the rest of the weekend making sure, drawing out his own children. He was most concerned about Patrick, whose severe asthma flared up with stress, and whom he had called every night he was out campaigning in 1982, reassuring him he was all right. The Senator's divorce from Joan Kennedy was an added strain on Patrick, who had moved to school in Boston to be near her. The next night the Senator and Patrick walked in together to the basement movie theater in the main house in the compound. They sat for a while, with Patrick's head on his father's shoulder, and then wandered off together before the movie was over. Everything was all right, surmised one family friend; Kennedy would not run.

On Monday, Kennedy told his staff that his children did not want him to run. He said his decision was not quite final because he still wanted to talk to his sister Eunice, the family's strongest advocate of another campaign. But when the meeting ended, Shrum started writing a withdrawal statement. The next day, when Kennedy told his aides that the decision was now final and a statement was needed, Shrum handed it to him.

No rumors of his plans leaked out until Tuesday afternoon. Then they flared because everyone whom reporters called in his office was unavailable. Dick Drayne, a former Kennedy press secretary now working for CBS, compared notes with newspaper friends and said that while he didn't know anything, the situation somehow "smelled" like a withdrawal to him.

Aides collected at his house in McLean, Virginia. Jack English, a Long Island lawyer and a fixture of Robert Kennedy's campaigns and the Senator's race in 1980, flew in. He told the aide who picked him up at National Airport, "I know what this is. I've been through this drill before with the other Kennedys. He's not running." After talking privately with Kennedy, he started telephoning people. So did Kennedy, as aides dialed calls and handed him the phone. He called Mondale, and joked that after campaigning with the former Vice President in Boston that fall, he knew he couldn't beat him. He called hundreds of people.

A reporter went to Kennedy's home in McLean, squeezing among ten parked cars to get to the doorbell. A maid asked his name, went away, and returned to say no one was home.

One of the operatives Kennedy called, pollster Patrick Caddell, reported the decision on ABC's *Nightline* and said, "I'm fairly sure the decision he made was not on political grounds but on family and personal grounds." Several newspapers had the story, with varying amounts of detail and conviction about his motive. The next morning, Wednesday, December 1, 1982, Kennedy held a press conference with his children on hand, to say that while he thought the political case for running was "strong," the family case against it was stronger.

Few doubted his explanation, although William Safire, a *New York Times* columnist, wrote that announcing he was not running was "a smart tactic" for someone hoping to keep his popularity up and prevail at a muddled convention, or positioning himself for a race in 1988. The argument was too clever; the only way to be nominated since the primary system developed in the seventies is to go out and run. But the Safire column reflected an enduring Kennedy problem. For years hardly anything he did, from his Senate votes to his travels, was taken seriously on its face, rather than as part of a grand strategy to recapture Camelot. Even when he remarried in 1992, that decision was sometimes derided as shrewd strategy to win a tough Senate race in 1994.

But people close to him had no doubts. Melissa Ludtke, a Cape Cod neighbor who was especially close to his children, wrote him Wednesday morning to say she shared the happiness his choice had given his family. She underlined how his children's opposition was based not just on fear, but on how they relied on him. Now, she wrote, "Patrick will have his ol' Dad to pal round with, to discuss the Magna Carta with, to find that shoulder to rest his head on when exhaustion and emotions bring the fears too quickly. Kara, of course, will have her Dad to point out all the cute looking wind surfers to, but more importantly will have her own time to gain a self confidence which makes her more radiant every time I see her. Teddy—well he needs someone to trim the mainsail and get some weight to windward if he's going to drink champagne from that trophy on Labor Day in 1984." She concluded, "It seems to me you've already won a lot more than the Presidency."

The statement he made on Wednesday morning did hold out the possibility that he might run in 1988, despite Patrick's pleading that it come out. But he never seriously approached another candidacy. After all, he had told three *Boston Globe* reporters in 1981 that life in the Senate was "fulfilling and satisfying and challenging," and that he could certainly spend his life there. For the first time since Robert F. Kennedy was shot in 1968, Edward Moore Kennedy was first of all a Senator, not a potential President: a lawmaker of skill, experience, and purpose rarely surpassed since 1789.

CHAPTER 2

Born to a Dynasty

When Edward Moore Kennedy was born on February 22, 1932, the Kennedys were far from being a leading political dynasty. If there was any remote thought that a Kennedy might run for President someday, it would have focused on the father, Joseph P. Kennedy. But in 1932 the family was just taking its first steps back from political exile, first imposed by the voters and then voluntary. It was not just that the family had abandoned Boston for Bronxville, New York, though Rose Fitzgerald Kennedy returned to Boston and a familiar doctor for the birth of her ninth child. More important, Boston's voters had last elected one grandfather, Mayor John F. "Honey Fitz" Fitzgerald, to office in 1910. The other grandfather, P. J. Kennedy, had not even been a candidate since being upset by an unknown rival for street commissioner in 1908.

After that defeat in 1908, P. J. Kennedy's son, Joseph P. Kennedy, had scorned the ward politics that had built those careers. He concentrated on making money in banks, movies, and stocks. He shrewdly got out of the stock market early in 1929, preserving his fortune. But as the Depression got worse, Kennedy was restless, and in 1932 he fastened on the Presidential candidacy of Franklin D. Roosevelt. He raised money, and he helped win over William Randolph Hearst, the newspaper baron who controlled Democratic National Convention delegates in California and Texas. He even traveled with Roosevelt, and he expected to be rewarded with a major job in

Washington. After a frustrating year of waiting—he kept busy manipulating stocks—in July 1934, Joe Kennedy was made chairman of the new Securities and Exchange Commission, which set out to regulate the stock market. In that post, then as head of the Maritime Commission, and finally as ambassador to Great Britain, he started growing Presidential ambitions of his own.

When Teddy was born in 1932, John F. Kennedy wrote to his mother from the Choate School: "Dear Mother, It is the night before exams so I will write you Wednesday. Lots of Love. P.S. Can I be Godfather to the baby." The earliest recollections about Teddy are of a chubby little boy, eager to make people laugh. His own first memories are of being taught to swim and sail by his oldest brother, Joe Jr., who was seventeen years older, of pillow fights and learning to ride a bike and throw a football with Jack, who was fourteen years older, of his father reading him the comics, and of his mother teaching him the lessons of the Sermon on the Mount and spanking him for walking home alone from nursery school instead of waiting to be picked up.

The exception in this bustling, competitive family was his oldest sister, Rosemary, who was mentally retarded. The other brothers and sisters, especially Eunice, the fifth child, made sure to include her at dances or sailing. Rosemary's littlest brother played dodge ball with her, and got a first lesson that some people were different and had special needs. Rose Kennedy once wrote that she thought her other children's experiences with Rosemary taught them "much of their concern and desire to help those less fortunate." As a veteran Senator many years later, Ted Kennedy said Rosemary's condition was the first influence pointing him to a political concern over health care.

In 1937, Roosevelt asked Joe Kennedy to become ambassador to the Court of St. James's, the first Catholic in that post. In March 1938, when the family arrived, the London newspapers were ecstatic. Six-year-old Teddy was a particular favorite, walking his dog Sammy, or costumed as a New England Puritan, or at the opening of the zoo. He patted a zebra, which clamped its teeth on his arm; "I thought I was being eaten by a zebra," he recalled.

Teddy and Bobby, who was six years older, went to two day schools, first Sloane Street, then Gibbs. Teddy solemnly got his father's permission to get into a fight, since he had been trying very hard to be good but a boy in his class named Cecil was picking on him. Teddy and Bobby took over the fun of running the elevator at the ambassador's residence, a grand six-story building at 14 Princes Gate, which the financier J. P. Morgan had given to the American government. Occasionally his father gave him a very special treat, taking him horseback riding before school.

In March 1939, when Teddy was seven, his father was designated as Roosevelt's representative to the coronation of Pope Pius XII at the Vatican. Mother, father, and eight children arrived, requiring twice as many seats in St. Peter's as any other delegation, but extra chairs were squeezed in for them. The next day the Pope, who had visited the family in Bronxville while still a cardinal, received the Kennedys and gave the children rosaries. Teddy told reporters, "I wasn't frightened at all. He patted my hand and told me I was a smart little fellow. He gave me the first rosary beads from the table before he gave my sisters any." Two days later, Teddy went to the Pope's private chapel and received his First Communion. Even though he was an American, he was the first young Catholic to receive that honor from this Pope.

Joe Kennedy's breezy, successful early months as ambassador became troubled as Europe approached war. He was close to Prime Minister Neville Chamberlain, an earnest, small politician. Chamberlain became synonymous with the word "appeasement" when he returned from a visit to Hitler in 1938 after abandoning Czechoslovakia and proclaimed "peace for our time." Chamberlain's hopes collapsed as Germany invaded Poland on September 1, 1939. Ambassador Kennedy stayed behind at the embassy when Joe Jr., twenty-four, Jack, twenty-two, and Kathleen, their nineteen-year-old sister, went to the House of Commons. They heard Chamberlain declare war, and Winston Churchill, not yet part of the government, say Britain was not "fighting for Poland. We are fighting to save the whole world from the pestilence of Nazi tyranny and in defense of all that is most sacred to man." Joe sent Rose and the children home, and then on a visit to Boston in November told a church banquet, "There's no place in the fight for us," a statement that finished his effectiveness as ambassador.

Roosevelt began a secret correspondence with Churchill and sent other officials to London to negotiate behind Joe Kennedy's back. In May 1940, when Churchill, the new prime minister, told the Commons he had nothing to offer but "blood, toil, tears and sweat," Joe Kennedy wrote a friend in Boston, saying, "The longer it goes on, the worse it will be for everybody." He lingered on as ambassador during 1940, resigning and returning for good just before Election Day. Earlier that year, Joe Jr., with Honey Fitz's help, was elected a delegate to the Democratic Convention. He was pledged to former postmaster general and Democratic Party chairman Jim Farley. When

Massachusetts's delegates were picked, Roosevelt had not yet announced that he would seek an unprecedented third term as President. After Roosevelt announced, Joe Jr. stuck to his promise and voted for Farley.

Teddy took a tiny part in that fall's campaign, not sounding nearly as isolationist or defeatist as his father. Back in Bronxville and a pupil at the Lawrence Park West Country Day School, he debated Westbrook Pegler II, the nephew of a very conservative columnist. "Today we are facing a crisis," Teddy solemnly proclaimed as he spoke for Roosevelt against Republican Wendell Willkie. "Across the seas, nation after nation whose belief in democracy has been as strong as ours, has gone down before the Nazi war machine. At a time like this, we must get things done whether it is good for private business or not." Then, sounding a bit like some of his father's London friends, he said, "We must borrow a little of the Nazi method. For this task, we must step on the toes of some people and institutions which Mr. Willkie favors." Even so, Willkie carried the school. But Roosevelt won a comfortable victory nationally the next day, on November 5, 1940.

A few weeks earlier, Teddy had received a remarkable letter from his father, lonely and frustrated in London and plainly missing his family. The Battle of Britain was on, with German bombers hitting London nightly, and his father wrote:

> I don't know whether you would have very much excitement during these raids. I am sure, of course, you wouldn't be scared, but if you heard all these guns firing every night and the bombs bursting you might get a little fidgety. I am sure you would have liked to be with me and seen the fires the German bombers started in London. It is really terrible to think about, and all these poor women and children and homeless people down in the East End of London all seeing their places destroyed. I hope when you grow up you will dedicate your life to trying to work out plans to make people happy instead of making them miserable, as war does today.
>
> I was terribly sorry not to be with you in swimming at Cape Cod this summer, but I am sure you all know I wanted to be, but couldn't leave here while I had work to do. However, I am looking forward with great pleasure to our swims at Palm Beach this winter.
>
> I know you will be glad to hear that all these little English boys your age are standing up to this bombing in great shape. They are all training to be great sports.
>
> I thought you might be interested to know, and you might tell this to

all your brothers and sisters, that the other night when I was going to the Concert to hear some music at Queen's Hall and afterwards going to have dinner with Duff Cooper, the Cabinet Minister of Information, I dashed home to 14 Prince's Gate, put on my dinner jacket and then left to go to the concert. When I got to Queen's Hall I found out the concert was cancelled, and then I went back to my office, and after sitting there three-quarters of an hour I noticed by the merest chance that I had forgotten to shave for a couple of days and I was going out to a dinner party without having shaved. So you can see how busy I am. I am sure everybody will laugh at this.

Well, old boy, write me some letters and I want you to know that I miss seeing you a lot for after all, you are my pal, aren't you?

Letters like this, which the children treasured, were one way of diminishing the impact of their father's long absences. When he was around, in London, or at Hyannis Port or Palm Beach or Bronxville, he could always find moments for one child at a time. He talked of the importance of public service, and he led the older children in discussions of current events at the dinner table. Teddy listened with awe to talk about Russia or Poland from brothers who had been there. "I remember a lot of exciting talk about these and other places, and I guess I knew that when I was old enough I'd get my chance to travel on my own, too," he recalled in 1962.

Joe stressed competition. He would say, "We don't want any losers around here. In this family, we want winners." In 1940, that first summer apart, letters from Hyannis to London record that the lesson had taken. In July, thirteen-year-old Jean wrote that eight-year-old "Teddy has gone out to race today. He is first in his series." And in August she reported, "Teddy has won one cup," and the two of them shared another for jointly winning a long-distance race.

The Kennedy children did not just compete with outsiders. The boys were spaced too far apart to compete with each other directly, but they could test themselves against the daredevil image of their older brothers. On the Riviera in the summer of 1939 the older boys teased Teddy until he followed them in jumping off a rock at least twenty feet above the water. "I was pretty scared," he recalled, "but they all seemed to be doing it." In Bronxville they taught him to jump off the garage roof with a parachute. In 1938, when Teddy was six, his big brother Joe threw him out of a boat into Nantucket Sound when he failed to respond to a sailing command he didn't understand.

Teddy was frightened for a moment, then Joe dove in and heaved him back into the boat.

If Joe Kennedy's values revolved around winning, Rose Kennedy's centered on her Catholic faith. She took the children into churches during the week, so they would learn that religion was not only for Sundays. She listened to their catechism lessons and emphasized the teaching of Saint Luke: "of everyone to whom much has been given, much will be required." She packed her teenagers, even if they would rather be sailing, off to weekend religious retreats once a year. And at Sunday dinner, after Mass, she would teach them the central truth of her religion; she wrote in her autobiography: "Faith, I would tell them, is a great gift from God and is a living gift, to sustain us in our lives on earth, to guide us in our activities, to be a source of solace and comfort, so we should do everything we can to strengthen its roots, to nourish it, and to help it grow and flourish, and try never to lose it."

Both parents set rules. If you were late for a meal, you started on whatever course was being served. Allowances were not lavish. Once Teddy was given permission to take some of his father's candy off to boarding school; when his father saw that he had helped himself to two large boxes, he was sent off without any. And though Teddy was something of a pet, there were continual efforts to keep the kids from getting stuck up. Eunice, the middle child of the five, once said that came naturally: "You can't be one of nine children and think you're the center of the universe." The older children, all but Jean and Teddy, had been to public schools, which his father praised for showing them that "some plumber's son will be smarter than you, and some carpenter's son will be tougher at football."

Of the Kennedy boys, Robert was closest to Teddy in age. Occasionally he was annoyed at being in charge of Teddy. At Portsmouth Priory, a strict Benedictine school where Rose had enrolled Robert in 1940, nine-year-old Teddy appeared in May 1941. He was too young for the school's lowest grade, but his mother did not know where else to send him. Teddy, as a fat kid, was a target for bullies. One day a boy named Plowden was twisting his arm, demanding that he say, "The Plowdens are better than the Kennedys." Robert happened by, and Teddy asked for help. Robert just said, "You've got to learn to fight your own battles," and went on his way. But a year or two later, when Bob had gone on to Milton Academy and Teddy was enrolled at the Fessenden School, not too far away, Robert would call him up to check on his studies. Or he would launch winter expeditions down to the Cape, where they stayed in the unheated garage of the boarded-up family house, cooking their own food and sleeping on cots, and having a wonderful time on their own.

Jack was a special companion at times during the war. Before his heroics in the South Pacific, when he led the crew of his sunken PT-109 to safety, Jack smuggled his ten-year-old brother onto a torpedo boat base. And after he returned to the Cape on convalescent leave in 1944, Jack read Stephen Vincent Benét's *John Brown's Body* to Teddy, and urged his godson to read political lives, like Edmund Burke's and Daniel Webster's.

But with Joe and Jack in the Navy and Robert itching to get in, Teddy spent more time with his sisters, especially Jean, who was just five years older. With the family, they would watch movies in the cellar projection room at the house in Hyannis, and staggered bedtimes would be arranged by sending Teddy clumping up to bed when characters first began to hold hands, followed by Jean at the first kiss.

After he returned from London, Teddy saw little of Joe Jr. He was twelve when Joe's munition-loaded plane blew up over the English coast. Ted's memory of August 12, 1944, was of two priests coming to the house at Hyannis Port and spending some time with his father. "Then he came out to the sunporch and said, 'Children, your brother Joe has been lost. He died flying a volunteer mission.' His voice cracked and as tears came to his eyes he said in a muffled voice, 'I want you all to be particularly good to your Mother.' "

But his oldest brother still played an important role in who Teddy came to be. President John F. Kennedy reflected on Joe's influence on all of them one day in April 1963 when Bobby was attorney general and Ted a freshman Senator.

The President told James M. Cannon of *Newsweek* magazine that Joe "set an example in an atmosphere which meant pressure. Whatever there was in me, that pressure helped to bring it out." He continued, "He set a standard of activity and commitment. I followed. In college I had something of a mediocre record, you might say, until that first book [*Why England Slept*]. The book, which I might well not have written except for pressure to do well, brought out latent ability that I had. . . . So if you say what was Joe's influence, it was pressure to do your best. Then the example that Joe and I had set put pressure on Bobby to do his best.

"The pressure of all of the others on Teddy came to bear so that he had to do his best," he said. "It was a chain reaction started by Joe, that touched me, and all my brothers and sisters."

The house in Bronxville was sold in 1941, and after that Joe and Rose Kennedy lived chiefly in Hyannis Port in the summer and Palm Beach in the winter. Those periods had no connection to school years, and Teddy was sent to one boarding school after another. Some of the most important lessons he learned in those years came not in an unpredictable series of classrooms but from his grandfather, Honey Fitz.

As kids, his two oldest brothers had seen a lot of their "Grampa Fitzgerald." Once the family moved away from Boston in 1927, the contact was less frequent. But in the fall of 1943, eleven-year-old Teddy was sent to the Fessenden School in West Newton as a sixth-grader. That year and the next he often made the one-hour trip into Boston on Sundays for lunch and a visit with his grampa. Before lunch, Teddy would go up to the suite the eighty-year-old former mayor and Congressman kept in the Bellevue Hotel near the State House. Newspaper clippings would be pouring out of his pockets, and Honey Fitz would be on the phone, offering condolences on deaths and hopes for quick recoveries to the sick. At lunch in the dining room, the former mayor was still a celebrity, and people would come up to talk about how business was in the North End, about discrimination against the Italians or the Irish, about a schoolteacher who had quit to go to New York for more money. They talked some about Ireland. After lunch, his grampa always took him through the kitchen to shake hands with the cooks, and then they would walk around Boston, seeing the place on the Common where British soldiers drilled and the church where William Lloyd Garrison had preached against slavery, and they would look at the decaying port, now far behind New York's in importance. He was learning about real problems from someone who was not preaching.

At Fessenden, Teddy was popular, played football, sometimes made the honor roll, and was frequently paddled when he got into trouble. Once he and a buddy, Danny Burns, tried to steal some candy from a geography teacher, Mr. Moore, who was given to cracks like "If Mississippi were in New Jersey, what would Delaware?" (The music-hall answer was "Idaho, Alaska.") Mr. Moore was nearly blind, so the boys thought he would be an easy mark. They snuck into his room through a shared bathroom one night. But he heard them and caught them. Teddy's punishment was a night sleeping in the bathtub.

He was among a handful of Catholics at Fessenden, sent off to church in a station wagon early each Sunday morning. But his mother felt strongly that the experience of Catholic schooling was important to building character and religious understanding, and so in 1945 he was sent to Cranwell, a Jesuit prep school at Lenox in western Massachusetts. He played football there, too. But though he played end, he was slow. He once challenged a priest in his robes to a footrace, and lost. He spent only the eighth grade there.

His longest stretch at any one school was the four years he spent at Milton Academy. He was there for the ninth through twelfth grades, starting in the fall of 1946. The school was just south of Boston, and Robert had spent two years there before going to Harvard and the Navy. Here Ted was an adequate student, always in the middle of horsing around if some was to be found. His letters to his parents reported falling through the ice playing hockey and struggling with Latin his first year, later serving Mass on Sundays and a call from "Kick," his older sister Kathleen, about three weeks before she died on May 15, 1948, in a plane crash in France. Ted went dancing at the Totem Pole, a round dance hall with tiers of couches and twinkling lights reflected off a revolving mirrored ball—"about the biggest dance place in the state for teen age people," he wrote. He was in the debating team, the drama club, and the glee club, and he belonged to a group called the "Boogies," which thrived on stunts like racing around the block in underwear.

Though he never held a school office, his senior yearbook ranked him a distant third as "Class Politician." He played on the tennis team, but his main sport was football. He was a strong end, a bit pudgy but hardworking. His big brothers had taught him always to run back to the huddle, not walk, so the coach would never think he was tired and send in a substitute. Milton's coach, Herbert Stokinger, recalled nearly a half century later that he held on to balls he could reach: "Ted really had sticky fingers. He was a good all-around football player. He was able to do his blocking assignments well." Ted caught a lot of passes, one for a touchdown in his final game, a 25–0 victory over Noble & Greenough.

In Ted's first two years at Milton, Robert was back at Harvard, and Ted would go to his football games, and then go down to the Cape with him and his pals. "I sort of hung out with them as a younger brother who was full of sort of stars in his eyes, respect for your older brother and also about Harvard and playing football," he recalled. Like his three brothers and many of his Milton classmates, Ted went on to Harvard. His Milton grades were ordinary, but in 1949 a diploma from a prep school like Milton was enough for admission to Harvard.

In the fall of 1950, he went out for freshman football, made the team, and made a lifelong friend in John Culver, from Cedar Rapids, Iowa. Culver was much more of a natural athlete than Ted, but he was impressed with how hard Ted practiced, and how unassuming Ted was, coming from a family whose famous patriarch paced the sidelines in a beret.

Ted remembers not having a very good time his freshman year, and one day in the spring of 1951 he cheated on an exam.

He was having trouble with Spanish, and worrying that if he didn't get at least a C-minus, he would be ineligible for football the next fall. His account is that he was going for a walk the night before the exam when a friend asked how he was doing. Ted said he was worried. They went over to the room of another friend, Bill Frate, and they asked Frate, who was good in Spanish, to take the exam for Ted. He agreed.

That night there was an element of joking and risk, something of a dare. But the next morning, Ted went by Frate's room again. Frate was sleeping in. He was awakened, and he cheerfully marched off to take the final exam in Ted's place.

But when Frate got up at the end of the exam and turned the blue books in, with Ted's name on them, the graduate student supervising the exam stared at him. He knew him as Frate, not Kennedy. Perhaps no more than an hour later they were both called to the dean's office and expelled. They were told they could apply for readmission in a year or two if they behaved themselves. That was not the most severe punishment Harvard imposed, but it was typical in serious cases.

His father, who had once warned Ted to be careful because he was the kind of person who would always get caught, was initially calm. The next day, though, Ted recalled, "He was absolutely wild and went up through the roof—for about five hours. From then on he was calm. It was just 'How do we help you?' " John F. Kennedy said a few years later that his father's toughness was critical: "If it hadn't been for that, Teddy might be just a playboy today. But my father cracked down on him at a crucial time in his life."

Ted spent three or four weeks figuring out what to do next. Then he did what Harvard often urged on students in trouble. He enlisted in the Army on June 25, 1951. The Korean War was a year old, precisely, for on June 25, 1950, Communist North Korea had invaded South Korea. The

United Nations condemned the attack. China entered the war that fall as UN forces, commanded by General Douglas MacArthur, swept north toward its Yalu River border with North Korea. On the morning Ted enlisted for four years, the newspapers reported truce proposals but also severe Communist attacks, along with twenty-eight New England casualties. Joe Kennedy, having lost one son to war, was furious again, asking Ted, "Don't you ever look at what you're signing?" In those days, most young men understood the laws about military service, even if they had a student deferment. But Ted did not know that he had the option of volunteering for induction, in effect getting drafted when he chose and for the draft's two years. After Joe intervened, Ted got the shorter term.

Once Ted had finished basic training at Fort Dix, New Jersey, he tried to get into Army Intelligence, even though it would mean longer service. He went to Fort Holabird in Maryland in November, but was dropped after a few weeks without explanation. He believed the cheating incident was the reason. Next he went to Camp Gordon in Georgia in January for three months of military police training. He set a camp record for pull-ups, chinning himself on a bar more than forty times. He got extra duty digging latrines after some Sterno cans he brought to the field to keep warm set fire to his sleeping bag. Another time he played basketball until dark and had his rifle disassembled, cleaning it by Sterno light, when an officer caught him and his tentmate, Matty Troy. More latrines.

After Camp Gordon, Ted went to Paris in June 1952 as part of the honor guard at the headquarters of the North Atlantic Treaty Organization. Joe Kennedy tried to get Troy the same assignment, but he was an inch too short, and ended up in Korea. Kennedy traveled a lot on weekends and had an easy tour of duty, returning to Fort Devens for discharge in March 1953. Troy, who became Democratic leader in Queens and frequently announced that he was leading movements to draft Kennedy for President or Vice President, was his only lasting Army tie.

Ted returned to Cambridge and took two summer school courses. He was readmitted to Harvard College in the fall of 1953. He was a sophomore and his original classmates were now seniors, though he kept up with them. The next spring, around Culver's graduation, the two of them drove to Washington to watch Robert's work in the Army-McCarthy hearings. They stayed with Jack and Jackie, and she asked Ted to coach her in some political questions she could ask her husband to show off her knowledge of his world.

This time Ted worked harder at his studies, getting an A-minus in the

Spanish course, earning a few more B's than C's, and thriving in a couple of courses. One was public speaking and another was a government course taught by Arthur N. Holcombe, a Harvard legend who had taught his father and all three of his brothers. The course's account of the struggles, debates, and compromises of the Constitutional Convention in 1787 fascinated Ted, and he took another popular but tough course on what the Constitution had become, taught by Robert G. McCloskey.

Outside of class, he attended Mass daily during Lent but also found time for stunts like bets about whether he could hit a golf ball across the Charles River behind Winthrop House. Once in Bermuda with the rugby team, he tied up a cow in Frate's room, with the predictable messy result. He coached basketball at a settlement house in Boston's South End. He made friends who were important all his life.

But football was his consuming interest. He was on probation his first term back, so he could not compete for Harvard. Instead he starred for his Winthrop House team. The day before the 1953 Harvard–Yale game in New Haven, the Harvard houses played against Yale's residential colleges. Ted asked Robert if he would come up to see the game. Robert was bored, working in New York City as his father's aide on a commission on reorganization of the executive branch headed by former President Herbert Hoover. He said he would rather play than watch. So Ted got an extra uniform for him, and Robert changed in his car. He played end and Ted played next to him at tackle, and Winthrop beat Davenport, 6–0. Ted's teammates were pleased at the spirit of an alumnus, class of '48, sneaking into the game and playing fiercely. John Ferry, the master of Winthrop House, was somewhat puzzled by the vaguely familiar face and asked Robert, "Aren't you graduating one of these years?"

The next fall, in 1954, Ted's probation was over and he was eligible to play for the varsity. For most of the year, he was a second-string end. He hoped to get in for at least four minutes in the Yale game, for that would give him sixty minutes for the season and win him a varsity H. Neither Jack nor Joe had lettered, although Robert, despite a leg broken early in the season, won his H in 1947. So when Ted was sent in just before halftime, he expected to play for a few minutes and earn his letter. But on his first play, Yale made a long gain around his end. He was taken out and did not play again. Harvard pulled out the game in a downpour, 13–9, with a touchdown with five minutes to play.

In 1955, things were different, better for Ted and worse for Harvard.

Jack came to many of his games. Ted started all season long, ensuring his varsity letter early in the season. He caught a touchdown pass in a victory over Columbia. Against Yale, in a blizzard, he scored when the ball bounced off the hands of Dexter Lewis and Ted caught it in the end zone for a touchdown. That cut Yale's lead to 14–7 in the third quarter. But then Yale then scored on an intercepted pass, and won, 21–7.

In 1954, Ted had been sure he had disappointed his father, but he knew he was supposed to be happy in the locker room because the team won. In 1955, Joe had brought a railroad carload of friends from Boston and another from New York to the game. He was euphoric over Ted's touchdown and impervious to Harvard's defeat. When his brothers and his father happily swarmed into the locker room, Ted had to restrain their glee out of consideration for his glum teammates, who had ended a 2-win, 7-loss season with a defeat in The Game.

Ted graduated in 1956. While Jack was nearly being nominated for Vice President at the Democratic National Convention in Chicago—for the opportunity to share with Adlai Stevenson a predictable landslide defeat at the hands of President Eisenhower—Ted was spending the summer in Europe. He also ventured down for a week in Algeria, where the rebellion against France was under way. He talked to his brother later about the inevitability of Algerian independence, a subject on which John F. Kennedy made a splash in the Senate the next year.

His college grades—and perhaps his expulsion—were not up to the Harvard Law School's standards, so he followed Robert to the University of Virginia Law School in Charlottesville. Jack urged him to go there, because it was near Washington and the school had an interest in public affairs. On registration day he met John Varick Tunney, son of the former heavyweight boxing champion Gene Tunney, who was a friend of his father's. They were instant friends, rented a farmhouse in the country, and were practically inseparable until each married in his third year. They threw two memorable parties for their entire class of about 150. Ted drove wildly and got several speeding tickets, once outrunning a police car. When Robert spoke at the law school he needled his brother, saying their mother had asked him to find out "what side of the court my brother is going to appear on when he gets out of law school, attorney or defendant." Ted spent hours studying, earning adequate, unspectacular grades.

Ted did well in two outside activities, matching Robert in one and outdoing him in the other. In his third year, he was president of the Student

Legal Forum, as Robert had been. The job was to bring important speakers to Charlottesville. Besides Robert, now investigating labor racketeering for Senator John McClellan of Arkansas, Ted got Supreme Court Justice William J. Brennan, Senator Prescott Bush of Connecticut, and Walter Reuther, head of the United Auto Workers. One notable guest was Senator Hubert H. Humphrey of Minnesota, the leading liberal in the Senate. Ted had come to Washington looking for speakers, and Jack brought Humphrey out of a Foreign Relations Committee session and asked him, in front of Ted, to speak.

Humphrey's appearance was on March 12, 1959, a year before the Wisconsin Presidential primary, and Ted's big brother wanted to know how his obvious rival had done in Charlottesville. "They're still cheering him down there, Jack," Ted replied, "and you couldn't get in the hall." His brother replied, "That isn't what I want to hear." Humphrey, who always treated Ted with an affection he did not hold for Bob, wrote him, "That was a splendid and enthusiastic audience. There is no finer compliment that can be paid to a speaker than to have an attentive and large audience. You provided both."

But his greatest distinction came when he and Tunney won the moot court competition over all their classmates. Tunney recalled the competition in 1995. "It had tremendous meaning in the law school because it was a competition that everybody in the class did participate in. You had to do it," he said, "outside of your studies, your normal studies, but it was something that was looked upon with great favor by the community. . . . We worked *extremely* hard at writing our briefs and preparing our arguments and practicing our arguments. And as a result of that, I think to the surprise of some people in the school, we won the first three competitions in our junior year, which then kept us alive for the senior year." Ted's syntax, which has confused reporters for decades, did not fail him.

Their last year, they won two more rounds and reached the finals, where they argued that a corporation was denied its rights of free speech by a provision in the Taft-Hartley Act that prohibited it from making campaign contributions. Stanley Reed, a retired Supreme Court justice, was one of the three judges, and Tunney and Kennedy found an old opinion of his that they used to help make their case. Reed voted with them. Lord Kilmuir, then Great Britain's chancellor (vaguely equivalent to chief justice and attorney general combined), voted against them. But they won with the deciding vote of Clement F. Haynsworth, a judge of the U.S. Fourth Circuit Court of Appeals. The victory earned Tunney offers from Wall Street law firms. For both of them, winning over all 150 classmates was a previously unknown level of academic distinction. And for Ted, there was also the satisfaction of winning a competition at which Robert had not excelled.

In the fall of his second year, Ted met Joan Bennett, a senior at the Manhattanville College of the Sacred Heart in Purchase, New York. The family had given a gymnasium to the college, where Rose, Eunice, and Jean had all studied. Ted filled in for Jack as the speaker on October 28, 1957. Joan skipped the speech to work on an English paper. But a roommate warned her that the nuns would give her a demerit if she avoided the tea that followed, and she hurried over to meet Francis Cardinal Spellman, and also the Kennedys. Jean had met Joan once before and said she wanted to introduce her to her little brother. Joan recalled, "I'll never forget that moment. I expected to see a small boy. Instead I found myself looking up at somebody 6 feet 2 inches and close to 200 pounds. And, I must say, darn good-looking." Joan caught his eye, too. Blond and statuesque, she had won beauty contests and modeled for television commercials.

They fell for each other, and had a few dates that winter and the next spring, brunch at the Sherry-Netherland Hotel in New York City, a couple of ski weekends and visits to Charlottesville. Joan singled out a Christmas trip to Stowe, when she skied for the first time, as the moment she fell in love with Ted because "he was always there to pick me up and urge me on, with such patient, sweet encouragement." Then Joan was invited to Hyannis Port in June. Rose, who thought it was time Ted settled down, was the only other Kennedy on hand, and they spent a lot of time together. The three of them ate together three times a day. Joan found the week cozy as she walked the beaches or a golf course with Ted and played the piano with his mother. Then Rose called Manhattanville to check out her grades and her character, and got good reports.

Joan got to meet the rest of the Kennedys on subsequent visits that summer, but not the father, who was in the South of France, where he often summered to avoid the noise of his many grandchildren. Joan organized a house party in Alstead, New Hampshire, where her family vacationed. Ted called square dances at a barn. When her mother bought easels for the young people, he did the best paintings of a mountain view. He had learned to paint in Palm Beach in 1955, competing with Jack when his older brother was recuperating from back surgery.

On a late-summer visit to Hyannis Port, Ted proposed, awkwardly:

"What do you think about our getting married?" After Joan agreed, she did meet Joe Kennedy later that day, just after his return from France. He asked, "Do you love my son?" She said she did, and Joe approved.

Though Joan had been growing up in Bronxville while the Kennedys were there, about a mile away, she lived in a different world, one where politics was never discussed by her Republican parents. Before she married, her only exposure to the political arena where the Kennedys thrived came one weekend that fall when she followed Ted around as he made speeches for Jack. "I didn't have a clue what I was getting into," she said years later.

They had fun together but really did not know each other very well. Between the proposal (and obtaining the permission of Harry Bennett, an advertising man, who quipped to the young millionaire, "Do you think you could support my daughter in the style to which she has become accustomed?") and the wedding nearly three months later, they saw each other only twice. The first time was at that political weekend. The next was at an engagement party at the Bennett home. Ted arrived late, bringing a ring Joe had purchased, in a box Ted had not yet opened.

That fall Joan got nervous, and she and her parents wanted to put off the wedding for a year. Her father met Ted and his father to tell them, but Joe Kennedy was furious and insisted the wedding proceed on schedule.

So on Saturday, November 29, 1958, the Kennedys and the Bennetts gathered at St. Joseph's Church in Bronxville. Francis Cardinal Spellman was there to officiate. The principals were wearing microphones for a movie a friend of Harry Bennett's had arranged for a wedding present. Jack Kennedy was his brother's best man, and apparently forgot about the microphone. As he and Ted were waiting at the altar, Jack told Ted that being married didn't really mean you had to be faithful to your wife. Joan discovered this when she watched the movie.

The audience of 475 guests was small for a Kennedy wedding but immense by Joan's standards. Everything went smoothly in church and as the guests went through the reception line at the Siwanoy Country Club. But a few minutes later, Rose looked up and could not find her sons. After a brief search, the groom and his brothers were discovered in front of the television in the men's grill, watching to see if Army could beat Navy and finish unbeaten for the first time since 1949. Before Army's 22–6 victory was clear, Rose hauled them back to the reception. Joan had not noticed they were gone.

CHAPTER 3

The Apprentice

One reason Joan and Ted had seen so little of each other was that Ted's political career had also begun in 1958.

After falling out with Roosevelt, Joe Kennedy transferred his Presidential ambitions to Joe Jr. When Joe Jr. was killed in the war, those hopes went to the oldest surviving son, John F. Kennedy. He was elected to the House in 1946, Teddy's first year at Milton. Jack was easily reelected in 1948 and 1950 and won a seat in the Senate in 1952, defeating Henry Cabot Lodge, Jr., despite the national and state landslides for Dwight D. Eisenhower for President. Ted was in Europe in the Army then, uninvolved except to round up a few absentee voters. After returning, he got to know aides like Ted Sorensen, the speechwriter from Nebraska, whom he taught the useful lesson that when Eastern drivers blink their lights, it means a policeman is waiting behind the next curve.

When his oldest brother was almost nominated for Vice President in 1956, Ted was traveling in Europe and North Africa, although about then Jack began talking him up to Massachusetts reporters as the family's best politician. But in 1958, when the immediate object for John F. Kennedy was to roll up a big victory in a certain reelection to the Senate, Ted was there.

He had the title of campaign manager, but he was not really managing anything. Larry O'Brien and Kenny O'Donnell were seeing to the campaign. What Ted did was go around Massachusetts, meeting people at factory gates, firehouses, police stations, wherever he could find a few voters to tell them that his brother cared, wished he could be there. In fact, John F. Kennedy spent a lot of 1958 going around the country, collecting IOUs for 1960 by campaigning for other Democrats. Ted's enthusiasm for the handshaking side of politics was much greater than John F. Kennedy's, or that of most Massachusetts politicians of the day. Once he was stuck on a bridge in a rush-hour traffic jam going out of Springfield. So as not to waste any time, he took a stack of bumper stickers, jumped out of the car, and went up to other drivers and introduced himself. Then he asked, "Can I put my brother's bumper sticker on your car?" It took half an hour to get off the bridge, and he found fifty or sixty willing drivers.

There was never any fear that Jack would lose to the Republican candidate, Vincent J. Celeste, who complained that he was running against the Kennedys' "financial steamroller." The steamroller's object was a margin of victory big enough to attract national attention. The press set that standard at the 560,000 votes, a margin once recorded by the state's Republican Senator, Leverett Saltonstall. In the fall, Ted came up from law school on weekends. He was there Election Eve for a classic political event, a rally at the G&G delicatessen in Dorchester, where all Democrats came to wind up the campaign. The Kennedys were happy and confident, so much so that even the more reticent Jack joined Bob and Ted in climbing up on a table to harmonize "Heart of My Heart," and, because of Ted's impending marriage, "Wedding Bells Are Breaking Up That Old Gang of Mine."

Ted was useful in the campaign, but the real point of his involvement was to prepare him to run himself one day. Joe Kennedy first put that idea forward in the summer of 1957, when he boasted of Ted to Harold H. Martin of *The Saturday Evening Post,* who was doing a gushy piece on the Presidential contender and his family titled "The Amazing Kennedys." Martin attributed to "fervent admirers of the Kennedys" this idea: "The flowering of another great political family, such as the Adamses, the Lodges, and the La Follettes. They confidently look forward to the day when Jack will be in the White House, Bobby will serve in the Cabinet as Attorney General, and Teddy will be the Senator from Massachusetts. If this should come to pass, of course, the joy of the senior Kennedy would know no bounds." The "admirer" was the senior Kennedy himself, Joseph P. Kennedy.

On Election Night, 1958, after a huge victory by 874,000 votes was secured, Ted toasted his brother: "Here's to 1960—Mr. President, if you can make it." Jack returned the toast: "And here's to 1962, *Senator* Kennedy, if *you* can make it."

When Jack ran for President in 1960, Ted went around the country speaking for him. He was given a modest strategic role, too, responsibility for states in the West. Presidential nominating politics was very different in those days. There were only a few primaries. Winning them was useful because it could impress political bosses who controlled the other delegations, but it would not guarantee nomination, as Estes Kefauver had found in 1952. Some delegations were chosen by governors or other state party leaders. In other states, including the sparsely populated Rocky Mountain states where Ted spent much of his time, national convention delegates were chosen in county caucuses or similar local gatherings.

The 1960 Democratic field was one of the strongest that ever competed for a nomination, including Kennedy; Humphrey; Lyndon Johnson of Texas, the Senate majority leader; Stuart Symington, a Missouri Senator with a strong reputation on defense; and Adlai E. Stevenson, the former governor of Illinois, who had lost twice to Eisenhower. Humphrey was the favorite of the party's organized liberals. Johnson, who labeled himself a Western (not Southern) Senator, was the main competition in the West, Ted's area.

New Hampshire's primary came first, as it always does, but no one ran against John F. Kennedy, the Senator from the state next door. The first real contest came in Wisconsin, the state bordering Humphrey's Minnesota. For John F. Kennedy, it was a place to begin to prove that in 1960 America was ready to elect a Catholic as President, as it had not been in 1928 when Governor Alfred E. Smith of New York was badly defeated by Herbert Hoover.

The Kennedys more than made up for Humphrey's geographic advantage with money, organization, effort, and family. They were everywhere. Humphrey said he "felt like an independent merchant competing against a chain store." Ted spent seven weeks in the state, with only a brief interruption when Joan gave birth to their first child, Kara Anne Kennedy, on February 27, 1960. He made speeches, handed out flyers, did whatever was needed or he could dream up. At Jack's urging, Bob Healy of the *Boston Globe* went out to watch as Ted stuck signs on the insides of car windows in a frigid parking lot, almost getting his arm taken off by a dog left behind to guard one car. Healy reported, "He bounces around at 5:30 in the morning the way most men do after the third martini of the evening." Another time, near Madison, Ted was promised a chance to speak to the crowd at a ski-jumping meet—but only if he made the 180-foot jump himself, something he had never tried before. He said later he went through with it because he feared he would be sent back East "licking stamps and addressing envelopes for the rest of the campaign" if his big brother heard he had given up. More to the point, it was the sort of daredevil stunt the brothers had been brought up to enjoy, and kid each other about. After Ted rode a bronco that summer, Jack called to say he was arranging to have a sharpshooter shoot a cigarette out of Ted's mouth the next week. Another time, Jack needled Ted to land the *Caroline,* the family's twin-engine Convair campaign plane, in Nevada at night. Ted, licensed to fly only single-engine planes, managed, but with a bump.

John F. Kennedy won the Wisconsin primary on April 5, 1960, with 55 percent of the vote. But it was an early example of a victory diminished because it was thinner than people expected just before the balloting. Critically, it was not clear whether Humphrey's votes were reflective of his own strengths or Protestant votes against a Catholic candidate.

So Humphrey chose to fight on in West Virginia, a 95 percent Protestant state. For several weeks Ted was there, learning something about poverty while handing out flyers at mine shafts or speaking wherever his brother could not be, and, for a couple of days, reading Jack's speeches while the candidate, silenced by laryngitis, looked on. Less than two months after Kara's birth, Joan joined in and traveled with Jack. But after her good looks distracted a crowd of wolf-whistling miners from the candidate's speech, she got other assignments. After the campaign, the President-elect gave her a cigarette case engraved "Joan Kennedy—Too Beautiful to Use."

The Kennedy campaign in West Virginia simply overwhelmed Humphrey. Money was lavished on county sheriffs and country preachers. The family brought in Franklin D. Roosevelt, Jr., the son of a genuine West Virginia hero, to accuse Humphrey of being a draft-dodger, a brutal attack in a state where patriotism was long measured in wartime casualty lists. That phony charge, and his belief that Robert Kennedy was responsible, left Humphrey bitter until Robert Kennedy's death.

John F. Kennedy won handily in West Virginia on May 10, with 61 percent of the vote. From then on the crucial campaigning shifted out of the hands of voters, and largely out of sight, too. Ted went back to the West, moving with Joan and Kara to San Francisco for a time. He was not on high-profile assignments like trying to win the allegiance of Governor Edmund G.

"Pat" Brown of California, which had a primary Kennedy had skipped to avoid a head-on fight with Brown. He would go to California to speak for his brother, impressing some later allies with his energy and leaving others thinking he was in over his head.

But he spent most of his time on the nickel-and-dime level of nomination politics, going from state to state, chatting up state chairmen and mayors and members of the legislature. One time he chartered a small plane and flew it all over Arizona for three days of meetings with county chairmen, in Globe for breakfast, Show Low for lunch, and Flagstaff for dinner, and then a few more little airports the next day. "I checked the plane in," he recalled in 1995. "I wrote all my little notes and went on to the next state. . . . And that was considered to be sort of high-powered and sophisticated." He was good at retail politics as the grinning, cheerful ambassador of the glamorous new political dynasty. When he made mistakes, he talked his way out of them. In Denver he told a group of Young Democrats that Hubert Humphrey would "not be a significant threat in the convention," and said Adlai Stevenson was "a brilliant but ineffectual man." When reporters heard that and caught up with him, he smiled and said, "Why, I can't imagine my saying" that.

Another time, flying to central Utah in pursuit of a delegate to the Democratic National Convention in Los Angeles who would cast half a vote, he found the town of Price fogged in. His pilot set him down on a highway about fifteen miles out of town, but there was no traffic and no way to hitch a ride. While Oscar McConkie, a companion, climbed a hill to get their bearings, Ted got into a car parked by the side of the road. When McConkie looked back, Ted had his head under the steering wheel, looking to see if he could hot-wire the ignition. Just then a menacing trapper wearing two revolvers appeared and demanded, "What are you doing with my car?" Kennedy's broad Massachusetts accent did not ease the trapper's reasonable fears of car theft. But it happened that McConkie's father, a judge, had once let the trapper off lightly, and so he drove them into Price, where the Carbon County Democratic convention awarded them the half delegate they had come to get.

The corollary of the minor role of primaries was an importance for conventions that is hard to imagine today, when it has been more than four decades since a nomination took a second ballot. But before 1972, primaries picked only a minority of all delegates. So when the people who chose and controlled the other delegates gathered at a national convention, their sense of what was good for the party—and especially their own local candidates—could be dominant. By the time the Democratic Convention began in Los Angeles in the first week of July, it was plainly Kennedy against the field. Though Johnson and Stevenson seemed to be the alternatives, Johnson had not demonstrated any personal appeal outside the South, and Stevenson, while his supporters had packed the galleries, had almost no support among politicians. They had seen him lose twice. Still, a second ballot would be an opportunity for a whole new round of deal-making and uncertainty. The Kennedys were anxious to get it over with by winning a majority, 761 votes, on the first ballot.

Ted's Rocky Mountain states provided a steady edge, a vote or two more than the Kennedys had been sure of. The total reached 710 when Washington was called. West Virginia's fifteen and Wisconsin's twenty-three were formalities to bring the total to 748. It was Wyoming's turn, with fifteen delegates. Ted had made seven trips to Wyoming that year, and was sure of ten votes. So that evening, before the balloting began, he went to the state's Democratic National Committee member, Tracy McCraken, and said, "If it comes down to where Wyoming can make the difference, would you be willing to commit all fifteen in your delegation?" McCraken thought the amiable young man must be dreaming to think five more delegates would even matter. But he said yes to the request. And when John F. Kennedy and the rest of the family watched the television cameras focusing on the Wyoming delegation, there was Ted in the middle, as McCraken announced, "Wyoming casts all fifteen votes for the next President of the United States."

Ted picked up where he had left off in the general election campaign against Eisenhower's Vice President, Richard M. Nixon. He and Joan toured colleges in Texas together. He still looked after the West. But in a general election Ted's retail campaigning didn't have much impact. Four televised debates, his brother's speech to Baptist ministers in Houston to reassure them about a Catholic in the White House, and a controversy over whether there was a "missile gap" with the United States trailing the Soviet Union were the issues that mattered. Not even the five seconds Ted stayed on a bucking bronco in Miles City, Montana, where he had gone to speak to a Democratic convention on August 27, made a difference. His brother lost Montana, though by just 6,950 votes. In the end, Ted's region had only Hawaii, Nevada and New Mexico in the Democratic column, supplying just eleven electoral votes of the 303 his brother received. Some other Western states were very close, as Nixon won California by 35,623 votes and Alaska by 1,144. Nationally, the popular vote was the closest ever, with Kennedy getting 34,227,096, or just 118,550 more than Nixon. But John F. Kennedy, the youngest ever, was the President-elect.

BEFORE ELECTION DAY, Ted and Joan had talked about moving to one of the Western states to launch a political career of his own. For Joan, Arizona had a particular appeal: still very sparsely populated and, she said many years later, a long way from the Kennedy family. Ted focused on New Mexico, whose two Democratic Senators were growing old. He thought Massachusetts would be ruled out, because his brother Robert would take Jack's seat in the Senate. But after the election they vacationed in Acapulco, and Robert told him he was not going for the Senate. So the "opening was there," as Ted put it years later, and he decided he would like to run himself.

By textbook standards, his ambition was breathtaking. Only one sitting Senator had attained the job with as little experience as Kennedy had—Russell Long of Louisiana also made the most of having a famous political family. But the Kennedys had little patience with tradition. Their view of qualifications was simple: if they could win it, they deserved it. The most revealing explanation came from the candidate himself, who told a reporter, "Listen, this thing is up for grabs, and the guy who gets it is the one who scrambles for it and I think I can scramble for it harder than the next guy." Ted's sense of seizing opportunity was in the family tradition. His father explained it in terms of his own passion for chocolate cake: "If there's a piece of cake on the table, you eat it." His oldest sister, Eunice, said, "As far as politics goes, there was an opening."

But it wasn't his decision to make. One day soon after Thanksgiving, Ted told John F. Kennedy of his ambition. Visiting the President-elect in his Senate office, he asked if there was some job in the new administration, perhaps in foreign policy, which he could use to attract some visibility before the next election in 1962. His brother disagreed with that idea, saying if Ted wanted to run he should get busy immediately getting known around Massachusetts. "Don't lose a day," he said. "Teddy, you ought to get out and get around. I'll understand, I'll hear whether you are really making a mark up there. I will tell you whether this is something that you ought to seriously consider."

If he was going to speak around the state, he needed something to talk about. So Jack telephoned Carl Marcy, the chief of staff of the Senate Foreign

Relations Committee, and asked if there were any trips his brother could get in on soon. Marcy said there was a group of Senators traveling in Africa, but Ted would have to leave that night to connect with the weekly flight from London to Salisbury, Southern Rhodesia. Ted protested briefly that this was rather sudden, and he hadn't even had time to talk to Joan. But it was settled by Jack's saying, "Well, I'd go."

Senator Frank Church of Idaho had organized the African trip with two other Democratic Senators, Gale McGee of Wyoming and Frank "Ted" Moss of Utah. They welcomed Kennedy, though they were a bit put off that local newspapers made him the celebrity as he assured reporters of his brother's interest in Africa. Ted visited nine countries in fifteen days, from the Congo to Nigeria to Senegal. He proved most useful at Abidjan, Côte d'Ivoire. President Félix Houphouët-Boigny gave a fancy dinner, seven courses on a gold service, two wines and champagne, for the Americans and the Presidents of three other former French colonies that had been made independent that year—Dahomey, Niger, and Upper Volta. But none of the African leaders spoke English, and the other Americans spoke no French. Kennedy spoke a little, so it was up to him to respond to their host's toast, telling the audience it was a great pleasure to be there.

While Ted was in Guinea on Sunday, December 18, 1960, the *Globe*'s Bob Healy reported that "the latest backroom word from the Kennedy camp" was that Ted might run for the Senate in 1962. Healy got that word from John F. Kennedy himself, as they lounged around the pool in Palm Beach. Not guessing Healy's source, most politicians did not take that report very seriously. One very eager aspirant for the interim appointment to succeed the President-elect was Massachusetts's attorney general, Edward J. McCormack. He got his uncle, John W. McCormack, then the House majority leader, to call the President-elect. Jack turned him down: "I'm putting someone in. I want to save that seat for my brother." Just as Ted had, the McCormacks assumed that Robert wanted the family seat available in the 1962 election.

Even if the Kennedy camp had been sure that Ted should have the Senate seat—and there were doubts expressed occasionally by President Kennedy and frequently by aides like Kenny O'Donnell—he could not be given the interim appointment. He would not be thirty, the minimum age set by the Constitution for Senators, until February 22, 1962. So, at John F. Kennedy's insistence, an old college pal of his, Benjamin A. Smith of Gloucester, was appointed by Governor Foster Furcolo. Although there was no promise to him that he would be allowed to run, Ted was being given a chance.

His father's support was constant throughout the two years, even though Joe suffered a serious stroke in December 1961. He was left almost unable to speak, though he delighted in his sons' company and could make approval or disapproval clear to them.

Ted's work for the nomination began at John F. Kennedy's inaugural. He asked a western Massachusetts union official he had impressed in 1958, Edward King, what he thought. King was enthusiastic, and the two of them cooked up a plan to have King drive around the state delivering Presidential inaugural medals to about four hundred volunteers from the 1958 Senate campaign. He thanked them on behalf of the President and got a good response to the idea of the youngest brother as a Senate candidate.

Ted Kennedy traveled too, telling audiences that Africa mattered to the world and to the United States because of its size, resources, and strategic location. He said its sixteen nations "are looking for leadership" and pointed out, "If we don't help them someone else will." He would say, "Nature has fashioned Africa in the form of a question mark, and to the world today, Africa is a question." He would also speak at luncheons and dinners on behalf of the state's Cancer Crusade. As Kennedy traveled he was often accompanied by Francis X. Morrissey, a municipal court judge and a retainer of his father's, and sometimes by a newspaperman reporting on his energy and the warm reception at Communion breakfasts, Polish veterans' meetings, or fund-raisers for retarded children. He seemed to be having fun, joining the community singing at a service club lunch, going to the kitchen to praise the coffee and cake at a PTA meeting. He gave the Fourth of July speech in historic Faneuil Hall, warning apocalyptically about Soviet threats to Berlin: "The fall of Berlin could be a prelude to the fall of Boston."

He also had a job, as an assistant district attorney at $1 per year. He filled a vacancy that came about that winter when Kevin H. White was elected Massachusetts secretary of state and left the district attorney's office. Suffolk County district attorney Garrett Byrne, a friend of the Kennedys, agreed to hire him. Byrne asked White to come by and swear in Kennedy. White snapped, "Let him come up here" to his office in the State House. But White went to Byrne's office and administered the oath on February 7, 1961.

While Ted was in Africa, Joe Kennedy dispatched Joan to Boston to find them a place to live. Joan was not yet sure just how lavishly she could spend the Kennedy fortune—she had no French dresses, and bought jewelry for next year's Christmas presents at a half-price sale. So she rented them a dumpy apartment with no heat as temporary quarters while a better place at the foot of Beacon Hill was decorated. And in March they bought a place on Cape Cod, on Squaw Island about a mile away from the family compound. The gray shingled house stood on a bluff, with a view of Nantucket Sound from every window, and Joan called it "a forever house, a home we bought to live in our entire lives." Joan decorated the bedroom in white and pink, telling an interviewer, "Teddy's away so much, he told me to decorate it any way I want to." It was across the street from a house President Kennedy would rent that summer from singer Morton Downey, a big place the Secret Service thought more secure than his house in the Kennedy compound. But the Downey place lacked a beach, so President Kennedy, preceded by two Secret Service agents, would often troop through Joan and Ted's house on the way to a morning swim. Joan would complain in mock outrage, "Jack, you don't call before you're through my front door." He would reply, "I don't have to, Joansie, I'm the President."

Early in his days as a prosecutor, Kennedy came up against one of the city's best defense attorneys, Robert Stanziani. A carpenter named Hennessy was accused of drunk driving after celebrating a Red Sox victory too enthusiastically. The evidence was pretty strong, from the waitress who served him to the policeman who arrested him. And the jury was fascinated by the President's younger brother, hanging on every word. Still, Stanziani suggested that the poor defendant's family would go on welfare if he was convicted, at an expense to the state of hundreds of dollars a month. It took the jury just twenty-two minutes to acquit. Kennedy also won some cases, and exulted over sending armed robbers to state prison. He got to know the police and heard their blunt descriptions of stabbings, robberies, and domestic violence. That may have influenced his own later stance on crime, which civil libertarians often condemned. But as he passed through the DA's office, he hardly made a ripple.

There was a general assumption in Boston that Kennedy was planning to run for something. There was talk about a House seat, but none were vacant. Some suspected he would seek McCormack's job as state attorney general, and McCormack's pals spread that story, just as the Kennedys praised McCormack as a potential governor. But the Senate was Ted's goal. He did well on the stump around the state, and the Kennedys had never been much impressed by the McCormacks. Ted's allies put out a story that before a decision was made, the Kennedys took a poll and it showed that Ted could beat any Republican but McCormack could not. Many newspapers reported it, but there was no poll. It was only propaganda his backers spread at a slow point in the campaign.

That summer he took a month-long trip in Latin America, visiting Costa Rica, Colombia, Brazil, Panama, Argentina, Chile, Peru, Bolivia, and Mexico. It provided fodder for more campaign speeches. In one he argued, as he had at dinner in Brazil, that ferment in Latin America had nothing to do with Communism, but with the hope of the poor to better themselves. He returned in late August. The *Globe* ran a five-part series on his findings, such as that "some 200 million human beings in Latin America are demanding membership in the 21st century." A few days later, on September 26, 1961, Joan gave birth to a son, Edward M. Kennedy, Jr. It had been a difficult pregnancy; she had stayed in bed much of the time. Ted told reporters he was trying to catch up with Robert, who already had seven.

By that fall, Ted was meeting with state legislators, buying them lunch at Locke-Ober, an elegant Boston restaurant most had never visited before, and asking them to back him for the Senate. By December, he was telling politicians he would definitely run. In February 1962, he finally told Ben Smith, who would have liked to run himself, that he would be a candidate. Smith offered to help, and he sent his legislative aide, Milt Gwirtzman, up to Boston every week to brief Ted.

After another whirlwind tour that took him first to Israel and then through the Berlin Wall, Ted made a national political debut on *Meet the Press* on Sunday, March 11. Three nights before, feeling ready, he had visited his brother in the Oval Office. The President sat him down behind his own desk and bore in on Ted with tough questions. Ted decided he needed to spend more time practicing.

The Kennedys had feared that the Harvard cheating issue would come up. It did not. On other topics, he laughed off the complaint of "too many Kennedys" by telling a questioner, "You should have taken that up with my mother and father," and he answered a question about the question of federal aid to parochial schools with a complexity that left listeners confused and President Kennedy pleased with his ambiguity on a sensitive issue. The President was too nervous to watch, but called Lawrence Spivak, the show's permanent questioner, the next morning and was assured that Ted had generally done well and left that issue muddy.

On Wednesday, March 14, 1962, Ted announced his Senate candidacy in his Beacon Hill apartment, assisted by an aide holding cue cards. He said he was running because he could win the nomination and the election and could provide "a voice of vigor and authority that will be heard." Soon his slogan became "He can do more for Massachusetts." He bluntly explained to a questioning columnist that if people accused him of running on nothing but his brother's name, they should stop and consider "that being the President's brother will enable me to do more for the people of Massachusetts than anyone else could do." McCormack, who had announced a few days earlier, sneered at his rival as a "kid just three years out of college" whose only qualifications were his relatives.

Then the Kennedys acted preemptively over the Harvard Spanish exam. Rumors were around in Boston about Ted's expulsion, but the university would not comment. Dick Maguire, the treasurer of the Democratic National Committee, was sent to Boston to see what the *Globe* knew. He called Healy, then the Washington bureau chief but at the Boston office that day, and asked him to visit him at the Parker House, Boston's political hotel. Healy told him it was widely known at the paper that Ted had been kicked out, but that Harvard was sitting on the details. In those days, newspapers did not print that kind of story without very firm confirmation. Maguire went into the next room and made a phone call, and then asked Healy to come in and handed him the telephone. President Kennedy came on and asked Healy to come and see him.

The next day, the President told Healy how Ted had been expelled and offered him the records of the case, if Healy would just slip the embarrassing incident into a profile. Healy told him that would mean that every other paper would have the cheating news in its lead and the *Globe* would look silly. Healy was supported by orders from Boston not to negotiate how the story would be handled. Eventually Healy was given all the details of the case except for Frate's name. Ted eventually spoke briefly to Healy, but it was the President who handled negotiations.

On March 30, the *Globe* ran a front-page story with the obscure headline "Ted Kennedy Tells About Harvard Examination Incident." The first paragraph said Kennedy "explained the circumstances surrounding his withdrawal from Harvard in 1951 when he was a 19-year-old freshman." It was not until the fifth paragraph that the explanation appeared: "I arranged for a fellow freshman to take the examination for me." And he said, "What I did was wrong" and "I have regretted it ever since. The unhappiness I caused my family and friends, even though 11 years ago, has been a bitter experience for me, but it has also been a very valuable lesson." Other papers followed up the next day, and some wrote it a bit more bluntly, using terms like "thrown out" and "cheated," but no one added detail to the story. Healy insisted in 1994 that there never was a deal with the White House about how the story would be written. He said he softened it because of pressure from within the *Globe*. The publisher, Davis Taylor, "didn't want me to kill this guy," Healy said, because he "believed in not hurting the Presidency."

The President insisted publicly that "my brother is carrying this campaign on his own" and avoided any public role. But he followed it intensely, as if reliving his own Senate campaigns.

He was not just a spectator. The President made it clear that a defeat would be not just Ted's loss, but his own, too, and would not be tolerated. He summoned all the White House aides with Massachusetts connections to a secret meeting on April 27, 1962. He even had politicians from Boston fly to Washington under assumed names. After hearing a summary of how things stood from Gerard Doherty, a young state legislator, the President made it clear that he expected everyone there to get busy and find ways to help. The President suggested discreetly using patronage. There is no evidence that many jobs were ever delivered, but the hope of them was certainly dangled before a lot of ambitious politicians. There were other kinds of help, too. Ted Sorensen, the President's speechwriter, sent the candidate quotations to use, and he was quizzed and coached through the summer at Hyannis Port. One administration aide, William Hartigan, effectively took an unpaid leave of absence to come up and help.

While neither the Harvard cheating nor the White House help seemed to bother most Massachusetts Democrats, they fanned the outrage at the candidacy felt by the establishments of Massachusetts and Washington. A leading Harvard Law School professor, Mark A. De Wolfe Howe, sent a scathing letter around the state calling Kennedy a "bumptious newcomer," opposing an experienced and able public servant in McCormack. "Teddy Kennedy seeks his Party's nomination for the Senate simply because he is the brother of the President," he wrote. "He knows as well as you and I know that were he not a coat-tail candidate his name would receive no consideration from any political body for such a high office as he seeks. His academic career is mediocre. His professional career is virtually non-existent. His candidacy is both preposterous and insulting."

Howe's apoplexy was matched by one of Washington's leading voices, James B. Reston, the *New York Times* bureau chief. He made criticism of the candidacy almost a crusade. In March 1962, he wrote that win or lose, Ted would embarrass the President: "In politics, nothing fails like success after a while. One Kennedy is a triumph, two Kennedys at the same time are a miracle, but three could easily be regarded as an invasion." In September, speaking for Washington, he said the candidacy "is widely regarded here as an affront and a presumption." Reston's editorial-page colleagues at the *Times* called Ted's success "demeaning to the dignity of the Senate and the democratic process."

It was a similar outrage that kept Edward McCormack going. Even though being John McCormack's nephew had surely helped, he had worked himself up. He served on the Boston city council and was completing four years as attorney general. He said years later he knew he was "tilting at windmills" when he took on the Kennedys, but did so because "it was wrong that that young man . . . who really had not worked a day in his life, had never been in the trenches, should get off being a United States Senator." Early on, McCormack had some reason to hope that the system would reward his approach, because the first stage in the process was a state convention in June, filled with ward and town politicians. He had two dozen assistant attorneys general and more than that many unpaid "special assistant" attorneys general around the state as the nucleus of a campaign organization.

But in the last six weeks before the convention, Kennedy's campaign outorganized McCormack's and Kennedy outworked his rival. Brother-in-law Steve Smith was in overall charge, and Doherty was the day-to-day campaign manager, checking every afternoon on how things were going around the state. The Kennedy campaign talked to almost every one of the 1,719 delegates. Kennedy himself probably spoke to three-fourths of them. He made as many as twenty stops a day, and his speaking style was a sustained shout. He was up at five to get to a factory gate, and climbed a ladder to get the support of a roofer in Lynn. He flew a small plane to some campaign stops. He would bound out of a car in a parade in Holliston to march the route, shaking hands with crowds on either side.

"Joansie"—her family nickname—was a big part of the campaign. She went to major events with Ted. But even with two small children at home, she went off on her own with home movies of her glamorous family, showing them at coffee hours, shaking every hand in the room, and taking an interest in all the people who volunteered to work for Ted. It was probably the closest time in their marriage.

Ted impressed people with the way he could bounce back quickly from a disappointing grilling by Brookline liberals, and could hold his ground and tell a church basement audience that whatever McCormack had told them, it was not constitutional for the federal government to buy books for a Catholic school. While a huge majority of academics were against him, he won the support of a few, like Sam Beer from the Harvard government department, Charles Haar of the Harvard Law School, and Robert Wood of MIT. He promised them reform of the state Democratic Party. Culver, a third-year law student, crowded them into a small television studio to look like the vanguard of an academic host.

By the time the state convention met in Springfield on June 8, Kennedy had the votes. Doherty's count was 1,196, or 70 percent of the 1,719 delegates. (McCormack claimed 917, or 53 percent.) Eddie Boland, the local Congressman and the only member of the delegation to buck John McCormack, nominated Kennedy, saying that while he respected Edward J. McCormack, "the nomination for the office of Senator from Massachusetts should not be given as a reward for services rendered. It should be given to the man who will best perform the work that must be done. The senator's office is the center of activity on behalf of our state, and the man who sits there must speak with a voice that will be heard." Once the balloting began, the beery delegates reflected Doherty's tally almost to the vote; it was 691–360 when McCormack conceded the endorsement but said he would now "take my case to the people" and contest a primary, as state law allowed a convention loser to do.

The main argument against Ted's candidacy within the administration had been concern over a rift with John McCormack, who had become Speaker of the House in January after the death of Sam Rayburn. The relationship had always been brittle, and White House aides feared that a rough primary would not help. They tried to find a job they could give the Speaker's nephew to get him out of the race. But the only job McCormack, a graduate of the Naval Academy, expressed any interest in was undersecretary of the Navy. Kenny O'Donnell, the White House appointments secretary, who was trying to broker the deal through Congressman Thomas P. "Tip" O'Neill, Jr., said that was out of the question. The President's old Navy buddy Paul "Red" Fay already had it.

As they had before the convention, the Kennedy forces easily outspent McCormack in the primary. They bought more television time, printed more flyers, paid more staff. They paid for polls, which showed a steady Kennedy lead. McCormack, now clearly the underdog, challenged his rival to a debate.

The two headquarters were next to each other on Boston's Tremont Street, and each day a sign in McCormack's front window would say how many days his July 25 challenge had gone unanswered. The tactic succeeded in goading Kennedy, who in early August wanted to rush out and have a debate right away. But his handlers held him back until they thought he was ready. It was August 13 before they agreed to two debates, first on McCormack's home turf at South Boston High School on August 27, and then out west in Holyoke on September 5.

The South Boston High School location, proposed by Kennedy's side, was designed to make him seem the brave underdog in enemy territory. But it also emphasized the image of McCormack as an old-fashioned politician who conjured up *The Last Hurrah,* Edwin O'Connor's novel of the declining James Michael Curley.

The one-hour debate was the campaign in brief. McCormack argued that Ted was simply not qualified by experience for the job. Ted, warned by Robert not to get into a brawl, never took the bait. Instead he flecked his answers with names and details of international, national, and local problems. He didn't always answer the questions posed by journalists (neither did McCormack), but he sounded knowledgeable. This was a line from his closing statement: "The hour is close to nine o'clock here in South Boston. It's close to three o'clock in Berlin, and only in a few minutes from now the convoys will start on their way down the Autobahn. I wonder tonight whether this convoy will be stopped like it was last night." And reporters thought he scored when he dismissed the idea of halting all production of nuclear weapons, which McCormack supported.

But the memorable words of the debate were McCormack's. In his opening, he slammed Kennedy, saying, "You never worked for a living." His younger rival seemed startled. McCormack sometimes called him "Teddy." Kennedy answered, "Mr. McCormack." The cramped, hot crowd was generally with the homeboy, sometimes cheering him like a prizefighter. Egged on, McCormack closed by pointing at his opponent as he tried to shock Kennedy into an angry response that might change the election's dynamic: "If his name was Edward Moore, with his qualifications, with your qualifications, Teddy, your candidacy would be a joke, but nobody's laughing because his name is not Edward Moore. It's Edward Moore Kennedy."

As the debate ended, Ted looked wobbly. He told Doherty he would have liked to sock McCormack for the closing line. The two candidates did not shake hands. Kennedy's supporters in the hall were worried, while Mc-

Cormack's seemed jubilant. But their attitudes soon changed. There was nothing debatable about McCormack's fundamental accusation; without the Kennedy name, his opponent would not have been in the race. Three decades later, much harsher blasts went unremarked. But in 1962 a rough personal attack jolted people watching television. McCormack had come across as a supercilious bully, and Ted as the polite underdog. Eddie Martin, a *Boston Herald* reporter on leave to work as Kennedy's press secretary, recalled: "When I got back to our headquarters, we had an elderly woman at this old switchboard. She said, 'What happened over there? I can't answer the calls. They're coming in from everywhere, saying "We're with Kennedy." ' We had a ward boss for McCormack, over in Cambridge, saying 'I'm with Kennedy.' They couldn't believe it. What we didn't realize, picture tubes in those days weren't that big, you know, so in the confines of the tube it looked like he was sticking his finger in Ted's face." He said, "Ted just stood there looking very hurt, with these relentless attacks on him, and he didn't answer back. And I think the people said, 'My, they take this young guy, and to get savaged by that pol!' "

But the candidate did not know that would happen, and neither did his brother the President, who had been calling Ted and Steve Smith all day, and finally got through to Ted at almost one in the morning. Milton Gwirtzman, the aide lent by Ben Smith, got on the phone when Ted told him to tell the President what had happened. Gwirtzman, who had watched on television, told him, "On points, McCormack probably won. He made a lot of the people take the things he said about Ted, and think about them, and he might have made some points. But on impression, on the general impression people get on television, Ted won, he was the good guy." The President answered sharply, "None of this on the one hand, on the other hand! He's the candidate. He has to get up in the morning and go out and campaign. Tell him that he did great. None of this objective shit, not with somebody who's running."

The next thing President Kennedy did was to order up an instant poll. Almost all polling in those days was done by personal interviews, and when Joe Napolitan in Springfield was told he had to have results for the President by the next afternoon, he did his first telephone poll. He reached 155 people around the state; sixty-six had watched the debate. Just 22 percent of them said McCormack made the best impression, while 44 percent said Kennedy did. Underlying that finding, only eight of the sixty-six found anything positive to say about McCormack's performance, either citing his experience or his support of an immediate tax cut, while thirteen found him rude or insulting. Sixteen had good things to say of Kennedy, splitting about evenly between manners and issues, while five were negative, citing inexperience. The margin of sampling error on a poll of this size is enormous; the percentages could easily be off by 20 points. But Napolitan's conclusion was sound, and McCormack's risky strategy had failed, even backfired.

Even before the news of the poll, the candidate was getting good reactions from callers to radio programs, where people called in to criticize McCormack. The next morning, shaking jam-smeared hands in a North End bakery, he heard one of McCormack's attacks turned in his favor. A burly worker said, "Ted, me boy, I understand you've never worked a day in your life." After a pause, he added, "You haven't missed a thing."

The second debate on September 5 had little drama. McCormack dropped his attack mode. One *Globe* reporter called it a "pillow fight." Napolitan polled again, and found that the eighty-four who had watched or heard it gave Kennedy a solid margin, with 32 percent saying he did best and 19 percent choosing McCormack. The spread on Primary Day, September 18, 1962, was even bigger. Kennedy got 559,303 votes, or 67 percent. McCormack got 247,403, and a few thousand more were scattered or blank.

The general election was never in doubt. The President and Robert were distracted from it by the Cuban missile crisis, but Ted Sorensen took care to warn Ted to avoid Cuba entirely because whatever he said would be overinterpreted in Moscow. Former President Eisenhower came to Boston to campaign for George Cabot Lodge and denounce Kennedy's candidacy. "All over the country," Eisenhower said, "I read and hear that in Massachusetts, one party, one candidate has boiled down the high qualifications long demanded of a candidate for the Senate of the United States into this crass, almost arrogant query: 'Who can get the most out of the United States Treasury for Massachusetts?' "

If that was the standard, it didn't bother the voters. On November 6, 1962, they gave Kennedy 1,162,611 votes, or 54 percent, to 877,669 for Lodge and 50,013 for Stuart Hughes, a Harvard lecturer running as an independent advocating peace and unilateral disarmament, and nearly 50,000 scattered or blank. But most of all, it was a victory for a Kennedy, not all that surprising in a state that had not only given his older brother six victories, but had also twice elected as its state treasurer an accountant from the Gillette Safety Razor Company with no more political experience than Ted Kennedy, but an even more unbeatable name: John F. Kennedy.

CHAPTER 4

The President's Brother

Soon after his election, before he was sworn in, Ted followed the President's advice and began paying courtesy calls on senior Senators. That was an easy assignment; the youngest Kennedy was used to treating his elders with respect. One visit left him laughing. He called on sixty-five-year-old Richard B. Russell of Georgia, the dominant Southerner. He passed on the President's warm regards and made the points the President had said would help. Finally, he observed that he had something special in common with the thirty-year Senate veteran because "you also came in your early thirties." Russell, a Senator at thirty-six, said there was a big difference; he had been governor of Georgia before coming to Washington. Kennedy was laughing as he left and told the story to John Culver, who had joined his staff as a legislative assistant.

The Senate that Ted Kennedy joined on January 9, 1963, was still very much the institution celebrated in William S. White's *Citadel,* a 1957 book Kennedy read before taking office, as "an Institution that lives in an unending yesterday where the past is never gone." A club of elders, mostly powerful Southerners and long-serving Republicans, ran the Senate. Ten of the sixteen committee chairmen were from the still solidly Democratic South, where they were hardly ever challenged for reelection and thus built up the seniority that guaranteed power.

This did not mean that the body was immutably conservative. The

Southerners, except on issues of race, were not alike. At the conservative end was Harry Flood Byrd of Virginia, chairman of the Finance Committee, which dealt with taxes and the national debt. In contrast, Lister Hill of Alabama, chairman of the Labor and Public Welfare Committee, came out of a populist tradition. Quite a few of the newer Senators in the 67–33 Democratic majority were strongly liberal supporters of President Kennedy's program, and they often found allies in a powerful group of moderate Republicans that included Jacob Javits of New York, Clifford Case of New Jersey, and Thomas Kuchel of California.

Lyndon Johnson and then Mike Mansfield, who succeeded Johnson as majority leader in 1961, encouraged newer Senators to take bigger roles. But the Senate was still a formal, private club. The anonymity of voice votes was standard and roll calls were the exception. Senate proceedings, except for an occasional major committee investigation, were not on radio or television. Newspaper and magazine reporters conveyed the institution to the country. Some came to think of themselves almost as auxiliary Senators, napping or playing cards in the press gallery and telling their juniors that the spectacle of a drunken member like Long, incoherent in the Senate chamber, was not newsworthy. And the press in the early 1960s was much more cautious about publishing attacks than it was even a decade later. Over the Senate's bean soup, reporters gossiped about how Johnson had stolen his 1948 election and enriched himself with a television station strengthened by federal favoritism, but they did not search out the facts and write them.

Because staffs were small, no more than a half-dozen in most offices, Senators did not have the sense they have now that their office had experts on anything that might come up. Instead they relied heavily on each other. When they came to the floor to vote, they dallied in clusters, searching out and seeking advice from a generally like-minded colleague who knew more about the day's issue than they did.

From the day he took his back-row seat, Kennedy had a better idea of what to expect of the Senate than it did of him. He had the President to give him advice. Even if John F. Kennedy had not been a devoted Senator, he understood the folkways. But the new Senator's colleagues knew little more of Ted than the arrogance that apoplectic columnists had seen in his candidacy, and the story of his jumping into Robert Kennedy's pool in 1961 in a dinner jacket and then changing into a dry one, so he could jump in again. And, thanks to the Harvard cheating case and his frequently tongue-tied syntax, they thought he was not very bright.

His arrival was hard to ignore. *Life* magazine, an eager promoter of the Kennedy dynasty concept then and of the Camelot analogy later, mentioned his arrival on the cover of its January 18 issue. The article, titled "Congress Opens, Ted's at Work," admiringly noted his efforts to be "as unobtrusive as a man who knows he's not only the latest Kennedy but in some circles the Last Straw. His time of arrival in Washington and even his means of transport were well-kept secrets." But the carefully posed pictures showed Ted talking to a reporter, Ted dictating to a secretary, Ted calling on his state's senior Senator, Leverett Saltonstall, and Ted telling Joan, Edward Jr., and Kara how his day went. If the pictures were taken the day Congress convened, three-year-old Kara needed the explanation, since she had been barred from the gallery to watch his swearing-in because she was not yet six. While Joan watched the ceremony with her sisters-in-law, a Kennedy aide soothed Kara's tears with a ride on the Senate's quaint little subway.

A few weeks later, Joan got equal attention in *Life*'s competition, *Look* magazine. Along with the gorgeous pictures, there was an innocent interview in which she annoyed her in-laws by revealing that Jackie wore wigs and the President's back was so painful that "he can barely pick up his own son." She said what she was looking forward to in Washington was "parties and an excuse for buying some new clothes. It seems an eternity since I've really dressed up for a party. We've been so busy moving around and campaigning, I've had to ask Ted for a date just to stay home and watch television. . . . I like to go glamorous at night." She said she got her clothes from Oleg Cassini: "He gives me 50 percent off which I think is terribly nice." She praised Ted as a father, adding, "We've been married four years, and Ted can't understand why we don't have four children. He wants nine. He says if his mother hadn't had nine, I wouldn't have him." But in May, she had a miscarriage.

Ted and Joan made a good first impression on his colleagues. Russell Long recalled Ted as a "quiet and unassuming sort of fellow" whom everybody liked. Birch Bayh of Indiana, also elected in 1962, said years later, "I don't know how you could not notice him. His brother was President. His brother was attorney general. Not an ordinary member of the United States Senate. A very personable, articulate, likable guy. I thought Joan was a beautiful woman." He and his wife, Marvella, made friends with Ted and Joan. "I think he treated everyone just as if they were longtime friends."

He also surprised the old pols of the Massachusetts delegation in the House, some of whom expected him to be as distant as the President was.

When John Kennedy entertained the delegation, Tip O'Neill recalled, "You were lucky, he would have a waiter pass a tray of drinks around and if you grabbed one when it went by you, you got your drink." But when Ted gave a party at the house he and Joan had rented in Georgetown, it was different. O'Neill remembered: "At eight o'clock Ted wasn't around and we all walk out. We get out in the street and we decide were we going to go here to eat or we're gonna go—Ted comes running out. 'Where you fellows going?' Phil Philbin looked at him and he says, 'Well, I'll tell you, when your brother Jack ran a party, it was over at five minutes of eight.' He says, 'I'm not my brother Jack and I've got a caterer and a lot of food in there,' and we went in and we were there until three o'clock in the morning."

In what the Associated Press called "a struggle for national anonymity," the new Senator tried to avoid drawing attention to himself. He turned down another appearance on *Meet the Press* and avoided non-Boston reporters. When he talked to the Boston papers, he stressed his belief in the tradition that freshmen should be "seen and not heard." He did join other new Senators in speaking at the Women's National Press Club in February 1963, but his message was that he would "stay out of the limelight, out of the headlines and out of the swimming pools."

One quiet place to learn the Senate was the weekly prayer breakfast. For Ted it was as much an opportunity to get to know the veteran Senators as it was for spiritual uplift. Jack would call his brother from the White House to get an idea of what was on his former colleagues' minds. In time, they guessed that Ted was a conduit and concentrated more on the religious business of the event.

Southern Senators were particularly taken with his deference. None was more important to Kennedy than James O. Eastland of Mississippi, the fifty-nine-year-old racist chairman of the Judiciary Committee and the plantation patriarch of Sunflower County. In late March, Eastland was ready to parcel out subcommittee assignments, and Kennedy was summoned to his office. A late-afternoon visit with Eastland meant serious drinking of Chivas Regal, and Kennedy's aides waited and waited. Finally he returned and told them that he had got the assignments he wanted, without even asking. Eastland had said, "You got a lot of Italians up there," and correctly assumed that Kennedy wanted the immigration subcommittee. He also put him on the subcommittee on constitutional rights, knowing that a Kennedy would want that position because of the civil rights bills in its jurisdiction. That was no problem for him, Eastland explained, because "we don't kill the bills in subcommittee, we kill them in full committee."

Eastland also gave Kennedy advice, telling him, "You want something, you come over and speak to me. I don't ever want to get a letter from you." They hardly ever voted alike on an important issue, but they genuinely liked each other, and Eastland would provide Kennedy with staff or money or travel authority for the subcommittees he later headed. As a chairman himself many years later, Kennedy said he really appreciated Eastland's advice about the importance of talking things over: "More often than not, you can work something out."

Kennedy was diligent in going to meetings of the Judiciary Committee and the Labor and Public Welfare Committee, and he sat in at other committees' meetings, too. But his focus was almost entirely local. The Soviet threat he worried about was the menace to the Massachusetts fishing industry from Soviet trawlers. He battled the Civil Aeronautics Board when it took the valuable right to fly to Florida away from Boston-based Northeast Airlines, and he looked for ways to protect the declining textile and shoe industries.

He did not get the sort of special help from the White House that his 1962 campaign slogan "He can do more for Massachusetts" had implied. Asking the President for help, he was told, "Teddy, those are your problems now." Sometimes the message was even blunter. Ben Bradlee, then *Newsweek*'s White House correspondent and a social friend of the President's, recalled seeing John F. Kennedy "roaring with laughter" while talking to Ted at a dinner-dance. "Some pipeline I have into the White House," the Senator explained to Bradlee. "I tell him a thousand men are out of work in Fall River; four hundred men out of work in Fitchburg. And when the Army gets that new rifle, there's another six hundred men out of work in Springfield. And do you know what he says to me? 'Tough shit.' "

In his first year, Kennedy brushed up against the three paramount issues of a momentous time: civil rights, nuclear arms control, and Vietnam. He would later have a major role in each, but in 1963 he did not influence what happened. Indeed, in that year the Senate itself largely ignored civil rights and Vietnam.

Of the three, the nation was most focused on civil rights. Since 1961, blacks had been demonstrating across the South, at lunch counters and on buses, for equal treatment. In May 1963, in Birmingham, Alabama, fire hoses and police dogs were turned on black schoolchildren demonstrating for the right to use snack bars and department-store dressing rooms, for nonracial hiring policies, and for the creation of a biracial commission to seek further progress. In Tuscaloosa, fifty-eight miles away, Governor George C. Wallace stood at the entrance to the University of Alabama on June 11, posing as though he were blocking a federal court order to admit two blacks, Vivian Malone and James Hood. Only after President Kennedy took control of the Alabama National Guard did Wallace step aside.

That night, June 12, 1963, the President went on national television to promise legislation. Until then, the administration had enforced court orders under existing law more than it had said what the law ought to be. That night the President chose sides: "We have a right to expect that the Negro community will be responsible, will uphold the law, but they should have the right to expect that the law will be fair, that the Constitution will be color blind." Four hours later, just after midnight in Jackson, Mississippi, Medgar Evers, the state secretary of the National Association for the Advancement of Colored People, was shot and killed in his front yard, ambushed by Byron de la Beckwith, who was hiding in a honeysuckle bush.* The President's bill was introduced on June 19. It promised equal access to employment and to motels and lunch counters. It broadened federal power to seek both court orders and conciliation. It tried to prevent states from denying the vote to anyone with a sixth-grade education.

The bill hardly moved at all that summer. And on August 28, 1963, a stunning demonstration, organized to the smallest detail, brought 200,000 people to the Lincoln Memorial to demand action. Ted had wanted to be part of the crowd. But the President, who was anxiously negotiating over the degree of White House involvement, said no. Without Ted, the throng heard the Reverend Martin Luther King, Jr., give his greatest speech:

"I have a dream that one day on the red hills of Georgia the sons of former slaves and the sons of former slave owners will be able to sit down together at the table of brotherhood. . . . I have a dream that my four little children will one day live in a nation where they will be judged not by the color of their skin but by the content of their character. . . . When we let freedom ring, when we let it ring from every village and every hamlet, from

---

*De la Beckwith survived two trials in 1964, when all-white juries could not agree on a verdict. He was tried, convicted, and sentenced to life in prison in 1994.

every state and every city, we will be able to speed up that day when all of God's children, black men and white men, Jews and Gentiles, Protestants and Catholics, will be able to join hands and sing in the words of the old Negro spiritual, 'Free at last! Free at last! Thank God Almighty, we are free at last.' "

Not three weeks later, Senator Kennedy proudly cited that day when he addressed the Interparliamentary Union at Belgrade, Yugoslavia. That international meeting was considering a resolution condemning racial discrimination, and Kennedy, though a freshman, was chosen to speak on it on September 16. As Russian delegates busily took notes, he began by saying that racial discrimination was not just an American problem. But, he acknowledged, it "certainly is most publicized in the United States. In a sense, my nation has asked to be judged in this area, because of the leadership we have taken in the cause of freedom and democracy around the world." Then he talked about the demonstration by "over 200,000 citizens, Negroes and white people who believe deeply in the cause of equality and dignity."

At the March on Washington, he said, "the Negroes in America were asking for protection of the right to vote, which is denied in some parts of the United States. They were asking for more jobs, higher wages, better education for their children; and most of all for the chance to enjoy, on an equal plane with white people, the accommodations of restaurants, motion picture houses and hotels."

He said America would succeed, as it had overcome discrimination before, observing: "Neither I, nor the President of the United States, would hold the positions we do, if America had not taken down the signs that said 'No Irish need apply.' "

But it was not by his speech that Kennedy made the most news out of Belgrade. His most publicized appearance was three days earlier at a September 13 lunch—attended by Joan and a New York Republican, Representative Katharine St. George—with Madame Ngo Dinh Nhu, the volcanic sister-in-law of the President of South Vietnam. The largely Catholic government in Saigon was fiercely opposed not only by the North Vietnamese and their Communist Viet Cong allies in the South, but by thousands of Buddhists. The United States had been demanding that President Diem ease up on the Buddhists and was considering support for a coup to oust him. For ninety minutes Madame Nhu harangued him, telling him that the Buddhists were Communist agents and that Americans were badly misinformed about her country. Both she and Joan later said the discussion was merely social. The Vietnamese leader said, "You can't talk politics with the ladies." Joan called her "very charming" and said, "She knew all about my children and we discussed our youngsters. We didn't discuss anything political."

That afternoon the State Department urgently cabled its deep concerns to the senior American diplomat in Belgrade, telling him, "You should attempt with utmost discretion determine how luncheon arranged and circumstances surrounding it." It urged him to discourage the members of Congress "from any association with Madame Nhu while of course avoiding any act of discourtesy or giving any offense." Eric Kocher, the chargé d'affaires, cabled back that Mrs. St. George had invited Madame Nhu, who drew the other Americans into the luncheon. But he took Senator Kennedy on a walk outside the embassy (to avoid any listening devices) to pass on the word from Washington, which he regarded as a direct warning from the President. He cabled that Senator Kennedy "recognizes his own vulnerability" and "agrees that an amiable but distant relationship should be maintained with Mme. Nhu."

Speaking to the Advertising Club in Boston on September 24, Kennedy recalled his meeting with Madame Nhu and made it clear she had not persuaded him, even though for ninety minutes "I listened—and looked." Then, in what seems to have been his first public discussion of Vietnam, he said:

"Our main interest in Vietnam is the military effort against the Communists. This is important because of the strategic position that country occupies in Southeast Asia. We have spent over two billion dollars in this effort. American soldiers have been involved, and some have lost their lives. Until the difficulty with the Buddhists, the war effort was going well. Listening to Madame Nhu did not change my feeling that no government which pursues a policy of religious persecution, a policy that sets group against group and causes so much strife, can hope to keep the confidence of the people that is so necessary to their fight against Communism. Our ambassador, Henry Cabot Lodge, has made it clear to Madame Nhu and her government what our policy is. If they expect our continued support, changes must be made. For we cannot advance the cause of freedom by supporting those who would deny freedom, anywhere in the world."

The coup to overthrow Diem, encouraged by the United States Mission in Saigon, came on November 2, 1963. Diem and his brother were murdered. But the new regime had its own ways of losing the confidence of the Vietnamese people—if it had ever had that confidence to begin with. The hope that the war was "going well" was voiced in one way or another for nearly twelve more years, until helicopters evacuated the United States Embassy in 1975.

The Vietnam speech was given on the day the Senate voted to ratify, by an overwhelming 81–19 vote, a treaty banning nuclear tests in the atmosphere. Kennedy and other Senators had left Belgrade earlier than planned to get back to Washington for the vote. With a public concerned about the dangers of nuclear fallout, the United States had been urging such a treaty on the Soviet Union for several years. But the negotiations started in earnest and were concluded with remarkable speed after a June 10 speech by President Kennedy, promising not to test any more atomic weapons in the atmosphere "so long as other states do not do so." At American University he had said, "A fresh start is badly needed," and a test ban treaty would check "the further spread of nuclear arms. It would increase our security. It would decrease the prospects of war." He said, "Let us persevere. Peace need not be impracticable, and war need not be inevitable."

Senator Kennedy supported his brother on September 9 when the Senate began debate on the treaty, which also prohibited tests underwater and in outer space. After rehearsing the assurances given by military experts that the treaty would protect United States security, he warned that further testing would lead to a "cycle of technology which has no end." He asked, "Should we not instead choose the other course—starting now to strive, cautiously and patiently, to come to agreement by which nuclear arms can be controlled?" He answered, "A limited test ban is better than an all-out arms race, and the time to make that choice is now."

In his first year in the Senate, Ted made a good impression on his elders and began to learn the institution. He took a briefcase of work home every night and usually got through it. He and Joan were a smart couple at the glittering parties at the White House and around the city that made Washington seem gay after the Eisenhower years. He played a lot of touch football, and he helped entertain his father at the White House, doing an imitation of Honey Fitz, complete with lisp. He would drive up to Camp David and visit Civil War battlefields with the President.

He saw more of his brother Jack than he ever had before, especially during the muggy summer when their wives and children were at Hyannis Port. If the Senate was not working late, and it rarely was in 1963, Ted would frequently go down to the White House for a swim, followed by dinner and a cigar. Sometimes Jack wanted the conversation light or focused on the family. But sometimes he wanted to talk about what was happening, matters like the civil rights movement, not in fine detail but in terms of moods and impressions. What did Khrushchev think he was accomplishing with the United States? How long would the House Rules Committee continue to tie up legislation? Wasn't it amazing that scientists believed the country could get to the moon? What about exploring the sea? One night, on the Truman Balcony, looking down toward the Washington Monument, the President told Ted that someday he might be here himself, as President. If that happened, he said, "Don't trust anybody with braid," for the more certain high-ranking military were, the more likely they were to be wrong.

AND THEN ON Friday, November 22, 1963, President John F. Kennedy was shot and killed in Dallas. The nation was dazed. Public events, except for National Football League games, were canceled. Television went on non-stop, instead of signing off in the early hours of the morning as it did then, and for many hours it showed no commercials. The country's instinct was to blame the killing on a right-wing nut; Dallas had many such. But the killer, Lee Harvey Oswald, had defected to the Soviet Union and then returned. Oswald was arrested within hours, and when his past was revealed, suspicion turned toward a Communist plot. Beliefs that he was the instrument of a grand conspiracy—managed by Communist Cuba, or by right-wing generals, or by the Mafia—built within weeks and are alive even today.

Ted Kennedy was doing his freshman duty of presiding over the Senate that Friday afternoon, expecting to celebrate his fifth wedding anniversary that night. After a brief but spirited argument over whether the Senate did its business in good time, with Senator Russell and Everett Dirksen of Illinois, the Republican leader, saying things were fine, a lackluster debate began over expanding federal aid to libraries.

Richard Riedel, a Senate aide, heard a shout in the lobby behind the rostrum from someone looking at the Associated Press printer, and he rushed over and saw AP's first report of the shooting at 1:41 P.M. Riedel ran onto the floor and told Kennedy that his brother had been shot. The Senate was in confusion. Kennedy left the chamber, and then the Capitol building, heading for his own office, hoping to get in touch with Robert.

But the phones were dead. The frenzy of calling in Washington had knocked out telephones all over the city. So Kennedy decided to go home to Georgetown and make sure Joan was all right. Milt Gwirtzman drove him and Claude Hooton, a college classmate who had come to attend the anniversary party. Gwirtzman ran red lights as Kennedy watched for oncoming cars and Hooton, a Texan, sat in the backseat, saying over and over, "The President was shot . . . and in my state." Joan was at a beauty parlor, getting ready for the party, and Gwirtzman got her while Kennedy continued trying to find a phone that worked. He finally found one, so he called Robert in McLean and was told, "He's dead. You'd better call your mother and our sisters." When Ted hung up, that phone went silent, too.

Gwirtzman drove Ted to the White House, where he found a functioning telephone and called his mother at Hyannis Port. Rose said she had heard the news of the shooting on television and was worried about her husband. Ted said he would come. Joan, overcome by grief, heard that he would not be home that night when he had a White House aide call. Ted and his sister Eunice were flown by helicopter to Andrews Air Force Base, outside Washington, and by jet to Otis Air Force Base on Cape Cod, and then driven to the house at Hyannis Port. Rose did not want her disabled husband told yet. Ted tore out the wires from his television set to shut out the news. He called Eddie Martin, who was back at the *Herald,* to find out for his mother if there was any doubt that the right man had been arrested. Martin recalled years later that he sounded "shattered, calm but shattered."

Ted and Eunice told Joe on Saturday. They returned to Washington with their mother early Sunday evening. For most of that staggering day they had been part of a communal national drama. Almost all Americans were listening to the radio or watching television.

Oswald had been shot in Dallas, on national television. Broad belief in a conspiracy became inevitable. The President's body had been taken on a gun carriage to the Rotunda of the Capitol. Senator Mansfield, Chief Justice Earl Warren, and Speaker McCormack offered eulogies. Mansfield's was memorable; he said the President "gave that we might give of ourselves, that we might give to one another until there would be no room, no room at all, for the bigotry, the hatred, prejudice and the arrogance that converged in that moment to strike him down." During the eulogies, Oswald died in Parkland Hospital, where the President had died two days before.

After dinner, Senator Kennedy found his cousin Joe Gargan, Gwirtzman, and another aide, Ed Moss. The Senator said, "Let's go up." As their car neared the Capitol, they began to see the immense line of people waiting in the chill midnight darkness to pay their respects, a procession that eventually numbered a quarter of a million. Many young people had driven hundreds of miles, and the line was about three miles long. The guards recognized the Senator and let him through. Then the other mourners saw him, and the shuffling line halted silently as he went up and prayed before the casket.

The next morning the dress suit Gargan had rented for him lacked a top hat. So he and his brother Robert walked bareheaded from the White House to the funeral at St. Matthew's Cathedral, seven blocks north, followed by family, friends, aides, and world leaders from Eamon de Valera of Ireland to Charles de Gaulle of France to Anastas Mikoyan of the Soviet Union, who took off their hats, too. Robert was on the right, Ted on the left, and the widow, Jacqueline Kennedy, between them as they marched up Connecticut Avenue to the cathedral.

When the coffin was brought into the church, the world paused. Trains were stopped. Traffic halted. Twenty-one-gun salutes were fired. Ships at sea cast wreaths overboard.

Richard Cardinal Cushing's raspy voice intoned the Latin of the low funeral mass. The family came to the rail to take Communion, and then the auxiliary bishop of Washington, Philip N. Hannan, a friend of the President's for more than a decade, read from Scripture, and from the President's inaugural address: "My fellow Americans, ask not what your country can do for you; ask what you can do for your country."

Then Cardinal Cushing resumed the service, praying in Latin, *"Libera me, Domine, de morte aeterna . . ."* (Deliver me, Lord, from everlasting death . . .), and suddenly switched to English, which not until the next year began to be regularly used as the language of the Catholic Mass. He said, "May the angels, dear Jack, lead you into Paradise. May the martyrs receive you at your coming. May the spirit of God embrace you, and mayest thou, with all those who made the supreme sacrifice of dying for others, receive eternal rest and peace. Amen."

John F. Kennedy, Jr., three years old that day, saluted the coffin. It was the most poignant image of the occasion. Nothing at the gravesite compared, not even the lighting of an eternal flame.

Most of the day, Ted was in the background as Robert stayed close to their brother's widow. At Arlington, the two brothers had planned to read selections from the President's speeches on peace and on civil rights. Cushing had urged against it, and they were not sure if it would seem suitable. Finally, after the flag from the President's coffin had been presented to Jacqueline Kennedy, as she told Manchester, "Teddy, Bobby and I looked at each other and the same thought was in all our eyes—and Teddy shook his head as if to say 'No,' and Bobby and I nodded—and as I was the one nearest to the Cardinal, when he turned to me, I said, 'No, your eminence.' "

That evening, after world leaders and old friends had paid their last respects at the White House, there was a birthday party for John Jr. Dave Powers, who first met the President in his 1946 campaign for the House, led the singing of "Happy Birthday," and then Robert and Ted joined him in "Heart of My Heart." The memory of singing it five years before at the G&G was too much for Robert. Hc rushed from the room.

CHAPTER 5

Crash

All the Kennedys were shaken by President Kennedy's death, a numbed Robert most of all. Ted pounded at his own grief through continual involvement in public commemorations of his slain brother, ceremonials that Robert was not up to. On Christmas Eve, when New York renamed Idlewild, its transatlantic airport, as John F. Kennedy International Airport, he was there, helping unveil the letters JFK. Two days before St. Patrick's Day, on March 15, 1964, he hailed the late President's oldest comrades—Dave Powers, Ted Reardon, Larry O'Brien, and Kenny O'Donnell—from Charlestown in 1946 to West Virginia in 1960, in Berlin and "in Dallas at the end. These are men whose loyalty and dedication know no bounds." Of the Irish, he said, "Let them say what they want about our people—and we have many faults—they cannot say we are not loyal to our chieftain."

In May, Ted traveled to Europe to raise money for the John F. Kennedy Library and Museum, which the family wanted to build near Harvard. He had met world leaders before, but then they had been doing a favor for the younger brother of a Senator or a President; now he was there for the family and as a Senator himself. The British prime minister, Harold Macmillan, wept when he recalled hearing of the President's death. The French premier, Georges Pompidou, told him of President Charles de Gaulle's reaction to the assassination: de Gaulle fixed on the location, Texas, and recalled Macbeth, saying the new President, Lyndon Johnson, would suffer from a modern "Banquo's Ghost," and would never recover from having a king die in his own home. The climax of the visit came in Ireland on May 29, when Ted retraced some of his brother's steps of eleven months earlier. Surrounded by the Irish public wherever he went, Kennedy told a crowd after mass in Dublin that he was joyful just to be in Ireland, but was sad, too, "because today is the President's birthday. My brother will not be able to come back and enjoy any more spring days."

Often his message was not just sentimental but political, too. On April 9, 1964, as the Senate debated the civil rights bill, he gave what he called his "maiden speech" in the Senate (despite earlier talks on the test ban treaty and Northeast Airlines and regional development). He cited President Johnson saying five days after the assassination, "No memorial oration or eulogy could more eloquently honor President Kennedy's memory than the earliest possible passage of the civil rights bill for which he fought so long." Ted went on to say, "My brother was the first President of the United States to state publicly that segregation was morally wrong. His heart and his soul are in this bill. If his life and death had a meaning, it was that we should not hate but love one another; we should use our powers not to create conditions of oppression that lead to violence, but conditions of freedom that lead to peace."

Joan watched from the gallery as Ted went on to argue that Massachusetts had overcome, for its minorities, the problems this bill sought to cure nationally for blacks. "In 1780, a Catholic in Massachusetts was not allowed to vote or hold public office. In 1840, an Irishman could not get a job above that of common laborer. In 1920, a Jew could not stay in places of public accommodation in the Berkshire Mountains." Those barriers had fallen, and Massachusetts was the better for it. The bill was not, as its opponents said, a harsh measure; it "abounds with reasonableness, with conciliation, with voluntary procedures." His voice broke as he spoke of his brother, but he finished, and three of the Senate's veteran liberals, Paul Douglas of Illinois, Wayne Morse of Oregon, and Hubert Humphrey of Minnesota, congratulated him.

No bill before Congress was more central to the legislative legacy of John F. Kennedy, nor to Johnson's crusade to prove himself a worthy successor. The House had passed it in February 1964, but when the Senate took it up on March 9, it was not clear that a Southern filibuster could be broken. Civil rights legislation had been talked to death eleven times before. Ending prolonged debate would require the votes of sixty-seven Senators. That meant that most Republicans would be needed. Everett Dirksen of Illinois, a florid orator and the minority leader, was the key player. But in April he was just beginning to show his hand. No Republicans heard Kennedy's speech, because they were caucusing to hear the amendments Dirksen was suggesting to get their support.

Ted also cited his brother in another kind of forum, going out to campaign for Johnson in Indiana and Maryland, where Governor George C. Wallace of Alabama had entered the 1964 Presidential primaries in order to denounce the civil rights bill. "Every Democratic voter who believed in my brother," he told Indianapolis Democrats on April 26, should back Governor Matt Welsh, who was running as a stand-in for Johnson. "What we are voting for is what President Kennedy believed in," he told steelworkers in Baltimore on May 17. Wallace did not run in Massachusetts, but Ted had encountered the same fears Wallace preyed on there. Marching in South Boston's St. Patrick's Day parade, he had sent his police escort to protect the float sponsored by the National Association for the Advancement of Colored People. The Irish crowd had responded to its comparison of oppression of the Negro and of the Irish by throwing rocks and eggs.

Another legacy of John Kennedy's death was the uncertainty about the future of the new head of the family, Robert F. Kennedy. He had stayed on as attorney general, at Johnson's urging, but there was never any warmth between them. Johnson had long blamed Robert Kennedy for the slights he felt he had suffered as Vice President, and the two neither liked nor trusted each other. Even so, Kennedy sometimes seemed to want the vacant Vice Presidency, as a way to maintain a career, even though Ted, for one, argued that the job wasn't worth having. By May 1964, Robert Kennedy had concluded that the Republicans would nominate their most conservative (and most beatable) candidate, Senator Barry Goldwater of Arizona, and so Johnson would not need his liberal image on the ticket. He told his brother and some longtime allies that he had decided instead to move to New York and run for the Senate. Steve Smith and Ted urged him to resign immediately and get to work. But Kenny O'Donnell, who had stayed on at the White House at Johnson's request, urged him to wait so pressure would build up on Johnson to choose Humphrey, whom they respected as a serious liberal. Robert Kennedy agreed, putting off his announcement until August 22. Johnson did pick Humphrey, who was probably his first choice anyway, after publicly ruling out Robert Kennedy and everyone else in the cabinet on July 31.

BUT LONG BEFORE those decisions were made, and Robert Kennedy appeared at the convention in Atlantic City to be cheered and applauded for sixteen minutes, another Kennedy career—Ted's—had almost ended.

After Dirksen finished working with Humphrey, the floor manager of the civil rights bill, and with the Justice Department to polish his amendments, Dirksen and Mansfield called for a vote on cutting off debate. As the last speaker before the June 10, 1964, vote, Dirksen pronounced civil rights legislation "an idea whose time has come." Twenty-six Republicans followed him, and the filibuster was broken, 71-27. That historic vote changed the nature of Russell's Senate; it was now an institution where a dedicated majority *could* work its will. On Friday, June 19, a year to the day after President Kennedy sent the legislation to Congress, the Senate passed the bill. The 73-27 vote banned discrimination in employment and in access to hotels, motels, and restaurants and tried to strengthen existing prohibitions on discrimination in voting and schooling.

The vote came late, at 7:40 P.M. Kennedy voted aye, and then hurried to National Airport. A private plane was waiting to fly him to West Springfield for the Massachusetts Democratic Party convention, where he was to be endorsed for his own full term. Joan, who had suffered a second miscarriage three weeks earlier, was already there. Birch Bayh was to give the convention's keynote speech, and he, his wife, Marvella, and Ed Moss, Kennedy's administrative assistant, were the other passengers. They took off after 8 P.M. By the time they reached the Springfield area, Barnes Airport was fogged in. Visibility just met minimum standards for an instrument landing, and Ed Zimny, the pilot, began an approach. Kennedy loosened the belt on his backward-facing seat so he could turn and watch the landing. "I was watching the altimeter and I saw it drop from eleven hundred feet to six hundred feet," Kennedy said that autumn. "It was just like a toboggan ride, right along the tops of the trees for a few seconds. Then there was a terrific impact into a tree."*

*The Civil Aeronautics Board reported in September. It blamed the pilot, who was killed when the plane crashed at 10:50 P.M. It found no malfunction in the plane or the Barnes radar. The probable cause was "an improperly executed instrument approach by the pilot in which improper altitude control resulted in a descent below obstructed terrain."

"The front of the plane opened as though a kitchen knife sliced through it," Bayh said the next day. "For a moment we were trapped. My wife crawled toward the opening and I pushed and somehow got her out. I tried to get some response from the others in the plane, and I could not raise anyone." Moss and Zimny were in front and did not answer. "Ted was crumpled up on the floor."

That fall Ted recalled, "He called my name a number of times. I could hear him but I couldn't respond." Then "Senator Bayh said he would go for help. And then Senator Bayh said he could smell fumes. Then he said there was the possibility the plane would catch fire. Up to that point, I was not sure I could crawl out. When I heard that, I was determined not to stay in the plane. I began crawling face down from the cockpit to the cabin in the rear. I pulled myself upon a seat by a window."

Bayh said, "I called his name and he answered. I reached my hand through the opening and he grabbed it."

Kennedy said, "I was half in and half out of the window. I told him I couldn't move from the waist down, he said to grab him around the neck, so I grabbed and he pulled me out."

Bayh said, "My wife and I struggled and scrooched on all fours and somehow scrooched him through the window."

Kennedy said, "I fell to the ground. I asked him to let me know if we were far enough from the plane. After what seemed an extraordinary time, he said he thought we were far enough away, and I let go."

Bayh said, "We dragged him over to the hill. My wife covered him with a raincoat. He was obviously in considerable pain and he certainly took it like a trouper. My wife and I were barefoot and must have been a motley looking pair. We walked toward some light in the distance. We came to a road and flagged down a car with a flashlight I had taken from the plane. Somehow the flashlight worked. I'll never know how. The driver [Robert E. Schauer] drove us to his home where we got blankets and pillows and went back to the scene."

State troopers came, and an ambulance. The two Senators, Mrs. Bayh, and Ed Moss were rushed to Cooley-Dickinson Hospital in Northampton. Eddie Martin rushed Joan to the hospital, arriving before Ted. When Martin saw Kennedy, he said later, "I thought he was dying. He was white." But doctors reassured him. Martin called the White House and was put through to Sargent Shriver; he told Shriver he could say Kennedy was badly hurt, but would live. Robert Kennedy drove through the night from Hyannis Port to Northampton. President Johnson had the White House Situation Room checking, and was advised that if all went well, Kennedy could get back to normal work in nine to twelve months.

Moss died on Saturday.

Kennedy's back was broken. One vertebra was crushed and two others fractured. Doctors could not tell immediately if the spinal cord was damaged. If it was, he might never walk again. But the first problem was that Kennedy was in shock, with very low blood pressure and uncertain internal injuries. One lung had collapsed and several ribs were broken. Dr. Paul Russell, the chief surgeon at Massachusetts General Hospital, was routed out of bed by the White House and drove the hundred miles from Boston to join Drs. Thomas Corriden and David Jackson of Cooley-Dickinson. They inflated Kennedy's lung and manipulated his ribs, but gave him no sedatives because they feared he had a ruptured spleen or kidney.

There were no spinal cord injuries, and after three weeks he was moved by ambulance to the New England Baptist Hospital in Boston. The next medical question was whether to operate to fuse the vertebrae. Robert McNamara, the secretary of defense, had sent Army doctors from Walter Reed Hospital, and they recommended surgery. But Joe Kennedy, visiting his son, raged at the idea, shaking his head fiercely and bellowing, "Naaaaa, naaaaa, naaaaa." The outburst stunned the doctors. But when Ted figured out what his father meant, that he feared that surgery would be as disastrous as it had been for Jack, he agreed quickly. Instead the treatment chosen was to keep Kennedy in a Stryker frame, a device rather like a waffle iron which kept his back rigid but could be turned over so that fluid did not collect in his lungs.

At the hospital, an office was set up in an adjoining room, and a certain amount of Senate business was conducted. McNamara visited to discuss the future of the Watertown Arsenal. Daniel Patrick Moynihan, then an assistant secretary of labor, came to talk about his department's business. Kennedy's aides arranged announcements of how he would have voted on issues on the floor, which were sometimes accomplished by announcing a "pair" with another absent Senator who would have voted the opposite way.

For most of the next five months, rehabilitation consisted of holding still and letting the vertebrae heal themselves. Ted was regularly wheeled out onto a sunny balcony off his fifth-floor room. His children came to play, and his room was decorated with their drawings. When they stayed on the Cape, he would telephone them and tell stories in which he played the roles of various noisy animals. He watched a lot of rented movies, especially mysteries and pictures starring Humphrey Bogart. He took up painting again. And just as Jack, while hospitalized for his back surgery, had assembled a privately printed book of remembrances about Joe Jr., Ted and Milt Gwirtzman put together *The Fruitful Bough,* a book of essays about Joe Sr., intended to give the grandchildren a sense of the patriarch before his stroke.

In his own essay, Ted wrote of his earliest memory, climbing into bed with his father to have the Sunday comics read to him, and of escapades like being caught helping himself grandly to his father's candy. And he reflected, "I always felt the greatest gift that Dad gave to each of us, was his unqualified support of any and all of our undertakings. . . . Dad was a taskmaster, too. He was also quick to admonish us for errors. He tolerated a mistake once, but never a second time. We were ashamed to do less than our best because of our respect and feelings for him. His standards were the highest for each of his sons, but they were different standards for each one—standards which recognized our individual strengths and weaknesses. Often, he compared us to each other, but only in a way which raised each of our expectations for what we hoped to accomplish. Comparisons between children are delicate to handle, but Dad could put each of us on our mettle, without creating resentment between us."

A stream of visitors, from Cardinal Cushing to entertainers like Ed Sullivan to President Johnson, kept coming by. Johnson's visit was the most dramatic. He had planned to stop off one evening on his way back from a speech in New Hampshire. But he kept falling farther and farther behind schedule, and the hospital was kept waiting. So Eddie Martin called Air Force One after midnight and told the President, "The Senator really appreciates your interest, but it's late and we would like to have you understand that it's a hospital and there are patients there, and you know the disturbance, I'm sure, is apt to bother them." Johnson was not to be put off, replying, "You tell Ted Kennedy that I'll be there. And don't worry about the patients, they all can sleep later in the morning." Johnson arrived around 1 A.M. and visited for twenty minutes. As he left, he kissed Ted on the cheek.

Joan, who had crumpled when Jack died, thrived in this season of adversity. She was needed. Joan went across Massachusetts speaking for Ted. She danced the polka and tried a sentence in Polish in Dorchester, showed home movies of Ted and the children in Lawrence, shook hands and appeared with visitors like Humphrey and Johnson. She collected cards pledging to vote for Ted. Thanking one audience for a pile of cards, she said, "The fun part is going back and telling Ted all about it. That's the best medicine, bringing a pile of signatures to Ted. . . . I'm no doctor, but when

I go back and tell Ted about this reception it'll take a few weeks off his recovery."

Ted fretted more about Robert Kennedy's race in New York than about his own. In November 1964, Johnson won in a landslide and Ted won his own term easily, with 74 percent of the vote over Howard Whitmore, the ex-mayor of Newton; Robert had 55 percent over Kenneth Keating. Another dividend was that Culver was elected to the House in Iowa and Tunney won in California. A couple of days after the election, Robert visited Ted and they posed for pictures. A photographer said to Robert, "Step back a little, you're casting a shadow on Ted." The youngest brother wisecracked, "It's going to be the same in Washington."

Ted spent much of those five painful, immobile months getting ready to be a more substantive Senator. Once or twice a week he would have professors from Cambridge lecture to him after having sent him a reading list a few days before. Sam Beer talked about state and local government. John Kenneth Galbraith offered a private survey course in economics. Jerome Wiesner, who had been President Kennedy's science adviser, talked science. Galbraith recalled years later, "The picture of it is so strong in my mind: strapped in with his injured back, having to read lying down on his stomach." The Harvard economist said Kennedy was a serious student, "more diligent" than he had been as an undergraduate. "He was very good. He had nothing else to do." It was more than that. Like many adult students, Ted had a motivation now that he had lacked a few years earlier at Harvard. Jack was gone. He and Robert were the Kennedys in politics. Robert would lead, but he would not be far behind.

Not all of the instruction was formal. Gerard Doherty recalled a visit and a long talk that focused on his own health. Suspected of having tuberculosis, Doherty had dropped out of Harvard and stayed in bed for two years. He said Kennedy asked how his family had handled the expense, since they were not wealthy. His father was a fireman, Doherty told him, but "he saved every penny. He spent it when it was important." He said they talked about catastrophic illnesses, and Ted became very interested in how the woman across the street from Doherty, who had a son with a health problem, handled the costs.

With his back in a brace, Ted walked out of the hospital on December 16, 1964. He actually walked out twice, first secretly at four in the morning to drive with Martin to Ed Moss's grave in Andover, awkwardly climbing an icy hill at the cemetery. Then they called on Moss's widow. It would have been Moss's forty-first birthday. Later, with reporters on hand, he left the hospital for good, to spend the holidays at the family's home in Palm Beach. The next year he reflected on his stay: "I never thought the time was lost. I tried to put my hours to good use. I had a lot of time to think about what was important and what was not and about what I wanted to do with my life. I think I gained something from those six months that will be valuable the rest of my life."

CHAPTER 6

The Lawmaker

Ted Kennedy walked stiffly into the Senate with his brother Robert, and they took the oath of office together on January 4, 1965. Robert did cast the shadow his younger brother had predicted, but this was the year when Ted himself began to be noticed in the Senate for more than knowing a freshman's place. He thrived on the institution, whose pace exasperated Robert. He spoke of setting a record for longevity and talked of great past Senators, like George Norris, whom he might someday rival.

He was looking for a tough issue as a way to prove himself as a lawmaker. He chose fighting the poll tax. That issue mattered for itself, but it was also a way to show Robert his skill. Robert had set an example as the attorney general who made civil rights the mission of the Justice Department. Ted served on the Judiciary Committee. Robert did not.

Voting was the civil rights issue of 1965. The most visible part of the 1964 law, the section requiring equal access to public accommodations from lunch counters to hotels, had taken effect with relatively little difficulty. But the law had done very little about the right to vote. Dr. King and John Lewis of the Student Nonviolent Coordinating Committee had been conducting demonstrations almost daily in Selma, Alabama, a town fifty miles west of Montgomery where local voting officials almost always found that whites met state literacy requirements and blacks did not. Police and state troopers broke up the demonstrations, using whips, cattle prods, tear gas, and clubs. In nearby Marion, Alabama, Jimmy Lee Jackson was shot on February 19 when a voting demonstration was broken up, and he died on February 27. The voting issue came home to Kennedy even more directly on March 9 when a white Unitarian minister from Dorchester, James Reeb, was savagely beaten after emerging from a Selma restaurant that served blacks. Reeb died on March 11. The Johnson administration had been planning to push for voting legislation sometime that year, but events in Alabama hurried the pace.

On March 15, President Johnson made the most commanding speech of his life. To a joint session of Congress, he compared Selma to Lexington and Concord. He called for swift action on legislation that would enable federal officials to register voters in the South. He said that while some civil rights issues might be complicated, "every American citizen has an equal right to vote. There is no reason which can excuse the denial of that right." And he went beyond the immediate issue of voting, saying, "There is no Negro problem. There is no Southern problem. There is only an American problem."

He said, "It is not just Negroes, but really it is all of us who must overcome the crippling legacy of bigotry and injustice." Then he borrowed the words of the civil rights anthem sung in Reeb's memory that day in Selma, and said with all the emphasis and authority a President can summon: "And we shall overcome."

Johnson was right when he reported, "Outside this chamber is the outraged conscience of a nation." There were no doubts that night, though some developed later, that a strong voting bill would be passed. Protected by federal troops, Dr. King and hundreds of followers marched from Selma to the state capitol in Montgomery without mishap later that month, but Ku Klux Klansmen shot and killed Viola Liuzzo, a Detroit woman who had come to help, the night the march ended.

Ted Kennedy was anxious to have a role, and the city's leading civil rights lawyer, Joseph L. Rauh, Jr., advised Kennedy's new legislative assistant, David Burke, that he should work on the poll tax. Its abolition was the one major legislative element which civil rights leaders wanted and the administration would not try for. Poll taxes were one of several devices used by Southern states to keep poor blacks—and sometimes poor whites, too—from voting. There was usually no effort to collect the tax if someone did not pay. Sometimes officials made themselves scarce at tax time so potential voters would not have the receipts they needed to vote. By 1965, only five states used poll taxes as a voting qualification—among them almost-all-white

Vermont, where the tax applied only to local elections. But the symbolism of the poll tax was monumental. As Ted said in a Senate speech, it was "the oldest and most infamous of the barriers to voting in the South."

He threw himself into the project over the next few weeks. He consulted often with Rauh. He called for advice to Thurgood Marshall, then a judge of the U.S. Second Circuit Court of Appeals in New York. Marshall, who had argued and won the great civil rights cases before the Supreme Court for the NAACP Legal Defense and Education Fund, suggested Kennedy get help from the professors at Howard University Law School who had quizzed him in preparation for his Supreme Court arguments. Kennedy also turned to Harvard, where Charles Haar, the law school professor who had backed him in 1962 when most other academics sneered, got students and professors together to research the issue. Mark De Wolfe Howe, the Senator's severest Harvard critic of 1962, wrote a memorandum, which Kennedy put in the *Congressional Record.* Paul A. Freund, the most eminent constitutional scholar of the day, came down from Harvard, and was joined at Kennedy's house in Georgetown by Clarence Clyde Ferguson, dean at Howard Law School, and by Howard professors Herbert Reid and Jeanus Parks to go over the constitutional arguments. This technique of collecting experts to brief him over dinner was one Kennedy was to use on issue after issue, from disarmament to paroling convicts to the meaning of the 1994 elections.

The constitutional issue was not clear-cut. The Supreme Court had upheld poll taxes in 1937, though not against a charge of racial discrimination. Even more recently, Congress had seemed to agree that it took a constitutional amendment to eliminate them. To prohibit the use of poll taxes in elections to federal office, it had sent the Twenty-fourth Amendment to the states in 1962, and it was ratified in 1964. On the other hand, the Fifteenth Amendment prohibited discrimination in voting, and its last clause carried this invitation: "The Congress shall have power to enforce this article by appropriate legislation."

The first battleground was the Senate Judiciary Committee. The administration bill was before it after an extraordinary Senate vote on March 18, allowing the committee just three weeks for action, less time than it usually took Eastland to call his first hearing on a civil rights bill. The chairman's own Sunflower County was a perfect example of the problem. In 1961, 13,524 voting-age blacks lived there, but only 161 of them were registered to vote. Eastland held hearings, and then the committee went behind closed doors to debate the bill on April 6. On April 8 it began to vote on amendments, and Ted proposed prohibiting states from basing registration or voting on payment of a poll tax. He cited a 1942 report by this same Senate Judiciary Committee saying the poll tax originated in "desire to exclude the Negro from voting." He said, "This is certainly one of the main impediments, I think we would have to recognize," to voting, because even $2 or $3 was a burden for Southern Negroes whose cash income might not exceed $25 a year. Everett Dirksen, who had voted to ban the poll tax while in the House, answered furiously: "There is nothing violative of the Fifteenth Amendment to the Constitution in a poll tax. . . . You are jeopardizing this whole bill if this thing stays in." But on a 9–5 vote, Kennedy prevailed.

A week later, Ted amplified his arguments in a Senate speech, largely written by Rauh. He argued that "the history of the poll tax is so entwined with racial discrimination that it can never and will never be separated from racial discrimination." He said that legislation could tip the balance in the Supreme Court, which would not lightly overrule Congressional action. His speech drew praise from Freund, who called it "magnificent."

Although President Johnson himself appeared ambivalent, his administration sided with Dirksen, who had to worry about his two Vermont Republicans because of their state's poll tax. Dirksen and Mansfield drew up a substitute for the Judiciary Committee bill and proposed it on April 30. Instead of banning the poll tax, it directed the attorney general to try to get the courts to rule against it. Nicholas deB. Katzenbach, Robert Kennedy's deputy who had succeeded him as attorney general, contended that Congressional action on the poll tax might slow the overall bill down, and that Dirksen's support was needed to get the votes to end a filibuster. And he thought that if the Supreme Court would not outlaw the poll tax in a case that was already on its way to the Court, it would not let Congress do so either. Others in the administration, like Lawrence F. O'Brien, who stayed on as chief of Congressional relations when Johnson became President, regarded efforts to amend the bill as naive attempts by "purists" who were risking everything. But Kennedy, Rauh, and others were angered at having their earlier efforts abruptly discarded, and they set out to round up votes for another try on the Senate floor.

Kennedy and the civil rights organizations worked hard, getting more support than Katzenbach had expected. He and Mansfield had to go all out to get the votes to defeat them. The dispute made both sides unhappy. By the time the vote came on May 11, 1965, Kennedy knew he would lose. Still, he made a stirring speech, arguing that the poll tax discriminated against both blacks and the poor: "The question is whether the Senate, in 1965, will say to the people of the country that the poll tax is being used to discriminate on the basis of the Fourteenth and Fifteenth Amendments. How can we ask the Supreme Court to make that declaration unless we are ready to assume our own responsibility?"

Kennedy lost, 49–45. A handful of Northern Democrats provided the margin. Most of those who spoke against him used Katzenbach's arguments of practicality, and they went out of their way to praise Kennedy's "lucid and compelling" case, as Mansfield put it. Vice President Humphrey, whose heart was with Kennedy, wrote him a few days earlier, "You have stepped in and taken leadership where it was badly needed." The newspapers were even more enthusiastic. Mary McGrory in the *Washington Evening Star* called it a "bar mitzvah," saying Ted "earned the right not to be called 'kid' any more." Andrew J. Glass of the *New York Herald Tribune* said it made Kennedy a Senate "heavyweight." Julius Duscha of the *Washington Post* called it a "legislative tour de force."

The House did include a complete ban on the poll tax in its version of the Voting Rights Act. But the Senate, where Southern seniority and the filibuster always gave it the decisive role on civil rights, insisted on dropping the ban when the two houses negotiated their differences. The next year the poll tax was found unconstitutional by the Supreme Court. The 6–3 decision on March 24, 1966, vindicated Kennedy's constitutional arguments in two ways, though without citing him. First, the majority opinion by Justice William O. Douglas adopted the argument that denial of voting opportunities to poor people violated the guarantee of equal protection of the laws. And in dissent, Justice Hugo L. Black said it was not the Court's role to throw out the poll tax, but the Fifteenth Amendment gave that authority to Congress: "I have no doubt at all that Congress has the power . . . to abolish the poll tax in order to protect the citizens of this country if it believes that the poll tax is being used as a device to deny voters equal protection of the laws."

Kennedy had not changed the law. But he had demonstrated that he could master a complicated issue and lead a Senate debate. He had impressed not only Senators but also people like Professor Parks, who said he never doubted Kennedy's sincerity and commitment: "He believed in the cause." Kennedy had not supplanted the Senate's leaders on civil rights, veterans like Hart and Javits. But he had started a fifteen-year path to that role, and many years later he would describe this effort as his first step toward what he regarded as the defining goal of his career, a nation where respect for civil rights was the "defining aspect of the American political experience: who we are or are not going to be—the helping hand to the dispossessed."

In 1965 that overwhelmingly Democratic Congress enacted dozens of measures that had been part of President Kennedy's program, such as Medicare to provide health care for the elderly, college loans and scholarships, and direct federal aid to elementary and secondary schools. President Johnson enlisted it in battles in his War on Poverty, too. Ted Kennedy was a reliable vote and an occasional craftsman, as when he and Senator Gaylord Nelson of Wisconsin won passage of a bill establishing a National Teacher Corps of specially trained teachers to go into slums and poor rural areas.

But his biggest personal victory was on an immigration measure that had been a personal cause of John F. Kennedy's since well before he became President. After 1924, immigration into the United States had been governed by a national origins quota system designed to replicate the ancestry of white Americans on hand in 1920. It parceled out tickets of admission in a way that strongly favored northern Europe, especially Great Britain, Germany, and Ireland. Britain was allowed 65,000 immigrants at the time and Greece only 306. Most Asian nations had a token allowance of 100, and that had come only from a liberalization of the law in 1952. The only way around these exclusions was to introduce a private bill affecting one potential immigrant or family, a system that was profitable for lawyers, and for many Congressmen. As a Senator, John Kennedy had denounced the quota system and called for a change to place all applicants on an equal footing, regardless of national origin. The existing system, he wrote in 1959, displayed an "indefensible racial preference." Eastland knew the Johnson administration had the votes to change the law, so he saw no reason to fight. He gave the project to Kennedy, who wanted it.

The family connection seemed to go back a lot farther than John F. Kennedy's years in the Senate. Ted told his staff how, as a member of the House in 1897, his grandfather Honey Fitz had made a speech leading the opposition to an effort of Massachusetts's leading Brahmin politician, Senator Henry Cabot Lodge, to impose a literacy requirement upon new immigrants. Representative John F. Fitzgerald argued, "Thousands of Irish and Jewish girls and women, owing to the injustice and barbarities that have been heaped upon them by the English and Russian governments and the lack of opportunity offered by these governments in the education of their subjects, are unable to read and write." He continued, "I am confident that if the provisions of this report had been enacted into law at the time of the arrival of my own mother from Ireland into this country she would have been denied admittance."

While it reads like a stirring oration, complete with a response to an anti-Italian argument voiced that very day, the speech was almost certainly never actually spoken but merely inserted later into the *Congressional Record.* (Fitzgerald, who was rarely in Washington, was probably out of town.) Nor is there any independent verification for Honey Fitz's claim that it was he, after the bill was passed, who persuaded President Grover Cleveland to veto and thus kill it as the session ended. But that was the family tradition, cherished as much as his Congressional desk that occupies a prominent place in the Kennedy home at Hyannis Port.

Ted seized the cause of ending the discriminatory national origins system, although Burke worried about the politics of ending discrimination that favored the Irish, and some Irish groups lobbied hard against the bill. Ted studied the records of past Congressional action and the various commissions that had studied the subject. He consulted not only with the new legislation's friends but with its enemies, like the Daughters of the American Revolution, trying to calm their fears.

The only real threat to the bill was in the House, where the subcommittee chairman, Representative Michael A. Feighan of Ohio, did not get along with the full Judiciary Committee chairman, Representative Emanuel Celler of New York. But after a long summer of disputes between them, the House passed the bill on August 25, 1965, by an overwhelming 318–95 vote.

The price of getting the bill through the Senate Judiciary Committee, demanded by Dirksen and Sam Ervin of North Carolina, was imposing an annual quota of 120,000 on Western Hemisphere immigrants, who until then had been subject to no limits at all. The House had narrowly defeated such a restriction. The Johnson administration, which was calling the shots on the bill, reluctantly gave in. The committee approved the bill on September 8.

Two weeks later, on September 22, 1965, the Senate voted 76–18 to pass the bill. Ted managed it on the Senate floor, allotting time to proponents and answering arguments against the measure. He deferred easily to Ervin when it came to explaining provisions he disagreed with, like the Western Hemisphere quota. No amendments were even offered.

THE FINAL DAY'S debate offered a telling comparison of the two Senators Kennedy and their approach to the Senate.

Though passage was a foregone conclusion, some of the Southerners attacked the bill anyway with speeches aimed at home state consumption. Senator Spessard Holland of Florida expressed shock that the bill would treat people from new African nations just as it treated people from "our mother countries." Could that really be true? he asked. Ted replied that it was because "this bill really goes to the very central ideals of our country. . . . We are the land of opportunity. Our streets may not be paved with gold, but they are paved with the promise that men and women who live here—even strangers and new newcomers—can rise as fast, as far as their skills will allow, no matter what their color is, no matter what the place of their birth. We have never fully achieved this ideal. But by striving to approach it, we reaffirm the principles of our country."

Holland said Ted had made his argument well, but he still disagreed. Ervin joined him in deploring the abandonment of the national origins quota system. Then Holland complained that raising the status of African nations was wrong because American Negroes did not even know where they came from in Africa. That was too much for Robert Kennedy, who asked if Holland did not understand that Negroes had been brought from Africa in slavery. Holland said of course he knew that, but still their sense of origins was not comparable, for example, to Robert's pride in his Irish heritage. Kennedy snapped, "The fact that I might know I came from Ireland does not make me any better than a Negro."

Holland replied, "The Senator may not be, but I shall let him be the judge of that."

The brothers reacted very differently to an argument they found repulsive. Ted debated principles; he had the votes and might need Holland's support another day. Robert had little use for the Senate's formal ways, especially its exaggerated personal courtesies in debate (Holland called him "my friend," but Robert Kennedy called him "the Senator"); he put the argument in personal terms, and left no doubt at all what he thought of Holland.

They reacted to the Senate very differently. They served together on the Labor and Public Welfare Committee, where Ted, with two years' seniority, outranked Robert. Once they waited several hours to question a witness, and

Robert whispered, "Is this the way I become a good Senator—sitting here and waiting my turn?" Ted said, "Yes." Robert pressed, "How many hours do I have to sit here to be a good Senator?" Ted answered, "As long as necessary, Robbie."

THE IMMIGRATION BILL was the first of many times Ted Kennedy fulfilled an unfinished dream of one of his brothers. But another family-related project turned out much worse. At about the same time, Ted took on a cause that quite deserved failure—an effort to get a federal judgeship for Francis X. Morrissey, a retainer of his father's and a loyal visitor to the Cape after other old pals stopped coming because Joe Kennedy's stroke left him unable to talk. As the *Washington Post* sneered, Morrissey "never let Joe Kennedy's coat hit the ground."

He was doing his duty by his father, but Kennedy drifted into the Morrissey battle unprepared. "After my father got sick, I accepted that as something that was important," he said years later. But he had not checked out Morrissey's weaknesses, which included failing the bar exam twice and possessing a degree from a dubious law school in Georgia. Nor did he measure political obstacles, from an intense effort by establishment lawyers wanting greater influence on judicial choices, to Senate Republicans trying to salvage something from a session that Democrats had dominated, to the duplicity of Lyndon Johnson.

In those days Senators could usually pick anyone they wanted to be a federal judge. True, Morrissey had neither the theoretical nor the practical experience to handle the complicated cases federal judges confront. In seven years as a municipal court judge he had tried perhaps a hundred misdemeanors and eight felony cases. But candidates not much better than Morrissey did get confirmed, and Senators were extremely reluctant to challenge a colleague's choice.

Ted pressed President Johnson to make the nomination, which John F. Kennedy had backed away from in 1961 after bar association protests. Katzenbach tried to talk him out of it. But he also told Ted and the President that Morrissey, however unsuitable, could be confirmed.

The issue was coming to a head at the end of the summer. On September 2, 1965, Katzenbach reported Ted's insistence to Johnson. Ted said he would drop the matter only if the President wanted him to. Hardly. While Johnson told Katzenbach the nomination was a favor to Joe Kennedy, whom

Johnson had known for a quarter century, he told J. Edgar Hoover, director of the FBI, to find enough dirt on him so he would not be confirmed. On September 13, Marvin Watson, one of Johnson's closest aides, urged the President to make it very clear that the nomination was made "because of Senator Kennedy's interest and belief that this man was qualified." Johnson also had a letter from Robert Kennedy, whom he could not stand, asking that he nominate the "worthy" Morrissey. Johnson never actually used the letter, because Robert Kennedy entered the fight on his own, but it was reserve ammunition. So Johnson brought Ted to the White House on Friday, September 24, and they called Joe Kennedy to tell him that his friend would be nominated. Two days later, the nomination was the only business when Joe Laitin, the assistant White House press secretary, briefed reporters who had flown to Texas with Johnson. On a slow Sunday afternoon, it produced headlines across the nation.

The *Globe,* which won a Pulitzer Prize for its coverage, launched a campaign against Morrissey. Other papers joined in. Succinctly, the *New York Herald Tribune* called the nomination "nauseous." The senior judge on the court where Morrissey would sit, Charles E. Wyzanski, sent the Judiciary Committee a letter complaining that Morrissey lacked both knowledge of the law "and the industry to achieve it." Albert E. Jenner of Chicago, a rising bar association figure, said the nominee had only judged "backyard fence fights." Morrissey did have some defenders, including Cardinal Cushing. And Speaker McCormack scoffed at bar association leaders from staid firms, like Boston's Robert W. Meserve: "Judge Morrissey is one who came up the hard way," McCormack said, "and in the opinion of some people this is a crime."

Dirksen lulled Ted along. After Ted discovered that Morrissey had twice failed the bar exam, he told Dirksen. But the minority leader said he had flunked the bar the first time himself, so that wouldn't matter. "I am very sympathetic," he said. Ted did not even stick around for the full hearing before the Judiciary Committee, leaving to march in a Columbus Day parade. After he left, Jenner accused Morrissey of getting a meaningless law degree from a "diploma mill" in Athens, Georgia, in 1933, and falsely claiming that he meant to practice in Georgia when he was admitted to practice there that September. Bernard Segal of Philadelphia, another bar association leader, called him the least qualified judicial candidate he had ever seen. And Morrissey made such an unconvincing witness that Katzenbach asked for a new investigation from the FBI.

Morrissey was called back the next day, October 13, 1966, for a closed hearing that was held in Dirksen's office in the Capitol. For seventy minutes, Dirksen hammered him. After an opening boast of his legal skills, exemplified by his getting someone acquitted of assault with intent to kill "even though he was guilty as sin," Morrissey tried to ingratiate himself with Dirksen. He went out of his way to agree that his law practice had been inconsequential, that the Southern Law School was not a law school but a diploma mill, and that it had been "very stupid" for him to attend. But he insisted he had gone there in the summer of 1933, and had read law every day. He said he had tried to get some legal business, but failed, and scraped by selling auto accessories for about six months before returning to Boston, ultimately attending Suffolk Law School and passing the bar in 1944. At one point Ted was seen giving him hand signals on how to answer. Two Kennedy allies, Senators Thomas Dodd of Connecticut and George Smathers of Florida, tried to slow Dirksen down. But he bored ahead, and then told reporters how he had made mincemeat of the witness. Even so, the Judiciary Committee voted 6–3 to recommend Morrissey's confirmation.

Then the *Globe* struck with a series of copyrighted stories that contradicted Morrissey's story. In the first, Bob Healy reported (after a reminder from O'Donnell, a White House aide to John Kennedy who shared his contempt for Morrissey as a spy for Joe) that Morrissey had run for the legislature in 1934. Such a candidacy required living in his district for a year which would have been impossible if he was in Georgia until March 1934, as he testified. In another, James Doyle and Martin Nolan wrote that he had appeared at City Hall in Boston to register to vote on July 26, 1933, while he said he was in Georgia.

Senators unconcerned about his qualifications began to be angered at apparent lies. And if a Senator expressed support for Morrissey, the American Bar Association got his home state lawyers to deluge him with complaints. Senator Joseph D. Tydings of Maryland, a friend of the Kennedys, thought Morrissey a liar and tried to talk them into withdrawing the nomination. He failed, and then made a speech in the Senate on judicial standards, saying, "By far the easiest way to remove an unqualified judge is not to have appointed him in the first place." Both Kennedys—for by now Robert was engaged, too—got angry at him.

The family was embattled. Robert pleaded for votes. He met with Dirksen, who was threatening to use reports that Morrissey and Ted Kennedy had met a deported Mafia don on the island of Capri in 1961. Robert told him, "You hate the Kennedys."

The Kennedy side supplied reporters with names of people to call to attest to Morrissey's Georgia studies. They were not very convincing. Katzenbach sent Eastland the second FBI report, which was not made public. His covering letter was released, and it said, "At least three former students identify Judge Morrissey as a former student who attended classes at the school during the period in question." Katzenbach wrote that there was "no basis whatsoever to question either Judge Morrissey's credibility or his recollection." But Katzenbach said years later that he had no doubt that Morrissey had lied and that his letter to Eastland reflected the reality that once a nomination was made, "you do what you can."

The administration had not lobbied on the nomination, at least not until Johnson was sure it would be defeated. Then he summoned his staff. With what Joseph A. Califano, then his chief domestic policy aide, recalled as a "cat who ate the canary smile," Johnson told them: "Work hard on this one. Teddy's gonna get his ass beat raw on this one and I don't want him to accuse me of not helping him, 'cause he ain't going to blame himself when he loses."

Johnson may have been wrong about the votes. A majority might have been found, though Robert Kennedy was told by a key Senate vote-counter that to do it "you're going to have to make an awful lot of good guys walk the plank." He asked for advice and was told it would be best to get Morrissey to withdraw, though that might be hard after Ted and the President and Morrissey had all said he would stay in. "If it's the right thing to do, we can get him to withdraw his name in ten minutes," Robert Kennedy snapped.

Ted Kennedy told the President he was backing down that night, October 20, 1965. But the secret was well-kept. The Senate gallery was full the next morning when he arrived for the debate, his desk piled with books. Fourteen Democrats were absent. But, at Dirksen's orders, all thirty-two Republicans were there.

At first, Ted challenged Morrissey's accusers, especially the ABA, whose opposition came "perhaps because he attended a local law school by night rather than a national law school by day." He said, "No one who knew Frank Morrissey could doubt that he was telling the truth." And he returned to the theme of class, sobbing as he said Morrissey went to Georgia because "he was young and he was poor—one of twelve children, his father a dockworker, the family living in a home without gas, electricity or heat in the bedrooms; their old shoes held together with wooden pegs their father made."

But then, insisting that he had the votes to win, Kennedy said he wanted any nominee of his to be confirmed without doubts, and Senators needed more time to study the record, which was impossible now with the "press of adjournment, the pressure of partisanship." He moved to send the nomination back to the Judiciary Committee. Relieved, the Senate agreed. Ted left that evening for Vietnam. A week later, with Congress adjourned for the year, Robert Kennedy had Morrissey write the President and ask that his nomination be withdrawn.

Fundamentally, the Senate was Ted's place, not Robert's. But this time it was the junior Senator—but older brother—who made the decision to cut the family's losses and to spare the institution.

CHAPTER 7

To Vietnam—and Back

After a thirty-hour flight from Dulles Airport outside Washington, Ted Kennedy arrived in Saigon on October 23, 1965. There he told reporters that it would be a "great mistake" to see much significance in the antiwar demonstrations that had begun in the United States. "The overwhelming majority of the American people are behind the policies of President Johnson in Vietnam," he said.

He was behind them himself, and he was hardly alone; an August 1965 Gallup Poll showed 61 percent of the American public backing the U.S. combat role and only 24 percent opposing it.

Kennedy went to Vietnam as chairman of the Judiciary Committee's Subcommittee on Refugees and Escapees. He had taken the position eagerly that spring because it gave him a chance to get involved in foreign affairs, Vietnam in particular. He threw himself into it, holding thirteen hearings in three months, mostly on Vietnam, quickly concluding that neither the American nor Vietnamese government had any real policy for dealing with the war's refugees. His aide David Burke said Kennedy reacted as a politician, seeing the refugees as a weakness that could bring defeat, and reasoning: "You cannot possibly be successful in this effort if you are destroying your constituency. . . . You can't win the hearts and minds of people if you are crippling them."

Somewhere around 400,000 Vietnamese had been driven from their villages by the war. Some fled the fighting around them. Others fled from government efforts to move them into "strategic hamlets," supposedly defended and loyal. They crowded the city slums, where the government ignored them, or were moved into refugee camps, where living conditions were nothing like in their villages, and where corrupt officials often stole their modest relief allowances.

Their circumstances reflected the increasingly chaotic situation in South Vietnam. Since Diem had been overthrown in 1963, there had been several new governments. Generals continued to fight themselves and the Buddhists more than they fought the Viet Cong. In 1965, it seemed the war was on the verge of being lost to the Viet Cong and the North Vietnamese, and the United States moved in Marines to prevent defeat.

That was not the first escalation of the American role in Vietnam. After the last French troops left in April 1956, the United States inherited a role of training and advising the South Vietnamese military. But that involvement was not deep. When President Eisenhower left office in January 1961, there were about nine hundred U.S. troops in the country. President Kennedy, seeing Vietnam as a crucial Cold War outpost, sharply increased American troop numbers and gave them more to do. U.S. helicopters, flown by U.S. personnel, were an important element in the war by 1963. At the end of that year there were about 16,300 American troops in the country; 78 had been killed.

The commitment deepened formally in August 1964. After reports of two attacks by North Vietnamese patrol boats against U.S. Navy ships in the Gulf of Tonkin, President Johnson ordered retaliatory bombing attacks against their bases. Johnson had already planned to ask Congress for greater authority to use U.S. forces in Vietnam. But after the second reported attack, which was no more than radar and sonar confusion doubted even by the captain of the presumably attacked destroyer *Turner Joy,* he moved swiftly. Johnson asked Congress to adopt a resolution calling the attacks part of a general campaign by North Vietnam against its neighbors and concluding: "The United States is, therefore, prepared, as the President determines, to take all necessary steps, including the use of armed force" to help Vietnam or another Southeast Asian nation "requesting assistance in defense of its freedom." Kennedy, stuck in his Stryker frame in the hospital, wrote Senator Ralph Yarborough of Texas, "Events of the past few days made me wish more than ever that I was down in the Senate in the midst of the goings on." He announced his support of the Tonkin Gulf resolution, which only

Senators Wayne Morse of Oregon and Ernest Gruening of Alaska opposed. On August 7, 1965, the Senate passed it, 88–2; the House vote was 414–0.

With that authority in hand, the United States began to bomb North Vietnam heavily in 1965. The bombing began in February, and U.S. combat troops waded ashore in Danang in March. The number of U.S. troops in Vietnam increased sevenfold during 1965, from 23,000 to 175,000. American opposition to the war was also beginning to form that spring, chiefly on college campuses. Vietnam was debated for hours at events known as "teach-ins," and Kennedy sent a letter to one at Boston University that drew two thousand people. To pull out, he warned, "would permanently undermine our credit with other nations in the area which are trying to remain independent of the historic and powerful influence exerted by China." He supported the bombing, saying its purpose was to strengthen South Vietnam's negotiating position.

Kennedy's acceptance of the administration's conventional wisdom was evident when his hearings began on July 13, 1965, more than six months before Senator J. William Fulbright's Foreign Relations Committee began nationally televised hearings that gave thoughtful critics like George Kennan a platform.

Opening the July 13 hearing, Kennedy said it was clear that "Communist forces are deliberately creating refugee movements to foster confusion and instability in the countryside, to overtax existing relief facilities, and to obstruct the movement of Government personnel and materials." He said the Viet Cong were using phony refugees to infiltrate government areas. He listened without protest when Rutherford M. Poats, assistant administrator of the Agency for International Development, told him that getting materials to refugees was a "very severe strain" and both Saigon and Washington looked on the problem of 380,000 refugees "as a temporary one."

But with the help of George S. Abrams, a Harvard classmate and Boston lawyer whom he brought down as subcommittee staff director, Kennedy soon became skeptical of administration satisfaction with refugee relief. He heard representatives of voluntary agencies complain that they could not transport supplies up-country, and tell of the shortage of doctors and nurses to treat civilians. He challenged the claim that there was a coordinated operation that reached all refugees. He suggested that the U.S. government was making no plans for dealing with the increased number of refugees that the increased fighting was guaranteeing. He began probing about corruption, but was assured by Poats that there had been only "a few isolated cases of diversion of goods to the black market," and that the refugee camps were "being remarkably well run."

It soon became obvious that the United States government had no clear grasp of how many refugees there were or what efforts were really being made to feed and house them. Poats said he did not know how relief allowances, financed by the United States, were paid out. As the year wore on, refugee estimates soared to one million, or nearly one South Vietnamese in ten. After his first hearings ended, Kennedy and Abrams began pressing the administration to do more. On August 30, the Agency for International Development announced that it was starting a refugee program. One official said Kennedy's hearings had shown that "in effect we have had no refugee program at all."

By then Kennedy was planning to travel to Vietnam. He invited Tunney and Culver, both then in their first terms in the House, and Senator Tydings to come along. They had an intense set of high-level briefings, which focused much more on the general military and political situation than on the refugees who were the justification for the trip. William P. Bundy, assistant secretary of state, told them that "the tide is running in our favor." Central Intelligence Agency officials said that the massive infusion of U.S. troops had stabilized the military situation, but that political circumstances still had not settled since Diem's fall. Averell Harriman, the undersecretary of state for political affairs, told them that the bombing had been effective in destroying enemy bases, in boosting South Vietnamese morale, and in showing United States firmness. He urged them to speak in Vietnam of American support for the war, predicting that such assurances would "encourage the North Vietnamese to abandon their conviction that they can win." General Maxwell D. Taylor, just returned from a difficult year as ambassador in Saigon, told them the war was going very well, and South Vietnam's new Vice President, Air Force general Nguyen Cao Ky, was a "flying George Patton" who could turn into the "George Washington of South Vietnam."

The trip itself was delayed by the Morrissey fight, which also left Kennedy and Tydings barely speaking to each other. When he arrived on October 23, Kennedy took Harriman's advice and scoffed at antiwar demonstrations in the United States. The most substantive of the delegation's meetings was probably the first one; shortly after arriving they called on Tran Ngoc Ling, minister of social welfare. He told them of the urgent need to find work for the peasant refugees: "All I need is the amount that is spent on one B-52 raid and I could finance this program." He dismissed the idea, argued by others in Saigon, that there was any risk of making refugee life too attractive. Peace would quickly get the people "back to their villages," he said. But during the war, once refugees moved out of his camps, "I do not have any more responsibility. Most of the provincial chiefs are militarily oriented and the military do not think the civilian is very important. If the chief of a province doesn't spend it, nothing we can do." The next day, Edward Marks, recently sent by AID to coordinate refugee programs, warned Kennedy that Saigon spent only a fifth of the money given it to help refugees.

After Kennedy, Tydings, Culver, and Tunney left Saigon on Sunday, October 25, 1965, their trip focused heavily on military matters. Most days they had a substantial press entourage along to record their meetings with home state servicemen. Kennedy awarded Purple Hearts to wounded GIs and gave out leftover tie clips from his 1962 campaign. In the mountain town of Banmethuot, children waved paper flags of the United States and South Vietnam. Blankets were distributed to refugees while the Americans were there, and taken away after they left. Reporters like Neil Sheehan of *The New York Times* asked the tough questions and found that the tidy huts in neat rows had been built only the week before, after Kennedy's visit was firmly scheduled.

While the itinerary was managed by U.S. forces and involved a lot of visits to military bases, Kennedy pushed it even farther in that direction, pressing unsuccessfully for permission to visit a besieged hamlet at Pleime or to go on some other combat operation. In Saigon, there were interviews and dinners with government officials, but Kennedy and his party also picked up dancers at nightclubs.

On October 27, their last day in Saigon, General William Westmoreland, the U.S. commander in Vietnam, told them he expected major Viet Cong efforts in the next two or three months. B-52 bombing raids had slowed the infiltration of regular troops from the North, he said, for otherwise there would now be four North Vietnamese divisions operating in the South, not just two. He did predict, "This will last a long time."

Kennedy conveyed that sober view to reporters when he left: "We recognize that the best we can possibly say about the situation is that we're hopeful about the ultimate outcome." He added, "But I would be forced to say that it certainly appears that this is going to be a long and enduring struggle." Back home, he began a series of visits to schools in Massachusetts. At Lowell Tech he called the war "the fundamental moral question facing the United States." He asked, "Are we concerned at all about people in a far and distant land? Do we want to defend freedom?" And answered, "We do, because this is our commitment, our heritage, our destiny." He told students at Peabody High School that many of them would fight in Vietnam. In visits to schools in the Merrimack Valley, he identified the ultimate enemy as China; he was as ignorant of the centuries of Vietnamese hatred for China as were most of America's strategic thinkers. He said, "The Chinese are noted for their patience. They think we will soon get tired of longer casualty lists and increased taxes to support the war." At Brockton High School, he asked, "If we do not support freedom in the world, who will?"

Kennedy recalled that his first doubts about the war were stirred that winter. He had agreed to write about Vietnam for *Look* magazine, and his plan was to take the same upbeat approach he had taken as he toured schools in his state. Abrams and Burke argued with him. Abrams called Tunney to ask what he thought. Tunney had just met Bernard Fall, the French expert on Vietnam, and Fall had systematically shown him that American claims of progress in reducing inflation and making the roads secure that they had heard in Saigon were false, using American statistics to contradict American claims. Tunney agreed with Abrams that it would be a mistake to write such a gung-ho article, and came out to Kennedy's house in McLean and talked him out of it.

When the article was published, on January 25, 1966, Kennedy chided Saigon and Washington for not fighting the civilian problems "with the same ferocity" as they fought the war. Kennedy wrote that he had been discouraged to hear from American officials "again and again that in Vietnam the problem of refugees is just that—a problem and a burden." But he found South Vietnamese apathy even worse: "While in Vietnam, I saw for myself the indifference of the Saigon government to the plight of their own. Government officials assured me that the refugee situation was well in hand—yet I inspected one camp of over six hundred people without a toilet. Construction was started on seven refugee camps in anticipation of my visit. Work stopped when my plans were temporarily altered. It began again when it was finally possible for me to go."

Following his brother Robert, he also sharply disagreed with another central element of South Vietnam's policy, its refusal to negotiate with the Viet Cong. Saigon's foreign minister, Tran Van Do, had told Ted in Saigon that his government "will never talk with Viet Cong." On February 19, 1966, Robert Kennedy issued a statement saying that the Viet Cong had to be a participant in any peace talks, and a coalition government in which they shared was one plausible result. Vice President Humphrey, traveling in the Pacific, said that amounted to putting "a fox in the chicken coop." With Robert under attack (the *Chicago Tribune* called him "Ho Chi Kennedy"), Ted backed his brother when he appeared on *Meet the Press* on March 6, saying, "If we are truly committed to the cause of negotiations, then we have to expect that the Viet Cong would come to the negotiating table, not just to preside over the surrender or the demise of the Viet Cong." He said, "They should be given assurances that, first of all, there will be free elections and that we will be willing to abide by the outcome of those elections."

But on questions where Robert had not committed himself, Ted often sounded supportive of the administration on Vietnam. President Johnson, who kept hoping that bombing North Vietnam would force Hanoi into negotiations and bring the Viet Cong along too, had halted the raids for several weeks at Christmas in 1965. At the end of January he heeded his generals' complaints that infiltration was speeding up and resumed the attacks. As Martin Nolan wrote in the *Globe,* Ted expressed sympathy for the President's "difficult decision" and said if bombing "served to convince Hanoi and Peking of our determination to stay in Viet Nam and defend the people there, this in itself could bring about the changed attitude on the other side that must come before they agree to meaningful negotiations." And Ted, on that same *Meet the Press* broadcast, began by saying, "I support our fundamental commitment in Vietnam. . . . It is fundamental and it is sound. I believe we have to utilize every resource in our power, whether it is military or diplomatic, to see that this commitment is fulfilled."

THE STEADY INCREASE in monthly draft calls to provide troops for Vietnam was making the draft itself an issue on college campuses, although in fact college and graduate students were almost all deferred from induction. One of the Harvard professors who had conducted seminars for Ted in the hospital, Samuel P. Huntington, argued that the system was unfair, and in the spring of 1966 Kennedy began to make that case. Appearing on the ABC television program *Issues and Answers,* Kennedy complained on June 12, "Those who have the economic resources to go to college are given a deferment." Instead, he suggested that draftees be chosen by a lottery so everyone had an equal risk. Lieutenant General Lewis B. Hershey, the autocratic seventy-two-year-old head of the Selective Service System since before World

War II, disagreed. He preferred the existing system, in which more than four thousand local boards had great discretion in deciding who should be drafted. "Fairness or anything else has to go," he said on the same broadcast, "because we have to have survival and therefore you have to have men."

The argument continued before the House Armed Services Committee, which heard General Hershey warn that a lottery had not worked during World War II and that any changes in deferment rules would cost the nation thousands of doctors and scientists. Kennedy argued that a random system worked perfectly well in West Germany, Australia, the Philippines, Bolivia, Chile, Colombia, and Venezuela. But the committee was satisfied with the system as it stood and proposed no changes.

AFTER THE MONUMENTAL civil rights laws of 1964 and 1965, there was a legislative lull in 1966. Yet civil rights problems were beginning to be felt in the North, too, and Boston's schools were almost as segregated as Birmingham's. The 1964 act had provided federal money to help plan orderly desegregation of school systems that were segregated by law. Kennedy tried in March to broaden that authority so that federal aid could go to school systems seeking to correct "racial imbalance." His proposal would have allowed money to be spent to plan busing of students from one area to schools in another. The plan went nowhere in Congress, as did his proposal to authorize Washington to withhold federal aid from communities with such de facto segregation in their schools. Massachusetts authorities were beginning to press the Boston schools to face that issue, and Kennedy urged the Boston School Committee (as the city's school board is called) to work with the state.

Kennedy reflected on those problems that summer when he traveled to Jackson, Mississippi, to address Dr. King's Southern Christian Leadership Conference. On August 8, 1966, he told the SCLC it was "on the firing line of freedom." He welcomed its announced plans to work in the North, saying, "The Negro in Boston, to our shame, goes to a segregated school, holds an inferior job, and lives in one of the worst parts of the city." The heart of his message was that the nation had to spend the money necessary on the poor, on education, on improving slum housing: "We have the resources many times over. The only thing we may lack is the will. We are spending two billion dollars a month to defend the freedom of fourteen million people in South Vietnam. Why shouldn't we make the same kind of effort for the twenty million people of the Negro race right here in America, whose freedom and future is at stake?"

IN 1966, TED KENNEDY found the cause that distinguished him in the Senate—health care.

It began in Boston at Columbia Point, a spit of land where the city had cleared away a garbage dump to build a six-thousand-person public housing project between two highways. The project was only about four miles away from Boston's center of teaching hospitals. But it took residents up to five hours to get there and back, adding up the time of waiting and riding on buses and subways, waiting in an emergency room, and then returning. Thinking of the comprehensive clinics in developing countries, in 1965 Drs. Count D. Gibson, Jr., and Jack Geiger of the Tufts University Medical School opened a health center on two renovated floors of an apartment building in the project. They staffed it with an internist, a pediatrician, a nurse, and a social worker, and soon added a dentist and hired community residents to work as aides and drivers. The effect was dramatic. Muriel Rue, a longtime resident, said thirty years later, "Before the health center, we used to have to call the police when we had a sick child. It was a blessing when this opened up."

Kennedy visited the clinic one day in August 1966, at the urging of Dr. Joseph T. English from the Office of Economic Opportunity. OEO was the headquarters of the War on Poverty, Lyndon Johnson's other great domestic crusade besides civil rights, and it had given a demonstration grant to set up the Columbia Point clinic. Officials thought it could be the model for a national system. Kennedy spent three hours touring the facility, talking with staff and patients, and hearing Dr. Geiger explain his ideas. What impressed him most was seeing women in the waiting room in rocking chairs, where they could look after their children or nurse their babies. He thought that recognized the patients' dignity.

Ted's brother-in-law Sargent Shriver was the head of OEO. Two of his aides, Julius Richmond and English, saw this sort of health care as a way for OEO to expand without running into the political opposition some of its other community projects were getting from traditional political leaders. OEO was already spending money on health, providing physical examinations as part of two of its most successful programs, Project Head Start for preschool children and the Job Corps for hard-to-employ young men. But those health efforts were frustrating and ineffective, because there was no provision for treating any medical problems that were discovered. Neighborhood clinics could provide treatment.

What happened after Kennedy's visit could not happen today. It probably could not have happened in any year after 1966. But Democrats still had overwhelming control of both houses of Congress. The budget deficit was just $8.6 billion, or 1.1 percent of the economy, and not a big worry. Most of all, the New Deal idea that government could solve problems had been revived. So within a couple of months, Kennedy got money for a program of community health centers through Congress.

After he visited Columbia Point, the actual legislation was drafted by English and others. Kennedy had English explain it to the Labor Committee, where he got the program adopted. Then he put in a word with Senator Lister Hill of Alabama and Representative John Fogarty of Rhode Island, who headed the appropriations subcommittees that controlled spending for the War on Poverty. Even the American Medical Association, which had opposed every federal health effort until then, was supportive, especially after Sarge and Eunice Shriver had its President, Dr. Charles L. Hudson, to breakfast at Timberlawn, their estate in Bethesda, Maryland. Ted was there, along with several of his and Shriver's aides. Eunice made a carefully staged entrance, saying, "Dr. Hudson, I want to meet you. Very nice to see you. And I am so delighted that you are here, because I think the American Medical Association ought to be helping on developing good health care for the poor. And I have to warn you that if you don't, my brothers will, and they don't know anything about it."

When the Columbia Point health center celebrated its first anniversary on December 11, 1966, Kennedy was there again as the main speaker, telling residents and staff, "You have not only assured the best in health care for your families and neighbors, but you have also begun a minor revolution in American medicine." He said they had demonstrated that "the old myth" that the poor were not interested in medical care had been shattered. He said the clinic's success in treating 5,400 of the project's 6,000 residents in a year had challenged "your government and the medical profession" to find ways to satisfy "the millions of people in this country who want and desperately need good health care" and are not getting it.

The $51 million appropriated that year got another thirty centers started, and the program grew steadily, though increased costs of the war in Vietnam got in the way. By 1995 there were 850 centers in urban and rural health areas, serving about nine million people. In the 1980s, Kennedy and Senator Lowell P. Weicker, Jr., a Connecticut Republican, fought off Reagan administration attempts to eliminate federal aid for the clinics. By the nineties some Republicans were advocating a vast expansion of the neighborhood system in order to assure health care for all Americans, as they fought a proposal by President Bill Clinton to provide national health insurance.

Nineteen sixty-six was a bad year politically for Democrats. The war was beginning to hurt. The nation's commitment to civil rights flagged as it became clearer that the problem was not just the South's. Republicans pulled themselves together after the disastrous Goldwater campaign in 1964. For embattled Democrats, Ted was in demand as a speaker. He traveled across the country drawing crowds and raising money, starting with an appearance in Omaha for Phil Sorensen, the Democratic candidate for governor of Nebraska and brother of Ted Sorensen, John F. Kennedy's chief speechwriter. Despite a flare-up of back pain, he campaigned all day through Illinois for Paul H. Douglas, from Chicago through the suburbs and Joliet and then down to Peoria and Moline, as teenagers screamed, "We want Teddy! We want Teddy!"—reminding Douglas of the excitement of 1960. Ted drew Douglas's best crowds of the fall and recalled that John F. Kennedy had called him "the greatest Senator of this century," adding on his own, "He still is." He also campaigned for Senators Lee Metcalf in Montana, Frank Moss in Utah, and Gale McGee in Wyoming, and in California for Governor Edmund G. "Pat" Brown. One veteran political writer, Bruce Biossat, summed up his efforts late in October: "Ted's campaign performance is probably the most commanding, captivating exercise any politician could wish for . . . he dominates the hall. You cannot be in it with him and not listen. He could very well be the finest political orator in the nation today."

Massachusetts politics never offered him the same satisfaction. Ted had kept his promise to Professor Beer and the others and chosen Gerard Doherty to be state chairman. But, like John F. Kennedy before him, he never had much enthusiasm for picking sides in intraparty fights. In 1966, he stayed neutral in a crowded Democratic primary for governor. He raised money for all the candidates, but was conspicuous for not endorsing Kenny O'Donnell, especially when his brother Robert came in to campaign for him. The Democrats lost the governorship.

Just as the Kennedys guarded their political capital in intraparty fights, they vigorously defended it against outsiders. Murray Levin, a political scientist who had written a solid book on the 1962 campaign that was less than worshipful, encountered intense Kennedy opposition to getting it published. Robert Kennedy took over that fight, forcing changes like the deletion of the phrase "the Kennedys succeeded in filling John Kennedy's Senate seat with his youngest brother Edward." He got the dust jacket altered, too, so that the title *Kennedy Campaigning* had this subtitle added: "The System and the Style as Practiced by Senator Edward Kennedy." Ted wrote his brother, using private nicknames and apparent amusement: "Dear Robbie, I am glad you were able to look after your own interests so successfully. You seem to forget who taught me Kennedy Campaigning. Your loving brother—Eddy."

Kennedy did not travel only on the campaign trail in 1966. In June he went to Texas with Ralph Yarborough to look at the conditions of migrant workers in the Rio Grande Valley. In December he met with Palestinian refugees in Jordan, where students told him Israelis had to be driven into the sea and an old man startled him by saying, "The Jews killed your brother just like they killed Christ." Kennedy snapped, "That's an inaccuracy."

His refugee subcommittee held a few hearings on Vietnam in 1966. But in general, Ted sought to press for improvements in the refugee program through private conversations. He traveled to Geneva to see if the Red Cross could do anything to help American prisoners of war, and on the Senate floor he joined other dovish Senators to warn Hanoi against holding show trials of POWs.

Ted returned to the draft issue that October, attacking the system's "class bias." On January 12, 1967, he told the National Press Club, "We have a system which allows professional athletes to join National Guard units which neither train nor guard. We have a system of local boards which apply widely different rules—which result in calling up married men in some states, while tens of thousands of single men in other states remain untouched; which conscripts nineteen-year-olds in one city and twenty-two-year-olds in another; which puts returning Peace Corps volunteers at the top of the list in one area, and at the bottom in another. We have a system which sends tens of thousands of young men into the Army because they cannot afford to go to college; one which lets seventy-five percent of those wealthy enough or bright enough to go on to graduate school escape military service completely."

His goal was a lottery system in which luck would decide who got drafted. After the flurry of attention the idea got in 1966, President Johnson appointed a commission to study it, with Burke Marshall, who had been assistant attorney general for civil rights in Robert Kennedy's Justice Department, in charge. When Marshall's commission recommended a lottery in March and Johnson followed with a message to Congress calling for one, Kennedy was elated. While a newly hired aide (and old sailing foe), Dun Gifford, thought that meant that their issue had been stolen, Kennedy told him that instead, "we've won, for Christ's sake. He's going to do it."

But it wasn't settled. Johnson needed Congressional support. Even though the draft was in the jurisdiction of Russell's Armed Services Committee, Kennedy held hearings to push the lottery idea in a subcommittee on manpower of the Labor Committee. By now, General Hershey favored a lottery and said it would be easy to accomplish. Marshall urged that all student deferments should be abolished, because they kept well-off kids out of the war. "Given the will, the means and the intelligence to do it," Marshall testified, "I think anyone can beat the draft."

The Armed Services Committee was dubious about a lottery, but wrote its legislation in a way that placed no obstacles to the President's putting one into effect. Kennedy worked with Russell on the floor, speaking repeatedly against the only serious challenge to the bill, an effort to move to an all-volunteer Army. He told of college appearances where students favored the idea, but made it clear they would not volunteer. He said the result would be "a black army." Russell thought the costs prohibitive, saying, "I do not know how much you would have to pay a man to pick up a gun and go over to the mud of Vietnam." The all-volunteer amendment was defeated 69–9, and the bill was passed, 70–2, on May 11, 1967.

But the House passed a version of the bill that forbade a lottery and barred a halt to student deferments. It also added sections narrowing the definition of conscientious objection and directed the Justice Department and the courts to speed up prosecutions for draft evasion. The Senate conferees accepted the House provisions, and when Russell brought them back to the Senate and asked for quick agreement on June 12, Kennedy fought him. His central complaint was over the lottery. He said, "Chance is the only way to choose fairly." Kennedy said the Senate should take a bit more time to consider the conference report. Russell complained that a July draft call of 39,000 men was imperiled by any delay. Kennedy said he wanted to ask Russell a few questions. Russell replied, "I do not like to answer questions under a threat of a filibuster." Mansfield intervened and got Russell to accept a two-day delay.

On June 14, 1967, Kennedy told the Senate he was going to seek to win three votes against the powerful Russell. First, to defeat the conference report. Second, to have new conferees appointed. Third, to instruct them to seek a one-year draft extension. He offered letters of support from all the members of the Marshall Commission, and from Cyrus R. Vance, deputy secretary of defense, Attorney General Ramsey Clark, and Kingman Brewster, president of Yale University. He laid out his objections to the conference report. Russell dismissed them as "nitpicking," saying a one-year proposal would be out of order, since both House and Senate had passed four-year extensions.

Russell won, as Kennedy knew he would. The vote was 72–23. But twenty-three votes against a conference report managed by a Senate elder was an achievement. Russell was impressed. Now that he had his victory, the Georgian was conciliatory, coming over to Kennedy and Gifford and saying, "Ted, that's a hell of a fight you put up. You did a great job." He went on, "I enjoyed it, I think I even learned a little bit."

They went off to have a drink, and Russell continued the flattery. He even promised to let Kennedy have at least one change in the law the next year. "I like you," he said. "I know you got problems in Massachusetts with the young people up there, and all those liberals. I don't like you to go away empty-handed." And then Russell asked for help—help the chairman of the Armed Services Committee and the second-ranking member of the Appropriations Committee hardly needed from a Senator who had not yet served five years. He told Kennedy he wanted to get some naval construction work in Savannah: "I'm looking to really expand my naval base in Georgia, and I know I can count on your support." Kennedy picked up on it immediately: "Yes, sir, we really do need a strong Navy. . . . We're a Navy state and a Navy family, and I'll be glad to help you with that."

CHAPTER 8

Alliances and Coalitions

Even as he was losing on the draft, Ted Kennedy was finding, on the issue of Congressional redistricting, the legislative method that made him effective. The Senate requires coalitions to function. And for all the millions of dollars the Republican Party has raised over the years by promising to protect the country from Ted Kennedy, there was hardly ever a major enterprise he accomplished without a partnership with one or more Republicans, who often ended up with much of the credit.

Kennedy made his first critical alliance with a Republican in 1967, teaming up with Senator Howard H. Baker, Jr., a freshman from Tennessee. They battled the Senate old guard and its efforts to slow the application to the House of Representatives of the Supreme Court's rulings that legislative districts had to be equal in population. The confrontation was delicious, at least for outsiders, because the leader on the other side was Baker's father-in-law, Senator Everett Dirksen of Illinois, the minority leader.

In 1962 the Supreme Court held that courts could consider cases challenging legislative districts of different populations as unconstitutional. In 1964, the principle of equality was applied to the House of Representatives when Justice Hugo L. Black wrote that as "nearly as practicable, one man's vote in a Congressional election is to be worth as much as another's." Over-represented rural interests (Republicans in the North and Democrats in the South) were on the defensive. Dirksen had led the fight for a series of measures designed to preserve the status quo. His chief antagonist in those fights in 1964, 1965, and 1966 was his fellow Illinoisan Senator Paul H. Douglas.

Kennedy had supported Douglas in those earlier battles. But Douglas was defeated in 1966, and Kennedy fell into the leadership of the 1967 fight unexpectedly. Representative John Conyers, a black whose own election in 1964 resulted from redistricting that gave Detroit a seat of its own, telephoned one spring day in 1967 with a warning. He said that the House was about to pass a bill that would allow a state's most populous district to be 30 percent larger than its smallest—a standard that meant twenty-two states could delay redistricting until the 1972 elections. (After that, variations by 10 percent would be the limit.) Representative Emanuel Celler of Brooklyn, chairman of the House Judiciary Committee, was pushing the bill, thinking it would postpone redistricting in New York. At Kennedy's office, Conyers reached Jim Flug, a new aide, whose approach was "If something needed doing and it wasn't clear that someone else was going to do it, we had to do it."

On April 27, 1967, the House passed the bill, 289–63. With Dirksen, Sam Ervin, and Jim Eastland behind it, the bill was rushed through the Senate Judiciary Committee. They increased the allowable differential to 35 percent after Celler discovered that 30 percent was not enough to preserve New York's districts, where the biggest had 471,001 people and the smallest had 350,186, a 34.5 percent difference. That was not as bad as California, where one district was 95 percent bigger than another, but not as good as Conyers's Michigan, where court-ordered redistricting had produced a map with the most populous district only 3.4 percent bigger than the smallest. Kennedy tried to delete the 35 percent provision in committee, but lost 10–5, on May 23.

Then Baker's support was enlisted, after Robert Kennedy, a friend of Baker's from the fifties, brought him and Ted together. For Baker, the alliance was politically easy but personally difficult. The first of the reapportionment cases, the one that opened the door in 1962 by holding that courts could consider the issue at all, was *Baker v. Carr,* a Tennessee case. (That Baker was Charles W. Baker, a Memphis voter and no relation to Howard Baker, who was representing the eastern end of the state in the House.) " 'One man, one vote' became a principal tenet of Tennessee Republicanism," Baker said years later. "I was hot for that; I preached that sermon all over Tennessee," both in his losing Senate campaign in 1964 and his winning race in 1966. As a great admirer of his father-in-law, Baker was embarrassed to go "from adulation to opposition in one fell swoop," as he put it. But he told Dirksen what he was doing and joined Ted in the fight with a May 25 Senate speech, complaining that the Judiciary Committee had rushed the bill forward without hearings.

Kennedy sent other Senators his minority report on the bill to read over the Memorial Day recess. Flug filled Kennedy's own holiday weekend on the Cape with readings of court cases and commentaries. Flug and Lamar Alexander, Baker's legislative assistant, called editorial writers and reporters, alerting them to the relatively little-noticed bill. Kennedy got organized labor to join the fight. On June 6, Kennedy told the Senate the bill was "an invitation to postpone vindication of a Constitutional right which the framers sought to protect 180 years ago." Baker followed up, pointing out that the Tennessee legislature had delayed its redistricting work when it heard that this bill was being considered. He said, "The State legislatures throughout the nation have a difficult enough dilemma in trying to bring about equal representation, without waiting to see or without being tempted with the idea that equality means only two-thirds equality."

The two Senators worked their colleagues on the floor and in the cloakrooms, arguing that while Celler's bill would probably ultimately be held unconstitutional, in the meantime it would only cause confusion. With Baker getting most of the moderate Republicans and a few conservatives to join Kennedy's contingent of liberal Democrats, they won a surprising 44–39 victory.

But the fight was far from finished. Most of the participants in the House-Senate conference chosen to settle on a final bill were senior members who wanted delay. The conferees' first effort seemed to put off all redistricting by state legislatures until 1972. Conyers called it worse than either the House or Senate versions. But the House feared it would lead to court-ordered, at-large elections with representatives running statewide instead of in a district where they were well known. At Celler's urging, the House rejected the first plan on June 28.

The conferees on the districting bill struggled for weeks. On October 9 they produced a new version, which said essentially that no state needed to redistrict until there was a new census, either the regular one in 1970 or earlier if a state wanted to pay for one. The House passed it on October 26, but Kennedy and Baker returned to the attack. Kennedy told the Senate on November 8 that the new version was "unconstitutional, unconscionable, unclear, unworkable and unresponsive." In a time of frequent appeals to law and order, he said, it was ironic that "we are being asked to reward and encourage disobedience to the Constitution and to court decrees." Baker followed, saying the issue was "whether the Congress may validly, or should desirably, enact a law which would in eighteen states, which include 259 Congressmen, delay for five years the enforcement of the clear constitutional mandate that each man's vote for his Congressman counts as much as the next man's vote." He said, "If a man has only a part of a vote, the candidate of his political party has only a part of a fair opportunity to compete for the right to speak for him in the House of Representatives, and his Nation has a Government which represents a part of the public more adequately than the rest."

This time they won, 55–22. Dirksen stayed away. Columnist David Broder praised Kennedy and Baker in the *Washington Post* for getting the Senate to reject a supposedly "untouchable" conference report. He wrote, "Defeating bad legislation, as Kennedy and Baker have done, is every bit as difficult as passing a good bill, but not nearly so long remembered. History is particularly careless about paying heed to those who managed to thwart legislative tinkering with constitutional processes."

It had been a busy summer and fall for the Kennedys, and not just in the Senate. On July 14, 1967, Joan gave birth to their second son. Patrick Joseph Kennedy was named for his paternal grandfather. For a time it eased a sense of inadequacy Joan felt among Kennedy women, especially Ethel, who had an easier time having lots of children. That doubt, and a sometimes lonely life in which she spent hours perfecting her beauty and clothes, had begun to lead to heavy drinking.

Another factor may have been the gossip in Washington about Ted and other women. Most of it was word of mouth. Even the supermarket tabloids did not print stories about politicians' affairs in the sixties, but they dropped arch hints. The *National Enquirer,* for example, ran the latest dirt about movie stars, but when it came to politicians used the device of quoting psychiatrists who offered generalizations that relied on readers' suspicions heightened by what they found in adjoining columns. Thus it asserted in June 1967 that Edward's future was dubious because "you get less responsibility of a real nature in younger sons. This is the descending order of siblings and is well documented." The only other regular source of printed gossip was *Women's Wear Daily,* with items like "Lots of beautiful young girls imported from New York . . . as always at those Kennedy parties" or "Ted leaves today for Bermuda. Returns next Tuesday. Maybe Joan will go there too," or a faux-innocent item from Geneva about how he "ran into what is surely the prettiest pair of American girls in town right now."

PATRICK'S BIRTH WAS also the occasion for planning to move out of Georgetown. The Kennedys bought a six-acre tract in McLean, Virginia, about a mile from Robert and Ethel's home at Hickory Hill. John Carl Warnecke, an architect who had done that house and President Kennedy's grave, designed the 12,500-square-foot house, with a soaring, 32-foot-high living room looking down toward the Potomac, a dozen bedrooms, a whirlpool bath, and a library with weathered wood from an old barn. Joan worked with Keith Irvine and Thomas Fleming on the interior design and furnishings.

Just the day before the last redistricting vote, on November 7, 1967, Kennedy and Boston had survived a scare at home when Kevin White was elected mayor. Kennedy had been silent during the September primary for mayor. But Louise Day Hicks, a rabble-rousing foe of school busing (which had not yet reached Boston), finished first in the primary ahead of White, who was still Massachusetts secretary of state. Kennedy then came out for White without naming him, calling for support of the candidate who dealt with problems "not from emotion and intolerance, but from experience and understanding." He spoke despite White's request that he stay silent. White feared that Kennedy's support would only win sympathy for Hicks, although he and his aides understood that a national figure could not avoid taking sides in a contest with such racial implications. Kennedy provided some fund-raising help, but White was most grateful that he encouraged Dick Goodwin, a veteran of the Kennedy administration, to come to write speeches.

Louise Day Hicks's strong campaign was a warning that passing the landmark civil rights bills of 1964 and 1965—whose impact was felt primarily in the South—had not taken race off the national agenda. Indeed, without the specter of police dogs and murders, passing legislation to deal with remaining racial problems proved much harder than advocates expected. When Northern schools were segregated it was largely because blacks and whites lived apart in Northern cities, much more so than in the South. That was the biggest remaining legislative issue—discrimination in housing sales and rentals. Johnson had made it the centerpiece of a 1966 civil rights message, but it went nowhere that year or for most of 1967.

Kennedy's role was not crucial. Javits was the liberals' leader on the bill, and Walter F. Mondale of Minnesota played a large role. But there was a critical moment when his commitment mattered. In the fall of 1967, after a year and a half, supporters of the bill thought they had made very little progress, and they were meeting in the Capitol to discuss what to do. The mood was gloomy. There was a sense that if the housing provision was abandoned, other elements of the bill could be passed. Kennedy was late, and when he walked in the others raised that idea. He said, "That's really not acceptable." He said they could not go back to Dr. King and the other leaders just to say "we can give them now, for dropping the housing title, what we could have given them a year and a half ago when they were first proposed together. We've got to do better than that."

They talked of seeing if they could get to Dirksen, then an opponent of the President's bill, and get him to take credit for a compromise that would bring enough conservatives along to break a filibuster. It took until February 1968, but ultimately it worked, as Dirksen found another idea whose time had come.

KENNEDY ALSO RETURNED publicly to the Vietnam issue. On March 4, 1967, speaking in Boston to the heart of the liberal movement Americans for Democratic Action, he said the war was diverting not only money but also moral energy: "The simple and brutal fact is that the liberal program, the concern for the unfinished agenda among our people, has been a casualty of the war in Vietnam."

In May he disputed the administration's casualty estimates, saying there were at least 100,000 civilian casualties a year, twice the official number. The administration's memos dismissed him for "agitating this medical business." He did win a pledge to build three U.S. military hospitals to care for wounded civilians. That promise was first made in April 1967, and was repeated regularly as though promises were plywood and commitments were construction. They were never built.

Kennedy's pressure over medical care did put enough heat on the Johnson administration so that it sent a group of prominent American doctors to Vietnam to study the situation in July 1967. They were led by Dr. F. J. L. Blasingame, executive director of the American Medical Association. In September, after their return, they called for a sharp expansion of U.S. spending on medical programs in Vietnam, especially surgical care. Their report said most civilian casualties were being treated in existing facilities, and that even the estimate of 75,000 civilian casualties a year was "too high." They opposed Kennedy's idea of building new hospitals staffed by Americans, saying that would slow the development of medicine in Vietnam. Most of their report was kept secret. The optimistic parts were summarized in a press release the White House gave reporters on September 21, 1967.

Kennedy was furious. He saw the doctors that afternoon and quoted the press release: "Hospitals appear to be sufficient in number and well located for our peacetime needs." Then he blurted, "We are not in peace over there, and as far as I myself have been able—I don't see what in the world that has to do with what the problems are." The doctors, especially John Knowles, general director of the Massachusetts General Hospital, defended their work and accused the administration of "manipulation." Kennedy was not satisfied: "We are the ones that are ripping that country up. And the question is what are we going to do about it."

Kennedy then called public hearings in his subcommittee. They began barely three weeks later, on Monday, October 9, 1967. He opened by lecturing the doctors: "To the sick and injured their only hope, for as long as this war persists, is that we are sufficiently concerned to bring our medical resources to their side." Dr. Knowles insisted that his passion was equal to Kennedy's: "Now, the final cure of this horrid disease in that country, again, is to stop creating the two million refugees that we welcome to democracy in the nice dust bins we stick them into, one eighth of the population, and stop shooting up the place and tearing up villages which creates 50,000 to 75,000 to 180,000 casualties; 180,000 casualties is even worse than 75,000 but it is like being two months pregnant instead of nine months pregnant. Any casualties as far as I am concerned, are awful. . . ."

But Knowles and the other doctors argued against the idea of building three shiny new U.S.-managed hospitals, saying it would be better to concentrate on improving the country's provincial hospitals. "How can you call a hospital a hospital if you have no sanitation, no hygienic measures whatsoever, almost no diagnostic facilities and generally unsanitary facilities, and overcrowding, and people lying out in tents waiting for up to a year for orthopedic surgery which never gets done because it is delayed constantly by the fresh influx of casualties and so on?" He said U.S. spending on civilian health in Vietnam should be doubled, not reduced as the administration planned. He complained that "the care of civilian casualties and the civilian population has been put at the lowest and brownest end of the stick."

On October 10, the emphasis shifted toward refugees. Don Luce, former director of International Voluntary Services, testified that the refugees he had dealt with when he arrived in Vietnam in 1958 were usually escaping the Viet Cong. But after 1965, he said, refugees were fleeing United States military actions, from bombs to the herbicides that killed the jungle cover that the Viet Cong used but killed crops too, to search-and-destroy missions and free-fire zones in the countryside.

On October 11, Kennedy put forward General Accounting Office reports on the failures of refugee efforts in Vietnam. He said they showed a "cavalier and almost disdainful attitude" toward the Vietnamese. On Friday the thirteenth, Roger Hilsman, a former assistant secretary of state in the Kennedy administration, argued that the Air Force's approach was counterproductive: "To bomb a village, even though the guerrillas are using it as a base for sniping, will recruit more Viet Cong than are killed."

Finally Kennedy gave the administration a chance to defend itself. On Monday, October 16, William S. Gaud, the head of AID, testified that his agency could report "substantial progress" since 1965. But, he said, while the United States was doing everything it could to help, it was still up to the Vietnamese to get the job done. Kennedy demanded to know why the United States could not take over the refugee effort as it had taken over the war. Gaud said that would be "contrary to our traditions" and that lasting progress could be made only if the Vietnamese government made a greater effort. "Does the Vietnamese Government care?" Kennedy asked.

Kennedy went back to Vietnam himself a few weeks later, arriving January 1, 1968. The trip was radically different from his 1965 visit. He wore civilian clothes and not Army fatigues. He avoided American reporters, telling them he wanted minimum or, if possible, no coverage at all. Instead of pressing to get close to battle, he rejected a suggestion that he fly near the siege of Khe Sanh.

The biggest difference was that this twelve-day trip was thoroughly prepared. Dave Burke organized a team of four advance men. They were Barrett Prettyman and John Nolan, two Washington attorneys who were close to Robert Kennedy; N. Thompson Powers, a partner of Nolan's; and John Sommer, who had spent several years doing volunteer work in Vietnam and spoke Vietnamese. For two weeks, they looked at hospitals and refugee camps in all parts of the country, found experts for Kennedy to talk to when he arrived, and assembled a huge briefing book telling him what he could expect to see and hear and what to ask.

Kennedy and a Wilmington doctor who had spent time in Vietnam, John M. Levinson, spent most of their long flight from San Francisco studying that briefing book. As they traveled in Vietnam, Burke and one or more of the advance team were with them. The advance team made the schedule, but Kennedy kept changing it, adding stops or rearranging the order. Levinson thinks that saved his life. A bomb went off at the home of David Gitelson of the International Volunteer Service in Can Tho just when Kennedy and Levinson were supposed to be there for dinner. At the last moment, Ted had switched the meeting to breakfast the next day.

One example of how the preparation worked came when Kennedy questioned an American colonel supervising the Chop Chai camp in Phu Yen about why the refugees' huts had no roofs. He asked if the Army could not just "put some tin on these huts." The officer answered, "We don't have any tin and we can't get any." Kennedy, alerted by Prettyman, told him where to find some two miles away. At other stops, Kennedy asked the refugees themselves whether they had received the money that chiefs or U.S. officials said they had been given, and often heard they had not.

He saw refugee settlements of all kinds. In Saigon itself, a friend of Levinson's, Dr. Thomas S. Durant of Boston, was running a clinic at a Catholic cemetery, where five thousand refugees had built their homes over the graves. They had taken caskets to use for bunk beds and slung hammocks between stone crosses. A chant went through the cemetery, "Ken-uh-dee, Ken-uh-dee," and refugees showed him pictures of his slain brother.

The United States Mission was less admiring. Robert Komer, a high-ranking U.S. official, refused to tell Ted of plans for the refugee effort, despite being urged to at dinner by Ambassador Ellsworth Bunker the night Kennedy arrived. "We have big plans" was all Komer would say. Bunker resented Kennedy's apparent disregard of his opinions, and Barry Zorthian, who headed the public affairs operation, complained that Kennedy was "unresponsive" to explanations it offered that many civilian casualties were really caused by the Viet Cong.

Kennedy heard differently from other Americans. Shortly after arriving, he had lunch in Saigon with veteran American correspondents. They largely reinforced his pessimism. At Danang, Dr. Gilbert Herod of Indianapolis and Dr. Alan Pribble of Covington, Kentucky, told him of delays in receiving supplies donated by the U.S. government. At Phu Cuoung, Herb Ruhs of the International Volunteer Service deflated a U.S. officer's satisfied report on attitudes among refugees by pulling out a Viet Cong flag and saying, "A whole bunch of these were found in the neighboring camp last night." Kennedy also heard from the legendary John Paul Vann, a former lieutenant colonel now back in the country as a civilian adviser. Kennedy listened while Vann and Burke argued over policy, with Burke challenging Vann's long-range optimism in the face of his list of failures. They traveled together by helicopter, and also in a breakneck midnight drive through Viet Cong–controlled forests in a vain attempt to gain access to a notorious insane asylum.

But it was civilian casualties and American responsibility for them that made the biggest impression on Kennedy. At Quang Ngai he met a forward air controller who told him, "My job is to go up there in this little plane, and I look around and if I see some guys running and they've got on black pajamas, I have to report it and they're going to go and aim some guns from someplace at them, and maybe air strikes and kill them. You know, they all wear black pajamas. That's what the peasant wears. They tell me that if they run in black pajamas I'm to report it because they're the bad guys. I am calling in air strikes and artillery on people and that's my job, and I can't do my job right and I hate my job." In the Mekong Delta, Kennedy visited a base near My Tho, where artillery fire was used for H&I, or harassment and interdiction, designed to make travel unsafe for the Viet Cong because they would never know just when the firing would start. Kennedy, shuddering at the gunfire, challenged claims by U.S. and Filipino officers that they knew there were no civilians where they were firing. Kennedy left convinced that precautions to protect civilians were a sham. He did not change his mind after General Westmoreland, still head of the U.S. forces, showed him a handbook that forbade indiscriminate shelling and told him the troops obeyed it absolutely.

Even Kennedy's departure from the Tan Son Nhut airport was delayed by a bomb threat. After the tension of the trip, he and the others drank heavily on the flight back. But when the plane landed at Guam and he was told the governor was waiting to greet him, he snapped to and performed.

Kennedy said years later that the 1968 trip provided the "final conclusion as to my own view about the war," leaving him troubled by the casualties the United States was causing and "the failure of the Vietnamese to fight for themselves."

That failure was what he emphasized when he first spoke publicly about the trip on January 25, 1968, at Boston's World Affairs Council, saying, "I believe the people we are fighting for do not fully have their hearts in the struggle. And I believe as well that the government that rules them does not have its heart in the cause of the people. So we are being forced to make the effort for them and take the risks they should be taking themselves."

He called officials in Saigon "colonialists in their own nation," caring little for their own people. The press focused on his extensive discussion of corruption. He said, "Corruption pervades all aspects of Vietnamese life, and it is brazenly practiced. . . . In the field of refugee care and in many other fields, the government of South Vietnam has been engaged in the systematic looting of its own people."

He suggested that the United States should threaten to withdraw if Saigon did not shape up: "It should be made clear to the elected government of South Vietnam that we cannot continue, year after year, picking up the pieces of their failures."

The day before, he and Dave Burke had called on President Johnson for ninety minutes to offer their impressions. Kennedy told Johnson the war was a disaster. "We're not winning hearts and minds," he said, "we're creating tons of recruits" for the enemy by causing casualties and forcing people to become refugees. Johnson was not pleased.

CHAPTER 9

The Last Brother

A few days later, on January 28, 1968, on the CBS News program *Face the Nation,* Ted endorsed the idea that the United States should tell South Vietnam, "Shape up or we're going to ship out." But any news he made was quickly drowned out. On January 31, during the Vietnamese New Year's holiday of Tet, Viet Cong and North Vietnamese forces attacked all over the country, shooting up the U.S. Embassy in Saigon, striking almost every city, seizing some, and startling the United States forces with their coordination and strength.

The attackers were beaten off, although it took twenty-five days to recapture the ancient city of Hue. In strictly military terms, the Tet offensive was a disaster for the Communists. They suffered severe casualties that battered their indigenous South Vietnamese forces hardest. But Tet's impact on U.S. public opinion, especially elite opinion, was equally devastating, because the offensive seemed final proof that the Communists were not on the run and there was no chance of winning the war. Clark Clifford, who was about to become secretary of defense, observed years later, "Its size and scope made a mockery of what the American military had told the public about the war and devastated American credibility." He wrote, "The most serious American casualty at Tet was the loss of the public's confidence in its leaders."

Tet also reopened the question of whether Robert Kennedy should run for President, if indeed it had ever really been closed. In the fall of 1967, some of the most intense opponents of the war urged him to enter the Democratic primaries against Lyndon Johnson. They were led by Allard K. Lowenstein, a New Yorker who had been involved in student politics, civil rights, South Africa (smuggling a student of mixed race out of the country in 1959 to protest apartheid), and now Vietnam. Robert Kennedy listened, but he refused. Lowenstein and Curtis Gans turned to Senator Eugene J. McCarthy of Minnesota. He agreed to run against Johnson. On December 2 he said he would run and enter five or six primaries, and that "the Administration seems to have set no limit to the price which it is willing to pay for a military victory."

When the news of the Tet offensive reached Washington, Robert Kennedy was at breakfast with a group of senior political reporters, discussing the arguments for and against running, and trying to make it clear that he would not run. During the meeting someone handed Frank Mankiewicz, his press secretary, a slip of wire service copy reporting the huge Communist offensive, and he showed it to Robert Kennedy. The ground rules were that Kennedy's remarks were generally not for quotation, but reporters could ask at the end to be allowed to quote a particular bit. The sentence they wanted was "I have told friends and supporters who are urging me to run that I would not oppose Lyndon Johnson under any conceivable circumstances." Mankiewicz argued for scaling it back to "foreseeable circumstances," and Kennedy, nodding toward the wire service copy, smiled archly and agreed. His disavowal was front-page news in the *Washington Evening Star* that day, though not played as prominently as the Tet story headlined "Reds Launch Mightiest Offense."

McCarthy and the Kennedys had never had much use for each other, dating back at least to McCarthy's 1960 convention speech nominating Adlai Stevenson. McCarthy was witty, devout, and cerebral, but was a lazy Senator known for leaving closed hearings early to get to the television cameras first. He was one of only two Northern Democrats to vote against Ted's poll tax measure in 1965. (Attorney General Katzenbach had asked him to.) McCarthy was not a comrade whose boldly staked claim the Kennedys would respect.

Any questions about Robert's running inevitably involved Ted. Since they had started serving together in the Senate, Ted and Robert had grown very close. They walked together from the Senate to their office buildings several times a day after votes. Ted remembered that Robert "would suggest that I speak at an early morning assembly at a local high school here in Washington, urging the students to stay in school and continue their studies. He would remind me to give whole-hearted support to a fund-raiser to be held in Boston for Cesar Chavez' farm workers. He would tell me of a recent trip to the Mississippi Delta, describing the extraordinary conditions of hunger, malnutrition and poverty he saw, and he would talk about the forgotten Indians in our country, and the injustices and indignities they suffered."

Ted found himself the central internal opponent of a Robert Kennedy candidacy. There were two main reasons. One he spoke of, in group meetings and alone with his brother. He argued that a run in 1968 would not succeed, that Johnson could not be denied the nomination, and that Robert's excellent chances for the Presidency in 1972 would be badly damaged if he tried now.

The other reason was more private. He did not say it in meetings of friends, relatives, and advisers. He did not even say it privately to his brother. He was afraid that Robert would be assassinated. "We weren't that far away from '63, and that still was very much of a factor," he said years later.

He asked people not to urge his brother on. One day in the Senate gym he encountered George McGovern and told him, "You know, George, I am not for this. And unless there are a lot of people willing to come out and endorse him if he decides to go, I don't think the thing would go anywhere. Well, you urge him to run, would you be willing to endorse him?" McGovern answered, "Well, you know, Ted, I am in a race for reelection to the Senate, and Hubert is from my state, and Gene McCarthy is next door. I am for Bob Kennedy, but we have kind of a tradition in South Dakota—if you are running for a position yourself, you don't weaken your position with anybody by public endorsements." And Ted said, "Well, shit, don't keep telling him to run then. If even you are not going to endorse him, don't tell him to run."

But there was an inevitability to it all. Robert thought the war should be ended, but he did not consider McCarthy worthy of the Presidency. And, unlike McCarthy, he was almost as firmly seized by the decay and despair of the cities. Johnson's War on Poverty was being cut back as the war in Vietnam increased; the President ignored the commission he had created under Otto Kerner, the governor of Illinois, when it demanded strong action to reverse the direction of a nation "moving toward two societies, one black, one white—separate and unequal."

Robert kept asking Ted what he thought. Ted would explain why this was the wrong year politically. And Robert would say, "I'm going to give it some more thought." Ted knew where things were going. On February 13, at Robert's suggestion, he had dinner in Boston with Dick Goodwin, who moved back and forth between the McCarthy and Kennedy camps but whose heart was with the Kennedys. Ted argued for waiting until 1972. Goodwin said he feared that four more years of Johnson would create a conservative revulsion that would defeat any liberal like Kennedy in 1972. As the evening came to a close, after a last drink at Goodwin's house on Beacon Hill, Ted conceded that his brother was inclined to run: "He usually follows his instincts and he's done damn well." Goodwin asked Ted what John F. Kennedy would have done. "I'm not sure about that," he replied. "But I know what dad would have said." Ted said he was sure what his father would have said: "Don't do it." Then he guessed about the late President: "Jack would probably have cautioned against it, but he might have done it himself."

On March 4, Fred Dutton, a former Kennedy administration aide who was close to Robert, came to Ted on an awkward mission. The day before, Robert Kennedy had asked Dutton to fly back from California to see him. When they met that morning, he asked Dutton to tell Ted that he had decided he had to run. Dutton said years later he tried to "wiggle out" of the assignment because it violated a basic principle: never get between Kennedys. But he went, and had sat down and cleared his throat when Ted broke the ice and said, "Fred, I know he is going to run. You don't have to tell me." Then they went over to Robert's office for a serious discussion of how to do it.

One political argument against running in 1968 had always been that it would seem ruthlessly arrogant to challenge Johnson, brother Jack's choice for Vice President. But the ruthlessness argument gained a new dimension after Tet as McCarthy's New Hampshire campaign suddenly took off. Hundreds of college students were shaving off their beards or cutting their hair to be "Clean for Gene" and not offend local voters as they went door-to-door, trying to reach every household in the state three times before the March 12 primary. Both antiwar voters and hawks impatient with a limited war joined behind McCarthy, or against Johnson (though the President was not on the primary ballot). Now Robert Kennedy would be seen as ruthless from the other direction, too, for jumping in after McCarthy had made the courageous challenge he had refused to attempt.

But the decision had been made. Robert thought of announcing before the New Hampshire primary, in California on Sunday, March 10. But he feared that would be taken as undercutting McCarthy in New Hampshire. On Friday, March 8, he spent ninety minutes with Haynes Johnson, a reporter for the *Washington Evening Star,* whose notes reported it "certain he will run." The day before, he had told Ted to let McCarthy know he was probably going to enter the primaries as soon as he could. But Ted, thinking McCarthy might use that against Robert, did not pass on the message. Robert finally sent word on Monday, March 11, through Goodwin, who was now writing for McCarthy.

Then McCarthy achieved the most successful defeat in the history of American politics. He got 23,263 votes in the New Hampshire primary. That was 42.1 percent. A write-in campaign for President Johnson got 27,520, or 49.8 percent. But McCarthy, who also won twenty of the state's twenty-four delegates to the Democratic National Convention, was the winner in the headlines. A typical lead was Robert J. Donovan's in the *Los Angeles Times:* "President Johnson defeated Sen. Eugene J. McCarthy in New Hampshire Tuesday, but with such a poor showing that McCarthy was a victor in defeat."

The next day Robert moved in abruptly. Reporters found him in a Senate corridor, and he told them, "I am reassessing the possibility of whether I will run against President Johnson." When McCarthy's student supporters heard, they were crestfallen and furious. In Manchester, a Connecticut College for Women student told Donovan, "Bobby didn't even give us twenty-four hours to enjoy the results of last night." Early that afternoon, Kennedy taped an interview with Walter Cronkite for broadcast later on the *CBS Evening News.* It was a calculated performance, begun by congratulating McCarthy and his followers. Then he complained about the Kerner report and about Vietnam, and said that McCarthy's victory had shown the party was already split, so his entry would not be divisive. "If I decided to run," he told the CBS anchorman, "it would be on the basis that I could win." Then back to the Senate, where he met McCarthy in Ted's office. McCarthy suggested that Robert stay out and wait until 1972, when McCarthy would not run for a second term.

Ted had summoned the usual crowd of advisers to meet in Steve Smith's New York office that afternoon, to discuss what should be done now in light of McCarthy's success. Then they adjourned to Smith's apartment to watch the Cronkite interview. When they saw Robert all but announce his candidacy, Ted said, "I don't know what we are meeting about, he has made all the decisions already, and we are learning about them on television." But by the time Robert arrived, there was nothing to do but to congratulate him on having decided.

There was reason to hurry. A public announcement by the end of the week was needed to get on the Nebraska primary ballot, and the deadline for naming a slate of delegates in California came the next week. But there was one more reason to wait: an odd, ill-formed idea that President Johnson might create a commission to find a new policy on Vietnam, a policy that would make a Kennedy candidacy unnecessary. Mayor Richard J. Daley of Chicago, who would control the critical Illinois delegation to the Democratic National Convention, had suggested it, both to Robert Kennedy and to the White House. Independently, Ted Sorensen suggested the idea to Johnson on Monday the eleventh. At Smith's apartment, Robert got a call from the White House saying the President wanted some suggested names. Then Mayor Daley, whom Robert called to say he was running, told him that Johnson wanted to use a commission to change policy in Vietnam.

So just before midnight, Ted called Clifford, who had succeeded Robert S. McNamara as secretary of defense just twelve days before, to say that Robert wanted to see him about Vietnam. An 11 A.M. meeting was arranged. It was futile. Kennedy wanted a commission of opponents of the war to change policy. Clifford, who wanted to change the policy himself, knew that the President could not do that, and warned that Kennedy would lose if he ran for President. The foreordained rejection came when Clifford met with Johnson. He spurned the idea immediately.

There was one last quixotic adventure that preceded Robert's formal announcement, and Ted was in the middle of it. Curtis Gans, the thirty-year-old ex-Marine who had run the McCarthy campaign in New Hampshire, started it. He talked to Blair Clark, a sewing thread heir who held the title of campaign manager and could talk to the quirky McCarthy when Gans could not. The idea was to see if some accommodation could be reached between McCarthy and Kennedy so that the antiwar vote would not be split, enabling Johnson to survive. The McCarthy campaigners knew their candidate had no prospects in Indiana, but it was not clear whether there was any state where Kennedy would give way to McCarthy. But Gans got Lowenstein to take the idea to Robert Kennedy. On Friday the fifteenth, Clark met with Goodwin in Washington, then called McCarthy in Wisconsin, and McCarthy agreed to meet with Ted that night. One reason Ted wanted to go was to give McCarthy a formal account of why Robert was running. He thought McCarthy deserved that much.

They had an awful trip in bad weather, and got from Chicago to Green Bay only when Clark chartered a plane. But since talking to Clark, McCarthy had changed his mind. His campaign had tipped off David Schoumacher of CBS News, who was waiting with a camera crew for this secret meeting. Instead of waiting up for Ted, McCarthy had gone to sleep, and his wife, Abigail, refused to wake him. Eventually, McCarthy arose and received them, well past midnight. Ted told him why Robert would run, and he suggested some states be the sites of Kennedy-Johnson races while others be McCarthy-Johnson. But McCarthy was not interested. He said, "I think we'd better go on as we were." About 5:30 A.M., Ted returned to Robert's house at Hickory Hill, waking Arthur Schlesinger, Ted Sorensen, and Bill vanden Heuvel and telling them, as he put it to vanden Heuvel, "Abigail said no."

A few hours later at the Senate Caucus Room, where John Kennedy had declared his candidacy eight years before, Robert Kennedy announced that he was running not "merely to oppose any man but to propose new policies." He said he felt obliged to run "because I have such strong feelings about what must be done." He said he sought new policies "to end the bloodshed in Vietnam and in our cities, policies to close the gap that now exists between black and white, rich and poor, between young men and old in this country and around the world."

Ted had seemed troubled that morning, walking around the yard at Robert's estate at Hickory Hill with Dave Burke. They walked for long periods without talking.

But he threw himself and his staff into the campaign. Gifford was charged with finding office space and working with the boiler room—the central communications hub of the campaign. With all of Robert's Senate staff on the road, Flug took over as the legislative assistant for both Kennedy offices during the day, then worked on student and youth organizing nights and weekends. Burke traveled with Kennedy as he sought support for his brother's campaign among big-city politicians and labor leaders and in some of those Western states he had learned in 1960. Ted had better union connections than Robert, whose past as a foe of racketeering compounded the chilly welcome an opponent of the war could expect from most labor leaders.

For the campaign's first trial, the May 7 Indiana primary, Ted recruited Gerard Doherty. Doherty went out for a long weekend in late March and found surprising enthusiasm for Kennedy, and Ted backed his view that Robert should run there. Doherty then brought dozens of Massachusetts pols to the Midwest. He had to. The Indiana Democratic Party organization was powerful and hostile. Local officials threatened not to certify the signatures a hastily assembled organization collected to put Kennedy on the ballot. On Wednesday, March 27, Governor Roger Branigin, who was running as a stand-in for Johnson, exploded at Ted over the phone, telling him he had deserted his natural friends, including Birch Bayh. He said it would be a long time before he ever got those friends back. "You probably can buy the state," Branigin said, "and I'll just call it Kennedy instead of Indiana. You bought West Virginia and you can buy this one."

On Sunday night, March 31, 1968, Ted and Doherty were in an Indianapolis hotel room after a daylong meeting with coordinators from all over the state. They were eating chicken sandwiches and watching President Johnson make another Vietnam speech on television. The President announced a sharp reduction in the bombing of North Vietnam and appealed for negotiations. But the bombshell came when he finished by saying:

"With America's sons in the field far away, with America's future under challenge right here at home, with our hopes and the world's hopes for peace in the balance every day, I do not believe that I should devote an hour or a day of my time to any personal partisan causes. . . . Accordingly, I shall not seek, and I will not accept, the nomination of my party for another term as your President."

The announcement startled even Vice President Humphrey, who had been given only a few hours' notice that the President *might* not run. He listened in Mexico City, where he was making an official visit. He agonized for a few days about whether to run himself, worrying: "Did I have a chance if I were to be available, or would I just be a punching bag for Kennedy, only to be humiliated and defeated in the convention. In other words, I wanted to have some time to think this through. I explained I am not anxious. Quite frankly, I am not sure I have the stomach for it, knowing the ruthless methods that are employed by both Kennedy and Nixon." But by the end of the week he had decided to run, urged on by Southern Democrats and by labor. It was too late to enter any of the primaries, and Humphrey made no efforts to test himself before Democratic voters, either through organized write-ins or by adopting the Johnson surrogate slate of delegates already on the California ballot.

The Kennedys went scrambling for supporters among Johnson backers. They got Larry O'Brien, who resigned as postmaster general, after first Robert and then Ted pressed him to rejoin the family's political enterprises, and he chose their side over Humphrey's. But Humphrey got big labor. George

Meany, head of the AFL-CIO, brushed off a call from Robert Kennedy and urged Humphrey to get in quickly to forestall a slide to Kennedy.

Another of that awful year's shocks came four days later. On April 4, Martin Luther King, leading a garbagemen's strike in Memphis, was shot and killed by a white racist, James Earl Ray, who escaped.

Robert was campaigning in Indianapolis that night, and he brought the terrible news to a black audience. In the most eloquent speech of his life, he appealed for calm, for compassion. He said vast majorities of both blacks and whites wanted to live together in peace. He said, "Let us dedicate ourselves to what the Greeks wrote so many years ago: to tame the savageness of man and make gentle the life of this world." There were riots in black slums all over the country that night and the next day, but it was calm in Indianapolis. Ted, in Colorado, said that looters "do not mourn a great leader, they mock him." He recalled King's speech at the Lincoln Memorial and said, "He will never see that dream, but the moment we realize his dream is our dream and work to make it ours, the nation can survive."

Ted and Robert talked almost every day, either by phone or in person. During that long Indiana primary campaign, Robert would soak in a bathtub after a day of campaigning, and Ted found that was the best time to talk. One day a New England Teamsters leader, whom they both considered personally honest, came to Ted with a proposition. If Robert would agree to follow the recommendations of the U.S. Parole Board in the case of Jimmy Hoffa, the corrupt Teamsters leader whose conviction he had made a personal crusade as attorney general, then the Teamsters would help him at the polls and give him $1 million in cash. Dutifully, Ted passed the message along, while Bob was in the tub. His brother answered, "Well, you tell so-and-so, that if I get to be President, that Jimmy Hoffa will never get out of jail and there will be a lot more of them in jail." In the fall election, the Teamsters supported Richard Nixon. In 1971, President Nixon commuted Hoffa's sentence and freed him.

Robert won the Indiana primary, with 42.3 percent to 30.7 percent for Branigin, who was now a stand-in for nobody, and 27.0 percent for McCarthy. A week later, on May 14, Robert won in Nebraska, swamping McCarthy with 51.7 percent to 31.2 percent.

Then, on May 28, a Kennedy of his generation lost an election for the first time. After his victorious defeat in New Hampshire, McCarthy had won on April 2 in Wisconsin, where he was on the ballot with Johnson, who had withdrawn just two days before. And he had won in Massachusetts on April

30, where his name was the only one on the ballot, though a variety of write-ins exceeded his total. In Oregon, there were no blacks, no labor, no natural Kennedy constituency. There was an early antiwar movement that had stuck with McCarthy. What little organization the state's Democrats had was that of Wayne Morse, one of the war's first critics, and it was behind McCarthy. McCarthy got 163,990 votes, or 44.0 percent. Kennedy got 141,631, or 38.0 percent. Johnson, still on the ballot because his withdrawal had come too late, got 45,174, or 12.1 percent. Write-ins for everyone from Humphrey to Nixon to George Wallace totaled 22,095.

There were two important primaries left, in California on June 5 and in New York two weeks later. California was winner-take-all. In New York the Presidential candidates' names were not even on the ballot; the race was for delegate, district by district. Humphrey, with his ties to labor and his inheritance of organization support in most states, was way ahead in the national delegate count. Robert Kennedy's task was a greatly compressed version of his brother's in 1960; Robert needed primary victories to impress the politicians whose main concern was having a winner who would help their local candidates. Those politicians might shrug off the Oregon defeat—it was a curious, prematurely environmentalist state with few electoral votes. But in order to persuade the professionals of his "winner" argument, Kennedy had to win both California and New York, two big states that decide general elections when they are together.

California was where the campaign finally came together for Robert Kennedy, where his appeal to Hispanics, blacks, and working-class whites began to connect. The Speaker of the state assembly, Jesse Unruh, and the Congressman who was the state's nearest equivalent of a political boss, San Francisco's Phil Burton, buried the hatchet and united behind him. In California, Ted finally found some labor leaders he could work with, and he campaigned hard. He also partied hard at the end of the campaign with Helga Wagner, the wife of a San Francisco businessman. Joan had gone to Paris to visit Sarge and Eunice Shriver. Sarge had been appointed ambassador to France by Johnson.

Ted was in San Francisco on primary night. John Seigenthaler, the Nashville newspaper editor who was Robert's close friend and had worked for him at the Justice Department, had asked for him, wanting a Kennedy on hand to thank the campaign workers who had made Northern California secure. The returns were slow coming in. They looked bad at first, but soon turned better. Shortly before midnight, Ted arrived at the party, where drinks had been flowing, and the workers went wild. "A great night," he said. "California is coming through for us." The crowd then watched Robert Kennedy on television, claiming victory and saying, "I think we can end the divisions in the United States. . . . What has been going on over the period of the last three years—the divisions, the violence, the disenchantment with our society, the divisions, whether it is between the poor and the more affluent, or between age groups—that we can start to work together again. We are a great country, an unselfish country, and a compassionate country."

Ted left then to return to the Fairmount Hotel. Three hundred and eighty miles south, in Los Angeles, Robert pushed through the crowd, out through a kitchen next to the ballroom at the Ambassador Hotel. At 12:13 A.M., Wednesday, June 5, 1968, Sirhan B. Sirhan, a Christian Palestinian outraged by Robert Kennedy's support of Israel, shot him in the head.

Ted heard the awful news on television back at the Fairmount. Burke raced down the ten flights of stairs at the hotel. He told an American Airlines clerk that he needed a plane immediately. The clerk looked at him, not comprehending the odd request. Burton, who had heard the news from Los Angeles, stepped in and said he would arrange a plane—and would go along. Burton reached an Air Force major, who was reluctant to order up a plane. The Congressman told him, "I am standing here with Senator Edward Kennedy, whose brother has just been shot and who may be the next President of the United States. You are at a point I call a career decision, Major. Either you get that plane or your career is over." He got the plane, but Burke saw to it that Burton did not come along. After Kennedy, Burke, Seigenthaler, Dick Drayne, Ted's press secretary, and a cousin, Robert Fitzgerald, piled into two police cars, Burke told the police to let no one else—meaning Burton—follow. Ted was silent on the drive and the flight to Los Angeles, where a helicopter waited for them. There wasn't room for everyone, but he made sure there was a place for his brother's friend, Seigenthaler.

When they arrived at Good Samaritan Hospital, Robert was about to be operated on to remove bullet fragments from his brain. The surgery began at 3:12 A.M. Wednesday, and finished about 7 A.M. Mankiewicz came out a few minutes later and told reporters that there might have been a loss of blood to the brain. Political reporters began to write obituaries. Medical writers explained the term "brain-dead." At 5:30 P.M., Mankiewicz told reporters, "Senator Kennedy's condition is still described as extremely critical as to life." By early evening, Robert's brain waves were flat. His relatives and his closest aides filed in to say goodbye. On his own farewell visit,

Mankiewicz recalled, he saw Ted "sort of leaning over the sink with the most awful expression on his face. Just more than agony, more than anguish. I don't know if there is a word for it."

Ted's worst fear for 1968 came true. At 1:44 A.M., Thursday, June 6, 1968, Robert Francis Kennedy was allowed to die. Edward Moore Kennedy's life changed forever.

Part Two

CHAPTER 10

A Summer of Entreaties

Robert Kennedy had not been dead an hour before Ted was urged to run for President in his place.

Ted and John Tunney were riding down in an elevator at the Good Samaritan Hospital to the autopsy room. Ted wanted a last look. Allard Lowenstein, who had flown out from New York, got in, embarrassed that he was intruding, and blurted, "Now that Bobby's gone, you're all we've got." He urged Kennedy to "take the leadership."

The pleas continued that afternoon. Ted flew back with the body to New York on an Air Force jet President Johnson had sent. Paul Corbin, a coarse Kennedy political operative, came up to him and insisted, "You gotta run, you gotta run for President." There was talk in Washington that day of him as a Vice Presidential candidate.

For the next fourteen years, he was a constant presence in Presidential politics. No one else ran for President in those years without weighing the likelihood and impact of a Kennedy candidacy. Over and over, Kennedy would think about running, put it out of his mind, and come back to worry the subject out again. Nor did this subject arise in his mind for the first time after Robert died. He and friends like Tunney had kicked the idea around for a few years, and assumed that "someday" he would run for President.

But it was completely beyond him in June 1968. He was devastated, more so than when John F. Kennedy had been killed less than five years before. Not only was Robert closer to him, in age, and especially since they had served in the Senate together, but now Ted was the only brother left.

On the flight back, Ted sat by the coffin with Dave Hackett, a chum of Robert's from Milton. He talked to the other passengers, family and campaign staff, to comfort them even as he was numbed by his own sorrow. He spent time talking with reporters who had traveled with Robert for those eighty-five days of his campaign. Jules Witcover of Newhouse Newspapers wrote that as he talked of the past and the future, "he stressed his increased family responsibility, both personal and political. It seemed clear even then he had dismissed thoughts of any active role for himself in 1968."

After Robert's oldest sons, Joseph P. Kennedy II and Michael Kennedy, helped him unload the coffin from the plane at La Guardia Airport, Ted joined Ethel in riding with it in the hearse to St. Patrick's Cathedral. John Rooney, a prominent New York funeral director, said he had never seen anyone do that. After a brief service, Ted walked Joan (who had flown back from Paris) to a car, but went back into St. Patrick's just past 10 P.M. He stayed for hours, gripping a rosary and missal. Early Friday morning, June 7, he went by the thirty-eighth-floor offices of the family business in the Pan Am Building, where the funeral was being organized. His face was puffed and his eyes red. Friday night he drove around the city with Culver, not going to sleep, talking and trying to muster the strength to deliver the eulogy that his friends feared he could not handle.

Saturday he persevered. At the cathedral, he said Robert "gave us strength in time of trouble, wisdom in time of uncertainty and sharing in time of happiness. He will always be by our side." He quoted from what Robert had written about their father's injunction to his fortunate sons to take responsibility for "others who are less well off." And he read at length from a speech Robert had given in South Africa: "Each time a man stands up for an ideal, or acts to improve the lot of others, or strikes out against injustice, he sends forth a tiny ripple of hope. . . . Our future may lie beyond our vision, but it is not completely beyond our control. . . .

"My brother need not be idealized, or enlarged in death beyond what he was in life. He should be remembered simply as a good and decent man, who saw wrong and tried to right it, saw suffering and tried to heal it, saw war and tried to stop it." Here his voice broke, but he went on:

"Those of us who loved him and take him to his rest today pray that what he was to us, and what he wished for others, will some day come to pass for all the world.

"As he said many times, in many parts of this nation, to those he touched and who sought to touch him:

" 'Some men see things as they are and say why. I dream things that never were and say, why not.' "

After the Mass, a funeral train took the body and the mourners south to Washington. It made a slow, sad journey. Crowds stood by the tracks. In Elizabeth, New Jersey, they were so thick that a man and a woman were hit by a northbound train on an adjacent track and killed. Ethel Kennedy and her oldest son, Joseph P. Kennedy II, walked through the train, thanking the mourners. Ted stayed mostly in the last car, with the casket, which was balanced on chairs so the crowds could see it through the windows. When the crowds were thickest, he would step out on the rear platform and wave, or at least nod, to acknowledge the mourners. Elsewhere, commencement ceremonies halted for silence. Longshoremen ceased their work in tribute. When the train arrived in Washington, it was raining. But there were more crowds. It was dark, except for candles held by mourners, when Robert Francis Kennedy was buried at Arlington National Cemetery on June 8, 1968.

For the next week, Ted sailed. With Gifford, he chartered a boat, the *Mira*, in Oyster Bay. Joan, not Ted, told Kara and Teddy about Uncle Bobby's death. Joan and the children joined Ted and Gifford at New London, and they cruised on to Hyannis Port.

Ted was out of sight, doing nothing to encourage the idea that he would run for anything. But the speculation continued. To some politicians, it seemed obvious that he would. To critics, he was just practicing the sort of manipulative calculation they expected of the Kennedys. But from neither perspective was anyone looking at Ted Kennedy as a person; he was just the embodiment of dynasty.

On Saturday, June 15, Ted and his mother appeared on television to thank the public for its sympathy. But he tantalized his followers with the conclusion: "And each of us will have to decide in a private way, in our own hearts, and in our consciences, what we shall do in the course of this summer, and in future summers, and I know we shall choose wisely."

Polls showed that if Ted ran for Vice President, he would help the chances of Hubert Humphrey against Richard Nixon. With Robert Kennedy's death, Humphrey had become the all-but-certain Democratic nominee. But even before the first polls were out, when Humphrey resumed his own campaign on Thursday, June 20, he was asked if he had invited Kennedy to run with him. He said he had not. Humphrey sent Ted a gentle condolence note, praising Robert Kennedy as "a brave man and a warm human being" who "lived a full life—giving and sharing."

Ted thought seriously about leaving politics entirely. He could not go back to the Senate. He tried one day in mid-July, but couldn't get out of his car, and turned back to McLean. To a Senate committee considering a new gun control bill, which Johnson had championed on the day of Robert's death, he wrote a letter but did not appear and testify.

Instead he concentrated on the family. He would take Ethel and her children on easy cruises to Martha's Vineyard. John Kennedy, Jr., nearly nine, was at the helm with Ted right behind him on a trip to Newport to watch the start of the annual Bermuda race. Ted took his nephew Joe to Spain to settle him for a summer away from reminders of his slain father. On one sailing trip off the New England coast with Bill vanden Heuvel, he grew a beard, which startled his father. He held a meeting with about two dozen of Robert's friends to discuss a memorial. But he never escaped speculation; that meeting was interpreted as a political strategy session. Later, Mayor Richard J. Daley of Chicago got word to him at sea to come ashore and talk politics. From a pay phone in Stonington, Maine, he said no.

In 1960, Daley had been a vital supporter, producing enough dubious city votes to compensate for Republican cheating in downstate Illinois and carry the state for John F. Kennedy. Daley hung back from supporting Robert Kennedy in 1968. But Robert expected an eventual endorsement, thinking Daley would see him as the candidate who could help his local ticket the most. Late in July, Daley, who controlled the Illinois delegation, started to generate a Vice Presidential draft for Ted. Governor Samuel Shapiro of Illinois, Daley's man, spoke up on July 21 at the National Governors Association meeting in Cincinnati. Governors Richard J. Hughes of New Jersey, John McKeithen of Louisiana, and Robert J. Docking of Kansas agreed. So did three New Englanders, Governors Kenneth Curtis of Maine, John Dempsey of Connecticut, and John King of New Hampshire.

On July 23, Governor John B. Connally of Texas, a Johnson protégé, tried to put a stop to it. He told reporters he did not see why Humphrey, "an outspoken liberal throughout his career," needed any liberal help. He said he doubted whether Kennedy had the "ability and capacity" to serve as President. Daley ignored him and joined in personally. The next day, at a press conference in his office at City Hall, he was asked if he thought Kennedy would help Humphrey win. He said Kennedy would help, and then muttered, almost under his breath, "And I hope the convention drafts him."

Kennedy ignored the talk from Democratic governors. But after Daley's announcement, on July 26, he issued a statement intended to kill the idea: "Over the last few weeks, many prominent Democrats have raised the possibility of my running for Vice President on the Democratic ticket this fall. I deeply appreciate their confidence. Under normal circumstances, such a possibility would be a high honor and a challenge to further public service. But for me, this year, it is impossible."

That shut the talk down, at least publicly. But after a Gallup Poll showed Humphrey trailing Nixon, Larry O'Brien, now at work for Humphrey, called Ted one midnight to ask if he was available. He was not. Kennedy sailed some more. Other things were happening that drew attention away from him. The Republicans nominated Nixon in Miami Beach on August 6, and he promised that "the first foreign policy objective of our next Administration will be to bring an honorable end to the war in Vietnam." The Soviet Union, fearful of the liberal regime in Prague headed by Alexander Dubček, was menacing Czechoslovakia, and eventually invaded on August 21, 1968.

On Tuesday, August 20, O'Brien announced that Humphrey now had 1,312 delegates pledged to him, a majority for the nomination at the convention in Chicago the next week. Kennedy called Humphrey to congratulate him, and they arranged to meet.

But the next day, before their meeting, Kennedy forced himself back into the public's eye. He went to Worcester to talk at Holy Cross College. The speech, his first since Robert's death, was nationally televised.

He had decided to stay in public life and carry on his brother's fight against the war, "the thing he really cared most about at the end, it dominated everything else," as he explained his decision a couple of years later, and to try to restore the unusual connection Robert had made both to poor blacks and to blue-collar working-class whites.

"Like my brothers before me," he said, "I pick up a fallen standard. Sustained by the memory of our priceless years together, I shall try to carry forward that special commitment to justice, to excellence and to courage that distinguished their lives." He would not retire; "there is no safety in hiding."

He focused on Vietnam. He said all his earlier hopes had "been buried by the incompetence and corruption of our South Vietnamese allies." He demanded that "we now resolve to bring an end to this war . . . as quickly as it is physically possible." He demanded an immediate halt to the bombing of North Vietnam, a reduction of U.S. military activity on the ground, and an effort to refocus the languishing Paris peace talks on withdrawing all foreign troops, including North Vietnam's, from South Vietnam. "As Hanoi withdraws her troops," he proposed, "we withdraw ours." He said, "The government in Saigon must not be given a veto over our course in Paris, our cessation of the bombing, or our mutual withdrawal of troops."

The speech, his sharpest ever on Vietnam, was an effort to influence what the Democratic platform said about the war, and the Humphrey forces quickly sought to accommodate him. But it also stirred Democrats who hoped Ted still might run that year, from Mayor Daley to rank-and-file delegates. They ignored his categorical description of himself "as one who will not run for office this year" and heard what they wanted to hear, the "no safety in hiding" line and a closing interpretation of the meaning of Robert's campaign, that there was more to citizenship than voting every four years. He said, "Each of us must take a direct and personal part in solving the great problems of this country. Each of us must do his individual part to end the suffering, feed the hungry, heal the sick, to strengthen and renew the national spirit." Lyndon Johnson heard it, and decided Ted was planning a coup. Daley heard it, and called to raise the idea of Ted running for President, not Vice President as he had urged before.

Humphrey himself was quickly reassured. The next night, August 22, he drove out to Kennedy's house in McLean and asked him to run for Vice President, perhaps suggesting that he would retire after one term so Ted could have the Presidential nomination in 1972. Ted said he had no interest in running for anything that year.

But for all his reluctance, Ted did take one step toward a candidacy.

The next day, Friday, August 23, he sent Steve Smith to Chicago to talk to Daley. Steve came back convinced that the mayor was playing it straight when he urged Ted to run for President, not trying to draw him into running for Vice President. Daley called again on Saturday, pressing Ted to come to Chicago, or at least say he would consider a draft. He did neither, but sent Steve back to stay in touch with Daley. And he also asked Tunney and Culver, neither of whom wanted him to run, to keep an eye on things there.

Emotionally, Kennedy almost certainly could never have brought himself to run. But he was curious, and tempted at least intellectually. He could reject a true draft, one that really represented what the party wanted. But it would not be as easy as turning down a few insiders just looking out for themselves. He wondered what was really going on in Chicago.

The convention was under siege. Outside, antiwar protesters camped in the park and fought with police, who routed them brutally. On television, it looked like revolution. Burke wondered how Kennedy, if he decided to go, could safely get from the airport to the convention hall. Where would a helicopter land? Inside, politicians battled over the platform and worried about their candidates. To a degree often missed on the scene, Lyndon Johnson was still in charge. He was being driven from office over the war, but still controlled what the platform would say about it. Humphrey could not budge him, and would not break. Johnson even planned to arrive on nomination day for a birthday celebration.

This was possible, despite the drama of New Hampshire, Indiana, and California, because the overwhelming majority of delegates had been picked not by voters but by other politicians. Even some who were in theory popularly elected were in fact chosen in the most autocratic ways. Voting dates were changed at the last moment, or never publicly announced at all. One township caucus in Missouri was held on a bus speeding around country roads so that no unwelcome McCarthy backers could get aboard. And most of the delegates picked in these odd ways were controlled by leaders formally committed to Humphrey but basically controlled by Johnson. Daley might buck the President with Illinois's 118 votes. But Governor Warren Hearnes of Missouri, with sixty, and James Gray, the Georgia state chairman, with forty-three, would not.

On Saturday, August 24, after Daley's call, Jesse Unruh, speaker of the California state assembly and leader of that state's 179-member delegation, arrived in Chicago. Bill Daley, the mayor's son and aide-de-camp, greeted him, and Unruh said, "I've got to talk to your father." Daley replied, "Fine, call him. He's at home." The son said years later that his father doubted that Humphrey could win, fearing that he lacked the needed independence. But he thought Ted could beat Nixon.

Unruh, whose delegation had been elected pledged to Robert Kennedy but since his death was now free to vote for anyone, made it clear he was looking around. He told reporters, "I want the California delegation to remain flexible and consider all alternatives."

On Sunday, August 25, Unruh and Daley talked politics at breakfast. Steve Smith arrived in Chicago a bit before noon. So did Humphrey, disappointed to find that Daley had sent a band but was not at the airport to meet him. Daley told Smith on the telephone that the Illinois delegation caucus that afternoon would not make a decision on a candidate, but would wait to see if anything happened. Daley did not tell Humphrey, who came to the Sherman House, made his pitch for endorsement, and only then discovered that Daley planned to wait a bit longer. Michael V. DiSalle, a former governor of Ohio, announced he was setting up a draft-Kennedy office. Lowenstein stirred the pot, encouraging delegates to drop in and state their availability. On Tuesday morning the *Chicago Tribune* reported that the leaders of the Michigan delegation, Senator Phil Hart and Sandy Levin, the state chairman, endorsed Kennedy; in fact, they were only praising him, but the report kept the pot boiling.

Smith stayed in a suite at the Standard Club, a Jewish club in the Loop, telephoning allies in various delegations. Late Sunday he insisted to Daley that Kennedy would not lift a finger to encourage a draft, and the mayor told his son he thought that meant Kennedy would not run. But DiSalle claimed 437 delegates had enlisted. Ted called DiSalle and asked him to shut down his office on Monday. But many of the pols brushed that off. Unruh told reporters he did not regard that action as final, saying he would "guess that Senator Kennedy might take a careful look at the situation if enough support is evident for him." And Smith asked friends from different states to get a solid estimate, delegation by delegation, of Ted's chances.

On Tuesday, August 27, Smith told Johnny Apple of *The New York Times,* "No one is going to find a shred of evidence that the Senator is working for the nomination." But by then Smith was convinced that it could be won. Some allies thought he wanted Ted to run; others were sure he did not. When John Seigenthaler, back working for the *Nashville Tennessean,* came up to Smith on the convention floor Monday evening and said he thought Ted could have the nomination if he wanted it, Smith told him to call the Senator at the Cape. He did, and then went back to tell Smith, "I don't think he is going to do it." Smith answered, "I don't think so either, but I'm glad you told him."

In the gathering confusion, Humphrey's key supporters were sure their votes were solid. Humphrey worried some, but Larry O'Brien assured him Ted would not run. On Tuesday the California delegation waved orange "We Want Teddy" signs. Various Kennedy allies were sure the nomination could be won. Senator Russell Long told vanden Heuvel that Louisiana, apparently locked up by a Johnson ally, Governor McKeithen, would break for Kennedy. Vanden Heuvel called Hyannis Port and told Kennedy of the offer: "You know, this is a long hill, the Presidency. It's a hard hill to climb, and all I'm saying to you, and I'm not trying to persuade you, that the nomination in my judgment is yours, if you're willing to be available to it. And I'm not pressuring you in any way, but it will probably be a long time before we're ever this far up the hill again." Kennedy laughed, but said, "No, I'm not going to do it."

McCarthy, who knew he could not win himself, was intrigued by the talk of drafting Ted. On Monday he had Goodwin call Steve Smith several times, suggesting that McCarthy might step aside. On Tuesday afternoon, Smith went to McCarthy's suite in the Conrad Hilton to talk. Goodwin and Peter Maas, who had traveled with Smith, wrote conflicting accounts, but they agreed on two essential points. Smith made it clear that he was not asking for his support, and McCarthy offered to drop his candidacy and urge his delegates to back Ted. The long-standing antagonism was underlined when McCarthy (according to Maas) said, "While I am doing this for Teddy, I could never have done it for Bobby."

Smith went back to the Standard Club. Unruh joined him. So did Abe Ribicoff, a Kennedy ally running for reelection to the Senate from Connecticut. They could not reach Daley, but did not take that for the grim sign it may have been. Smith's distrust of McCarthy boiled over when CBS's Schoumacher went on television to report that Smith had spent two hours with McCarthy seeking his backing. He tried and failed to get the McCarthy forces, who were obviously Schoumacher's source, to deny it.

Steve Smith called Ted frequently in Hyannis Port. After an evening meeting with the friends he asked to check the delegations, he told Ted he had the votes to be nominated. Ted said no. Steve told the group it was over. Lester Hyman, who had succeeded Doherty as Massachusetts party chairman, could not believe it. He had always known the Kennedys to seize any opening they could. So when he got back to his hotel, he called Hyannis Port and asked Ted why not. He recalled that Kennedy gave him two reasons: "One, because if the delegates did this, they'd be doing it out of sympathy for Bobby. And number two, 'I am not yet qualified to be President of the United States.' " If he got the 1968 nomination, he would not be getting it as himself, but as the kid brother. Kennedy explained it a bit more to Warren Rogers of *Look*, saying, "How could I conscientiously combat allegations by Nixon—and we had to anticipate he would make them—that

I was too young, that I had no record in public life strong enough to recommend me for the high office of President, that perhaps I was trying to trade on my brothers' names."

On Wednesday morning, Ted called Humphrey to say he would not allow his name to be put in nomination. He had Culver carry the same message to Carl Albert, the majority leader and the convention's presiding officer.

The next evening, watching the convention on television, Kennedy saw alternating pictures of delegates voting for Humphrey (Kennedy got twelve and a half diehard votes himself) and of the Chicago police going wild, beating up McCarthy supporters and anyone else within reach of their nightsticks.

He turned to Dave Burke. In a manner he uses when he is satisfied that he has made the right decision, he started riding Burke mercilessly. "You know, someday I'm going to be in Coos Bay, Oregon, or someplace like that, and I'll just have blown my last chance at the nomination and I'm going to call you and remind you of what we decided here." He went on, "You'll regret it ten years from now, when you're nobody and you could have been and this could have been, and I could have been, and God, we would have taken over the country, and God, we would have done this, and we would have done that. Oh, no, no, no, Dave says, 'No, I'm not going to hear it.' Oh, no, get Dave another drink, he's going to be a lot of help tonight."

CHAPTER 11

A Leader with Robert's Causes

Ted was barely involved in the rest of the 1968 campaign. He asked Kenny O'Donnell to campaign with Humphrey, and Hyman to tour with Senator Edmund S. Muskie, the Vice Presidential candidate, as evidence of the family's involvement. Muskie had been chosen by Humphrey after Ted rejected the nomination himself and made it clear that his brother-in-law Sargent Shriver would not be acceptable to him. When Shriver went to France as ambassador for Lyndon Johnson and then stayed instead of resigning to come home and campaign for Robert that spring, his political future, at least among his in-laws, was over.

Joan went out and campaigned for Birch Bayh in Indiana and John Culver in Iowa. But Ted made only three public appearances in the campaign himself, and they were all in Massachusetts. The first was on September 9, when he went west to Chicopee to raise money for Eddie Boland's House race. His rousing endorsement of Boland drew cheers, but only the reporters in the crowd of twelve hundred appeared to notice when he said that Humphrey and Muskie were "best equipped to handle the problems facing the nation" and "to provide the kind of programs and plans we need to move this country ahead."

Humphrey, desperate for support and running well behind Nixon in the polls, wrote a cloying response, saying, "I know of no one who can be of more help to me than you. . . . I do want you to speak up for me. It can mean the difference in this election. I shall try to do the things that will earn and merit your respect and confidence." On September 19, Humphrey came to Boston, and he and Kennedy met a screaming, abusive crowd. Humphrey was getting used to it. But Ted was not, and the cries of "Sellout" and "Shame on Teddy" and "Dump the Hump" often drowned him out. Joan sat grimly on the platform. The *Herald*'s Peter Lucas wrote, "It was probably the worst Massachusetts reception Kennedy ever has received."

All across the country the Humphrey campaign was dogged by demonstrators furious about Vietnam. Humphrey had been unwilling in Chicago to break with Johnson over the war. O'Brien kept pressing him to prove that he was his own man. The campaign manager scraped together $100,000 for a televised half-hour speech on September 30, 1968. Humphrey told the nation from Salt Lake City, "As President, I would be willing to stop the bombing of North Vietnam as an acceptable risk for peace, because I believe that it could lead to negotiations and a shorter war." That speech revived Humphrey's standing among many of his traditional liberal supporters, who started sending money. Kennedy sent an enthusiastic telegram praising him as "the only candidate for the Presidency who has offered the American people proposals for ending the war in Vietnam, a war which must end before we can meet effectively our grave domestic needs." Humphrey was delighted. O'Brien wrote later: "He was beaming as he showed it to us. He was a new man from then on."

Ted made a couple of television commercials for Humphrey. He praised his record on everything from arms control to education, saying that "in the past he was always considered a man of the future." In an Election Eve program, Kennedy said that "no one worked harder" than Humphrey had worked for John F. Kennedy in 1960 after losing the nomination to him. Kennedy criticized Nixon for refusing to debate Humphrey, and for not speaking out on what he would do as President.

Kennedy was much tougher on Wallace, who was running as an independent. Polls showed Wallace getting the support of blue-collar workers who had favored Robert Kennedy before his death. Wallace campaigned for "law and order," while his running mate, retired Air Force chief of staff Curtis LeMay, said the way to end the war was to "bomb North Vietnam back to the Stone Age." Many politicians feared that Wallace would carry enough states to deadlock the Electoral College and force the decision into the House of Representatives, where each state would have one vote. (Dun Gifford, one of Kennedy's aides, even calculated how this bizarre process might result in Kennedy's becoming President after all, and drew an amiable rebuke from the Senator, who suggested that Gifford had been enjoying a too-liquid lunch.)

Ted attacked Wallace as a supporter of discrimination at the Boston appearance with Humphrey, in an interview he gave the *Globe* and the *Worcester Evening Gazette* on October 10, and then, in a major way, in New Bedford less than two weeks before Election Day. On October 24, in a speech whose importance was promoted in advance to the national press, Kennedy spoke to a meeting of New Bedford's labor council and its service clubs.

A strong Wallace vote, he said, "would lift the haters and the wreckers to positions of formidable influence in our country." Addressing those Wallace backers who had supported his brothers, he said that "nothing could be further from the principles of these men." He elaborated:

"President Kennedy upheld the Constitution and the laws of the United States. George Wallace defied them.

"President Kennedy defended America against the extremists. George Wallace is in league with them.

"President Kennedy fought for legislation that would ensure every citizen the right to vote. George Wallace would repeal it.

"Robert Kennedy stood for reconciliation between the races. George Wallace stands for division and suppression. He was elected and he runs today on the slogan of 'Segregation Forever.'

"My brothers believed in the dignity of man. How can those who stood with them support a man whose agents used cattle prods and dogs against human beings in Alabama?"

The audience listened quietly for twelve minutes and gave him a one-minute standing ovation when he finished. On Election Day, November 5, 1968, Wallace got 13.5 percent of the vote and five states with forty-six electoral votes. Humphrey, who had been gaining fast at the end, got only half a million votes fewer than Nixon (31,275,166 to 31,785,480) and 42.7 percent, but only 191 electoral votes, while Nixon had 43.4 percent and 301 electoral votes and was elected.

The biggest news any Kennedy made that fall came when Jack's widow, Jacqueline Bouvier Kennedy, married the Greek shipping billionaire Aristotle Onassis, on the island of Skorpios on October 20. The marriage was a dismaying surprise to the country, which had made her into an icon. But except for Rose, the Kennedys supported her. Jean and Eunice attended the ceremonies off the Greek coast. Ted stayed behind because Ethel, the other widow, had almost lost her last child, Rory, a few days earlier in her seventh month of pregnancy. He did not want to be several thousand miles away if anything happened. But he had visited Skorpios with Jackie in August. Her biographer Willi Frischauer wrote that Onassis told him Ted made the trip to negotiate a prenuptial agreement. Ted says he did not, and certainly the details were left to lawyers.

Kennedy also took a small role in the Senate for the rest of 1968. With other Democrats, he tried to get a vote on President Johnson's nomination of Abe Fortas, then an associate justice of the Supreme Court, to be Chief Justice. But the Republicans, expecting to win the White House and get their own appointment, filibustered the nomination to death. The one issue that captured Kennedy was the civil war in Nigeria, where starvation threatened the people of the breakaway state of Biafra. On September 23, 1968, in his first Senate speech since Robert's death, Ted said that Biafrans were dying at a rate of well over seven thousand a day while "our own government and the international community stand paralyzed." Russell joined in, calling the situation "genocide," and Mansfield followed, saying Washington had tried to treat Biafran starvation as a matter to "sweep under the rug and ignore as if it did not exist."

Kennedy kept at it all fall. He visited U Thant, secretary-general of the United Nations, on September 25, and was told the UN could do little because its members were more concerned about sovereignty than starvation.

He wrote Dean Rusk, the secretary of state, urging greater relief efforts and complaining of a lack of urgency in the American response. He wrote Clark Clifford, the secretary of defense, asking for greater airlift support to get food to isolated Biafra. And in his first contact with the new President, he wrote Nixon on November 15, asking him to give the issue a high priority and to work to send food and halt the arms flow.

On December 6 he blamed both Nigeria and Biafra for shooting down relief planes, the other African nations for backing Nigeria as a generic warning against secessionism, and European countries for arming both sides. "You may ask why," he said, "if Nigeria-Biafra and the people of Africa care so little about the suffering, why should we?" He answered, "We hold ourselves out as something different on this globe; a nation that has responded to suffering, has helped the helpless, has looked beyond our narrow interest. . . . Perhaps the starvation of people in Nigeria-Biafra is not in our vital interest. But it is in our conscience."

A few weeks later, at Christmastime, he flew from a visit to his father at Palm Beach to a ski trip at Sun Valley, and made the only impulsive political decision of his life. Ed Muskie, whose campaign dignity distinguished him in 1968, had thought about running for whip, or deputy majority leader, against Russell Long of Louisiana. Democrats like Phil Hart had urged Muskie to run, wanting a consistent liberal in the job. But Muskie had decided against it after counting the votes and coming up short. After a phone call from Charles Ferris, Mansfield's chief aide, Kennedy jumped in. Over a couple of days and without the sometimes agonizing weighing of arguments pro and con he usually went through, Kennedy decided to run.

This was an opportunity to reassert Kennedy leadership, and just as in 1962, there was an opening.

There was an opening because Long had not paid much attention to the job, which involved largely the tedious business of salving Senatorial egos by scheduling votes when they were most convenient and handling some of the most trivial parliamentary duties. There were two reasons Long had slacked off. First, he had become chairman of the Finance Committee in 1965, not long after being elected whip. That chairmanship was a major job, especially for an oil-state Senator. Second, Long was a drunk, sometimes incoherent on the Senate floor. Newspapers did not report it, but he embarrassed the elder Southerners.

While a couple of friendly Senators asked why Kennedy would want the detail-ridden job, he told them it could be made into something grander. He could lead Senate Democrats in confronting the new Nixon administration and its expected threats to Democratic programs, fights the dignified Mansfield would not go out of his way to pick. Kennedy's staff did not argue. They were delighted at the sign of life, at the evidence that his depression after Robert's death was lifting.

Kennedy called Muskie on Saturday, December 28, from Sun Valley and confirmed that Muskie was not running. He enlisted his support. He got Humphrey's too. But he did not promptly announce his candidacy. Kennedy called his friends, first, then other liberals, and finally Senate elders. Eastland was puzzled, telling him, "Ain't no vacancy there." But Russell's answer was more important. He told Kennedy, "I will put no stone in your path." That meant, first of all, a license for Southerners like Bill Spong of Virginia to vote for Kennedy if they wanted to. But it also meant that when Kennedy finally got around to calling Long on Monday, December 30, he found that none of the nearly fifty Democrats he had called first had tipped Long off. He startled Long not just by his announced intentions, but by asking for an assurance from Long that if he won, Long would cooperate with him on issues—which meant taxes, the Finance Committee's chief jurisdiction. Bruised, Long recalled years later, "I just didn't feel like making any commitments at all."

The operation moved very quickly. Kennedy was back home in Virginia when he announced his candidacy late in the afternoon of December 30, 1968, after calling Long. He cited the backing of Muskie and Humphrey and argued that Congressional Democrats had greater responsibilities now that Republicans were in the White House, and he would not be distracted by the chairmanship of a major committee.

Mansfield took no public stand, but Ferris worked with Burke in lining up support for Kennedy. So did Muskie and two other liberals, Walter Mondale of Minnesota and Tom McIntyre of New Hampshire. His backers reached outside the Senate, drawing on the last Kennedy's national standing and getting help from I. W. Abel of the Steelworkers and from within state parties in New Mexico and Connecticut. Ferris was the source of a story in the *Globe* that said Mansfield "dislikes Long and considers him disloyal." Kennedy's announcement played on that idea, saying that "the primary responsibilities of the occupant of this position are to assist the majority leader faithfully, to serve each Democratic member of the Senate equally and to inform them fully and fairly on all matters coming before that body."

Long had lined up enough votes to defeat Muskie. But, as he said years later, running against Kennedy was different: "At that point, nobody could beat Ted Kennedy for anything. . . . Some of the best-regarded Senators in the Senate came to me and said, 'Look, I am sorry, but I told you I would vote for you and I had every intention of doing so, but I never anticipated that I would be facing this. I am sorry but I just can't stay with you.' " Long held on to some old allies, like Eugene McCarthy, and almost all the Southerners except for Spong, Yarborough of Texas, J. William Fulbright of Arkansas, and Albert Gore of Tennessee. Muskie nominated Kennedy in the Democrats' caucus on January 3, 1969, praising him for commitment to the Senate and saying his region, the East, was "too seldom represented in the Senate leadership." Gore and Henry M. Jackson of Washington seconded him. Kennedy won a clear victory, 31–26. Long emerged to tell reporters he had been beaten "fair and square." He added, "I couldn't have lost to a better man and I don't think I would have lost to any other man who would have been likely to run."

Even though Kennedy insisted the whip's job would be time-consuming and had no implications for Presidential politics, hardly anyone in Washington was listening. Long spoke for them when he observed, "Having had the experience that Lyndon Johnson experienced in 1960, Richard Nixon experienced in that same year, and a number of other people have had in taking on a Kennedy, I can say it is a very interesting experience. I would suggest that Mr. Nixon himself should be very careful and watch himself for the future, because in all probability he has a very able opponent ready for him."

White House dreams, for Kennedy if not by Kennedy, certainly helped him win. Senators may be used to constituent pressure on issues, but calls telling them how to vote for whip were something new. Those calls were generated not by a sudden awareness among Hispanics in New Mexico or liberals in Connecticut of the importance of Senate scheduling, but by a craving to see the Kennedy dynasty restored.

And even with Nixon not yet sworn in, Washington started to focus on Kennedy as his likely opponent in four years. The first paragraph of *The New York Times*'s story on his victory used the phrase "stepping stone to the Presidency." *Time* and *Newsweek* put him on their covers, each describing his Presidential candidacy as all but inevitable. Phil Hart dropped him some information on a friend who would help with fund-raising. From Paris, Pierre Salinger, President Kennedy's press secretary, promised help, telling *The New York Times*, "We started something with Bobby and we've got to finish it. I'll drop anything, do anything to help. Today's Democratic party is a Kennedy party." In the Capitol itself, someone scratched "Mr. President" in the wood of the front-row desk that was now his.

Kennedy sought to brush off that speculation, saying he expected to be kept busy in the Senate. Mansfield praised him, saying, "Of all the Kennedys, the Senator is the only one who was and is a real Senate man."

In 1969, that was more true of his instincts than his accomplishments. After six years, he enjoyed the ways of the Senate and had an easy rapport with Senators twice his age. He had been comfortable in the supporting roles that junior Senators get more often than star parts. His legislative record was short. With Howard Baker he had blocked the redistricting bill, fighting off the Senate leaders. Otherwise, the poll tax had been an admirable defeat and the immigration bill a sure thing. Community health centers and the teacher corps had been accomplishments, but modest ones achieved without battles.

And he was not really familiar with the quirky procedures of the Senate, a parliamentary thicket offering dozens of ways to stop anything.

Yet if his sense of picking up the family standard did not necessarily require a Presidential candidacy, it did seem to demand leadership on issues. Aware that half the Democrats in the Senate were from the West, he hired Wayne Owens, a thirty-one-year-old Salt Lake City lawyer who had worked with him on Western states for Robert Kennedy, and set him and Dave Burke to work trying to define the job. They brought in outside experts to brief groups of Democrats. Senators came by the chandeliered whip's office for a late-afternoon drink.

Two days after Nixon was sworn in, Kennedy told the Senate the President should appoint an American relief coordinator for Biafra. He talked to the President and William P. Rogers, the secretary of state, and told a reporter they "both want to do something about stopping the fighting and getting relief in. Mr. Rogers's first day in office was spent on the humanitarian aspects of this thing. I think we've all got to help." When the President appointed Clarence Clyde Ferguson, dean at Howard Law School, as coordinator, Kennedy praised the choice.

But he was working against the administration on an issue that mattered more to it—antiballistic missiles. The Johnson administration had proposed protecting major American cities against the possibility of a Chinese attack, and while Ted was mourning, away from the Senate, legislation had been passed to begin preparing the missile and radar installations that were part of the complicated and very theoretical plan for spotting and shooting down intercontinental missiles.

He got involved that winter when residents of North Andover complained about imminent construction, and when Jerome Wiesner, who had been President Kennedy's science adviser and was now provost of MIT, called to tell him how the system would not work. Two of the Senators Ted admired most, Phil Hart and John Sherman Cooper, a Kentucky Republican, had already fought and lost against the Johnson plan in 1968 and were fighting again. But Kennedy's name immediately drew attention, and so when he wrote to Secretary of Defense Melvin R. Laird on February 1, 1969, calling the system "folly" and a threat to peace that invited Soviet escalation, he was instantly elevated into the front rank of opponents. He urged Laird to halt the work. A few days later, Laird did, saying he would review the issue. But, suspicious that Laird's study was no more than a propaganda exercise, Ted announced in mid-February that Wiesner and Abraham

Chayes, the State Department legal adviser under President Kennedy, would publish a study of the project—for which he would write an introduction.

But on March 14, President Nixon announced a crafty political compromise. He changed the purpose of the ABM system to protecting American missile-launching sites against a Soviet first strike. That scaled it down, reduced costs, and removed the system from populated areas to Montana and North Dakota. Most of the opponents were unmoved, seeing the plan as an entering wedge for the huge, costly system the Pentagon really wanted. And so they battled on.

Gifford, with a security clearance that enabled him to see the Pentagon's documents, met regularly with aides from other Senate offices as they plotted strategy. Kennedy himself called around the country to stir up opposition. The administration's response was to call his efforts part of a 1972 Presidential campaign, and to berate Republicans like Illinois freshman Charles H. Percy for working with him. When that happened, and Cooper warned him that the coalition was endangered, Kennedy stepped back and stopped making speeches.

The book, *ABM: An Evaluation of the Decision to Deploy an Antiballistic Missile System,* came out in May with an introduction by Kennedy. He charged that the system would cost $20 billion, three times the Pentagon's estimate. He called the pending ABM decision a crossroads, with one path leading "towards an upward spiral in the arms race" and the other "towards some reduction in international tensions and nuclear arsenals."

In the short term, the opponents lost. The program got through by a single vote in August, and two missile sites were built. But in the long run, the effort made a difference to the way Congress looked at Pentagon requests, at both their feasibility and their costs. The case was argued in Senate hearings with a degree of authority the Pentagon had never confronted before. John W. Finney, a veteran *New York Times* correspondent, wrote, "A new sophistication is entering the chamber's debate on military matters," and Philip M. Boffey of *Science* magazine saw an end to the "old system of having costly weapons systems devised by groups of experts and then approved by docile Congressional committees which seldom ask questions."

This was only one of several areas where Nixon worried about Kennedy. In February he demanded that John Ehrlichman, his chief domestic aide, set up a small group to challenge Kennedy on the draft and on amnesty for deserters. In March, when Ted held a hearing on job discrimination, the White House and Dirksen coordinated a counterattack. In April, when Kennedy spoke out against student demonstrations, Nixon ordered one group of operatives to come up with a way to put him "*squarely* on the spot." When Nixon's news summary reported that Kennedy had been on television, he demanded, "Give me a report on time Teddy has received on Networks over past 30 days." So when Kennedy challenged the administration over its plan for an antiballistic missile defense system, Nixon's chief of staff, H. R. Haldeman, wrote in his diary for April 8: "This is the first battle of '72, vs Teddy Kennedy, and we *must* win."

That same sense of an inevitable candidacy could be found both among his potential rivals and in the press. Muskie had gone out to test the waters and returned in March to tell Stewart Alsop of *Newsweek,* "When everybody begins saying this early it's going to be Teddy, why, it's going to be Teddy—almost sure to be." In the *Globe,* Bob Healy, though noting that Kennedy might not want to run in 1972 if Nixon was strong, wrote in March that "right now he appears to be the inevitable candidate." James Reston of *The New York Times,* his severe critic of only a few years before, wrote in May, "His principal Democratic opponent four months ago was Senator Edmund Muskie of Maine, but in the last few weeks, Kennedy has clearly dominated the Democratic scene."

Joan was plainly troubled. In March, she told a *Globe* reporter that she shared Ted's dreams. But when Ann Blackman asked her about his running for President, her jaw tightened and her smile faded. "I can't say I'd urge him to run," she said. "Anyway, it's his decision, not mine." A few days before, in what she later acknowledged was something of a cry for attention, she had worn a low-cut silver dress with a short skirt to a formal White House reception. And she was wearing miniskirts to mass at Holy Trinity. But she also had moments of striking self-confidence, as when she narrated *Peter and the Wolf* with the Boston Pops Orchestra on May 11.

Ted sounded dubious, too. George McGovern, who was thinking of a serious, antiwar candidacy for himself, asked Kennedy if he was going to run, and Ted answered, "I don't think so." At lunch with William vanden Heuvel, who told him the 1972 nomination was his if he wanted it, Ted said he knew that everyone assumed he wanted to be President, but they forgot what had happened to his brothers. Jack's death might have been an accident of history, but Robert's was a clear warning that the same irrational forces that had killed his brothers would get him, too, if he ran. Or, as he put it crudely on a long flight home from Alaska in April, "They're going to shoot my ass off the way they shot off Bobby's."

He did not stay out of harm's way. On Good Friday, he went to Memphis, where Dr. King had been murdered one year before. Several thousand blacks were marching to City Hall for a memorial meeting, and there was some looting along the route. As he was coming in from the airport, gunfire was heard not far from City Hall. A young black aide, Bob Bates, had to decide whether Ted would be safe, and since the chief of police had refused to shake his hand the day before, he worried about it, but finally decided Ted should speak as planned. As Kennedy reached the speaker's stand, Mayor Henry Loeb signed an order imposing a curfew to take effect at 7 P.M.

The Reverend Ralph Abernathy, King's old comrade, introduced Kennedy to the crowd of ten thousand as "the President of the poor people of America." Thunder and intermittent rain punctuated Ted's tribute to Dr. King: "He taught the way of peace through social justice for men of all races. He gave all Americans a deeper understanding of the meaning of freedom and the importance of nonviolence. . . . He showed us that bigotry and violence are just as corrosive to the lives of those who practice them as they are to the lives of those who suffer from them." He said that "the promises we made a year ago, to finish up his work, have not been kept." He quoted Robert's speech in Indianapolis on hearing of King's death, the call for "love and wisdom and compassion toward one another." Speaking of Dr. King and his own slain brothers, he said, "If their lives and deaths had a meaning, it was that we should not hate but love one another."

The Alaska trip the next week was another hectic memorial, this one to Robert Kennedy, who had canceled his own visit to Alaska in 1968 to attend Dr. King's funeral. Ted flew into Anchorage on Tuesday night, April 8, as chairman of the Subcommittee on Indian Education, a position Robert had held. He had an intense three-day schedule of visits to Eskimo and Indian villages, though some of it was scrubbed because of bad weather. He was joined by Mondale; four Republican Senators, Henry Bellmon of Oklahoma, George Murphy of California, Bill Saxbe of Ohio, and Ted Stevens of Alaska; and one Republican Representative, Howard Pollock of Alaska. But after one fifteen-hour day in the bush when the photographers and reporters predictably clustered around Kennedy, most of the Republicans pulled out. Seizing on a *Chicago Tribune* story reporting that Kennedy planned to make political capital of the trip, Murphy called Kennedy Thursday morning to say they were dropping out. He issued a statement calling the trip an "unfortunate political publicity junket at the taxpayers' expense."

Kennedy, Mondale, and the press went on with the trip, by transport, ski planes, and dog sleds. So did Stevens and Pollock, although Senate Republican leaders pressed Stevens to drop out too. But he continued, saying, "My feeling is that the trip will be beneficial to the state." (Several months later, he and Kennedy got legislation passed to build water treatment plants in Indian villages, where ear infections from polluted water were common.) The visitors heard about the Bureau of Indian Affairs sending children thousands of miles to high school. They saw federal installations with pure water not shared with natives. They crouched to get into one-room shacks that housed eleven people.

The visit wound up with a four-hour hearing in Fairbanks on Friday night. Fourteen witnesses added complaints about public health and insensitivity to native cultures and languages. Mondale told that hearing, "We haven't seen a single Eskimo or Indian teacher. I can't believe that would ever be true if Indians and Eskimos ran their own schools." Kennedy told of a teacher in Pilot Station who said that "the warmest they could ever get their classroom up to was zero. I wonder how you expect children to even think about reading any kinds of books or concentrating or studying at zero degrees in the classroom."

But he brooded. They were delayed one night in Arctic Village because of an engine problem. Walking back to the plane from a stop in the village for coffee, he turned to a pair of photographers with him and said, "Why don't you two walk on ahead? I'd like to be alone." The *Boston Herald* reported that "with the northern lights putting on a brilliant display in the sky overhead, Kennedy, parka open at the neck in the chilly night air, walked down the wilderness trail alone with his thoughts."

And he got very drunk. When the Fairbanks hearing ended at midnight, the rest of the party got something to eat. He did not, but started drinking in the hotel bar, then moved to the airport bar. As soon as a rowdy group of aides and reporters boarded the Pan American flight to Seattle late, about 3 A.M., flight attendants started serving drinks, and Kennedy knocked them back one after another. He was talking loudly, making silly jokes, leading chants of "Eskimo power! Eskimo power! Eskimo power!" Northwest Orient Airlines delayed its flight to Washington so the Kennedy group could make its connection. Reporters started a pillow fight before takeoff. Kennedy avoided that, but was drinking or wandering unsteadily in the aisle for most of the trip, oblivious to other passengers. In the convention of the time, no one wrote about it for publication.

That was only one of several bursts of intense activity.

Another revolved around his brothers and their causes. During the third week of May, Cesar Chavez, the charismatic farm workers' union leader, pleaded with Kennedy to come to California for a celebration at the end of a ten-day, one-hundred-mile march by one hundred grape workers protesting the use of Mexicans as scabs. Chavez had been very close to Robert Kennedy, and Ted admired his cause and his nonviolent approach. His staff argued that security would be terrible, made worse by death threats against Chavez, and the politics needless, for he was already identified with their cause. Others, like Tunney and Mondale and Yarborough, were going, so Chavez would not be abandoned. He told Chavez no on Thursday, May 17. But on Saturday, Ted went to Hickory Hill for Ethel Kennedy's charity pet show. He ended the afternoon playing touch football. And then he changed his mind, telling Burke to meet him at Baltimore's Friendship Airport. It was about four-thirty. They just made the six-o'clock plane for Los Angeles.

He was returning for the first time to the city where Robert had been murdered. That night at the Beverly Hilton, he wrote to urge clemency for the killer. A jury had convicted Sirhan and recommended the gas chamber, and the formal sentencing was to come the next week. Kennedy wrote a five-page longhand letter to Evelle Younger, the Los Angeles district attorney, telling him he and his sisters felt Sirhan should not be executed: "My brother was a man of love and sentiment and compassion. He would not have wanted his death to be a cause for the taking of another life. . . . If the kind of man my brother was is pertinent, we believe it should be weighed in the balance on the side of compassion, mercy and God's gift of life itself." When the letter was read in court the next Thursday, Judge Herbert V. Walker was unmoved. He sentenced Sirhan to death.*

On Sunday the grape workers celebrated the end of their trek in the cool night air of the border town of Calexico. The crowd reached three thousand. Kennedy's appearance came as an exciting surprise. The other politicians, especially Mondale, who had walked the last five miles, were eloquent; Mondale said, "They tell us that grapes are perishable. Well, children are also perishable. The country is hearing your cause and is responding to the boycott." But Kennedy was the one they were waiting for. In his best

*Sirhan's sentence was commuted to life in 1972 after the California Supreme Court held the death penalty unconstitutional.

rousing stump style, he told them that "a nation that can afford to send men to the moon, that can spend thirty billion dollars a year on a senseless war, and that can provide hundreds of millions of dollars to farmers for not growing crops, can afford to provide decent wages for hundreds of thousands who want to work for a decent wage." Speaking from the back of a truck, he hailed their leader, recalling Robert's feeling for him: "He called Cesar brother. He is your brother and mine. *Viva Chavez! Viva La Huelga! Nosotros Venceremos!*"

On Monday, May 19, he flew from Los Angeles to Boston and drove forty miles to Worcester for a speech at Clark University honoring Robert H. Goddard, the pioneer of American space science. John Kennedy's crash program to land a man on the moon before the end of the decade was on the verge of success. Three astronauts were making the final prelanding flight as he spoke, and one who would travel to the moon, Colonel Edwin "Buzz" Aldrin, was in the audience.

Ted quoted his brother as saying that the lunar goal "will serve to measure the best of our energies and skills." But he called for a substantial cut in the space budget once the four planned moon landings were concluded, so that the nation would have the money for other urgent needs such as housing, crime control, education, and the environment.

"I am for the space program," he said. "But I want to see it in its right priority, one which will let it continue into the future and not have to be cut back or abandoned because the nation that supports it is hobbled by internal disorder."

"We should develop a plan for an orderly, programmed exploration of outer space, but we no longer need an accelerated program," he said. "We need not try to get to Mars or Venus merely because the Russians might get there first. We need not pay the overtime and incentive costs necessarily involved in a program with an urgent deadline."

Thomas O. Paine, head of the National Aeronautics and Space Administration, immediately accused Kennedy of wanting to ground American astronauts as part of a "dispirited vision of the Nation's vigor and destiny in space." Nixon scrawled, "Good job, Paine," in the margin of his news summary.

That wasn't all. On Tuesday morning, May 20, Kennedy came boiling into the office, furious over what he had seen on morning television from Vietnam. Two months before, he had joined Mansfield in chiding George McGovern for being too quick to condemn the President over the war in

Vietnam he had inherited. But this day, for the first time, Kennedy took the lead in attacking Vietnam policy.

The news was that U.S. forces, in their eleventh uphill assault, had captured Apbia Mountain. The soldiers called it Hamburger Hill because of the way it chopped them up. Hamburger Hill was a ridge in the Ashau Valley near the border of Laos, captured after a ten-day battle labeled a victory with the claim that 426 North Vietnamese were killed, but only 43 Americans. Major General Melvin Zais, commander of the 101st Airborne Division, said, "This was a tremendous, gallant victory and we decimated a large North Vietnamese unit." The battle was part of an operation designed to kill enemy troops; the hill's tactical value was demonstrated by the fact that it was abandoned soon after, as U.S. officers had told reporters it would be.

As the Senate convened, Kennedy attacked. In a little less than three minutes, he argued that casualties, bomb tonnage, and battles had all increased since the bombing halt of the previous November, and to no clear purpose. He argued that offensive actions undercut the peace negotiations in Paris and quoted President Nixon as saying the United States did not seek a military victory.

"How then," he asked, "can we justify sending our boys against a hill a dozen times or more, until soldiers themselves question the madness of the action. The assault on 'Hamburger Hill' is only symptomatic of a mentality and a policy that requires immediate attention. American boys are too valuable to be sacrificed for a false sense of military pride."

Senator Hugh Scott of Pennsylvania, the Republican whip, who was no great enthusiast for the war himself, shot back, "I shall not try to second-guess the President on the conduct of the war." He compared Apbia to Little Roundtop and Cemetery Ridge at the Battle of Gettysburg. "If our military are told to contend for a hill," he said, "it is part of the strategy which is essential to maintaining the military posture which we take for peace." Dirksen chimed in two weeks later, saying Kennedy's speech would damage morale. Mansfield defended Kennedy and called the war a tragedy, demanding to know what 179 Montanans had died for.

TED'S CONCERN FOR Robert's children was another force in his life. He took Kathleen, Robert's oldest daughter, with him when he spoke to honor

Robert's concerns about Indians at graduation at Rough Rock, an experimental Navajo school in Arizona. She had worked there the summer before, after her father's death. Ted quoted Robert's saying Indian education should resist "total assimilation" and enable its students to "meet the complexities of life with versatility and grace." And he quoted from the "Song of Black Mountain"—"Who learns our songs shall be our child"—before telling the twenty-eight eighth-graders, "You have learned the songs of the Gods—now you must teach us to listen and to sing."

Kathleen said years later, "He made it his mission, I think, to make sure we felt connected to my father, and to John Kennedy." But he was also available for personal advice. Kathleen, who as the eldest had been expected to supervise the others, got called away from the dinner table at the Putney School in Vermont often that year, to go to the basement pay phone and consult with her uncle about her troubled brothers.

A friendly reporter, Joe Mohbat of the Associated Press, wrote that Kennedy was submerging his grief in "frantic activity." He called him a "fearsome driver" who proceeded at breakneck speed on his way to a day of dozens of events. Ted traveled a lot that spring, frequently to family events, like Kathleen's graduation from Putney, when his small plane flew over the terrain where he had crashed five years earlier. But often he was traveling to colleagues' fund-raising events, and he could hardly go anywhere without the question of a Presidential run coming up. In June he told Mohbat he would like to see someone else try, someone like Muskie. He said he would have no trouble rejecting party demands that he run: "Damned right, I could, in an instant. I honestly don't feel any obligation to pick this one up." He told Mohbat that campaign events "pretty much turn me off now. When I first came into this in 1962, it was really good, easy. But the kicks aren't . . . I mean, meeting Molly Somebody and hearing about her being Miss Something."

He continued, "What's it all for? I used to love it. But the fun began to go out of it after 1963, and then, after the thing with Bobby, well . . ."

On Friday, July 18, 1969, Mansfield told reporters at breakfast, "My judgment is that he will wait until 1976. He's in no hurry. He's young. He likes the Senate."

And that afternoon, after a couple of hearings, Ted caught a plane for Boston and then another for Martha's Vineyard. He flew to Boston with Tip O'Neill, who remembered him complaining, "I've never been so tired in my life."

CHAPTER 12

Chappaquiddick

The trip to Martha's Vineyard had two purposes. First of all, it was the weekend of the annual Edgartown Regatta. Ted was going to race Jack's boat, the *Victura*.

And there was a reunion scheduled that evening with several of the women who had worked at the nerve center of Robert's campaign, the "boiler room" where they kept continual track of developments, identifying delegates who might be enlisted or arguments that could help in a particular state. They were smart, professional women, interested in politics. Most of them now held other Washington jobs. They met from time to time, but Ted had missed the last gathering. Nance Lyons, the only one of them on his staff in 1969, chided him for his absence, so he felt obliged to be there, though he did not enjoy organized reminiscence.

It was a hot day, and when Ted's plane from Boston landed about one o'clock he asked Jack Crimmins, his driver, if there was time for a swim before his race. Crimmins said there was, and, after a stop for some fried clams, took Kennedy on a two-car ferry to Chappaquiddick Island, a modest sand spit five hundred feet across from Edgartown. Crimmins drove the Senator's Oldsmobile to a modest two-bedroom cottage that had been rented by Kennedy's cousin Joe Gargan, a lawyer and vice president of a Hyannis Port bank. After Ted changed, Crimmins drove him to the beach where the women were swimming. To get there, Crimmins turned right from the cottage on Schoolhouse Road, then right on Dike Road and across a narrow wooden bridge without railings, the Dike Bridge.

Ted swam, returned to the cottage to change, and was driven back to the ferry and crossed to Edgartown. The race began a few minutes before three. Gargan and a local young man, Howie Hall, crewed for him. The breezes were very light, not the heavy weather a restless sailor like Kennedy does best in, and after about two hours, he finished ninth in a field of thirty-one Wianno Senior sailboats. The women, Nance Lyons and her sister Maryellen, Rosemary (Cricket) Keough, Mary Jo Kopechne, Esther Newberg, and Susan Tannenbaum, watched from a boat hired for them by Paul F. Markham, a friend of the Kennedys and a former U.S. attorney for Massachusetts.

After Friday's race, Kennedy checked in at the Shiretown Inn, near the Edgartown ferry slip, where he was to share a room with Gargan. He changed and then shortly after seven returned to the cottage on Chappaquiddick, where he soaked his bad back in a hot bath. Crimmins fixed him a rum and Coca-Cola. The women arrived about eight-thirty or nine, and he served them drinks.

It hardly seems to have been a wild party, though some neighbors thought it fairly loud. Gargan heated frozen hors d'oeuvres in the oven and grilled steaks over charcoal outside. People told campaign stories and sang and kidded the Senator about his poor finish in the race. They wandered in and out of the small cottage. Charles Tretter, another lawyer who had driven four of the women down from Boston on Thursday, left with Cricket Keough about nine-thirty to borrow a radio from the Shiretown so there could be music for dancing. There is no way of knowing just how much drinking took place. Newberg, the only one at the party who talked to reporters soon after, said no one had more than two drinks. Kennedy testified at an inquest in January that he had two. But Mary Jo Kopechne, who normally hardly drank, probably had three or more drinks.

Sometime after eleven that Friday night, Kennedy left with Mary Jo Kopechne. He and Crimmins both testified that he asked Crimmins for the car keys, saying he was going back to the Shiretown Inn and was taking Kopechne back to the motel where she was staying. Crimmins, but not Kennedy, testified that the Senator said Kopechne was not feeling well because of exposure to the sun. Several of the others there testified they had noticed them leaving together, but said nothing was said by either of them about why they were going. And no one said Kopechne had complained about feeling unwell.

Before the 1968 campaign, Mary Jo Kopechne had been a legislative aide in Robert Kennedy's Senate office. In 1969, she was working for one of the first political consulting firms, Matt Reese and Associates. She was a devout Catholic with a serious manner and no reputation for playing around. Kennedy did have that reputation. His name had been linked with various women, generally glamorous, unlike the trim, pretty Kopechne. His girlfriend at the time was Helga Wagner.

As Crimmins had done earlier when driving Kennedy to the beach, the Senator turned right from the cottage onto Schoolhouse Road. But instead of turning left on Chappaquiddick Road toward the ferry, Kennedy turned right on Dike Road toward the beach.

When he came to the unlit humpbacked bridge, which angled to the left, he drove off the right side of the bridge into the pond it spanned.

The bulky Oldsmobile overturned in the water. The windows that were not already open blew open from the crash, and the car started filling with water.

The next day, in a statement to police, Kennedy said he did not know how he escaped from the car. And he said he "repeatedly dove down to the car and tried to see if the passenger was still in the car," but was unsuccessful. But when he testified in January at an inquest, he said, "I knew that there was a girl in that car and I had to get her out." He testified that he became exhausted and stopped diving after seven or eight tries, rested for perhaps twenty minutes, and then went for help.

Kennedy suffered a moderate brain concussion in the accident, an injury which causes confusion and memory failure. That undoubtedly affected his actions and his subsequent accounts of them.

But his behavior was also consistent with a clumsy attempt to cover up his role in the accident.

He stumbled back the mile and a half to the cottage, passing two other houses near the bridge. He testified he saw no lights, but he did not bang on doors to see if he could summon help. When he reached the cottage, he slumped into a parked car. He saw Ray LaRosa, a federal civil defense official in Andover and the sixth man at the party, and asked him to get Markham and Gargan. Some of the others saw them leave.

Kennedy, Gargan, and Markham all testified that they returned to Dike

Bridge, where Markham and Gargan dove to try to reach the victim, but failed. Then, they testified, they returned to the ferry slip, where Kennedy assured them he would report the accident. Instead of calling for the ferry, which made night runs on request, he dove into the water to swim back to Edgartown. There was a pay phone at the ferry slip, but no one used it. There was no phone at the cottage.

About 2:25 A.M. on Saturday, a dry and neatly dressed Kennedy emerged from his room at the Shiretown and asked its co-owner, Russell Peachey, what time it was. The next morning, he spoke to some other guests before the breakfast bell rang at eight. He never mentioned the accident.

Then he met in his room with Markham and Gargan, who had crossed by ferry after telling the other women that there had been an accident and Kopechne was missing. Then the three men took the ferry back to Chappaquiddick and made some telephone calls, trying to reach Dave Burke, Steve Smith, and Burke Marshall, an assistant attorney general in Robert Kennedy's Justice Department. Another early call—before he told Joan—went to Helga Wagner; Kennedy told *The New York Times* in 1980 he called to see if she knew how to locate Steve Smith, who was in Spain.

By this time, two early-morning anglers out for bluefish had spotted the wrecked car, and the police had been called over from Edgartown about eight-twenty. Chief Dominick J. "Jim" Arena tried to dive to the car, but the tides that swept through the narrow entrance to the pond spanned by the bridge were too strong. He got the license plate checked, and learned it was the Senator's car. His reaction was that another tragedy had hit the Kennedy family.

John Farrar, a trained scuba diver for the Edgartown fire department, got to the car and saw Kopechne in it. At first he thought she might be alive, but when he reached in and touched her, she was stiff with rigor mortis. He put a rope around her neck to be sure the tide would not sweep her out to sea and pulled her to shore. Arena's first question was whether the victim looked like one of the Kennedys.

Dr. Donald R. Mills, an Edgartown general practitioner and a relief medical examiner, was called to the scene about 9 A.M. With him was a local undertaker, Eugene Frieh. Dr. Mills examined the corpse and concluded she had died of drowning. Frieh then took the body to his funeral home and prepared it for burial. He removed her clothes—a blouse, slacks, and a bra. She was not wearing panties. At Dr. Mills's direction, he also took a blood sample, which state police later found to have a blood alcohol content of 0.09 percent.

Chief Arena heard that Kennedy was by the ferry slip, and went to the nearest house to call the station to tell the police to meet the ferry and find the Senator. But when he made the call, a bit after nine, Carmen Salvador, a policewoman on duty, told him Kennedy was there and he came on the phone.

According to Arena's inquest testimony, he said, "I am sorry. I have some bad news. Your car was in an accident and the young lady is dead."

Kennedy replied, "I know."

Arena said, "Can you tell me if there was anyone else in the car?"

Kennedy said, "Yes."

Arena said, "Are they in the water?"

Kennedy said, "No."

Arena said, "Can I talk to you?" and when Kennedy agreed, he returned to the station in Edgartown. He found Kennedy on the phone. He had been making calls steadily since he arrived, to Mary Jo's parents among others. He hung up and told the chief, "I was the driver," and told him who the victim was.

Word of the accident was spreading, and reporters began turning up at police headquarters. Arena shooed them away. He did not question Kennedy. Instead, he said he would need a statement from him, and sent Kennedy and Markham off into a quiet room to write it out. Kennedy dictated and Markham wrote. This is the statement:

> On July 18, 1969, at approximately 11:15 P.M., I was driving my car on Main Street in Chappaquiddick on my way to get the ferry back to Edgartown. I was unfamiliar with the road and turned right on Dike Road instead of bearing left on Main Street.
>
> After proceeding for approximately one-half mile on Dike Road, I descended a hill and came upon a narrow bridge. The car went off the side of the bridge.
>
> There was one passenger in the car with me, Miss Mary Jo Kopechne, a former secretary of my brother Robert Kennedy. The car turned over and sank into the water and landed with the roof resting on the bottom. I attempted to open the door and window of the car but had no recollection of how I got out of the car. I came to the surface and then repeatedly dove down to the car and tried to see if the passenger was still in the car. I was unsuccessful in that attempt.
>
> I was exhausted and in a state of shock. I recall walking back to where my friends were eating. There was a car parked in front of the cottage and

> I climbed into the back seat. I then asked for someone to bring me back to Edgartown. I remember walking around for a period of time and then going back to my hotel room. When I fully realized what had happened this morning I immediately contacted the police.

It was a bizarre use of "immediately." He had been up for two hours before he did so. And the statement said nothing about returning to the car with Markham and Gargan; Markham testified that Kennedy told them at the Shiretown that morning, "I'm not going to involve you. As far as you know, you didn't know anything about the accident that night."

Arena read the statement, typed it, and let the Senator go without asking him about its contents or its omissions. Arena did not ask about speed, or whether Kennedy had been drinking. He did not ask Kennedy why he had taken hours to report the accident. Questions might have been futile, because by the time he had finished the statement, the Senator was saying "No comment" to a supervisor from the Motor Vehicles Registry, George Kennedy.

The Senator told the chief he wanted to contact Marshall, whom he called his family attorney, and would get back to him. Markham asked the chief not to release the statement until they had reached Marshall. Then Arena helped Kennedy escape the press, including James Reston of *The New York Times*. He owned the *Vineyard Gazette* and was outside. Around noon, George Kennedy drove the Senator, Markham, and Gargan to the airport and they flew in a private plane to Hyannis. On the drive to the airport, the Senator was sitting in the front seat mumbling, "Oh, my God, what's happened? What's happened?" When he got home, he went to see his father and told him there had been an accident and he was in trouble. "I'm telling you the truth," he said. "It was an accident." That evening he finally saw a doctor. Dr. Robert Watt diagnosed a concussion.

The other five women returned to their motel on Martha's Vineyard after a cramped night of sleep in the cottage and the cryptic comment from Gargan that Mary Jo was missing. Then about ten o'clock Gargan telephoned to say that she had drowned, and insisted the Senator had tried to save her. At his urging, they took the ferry to the mainland at Woods Hole about 3 P.M.

Arena was not sure what to do about the case. Walter E. Steele, a prosecutor on the Vineyard, told him he had to issue a complaint charging Kennedy with leaving the scene of an accident, and to inform Edmund Dinis, the district attorney, who lived on the mainland in New Bedford. Arena was reluctant to disturb Dinis on the weekend, but did call him. Dinis immediately announced that he was taking over. But a few minutes later, Dinis's investigator, Lieutenant George Killen of the state police, called to say they were not getting into the case.

Dr. Mills had been sure Kopechne had died by drowning and did not order an autopsy. But he could not get Dinis's office to answer when he tried to find out if it wanted one. No autopsy had been done when the body was flown on the afternoon of Sunday, July 20, to Mary Jo's girlhood home in Pennsylvania for burial.

Besieged by reporters, Arena decided in midafternoon to release Kennedy's statement. He read it through twice while they took notes.

The Sunday papers were full of the Chappaquiddick story. In the *Boston Globe,* it got more dramatic play than the fact that on Sunday afternoon, John F. Kennedy's vision of men landing on the moon would be fulfilled. *The New York Times* headline was "Woman Passenger Killed, Kennedy Escapes in Crash: Senator Tells the Police He Wandered About in Shock After Car Ran Off Bridge Near Martha's Vineyard."

Many newspapers reported the story as another Kennedy tragedy, but the political implications were heavy. Another *Times* headline was "Kennedy's Career Feared Imperiled." Walter Mondale recalled telling his wife that "he took a lot of us with him, because there were no other stars in the sky." At the White House, President Nixon reacted with chilling calculation. On Saturday he had John Ehrlichman send investigators to Chappaquiddick. That evening the President told Haldeman he wanted to make sure Kennedy did not get away with it. Nixon was sure, as Haldeman wrote in his diary, that Kennedy "was drunk, escaped from car, let her drown, said nothing until police got to him. Shows fatal flaw in his character, cheated at school, ran from accident."

On Sunday, at the triumphal moment of Neil Armstrong's steps on the moon, when Nixon could speak to the nation on television and announce that his administration had accomplished what his first rival for the Presidency had set out to do, he was distracted. Waiting to congratulate the astronauts, he told Haldeman, "It marks the end of Teddy." After making brief remarks and accepting congratulations from Bill Safire, a speechwriter, Nixon said, "You know, this is quite a day on another front, too." He said it would be hard for the Kennedys "to hush this one up; too many reporters want to win a Pulitzer Prize."

Advisers descended on Hyannis Port. Burke Marshall and Dick Goodwin came first, on Saturday. Marshall's lawyer's instinct was to say nothing. Reporters who gathered outside the compound were left with nothing more to report than comings and goings. Even so, Kennedy eluded them on Monday to go to the Cape Cod Hospital for a more thorough medical examination. A neurosurgeon, Dr. Milton F. Brougham, was called in. The diagnosis of concussion was unchanged, and Kennedy was also fitted with a cervical collar for his neck injury.

On Tuesday, Mary Jo Kopechne was buried. Ted, Joan, Ethel, and various family friends flew to Wilkes-Barre on a DC-3 owned by one of Ethel's family's businesses. At the funeral at St. Vincent's Church in Plymouth, Pennsylvania, Ted talked to the Kopechnes and to the women from the boiler room, but told them nothing more about the accident.

When his plane got back to Hyannis that afternoon, reporters were clamoring for a statement. Kennedy told them, "This is the day of the funeral. This isn't the appropriate time. But I will make a full statement at the appropriate time." One reporter pressed on, asking how all this would affect his political career. He snapped, "I've just come from the funeral of a very lovely young lady, and this is not the appropriate time for such questions. I am not going to have any other comment to make."

Kennedy drifted in and out of meetings, inattentive, sometimes confused, in a stupor. More advisers arrived. Robert McNamara, Ted Sorensen, and ultimately Steve Smith took over. Smith shared Marshall's view that nothing should be said until any charges had been dealt with. He was still concerned about a more serious charge, perhaps manslaughter. Steele had told Arena they had no case stronger than leaving the scene.

Marshall hired a local expert in motor vehicle law, Robert Clark. On Tuesday night, after the funeral, he called Arena to arrange a meeting. They met on Wednesday, at Steele's out-of-the-way fishing cottage. After some fencing, they agreed that Kennedy would plead guilty to leaving the scene of an accident and Steele would recommend a two-month suspended sentence. But first Clark had to return to the mainland and check with Steve Smith. Two more meetings on Thursday were needed to settle all the details, including what dock Kennedy would arrive at by boat. On Thursday the twenty-third, Dr. Brougham examined Kennedy again at his home and did an electroencephalogram, measuring his brain waves. By that day, Dr. Brougham found Kennedy alert and his speech normal.

Just before nine on Friday morning, Kennedy appeared before Judge

James A. Boyle in district court in Edgartown. He said "Guilty" when asked how he pleaded to the charge of leaving the scene of an accident. He said it softly, then swallowed and said it again so the whole courtroom could hear. That was all he said.

Chief Arena then described the accident and how he discovered it. He volunteered, "Investigation of the accident and accident scene produced no evidence of negligence on the part of the defendant." But, he said, Kennedy had opportunities to report the accident and did not.

Judge Boyle asked one question: "I would be interested in determining from the defendant or the Commonwealth if there was a deliberate effort to conceal the identity of the defendant." Chief Arena answered, "Identity of the defendant—not to my knowledge, your honor."

Richard McCarron, a local defense attorney, and Steele then asked the judge to impose a suspended sentence. He agreed, saying, "Where it is my understanding, he has already been and will continue to be punished far beyond anything this court can impose, the ends of justice would be satisfied by the imposition of the minimum jail sentence and the suspension of that sentence."

The whole proceeding took nine minutes.

As he left the courthouse, Kennedy told reporters he had asked for television time that night to "make my statement to the people of Massachusetts and the nation."

What he would say was intensely debated within the compound. At one point he wanted to say that he would never run for President. His big sister Eunice had that taken out. More broadly, the younger advisers argued for a simple, contrite admission, and they lost.

Instead the speech tugged every emotional string Sorensen and the other writers could think of.

Kennedy began with a tribute to Mary Jo as "one of the most devoted members of the staff of Senator Robert Kennedy." He denounced the "ugly speculation about her character" that would lead anyone to harbor "suspicions of immoral conduct that have been leveled at my behavior and hers that evening." He said Joan would have been there except for reasons of health, which his press secretary later translated for reporters to mean that she was pregnant.

Kennedy said he was not "driving under the influence of liquor," and described the accident flatly, noting the road was not lit and the bridge lacked guardrails and angled away from the road. He said he thought he was drowning, somehow escaped, and then dove repeatedly to try to save Mary Jo.

Using a confessional style that makes excuses while disowning them, he said that although doctors "informed me that I suffered a cerebral concussion as well as shock, I do not seek to escape responsibility for my actions by placing the blame either on the physical, emotional trauma brought on by the accident or on anyone else."

He called his failure to report the accident immediately "indefensible." For the first time he described the efforts by Gargan and Markham to rescue Kopechne.

Then he turned florid:

"All kinds of scrambled thoughts—all of them confused, some of them irrational, many of which I cannot recall, and some of which I would not have seriously entertained under normal circumstances—went through my mind during this period.

"They were reflected in the various inexplicable, inconsistent and inconclusive things I said and did, including such questions as whether the girl might still be alive somewhere out of that immediate area, whether some awful curse did actually hang over all the Kennedys, whether there was some justifiable reason for me to doubt what had happened and to delay my report, whether somehow the awful weight of this incredible incident might in some way pass from my shoulders.

"I was overcome, I'm frank to say, by a jumble of emotions, grief, fear, doubt, exhaustion, panic, confusion and shock."

Then he used a device reminiscent of Richard Nixon's call for the public to say whether he should drop off the Republican ticket in 1952 because of revelations about a secret fund maintained for his office expenses, the "Checkers Speech." Kennedy asked the people of Massachusetts to tell him whether his standing had been so impaired that he should resign from the Senate. He said that if the people lacked confidence in a Senator's "character or his ability, with or without justification," he should resign.

The Massachusetts reaction, like the national response to Nixon seventeen years earlier, was overwhelmingly favorable.

NO OTHER AUTOMOBILE accident in history has stirred as much suspicion and speculation. There is no question that the authorities, simultaneously sympathetic, in awe, and in fear of a messy case involving a hugely popular Senator, did a terrible job of investigating the accident. The Kennedy camp did not help, but in legal terms, it did nothing to obstruct, either. Absent different evidence, it is almost impossible to see how the authorities could have proved a more serious crime, like the manslaughter charge Steve Smith feared.

There may have been some tinkering with the timetable of events (if not by Kennedy, whose concussion may have induced partial amnesia, then by others) in an effort to show that Kennedy and Kopechne did not have time to do anything but drive straight from the party to the bridge. However they used the time, they probably had some. Kennedy's testimony that he saw no house lights as he returned to the cottage fits best with testimony by a local deputy sheriff, Christopher Look, that he saw a car that night about 12:45 A.M., and that it sped away from him to go down Dike Road. He said it appeared the next morning to have been Kennedy's car. And, as James E. T. Lange and Katherine DeWitt, Jr., argue in *Chappaquiddick: The Real Story,* the calmest examination of the accident, the later timetable fits the tides better.

The accident itself is consistent with the fact that Kennedy was a terrible, easily distracted driver. Moreover, even if he had only the two drinks he reported, liquor probably mattered. He was almost certainly not legally drunk in 1969, when the blood alcohol standard was 0.15 percent. He might not even have been over the current Massachusetts standard, which is 0.08. But any alcohol consumption diminishes driving ability.

There is no direct evidence that Kennedy dove to the wrecked car. Some critics have dismissed his account, claiming that his bad back and the tides made this impossible. But since childhood, Kennedy had taken physical risks. After he broke his back, he was skiing on difficult slopes again the next year. Adrenaline would have boosted his energy. There is no reason to doubt that he tried as hard as he could to rescue Mary Jo immediately.

Just as there was no good excuse for Kennedy's not knocking on doors until he found a house with a telephone, there was no excuse at all for Markham and Gargan's failing to summon help. Each was a lawyer, and neither had a concussion. Their instinct, and perhaps Kennedy's too, appears to have been to prevent disclosure of his having been in the car with a pretty young woman under circumstances that invited suspicion.

Nevertheless, it is very unlikely that faster action to summon help would in fact have saved her life. Considering that it took more than an hour to get her body out the next morning when everyone involved was already awake and dressed and visibility was good, Lange and DeWitt argue convincingly that she had to have died before help would have arrived in the middle of the night.

But Markham, Gargan, and Kennedy didn't know that. They should not have worried about how it would look and should have called the police without delay, without going back to Edgartown.

CHAPTER 13

A Staff to Lean On

For the second summer in a row, Ted Kennedy was trying to survive. When he returned to work, on Thursday, July 31, 1969, Mansfield greeted him on the Senate floor, saying, "I'm glad you're back. This is where you belong." Democrats and Republicans streamed by his front-row whip's desk, shaking his hand or patting his shoulder, and Kennedy thanked them. Even President Nixon, who had constantly badgered his aides for news or rumors while touring Asia, wrote to thank him for waiting in the rain at Andrews Air Force base on his return. On August 4, Nixon took him aside after a Congressional leadership meeting to offer sympathy, especially for how tough the press had been on him. He warned Kennedy that the press, even if it seemed to like him, was always the enemy, because the story mattered more than anything.

Chappaquiddick changed the press, too, offering a justification for writing about politicians' private lives. Kennedy was the first target in *Newsweek*'s and *Time*'s reports on the accident as they carried gossip they would not have touched the week before. *Newsweek* said Kennedy's friends worried about his "ever-ready eye for a pretty face," and *Time* said, "As for women, there are countless rumors in Washington, many of them conveyed with a ring of conviction." But he was not the only subject, as a *Washingtonian* magazine article titled "Washington's Biggest Male Chauvinist Pigs" (citing Ted's "roving eye") showed in 1972. In the next few years, Speaker Carl Albert and committee chairmen Wilbur Mills and Wayne Hays would find their failings widely reported: Albert for booze, Hays for sex, and Mills for both.

Politics intruded on Ted immediately. Reporters asked him that same day he returned if he had abandoned any thought of running for President in 1972, and he said, "That's right. I intend to fill out my Senate term if re-elected" in 1970. It was soon clear that he could no longer set the issues agenda from the Senate. He tried, attacking Nixon in September for cutting medical research money while continuing the war in Vietnam. Nixon flared up, but an aide, Alexander Butterfield, told the President that the speech had not got much attention and should be ignored.

So Kennedy tried to be a presence within the Senate instead. He would ask Democrats if he could help gather support for their bills, and if they wanted him to, he worked at it. In October, Muskie was fighting the administration over how much to spend fighting water pollution. So Kennedy appealed to other Democrats for help, saying it was time for a true commitment to a cleanup. His staff looked not only for things for him to do, but for ways to make him seem to be doing more than he was. Gifford suggested that during roll calls, Kennedy move around the Senate floor talking to Democrats so reporters watching the voting from the press gallery would think he was rounding up votes.

He tried to learn the Senate's rules, having regular seminars with Owens and the Senate parliamentarian. He needed to because Robert C. Byrd of West Virginia, a procedural wizard, would fence with Kennedy to show him up. Byrd bitterly resented Kennedy as whip because Byrd, as secretary of the Democratic caucus, was next in the leadership line after Long, and Kennedy had jumped the line to usurp a position Byrd regarded as his to inherit. After one painful procedural embarrassment, Kennedy turned to Owens with a grin and said, "You know, I really feel sorry for you." His Mormon aide was surprised and asked why. Kennedy said, "Because you can't go drink a huge glass of Scotch in a hot bath and forget what happened today."

In November, Ted's father died. This death was more of a relief than a shock. Joe Kennedy lived almost eight years after his first stroke, but he never walked or talked clearly again. Toward the end his decline was steep, and he could no longer feed himself. On November 15, 1969, he had a heart attack and slipped into a coma. Ted and Jackie spent hours by his bed. He died that morning, eighty-one years old, with his wife, his children, and their husbands and wives around him, saying the Our Father. Eunice began, "Our father, Who art in heaven, hallowed be Thy name." Ted continued, "Thy kingdom come. Thy will be done."

The funeral was held at the Church of St. Francis Xavier in Hyannis, where Rose went to Mass every day, and where the family had endowed the altar as a memorial to Joe Jr. Ted spoke for the Kennedys, saying, "This is not so much a final prayer to Dad as a reminder to those of us he left behind of his deep love for us and our obligation and responsibility to lead the kind of lives he would want us to lead." He thanked Ann Gargan for the "love and affection she gave Dad."

Otherwise, the lone surviving son did not rely on his own words.

Ted's throat caught as he read a prayer his mother had written for *The Fruitful Bough,* the book of reminiscences of his father he had assembled while recovering from his broken back. And he read Robert's essay for the book, too. Where Ted's chapter was joyous and talked of happy moments—and some stern reprimands—Robert's was analytical and almost somber, as it described the father's role in the background of his children's lives and his basic lesson:

"He has called on the best that was in us. There was no such thing as half-trying. Whether it was running a race or catching a football, competing in school—we were to try. And we were to try harder than anyone else. We might not be the best, and none of us were, but we were to make the effort to be the best. . . . If we were racing a sailboat, he was there in his cruiser. One time we did badly. He felt it was because we were not paying attention. There was absolute silence at dinner that evening."

The day of the funeral, November 20, would have been Robert's forty-fourth birthday, and it was three days before the sixth anniversary of Jack's death. After Cardinal Cushing, himself frail at seventy-four, celebrated Mass at the church in Hyannis, the family drove away in a downpour. The sun was out when they reached Hollywood Cemetery in Brookline, for the final brief service before a huge granite gravestone engraved only with the word "Kennedy." Ted took his mother back to her limousine, then returned for a moment and knelt by the casket to say a final prayer for the soul of the father he had tried so hard to please.

And Chappaquiddick did not go away. Owens's chores included checking the wire service printers to make sure that Dinis had not done anything. The district attorney did seek to get Kopechne's body exhumed for an autopsy. But a court in Wilkes-Barre denied the request after a two-day hearing in October. Her parents had opposed exhumation, saying it would be like a second funeral for them.

Kennedy was not needed for that hearing, but was the first witness when a four-day inquest was conducted in Edgartown in January 1970. The sessions were closed to the press. The testimony, the exhibits, and Judge James A. Boyle's conclusions were made public only the next April. Boyle said he believed that "Kennedy and Kopechne did not intend to return to Edgartown" and that "Kennedy did *not* intend to drive to the ferry slip and his turn onto Dike Road was intentional." And he announced that he thought there was "probable cause to believe that Edward M. Kennedy operated his motor vehicle negligently . . . and that such operation appears to have contributed to the death of Mary Jo Kopechne." Even though the Edgartown grand jury showed some interest in pursuing the case, Dinis did not, and the legal proceedings ended there.

Not long after, Ted met with Mary Jo's parents in the Central Park South apartment his mother kept. Shortly after the funeral, they had asked to spend some time alone with him. Rose and Dun Gifford were there at the start, then left the other three alone. Leaving, the Kopechnes told Gifford they were satisfied.

THERE WERE STILL two issues on which Kennedy was a national leader. One was a familiar one, Biafra. At the beginning of 1970, the rebellion was collapsing. The U.S. Embassy in Lagos said the Nigerian government had assured it there was no starvation problem. Elliot L. Richardson, the undersecretary of state, said that "we should at this point accept at full face value the sincerity" of the central government's desire for reconciliation. When Kennedy called a hearing of his refugee subcommittee, the State Department censored out the candor in the prepared testimony of Clyde Ferguson, the relief coordinator. But sympathizers in the department slipped Kennedy and his aide, Dale de Haan, official cables reporting that the Nigerian forces, far from seeking reconciliation, were completely out of control, looting and raping in Biafra. "A disaster of major proportions appears to be developing here," the cable said. "At least one million people are in acute need now, and the situation grows worse daily. . . . Problem particularly explosive because press have seen enough during recent escorted visit to realize how inadequate relief effort is." Kennedy also learned of a study that the U.S. government had paid for and then buried, reporting the food needs in Biafra at 260 tons per day, far more than the U.S. ambassador, William Trueheart, who was afraid of irritating the Nigerian government, would acknowledge.

Kennedy had two important allies when he pressed for greater relief efforts, President Nixon and Henry A. Kissinger, his national security adviser. Kissinger feared that somewhere between 1.5 and 6 million Biafrans would starve or be massacred. Because of Kissinger's alarm, and spurred by the publicity Kennedy generated, Nixon stepped in. He telephoned Prime Minister Harold Wilson of Britain to coordinate relief efforts and even met with Kennedy to tell him more food would be sent. Kennedy told reporters it was "more than just another problem" to the President.

The other issue was health care. Walter Reuther, the head of the United Auto Workers, was trying to get the United States to adopt a system of national health insurance, paid for out of a specific tax, like Social Security. Around the end of 1968, Reuther had persuaded Kennedy to join a national leadership group called the Committee of 100 for National Health Insurance. Kennedy was central to Reuther's hopes for Congressional action, and the labor leader went to Boston early in 1969 to talk privately with Kennedy and persuade him to take a leading role on the issue. Kennedy was flattered, but uncertain. He told Reuther he had expected to concentrate on education because Claiborne Pell of Rhode Island was senior to him on the health subcommittee. Reuther promised to talk to Pell, and persuaded him to drop off the health subcommittee and leave a clear field to Kennedy.

Kennedy's advocacy, though not his leadership, became clear on December 16, 1969, when he spoke at the Boston University Medical Center and said, "We must begin to move now to establish a comprehensive national health insurance program, capable of bringing the same amount and high quality of health care to every man, woman and child in the United States." He complained that the United States was unique among industrial nations in lacking universal health care. Acknowledging that the cost would be high, he said the program should be phased in gradually. He proposed starting by covering all children in 1971, with all Americans covered by 1975.

At the beginning of 1970, Kennedy was the third-ranking Democrat on the subcommittee, behind Yarborough and Harrison Williams of New Jersey, but still a leader in the growing campaign to achieve universal health insurance. But on May 2, Yarborough, who headed both the full Labor Committee and its health subcommittee, was defeated in the primary for the Democratic Senate nomination by Lloyd Bentsen. One week later, on May 9, Reuther died in a plane crash. Kennedy replaced him five days later at Reuther's next scheduled health care appearance, saying that "the legislation we enact for national health insurance will be a living memorial to Walter Reuther."

By August 27, 1970, when the Committee of 100's bill was introduced in the Senate, events had propelled Kennedy into a position as the nation's leader on the issue. He introduced the bill, for himself, ten other Democrats, and two Republicans, and told the Senate that most Americans could not assure themselves of quality health care: "We therefore offer today a health security program that will enable our nation to make the right to health care not merely a principle or a social goal, but a living and functioning reality."

That was the beginning of a career-long crusade.

IN THE SHORTER term, though, Kennedy's personal influence had plainly diminished. For example, the tax reform effort he had persuaded Mansfield to make a central Democratic priority ran into full-scale opposition from Russell Long in the summer. By the time the legislation limped to the floor in December 1969, there was little to it except one idea that Robert Kennedy had urged—a minimum tax designed to make sure that even the richest Americans paid a little. Kennedy had pushed hard for that proposal in leadership meetings. But when Kennedy tried to strengthen Long's bill with floor amendments, he lost three times out of three.

Yet Kennedy in eclipse still had a central, even decisive role in two major Senate actions, the defeat of G. Harrold Carswell, an absurd Nixon nominee for the Supreme Court, and the passage of the eighteen-year-old vote. He did not have the leading role personally, but two of his staff members were singularly responsible for those Senate actions.

Carswell was nominated only after the Senate had rejected Clement Haynsworth, a conservative appeals court judge from North Carolina. The arguments against Haynsworth were that he had often been reversed in labor and civil rights cases, and that he should have disqualified himself from a case involving a company in which he held stock (a standard that did not then appear in codes of judicial ethics). But the underlying cause was retaliation for Republican efforts to filibuster Abe Fortas's nomination to be Chief Justice in 1968 and then in forcing him to resign from the court in May 1969. Labor and civil rights groups went after Haynsworth and got Bayh to lead the fight. Kennedy, who had doubts years later, joined in. The nomination was rejected on November 21, 1969, by a 55–45 vote. Haynsworth had voted for Kennedy in the moot court final in Charlottesville. Kennedy voted against him in the Senate.

Supreme Court nominees had been rejected frequently in the early years of the Republic, but the Senate had not voted down a justice's nomination since 1930. Nixon, with no small sense of Presidential prerogatives, was furious. Determined to show civil rights groups who was in charge and confident that the Senate lacked the will for another bruising battle, he proposed Carswell on Monday, January 19, 1970. A judge of the Fifth Circuit Court of Appeals for only a few months, the Floridian had none of Haynsworth's quality. Indeed, until Haynsworth's Supreme Court nomination, Carswell had been the only judicial nominee ever opposed by the Leadership Conference on Civil Rights.

At first, the Nixon strategy seemed to work. Hugh Scott of Pennsylvania, the Republican leader, who had damaged his ties to the administration by opposing Haynsworth, came out immediately "without qualification" for Carswell. The *Los Angeles Times* bureau chief, Robert J. Donovan, wrote that "Senators are mostly in no mood for another such donnybrook." Three days later, though, the *CBS Evening News* carried a report by Ed Roeder, a stringer for its Jacksonville affiliate, WJXT-TV, that made Carswell's early racism clear. In 1948, running for the Florida legislature and twenty-eight years old, Carswell told an American Legion rally that he would be "the last to submit to any attempt" to end segregation. "I yield to no man as a fellow-candidate, or as a fellow-citizen, in the firm, vigorous belief in the principles of white supremacy, and I shall always be so governed."

The next morning, Jim Flug got involved. He had been with Kennedy for five years, and was now in charge of a subcommittee with a broad hunting license, the Judiciary Subcommittee on Administrative Practice and Procedure. On Friday, January 23, 1970, Flug sat down with aides to Senators Hart, Bayh, and Tydings and civil rights and labor lobbyists. Some of the others felt it was important just to make a showing against a nominee as bad as Carswell. Flug told them that he could actually be defeated. But in either case, they needed a Senator to lead the fight. Bayh did not want to take on a second challenge. Hart felt disabled by having led the battle to confirm Fortas as Chief Justice—a cause that looked worse in the hindsight of his resignation. Tydings had a tough election campaign coming up, and Kennedy had the Chappaquiddick cloud. So they broke up without finding a leader. The next day, Flug sent Kennedy a memorandum calling the nominee "a mediocre candidate with no indications of particular intelligence, leadership, insight or respect among his brethren." He called him a segregationist and a white supremacist and told Kennedy that evidence of that was being compiled.

When Eastland opened Judiciary Committee hearings on January 27, Kennedy questioned Carswell sharply. In 1956, while serving as United States attorney in Florida, Carswell had drawn up the papers to transform a public golf course into a whites-only private club. Carswell, though he had been shown the papers the night before, said he could hardly remember them. Kennedy asked if he had read them before signing them. The hearings also disclosed that as a judge, Carswell had treated civil rights lawyers rudely in court. Another Flug memo to Kennedy, during the hearings, was titled "How to Beat Carswell" and said, "I smell blood. I think it can be done if we can get the full civil rights apparatus working, which it's beginning to do."

But Flug's determination was outrunning Kennedy's. Suspecting that the judge's unpublished opinions might be even worse than the ones which were published, Flug ordered the unpublished opinions shipped to Washington. When several pallets loaded with them arrived at the Judiciary Committee, Eastland's staff was baffled and annoyed. Eastland called Kennedy sharply and asked him what was going on: "Keny, what you doing to me? Keny, what's that Flug doing?" Kennedy wavered, and told Burke to tell Flug that Eastland's unhappiness was an unwelcome problem: "We don't need this. So tell Jim that we're sort of slowing down on this stuff." Burke went over to the floor, sitting in the back, to tell Flug, who replied that "this man is worse than Haynsworth." Burke tried to tell him, "I'm the administrative assistant and you're not." Before long they were both in tears. In the end, Burke told Kennedy that Flug wouldn't slow down, and Kennedy accepted that.

It was a tough fight. On March 19, with Kennedy off the floor in the cloakroom, Senator Bob Dole of Kansas mocked the opposition to Carswell from onetime backers of Frank Morrissey. Kennedy emerged and said, "At that time the proponent of the nomination had, I think, the wisdom to withdraw the nomination. Some of us who have expressed reservations about this nomination hope that the same judgment would be expressed by the administration on this nomination." The next day, Kennedy said confirmation would tell Nixon that "you can appoint anyone you like, no matter how pedestrian, no matter how undistinguished, no matter how unworthy of respect, no matter how abhorrent to our ideals, our traditions and our liberty." On April 6, he pointed out that the issue was broader than Carswell himself. "Can we accept the proposition that the appointment of Supreme Court justices is the President's own to make, unfettered by Senate review? The Constitution says no; logic says no; history says no; past Senates have repeatedly said no; and I believe this Senate will say no."

Other Senators were clearly more central to the campaign to defeat the nomination. On March 10, Bayh agreed to take the lead for the Democrats, and he was on the floor from then on. The Senate's only black, Republican Edward W. Brooke of Massachusetts, took on the chore of rounding up Republican votes. Tydings was prominent, too. Lobbyists from the civil rights community, like Joe Rauh, Marian Wright Edelman, and Clarence Mitchell, mattered a lot. But among Senate staffers, from the first critical day when he told other aides Carswell could be defeated, no Senate aide had a bigger role than Flug. At every level, from collecting rumors and checking them out, to organizing students to read those unpublished opinions, to passing on credible vote counts and figuring out who needed to be persuaded, Flug was tireless.

The vote was one of those rare Senate moments when the result was in doubt as the roll was called. The White House struggled to save Carswell. It overreached. Senator Margaret Chase Smith, a tight-lipped Maine Republican, had never said how she would vote. But when she learned that the White House was telling other Republicans she was for Carswell, she was furious and voted against him. On Wednesday, April 8, when the nomination was beaten 51–45, there was cheering in the gallery and fury at the White House. Flug was sitting at the back of the Senate chamber, laughing and weeping, saying, "I just can't believe it. It's too good to believe."

The other issue on which Kennedy's staff played a dominant role was the eighteen-year-old vote, which arose in the midst of the Carswell battle and was settled, at least in the Senate, before the nomination was defeated.

The aide for this issue was as cerebral as Flug was passionate. Carey Parker, a thirty-four-year-old Harvard Law School graduate who had clerked for Supreme Court Justice Potter Stewart, joined the staff in February 1969.

The argument was being made that if young men were old enough to fight in Vietnam, they were also old enough to vote for, or against, the politicians who sent them. But states set the voting age then, and most set it at twenty-one. Only Georgia and Kentucky allowed eighteen-year-olds to vote. And while there was national support for amending the Constitution to set an eighteen-year-old standard, that path seemed both time-consuming and uncertain.

But Parker had read an article by Archibald Cox, President Kennedy's solicitor general, in the *Harvard Law Review*. It offered an intellectual spark for doing it another way. In a single paragraph in a review of Supreme Court decisions in its 1965–66 term, Cox saw broad implications in *Katzenbach v. Morgan*. That 1966 decision upheld the constitutionality of a provision of the 1965 Voting Rights Act, a section written by Robert Kennedy prohibiting an English literacy requirement for anyone who had completed the sixth grade in Puerto Rico. "If Congress can make a conclusive legislative finding that ability to read and write English as distinguished from Spanish is constitutionally irrelevant to voting, then," Cox wrote, "Congress would seem to have power to make a similar finding about state laws denying the franchise to eighteen, nineteen and twenty-year-olds even though they work, pay taxes, raise families and are subject to military service."

Parker drafted a memorandum marshaling the arguments for lowering the voting age. It cited the greater education and sophistication of young people, the hope that the vote would encourage civic responsibility, and the classic argument of "old enough to fight, old enough to vote." It went on to discuss tactics, saying that a constitutional amendment would take "many years." Then, expanding the Cox argument from one paragraph to three legal pages, Parker explained the constitutional basis for acting quickly, by statute.

His constitutional arguments persuaded Kennedy that it could be done, but there was still a major tactical problem. The obvious place to legislate on the voting age was the renewal of the Voting Rights Act. The Nixon administration had sought to weaken the act, which was to expire in 1970, and Southern states were fastening on devices like changing the boundaries of elective districts or making offices appointive rather than elective to dilute the weight of the new black vote. Civil rights leaders and Senate allies like Phil Hart were worried about doing anything that would jeopardize renewal of the voting act, and they feared that the opposition of Emanuel Celler, the eighty-one-year-old chairman of the House Judiciary Committee, would endanger the whole bill if the Senate added the eighteen-year-old vote.

Kennedy was not ready to act, but he circulated Parker's memo among friendly Senators. Then John W. Finney of *The New York Times* got a copy and wrote about the plan. His story appeared on February 23, 1970, and that day Kennedy issued the memorandum as a press release. Along with the analysis, there was a promise that if the eighteen-year-old vote provision threatened renewal, that provision would have to give way. Kennedy was not sure that Celler would actually block the voting bill over this issue, and he talked with younger members of the House, like Lowenstein (elected from Long Island in 1968) and Tom Railsback of Illinois. They reported that while many older members, especially those with colleges in their districts, were uncomfortable with adding younger voters, they would find it hard to vote against if it got to the floor.

The tactical question was not resolved when Kennedy left for Ireland in early March to give a lecture on Edmund Burke. But while he was gone, Warren Magnuson of Washington, who had been trying to lower the voting age for more than a third of a century, told Mansfield that he thought Kennedy had a good idea. So Mansfield and Magnuson introduced the proposal on March 4, without Kennedy. His post-Chappaquiddick image was a reason not to have him front and center, but he joined as a cosponsor, and Mansfield gave him plenty of credit for the idea.

Mansfield's prestige and intense interest (he said in later years this was his proudest achievement as a Senator) carried the issue in the Senate. It moved very quickly. On March 11, Kennedy spoke: "I think one of the significant arguments for a finding by Congress that the voting age should be lowered is that if young people are old enough to fight, they are old enough to vote. Thirty percent of our forces in Vietnam are under twenty-one years of age. Tragically one half of the deaths in Vietnam are of young Americans under twenty-one." He argued that setting the age of majority at twenty-one was archaic, not only because of improved education but because it had been established in the eleventh century because "it was believed at that time that a young man had to be twenty-one years old to carry the heavy armor of a knight."

The next day the measure passed overwhelmingly, by a 64–17 vote. And while President Nixon, the *Washington Post,* and most of the Yale Law School faculty argued that the approach was unconstitutional, Parker and Kennedy kept the heat on the House. Parker drafted a letter to the *Post,* enlisting Barry Goldwater for the proposition that the Senate "found unfair discrimination in the fact that 18-year-old Americans who die in Vietnam and who work, marry and pay taxes like other citizens are denied the most basic right of all—the right to vote."

The pressure, including support by Speaker John W. McCormack, who was retiring, persuaded Celler to go along. Nixon, worried about younger voters, hoped Celler would bottle the bill up. But Celler told the House the voting age provision no longer worried him, because the bill enabled the Supreme Court to review the issue quickly. So the House passed the voting rights bill, with the eighteen-year-old vote, on June 17. Nixon, still questioning its constitutionality, signed it six days later.

Kennedy was not finished. That fall, when the issue came before a federal district court in Washington, Kennedy argued as a friend of the court that the law was constitutional. His side won in the district court, but in December the Supreme Court split three ways. Four justices said Congress had all the power it claimed over voting ages, and four said it did not. Justice Hugo L. Black had the decisive voice, saying Congress could control the voting age for federal elections, but not for state elections. The decision was widely seen as a victory for Kennedy. The states had no interest in the expense of maintaining two separate rolls for state and federal elections. They urged Congress to amend the Constitution to make the voting age a uniform eighteen. That amendment was sent to the states on March 10, 1971, and ratified by thirty-eight of them in record time, becoming the Twenty-sixth Amendment to the Constitution on July 1, 1971.

Those events in 1970, when Kennedy himself was weak, were a measure of the importance he placed on staff and the empowerment his aides felt. For years it has been a common Senate observation that Kennedy's staff is among the very best on Capitol Hill. His aides stay longer than most able assistants in other offices. Kennedy, who heard his father preach about the importance of having assistants like the young Bill Douglas before he went on the Supreme Court, has occasionally kept topflight aides from leaving by supplementing their salaries from his own funds. He has said that he gets good people because they know he gives them responsibility and will hear their arguments. The personal, Senator-to-Senator institution was changing. Senators had more to do and staffs were getting bigger. Kennedy never gave up his role of dealing with the other ninety-nine Senators, one on one, but in 1970 he showed that he could affect events through his staff, too.

CHAPTER 14

Victory and Defeat

Kennedy's opposition to the war and his support for the eighteen-year-old vote sometimes led him to be stereotyped as a backer of student protests. He was quick to condemn harsh crackdowns, such as the May 4, 1970, killing by Ohio National Guardsmen of four Kent State University students protesting the invasion of Cambodia. "Discontent," he said at a memorial service, "must either be met or suppressed; . . . to meet it is liberation and to suppress it is the end of liberty. . . . Dear God, help us, this war must end."

But his heart seemed more with working youths, who were much more likely to serve in Vietnam than young men who could afford college. Shortly after Kent State, while telling students at Johns Hopkins University in Baltimore that he would work to cut off funds for Vietnam, he insisted that whatever the cause, "violence has no morality to it." Attacking the vernacular of student protest, he said, "To call a police officer a 'pig' is sheer malice and a hateful act."

At an Independence Day celebration in Wakefield, a small blue-collar town a few miles northeast of Boston, he condemned both hawks and doves for the twisted symbols they were making of the American flag. He said the flag was too often worn "on the outside instead of the inside, on our sleeve instead of in our heart." Speaking from the library steps at the conclusion of a parade whose winning float showed servicemen and villagers together in Vietnam, Kennedy said, "I love the flag no less because I believe America has lost its way in Vietnam." But he scorned those who burned or tore the flag to protest the war. He said they committed "a crime against everything for which we stand, a crime against our past, a crime against our future." A few weeks later, he warned Democrats not to let a "love affair with campus youth" make them forget about "young hard hats." He said the party should be broad enough for supporters of the war as well as its enemies. "We have become so preoccupied with the generation gap," Kennedy said, "that we have failed to see the class gap that is opening and that threatens us far more seriously."

A different threat was on his mind, and his family's. Every afternoon he called home after school to talk to Teddy, who turned eight in September, and reassure him that nothing bad had happened. As Joan put it, "Teddy is the one who asks. He wants to know why all these things happened to Uncle Jack and Uncle Bobby, and will they happen to Daddy? What do you tell an eight-year-old when he asks you why that man shot his uncle? I tell him he was just a bad man who didn't know Uncle Bobby."

Reassurance for Teddy may not have come all that easily from his parents. Ted would sometimes crumple when a metal door slammed or a balloon popped. He even speculated that he was not perfectly safe in his home by the Potomac, for someone with a rifle could fire from several hundred yards across the river at him. Joan was even more open about her fears. She told the *Ladies' Home Journal,* "Frankly, I worry all the time about whether Ted will be shot like Jack and Bobby." She said he "tries to keep things from me—serious threats against his life—that kind of news—but I know what's going on."

When that interview was published in late June it appeared as though Joan was trying to talk to Ted through the magazine. She told of her fears for Ted's life and said, "This is such a painful subject with us that we can't even discuss it." Joan was insisting that their marriage was stronger than ever, too, now that "we know our good and bad traits, we have seen one another at rock bottom, and we still love each other." But in person, the awkwardness was obvious. Joan flew to Boston and drove to join Ted for a parade in 1970, but when they met, they just chatted casually, without a kiss or even a squeeze. A reporter asked why and Joan said, "Neither of us thinks it's in good taste" to show affection publicly.

Ted sometimes talked about quitting politics entirely. With Dun Gifford he joked about going into the boat business, leasing yachts. They talked about it occasionally, fleshing out details, but never in earnest. One day he had Fred Dutton come by for a sandwich in his Senate office and asked, "What if I didn't run again? What if I retire?" Dutton asked, "Well, do you have anything in mind?" Kennedy said, "Maybe return to law practice in Boston. Maybe see if I can buy the *Boston Herald*." He laughed. "Maybe sit in the South of France."

He did firmly take himself out of the 1972 Presidential race, repeatedly and on national television. In May 1970 he said on NBC's *Today* show that he would not run but would "remain in politics, or public life, as long as I felt that I could be effective." Asked about 1976, he said, "I don't make any plans that far, certainly, into the future." In early October, on *Meet the Press,* he made his reasons clearer: "The uncertainties of higher office would place a great burden on my family."

But he was running for reelection to the Senate. He started with appearances in St. Patrick's Day parades. The first was in Lawrence on Sunday, March 16. There had been death threats, and police sharpshooters were visible on the roofs as he and Joan marched and waved to the crowd. The next day the atmosphere in South Boston was more festive. The police estimated that a quarter of a million people turned out, sloshing beer and whiskey on marchers, shouting "We love ya, Ted!" while trying to hug him, kiss him, or shake his hand.

At the beginning of the campaign, no one was quite sure if Chappaquiddick would hurt him. The South Boston reception was an answer to that question, especially when the Nixon administration tried and failed to persuade a Republican officeholder to take him on. Cy Spaulding, a former state party chairman, inherited the nomination.

Kennedy campaigned as hard as he had against Eddie McCormack. He sometimes made more than twenty stops in a day. The intense schedule was designed to show voters that nothing had changed about his concerns for them and the state, and to let them ask him anything they wanted. "The voters need reassurance," he told Johnny Apple of *The New York Times*. "They need to see me, to be convinced that I'm reliable and mature. You can't counter the Chappaquiddick thing directly. The answer has to be implicit in what you are, what you stand for and how they see you."

Another way to dull the political implications of that tragic weekend was high visibility for Joan, who had made headlines of her own by another striking White House appearance, wearing a see-through blouse and a blue bra to a luncheon in September, and then by making her concert debut as a pianist at a Democratic fund-raiser in Philadelphia, playing Mozart and Debussy. She was a frequent campaigner, but while she had thrived on the 1962 and 1964 races, this time she resented it. "I felt used, rather than needed," she said years later. And, she told another interviewer, "That's when I truly became an alcoholic."

The tactics worked. While reporters sometimes brought up Chappaquiddick, ordinary voters never did. Kennedy gave few formal speeches. He listened more than he talked, visiting newspaper offices, service clubs, Italian festivals, factories, homes for the aged. Vietnam did come up repeatedly, and he had a standard answer: "I'm opposed to our involvement in Southeast Asia. I think we should get out of Southeast Asia, lock, stock and barrel."

He was often asked about the new women's liberation movement, and the Equal Rights Amendment to the Constitution which it championed. In Belmont on August 31, he told an audience of seven hundred that he favored equal rights for women but opposed the amendment because it would cause confusion about many state laws, Social Security, and even the draft. Kennedy, who had no women on his staff in policy positions, prepared for the speech by meeting the secretaries in his office. "During the discussion we all pretty well decided we were not members of the women's lib movement," one of them told the *Herald*.

Kennedy was even further away from the women's movement on abortion, the other issue that came to define it. In the spring, he had scoffed at the efforts of Oregon's freshman Senator, Bob Packwood, to pass federal legislation guaranteeing the right to an abortion. "I wonder how you can even stretch the commerce clause that far," he said. Campaigning in Cambridgeport that fall, he said, "I don't believe in abortion on demand. . . . The day that we can solve the world's population problem, the problem of browns in Central America, the problem of blacks in the ghetto, by aborting them, that's unacceptable to me. How about the kids in mental hospitals: they're parasites on the environment. How about the old people in institutions: they're cluttering up the landscape. Do you want to exterminate them, too?"

Abortion was about the only issue on which Kennedy and Spaulding differed, with the Republican taking a more liberal position. Spaulding, who never raised Chappaquiddick, never threatened the incumbent. The race was also notable as a rare moment of amiable political cooperation between Kennedy and Kevin White, who was running for governor. They shared headquarters and some new ideas about using computers to boost registration and turnout. And when emergency ulcer surgery laid White up for the campaign's last three weeks, Ted took his wife, Kathryn, along to his campaign stops.

White lost anyhow, getting only 43 percent of the vote against Francis W. Sargent. But Ted won easily with 62 percent. His victory prompted Charles Colson, one of the sleazier political operatives in the Nixon White House, to send a private detective to follow him in Paris, where Kennedy went in November for de Gaulle's funeral. The detective photographed him dancing with a woman identified as Maria Pia, the daughter of the deposed King Umberto of Italy. Colson got the picture and a line about insulting de Gaulle's memory into the 5.5-million-circulation British tabloid *The People,* and then he and Nixon spent hours plotting ways to get the picture circulated widely in the United States. Dick Drayne, Kennedy's press secretary, said, "It's preposterous, and it's a phony." He said Kennedy was not dancing with anyone that night.

But while Colson was preparing for a race that never happened, Kennedy suddenly found himself in a race he was not ready for. On November 7 the *Globe* reported that Byrd was going to run against him for whip and might win. Kennedy was on the phone right away to argue with the reporter, Marty Nolan.

Byrd did not say he was running, but told reporters that if he was elected whip, the only change would be that "I would have the title." Byrd said he had been doing the work all along.

Kennedy took the threat casually. He went to Europe for a NATO parliamentarians' meeting. At dinner one evening, he was called away for a telephone call from his office in Washington, and returned to say it was now definite, Byrd was running against him. His friends there said he should go back and start campaigning. He replied, "Oh, I've got the votes."

Later, Kennedy called Byrd and asked, "Are you going to run for whip?" He said he was leaving the country for a few days and wanted to know before he left. Byrd replied, "I don't know." Kennedy said, "Well, of course you know," and Byrd answered, "Well, if I did know, I wouldn't tell you."

When Ted finally got around to canvassing his colleagues, after Christmas, he began to understand his difficulties. Senator Henry M. Jackson of Washington, a friend who had been Democratic national chairman when John F. Kennedy ran for President and would later defend Kennedy to groups who doubted his commitment to national defense, kept him waiting for half an hour. On December 4, Kennedy had voted against federal financing for construction of a supersonic transport plane, thinking it a waste of money. But it was money that mattered to the state of Washington and its dominant employer, the Boeing Company. After Kennedy finally saw Jackson, he told Owens, "He won't commit. I can't believe he'll vote against me, but the fact is, he won't commit."

He probably lost another 1969 supporter, J. William Fulbright, of Arkansas, over a question of turf. Just before Christmas 1970, Kennedy got a cable from the North Vietnamese in Paris, offering him a list of American prisoners of war. Kennedy sent John Nolan to get the list. On the Senate floor, Kennedy told Fulbright about the offer and his sending Nolan. Fulbright had received a letter, too, and said firmly, "That's a matter for the Senate Foreign Relations Committee." The list added no new names to the known prisoners, and the exercise accomplished nothing, except to annoy Fulbright.

Byrd never did formally announce. Kennedy, in the absence of an open challenge, did not ask for the help from unions and liberals that had won him the whip's job in 1969. And some of his cannier allies, like Tydings, had been defeated. Finally, when the vote came on January 21, 1971, Byrd checked with Walter Reed Army Hospital to make sure that the dying Richard Russell, who had given him his proxy, was still alive. Only then did he tell his allies in the caucus he was running. He didn't need Russell's vote, winning 31–24. Two hours later, Russell died.

The press verdict was almost as dismal as it had been euphoric two years before. A *Washington Evening Star* editorial called it "a humiliation for Kennedy and a setback for any future national political plans he may have." The *Boston Herald-Traveler* said it was an "astonishing upset." But when Nixon heard the news, he predicted the press would exaggerate. Haldeman wrote in his diary, "He made the point that you don't kill a man who has been built up the way Teddy has, by a defeat, any more than you killed Nixon by his defeat in California. He thinks it will provide a momentary setback, but that Teddy will move ahead in spite of it, with considerable strength."

But defeat was galling, and Kennedy joked bitterly about it. From time to time he thanked "the 28 Democratic Senators who pledged to vote for me . . . and especially the 24 who actually did." He spoke at that winter's Gridiron Club dinner, a white-tie-and-tails affair where senior Washington journalists did musical skits for the capital's bigwigs, saying, "The Secret Service says I receive more anonymous threatening mail than anyone else on

Capitol Hill. It wasn't until January that I realized that most of them came from my colleagues in the Senate."

But, as Kennedy told Byrd several years later, the loss was one of the best things that ever happened to him. It pushed him toward committee work, something he did far better than he would ever do as a Senate leader. And the first opportunity was coming to a head even as Byrd counted votes.

FOR SEVERAL YEARS, Mary Lasker, a New York philanthropist and a liberal Democrat, had been promoting the idea of a government war on cancer. In the spring of 1970 she persuaded Yarborough to form a committee of consultants to his Labor Committee, to report on what should be done to treat, cure, and eliminate cancer.

But then Yarborough lost in Texas's Senate primary to Lloyd Bentsen. When the consultants reported back on December 4, he had only a month left to serve in the Senate. He welcomed their key proposals: much more money for research and for spreading what was already known about cancer around the country, and, dramatically, taking cancer research out of the National Institutes of Health and establishing a new agency to give it more influence. Yarborough urged Kennedy and Javits, the senior Republican on the Labor Committee, to carry on.

Kennedy's ascension to the health subcommittee chairmanship was now assured, because Harrison Williams, who was moving up to the full committee chairmanship after Yarborough's defeat, did not want to give up its labor subcommittee. Ted asked Benno Schmidt, a New York investment banker who had chaired the panel, if he thought the country could afford the recommendation of quadrupling spending, to $1 billion annually within five years. Schmidt replied, "My strong personal view is that not only can we afford this effort, we cannot afford not to do it."

Over the next few months, Schmidt and Kennedy were out front in Washington, and Lasker stirred up the public. They had a critical, quiet ally in Elmer Bobst, the eighty-five-year-old chairman of the Warner-Lambert drug company. He had founded the American Cancer Society with Mary Lasker's late husband, Albert, and had served on the panel of consultants. But most of all, he was very close to Nixon and a virtual uncle to the President's daughters.

On December 16, 1970, Nixon threw him a birthday party at the White House. Bobst enthusiastically told him of the report. John Ehrlichman advised Nixon that the planned increase from $173 to $184 million for the National Cancer Institute was enough. But Bobst was more persuasive. Nixon, in his State of the Union address on January 22, proposed spending an extra $100 million on cancer. He said, "The time has come in America when the same kind of concentrated effort that split the atom and took a man to the moon should be turned toward conquering this dread disease. Let us make a total national commitment to achieve this goal."

Kennedy then seized back the initiative. He and Javits introduced Senate Bill 34, which called for four times as much money as Nixon had proposed, $1.2 billion over three years. He took Nixon's "moon shot" analogy a step further by saying a new agency, modeled on NASA, should take charge of cancer.

Except among cancer researchers, the idea drew skepticism from medical schools and scientists. Although Schmidt would continually explain in his testimony that they expected no instant breakthrough and that the analogies everyone made to the moon or the atom bomb were not literal, those headline-grabbing phrases sounded like promises of quick success. Many scientists argued that before cancer could be overcome, much more basic research in molecular biology was required. Schmidt, Javits, and Kennedy agreed and felt their bill would do just that. But they would not say so very loudly, since spending on cancer was vastly more popular than spending on basic scientific research.

In early March, Kennedy ran two days of hearings which focused less on money than on creating a separate agency outside the National Institutes of Health. Administration witnesses and some outside experts argued that the mix of disciplines and interests within NIH was essential to progress. They warned against splitting off first one disease, then another. Kennedy argued with them, saying that NIH was unsuited to coordinating and expanding cancer research because it would not let one of its component institutions grow much more quickly than the others.

"The American people want action in respect to the conquest of cancer and they want that action to be effective," he said. "The average American is not concerned with the multiple layers of bureaucracy which rest upon the National Cancer Institute. The average American is probably not aware of the intense competition for limited funds within HEW or the inevitable waste of effort and manpower which that bureaucracy and competition exact." Schmidt, speaking for the consultants, said the layers of bureaucracy in NIH "bring about inordinate delays."

By mid-April the Nixon administration was worried that Kennedy was going to get his bill through the Senate and get all the political credit. Mary Lasker had persuaded Ann Landers to write a column, not of her usual witty personal advice but of earnest political counseling, telling her readers to write their Senators and tell them to pass Senate Bill 34. Millions did so. James Cavanaugh, a White House aide who dealt with health issues, warned, "The Lasker forces are working very heavily to not only secure enactment of the bill but to see to it that it is Senator Kennedy and not the Administration who receives the credit for pushing the cancer research activities ahead."

Bobst pitched in again. Nixon's secretary, Rose Mary Woods, warned other aides, "Elmer says it is imperative for him to see the President." On April 26, Ehrlichman sent the President a memo predicting what Bobst would say about the need for an independent cancer agency. He suggested the President tell him "it would only mean more bureaucracy and bureaucracy is not the answer."

But even before the meeting, the President bought into Bobst's approach, which played to his own suspicions of the bureaucracy. On April 28, he told Haldeman the administration should not use NIH to fight cancer but "should set up our own agency to do it, so that we can get some credit and get some steam behind it with a real manager and PR type running it, instead of a doctor." After the May 5 meeting between Bobst and Nixon, Ehrlichman told his staff that Bobst had prevailed and they should now devise a plan "to organize a breakthrough in finding a cure for cancer and a way which will also achieve maximum credit for the Administration."

On May 11, 1971, the President came to the White House press room to announce the shift in administration policy, saying that while the cancer program would remain in NIH, it would be "independently budgeted" and "directly responsible to the President of the United States." He recalled that his aunt Elizabeth had died from cancer at the age of thirty-eight, and promised that the effort "will not fail because of lack of money. If $100 million this year is not enough, we will provide more money."

What the President called for satisfied Kennedy and the consultants. Ted had not been nearly as confident of passing a bill without administration support as the White House was fearful that he could. But then the legislation the administration proposed did not go as far as the President said, and in late May it also appeared that the $100 million he had promised for that year was not real.

Kennedy urged Schmidt to talk to Nixon, and Bobst got him an appointment. The banker persuaded the President that the Kennedy–Javits bill, altered just a bit to adopt his formula for budgetary independence within NIH, was the way to get what he wanted. "I'll support it and I'll take care of it," Nixon said. But there was a price. As Schmidt left, Kenneth Cole, another White House aide, told him, "You can't ask this President to support a Kennedy–Javits bill." Schmidt said, "Well, I just did, and I understood him to say he would. Didn't you understand him that way?" Cole said, "Oh, yes, he will support the bill, but you have got to get the names changed. Nixon could be running against Kennedy when this term is over." He also wanted Javits off the bill; the liberal Republican was no Nixon favorite.

Schmidt told Kennedy that there was good news, that Nixon would back the bill, and bad news, too. Kennedy asked what the bad news could be. Nervously, Schmidt told him. Kennedy laughed and said he would take his name off. Javits did, too, a bit more grumpily.

On June 16, the health subcommittee met and agreed on a bill. It took the latest text of the Kennedy–Javits bill, but under the bill number, S. 1828, of the administration measure which Peter Dominick, a Colorado Republican, had introduced. The full committee met the same day and approved the measure, and Kennedy said, "Peter, why don't you report the bill?" That produced a committee report hardly ever seen in the Senate, one that cited a member of the minority party as bringing a bill out of committee.

The Senate passed the bill, 79–1, on July 7. The day before, Kennedy said, "The conquest of cancer is a special problem of such enormous concern to all Americans. We can quote statistics, but I think every one of us in this body, and most families across the country, have been touched by this disease one way or another."

The House took its time. Representative Paul Rogers of Florida, chairman of the House Commerce Committee's subcommittee on health, was much more cautious than Kennedy. He was dubious about creating a separate agency, and his bill provided less independence than the Senate measure, though in the end it authorized more money, $1.5 billion over three years. Kennedy did not fight in conference over the independence issue. When the final version was adopted, 85–0, on December 10, he told the Senate it would "give the President of the United States the necessary tools so that he can direct his very considerable sense of urgency toward meeting the problems of cancer."

Nixon was not quite as gracious to Kennedy. Reluctantly, he held a signing ceremony at the White House on December 23. Kennedy and Rogers were there, but neither was singled out for credit.

Kennedy and the Nixon White House worked together, if occasionally at some distance, on cancer as they had on Biafra. Kennedy startled Cole after the bill was enacted by calling him at the White House and thanking him for doing "a great job."

But on other issues, especially national health insurance, they scrapped. Nixon called Kennedy's bill a prescription for "complete federal domination of our medical system" that would "actually do the most to hurt American health care." Kennedy called the administration's plan "poorhouse medicine," a result of a "marriage of convenience" with the American Medical Association. At the White House, Ehrlichman told Cole to find people to start "kicking" Kennedy around.

The war in Vietnam dragged on in 1970. Although American troop strength was reduced, fighting spread to Cambodia and Laos. In March, Kennedy complained to William P. Rogers, the secretary of state, about the devastation of civilians in Laos. When the President announced an "incursion" into Cambodia in pursuit of a Viet Cong headquarters, Kennedy on April 29 called it a "sad and tragic step." A week later, he told the Massachusetts Dental Society, "The strange and tragic fascination of military victory in Vietnam has cast its mad spell over two successive Presidents and thousands of young Americans have gone to their death." The administration then sent Vice President Spiro T. Agnew to Cleveland, where he said Kennedy and several other war critics had "developed a psychological addiction to an American defeat."

Kennedy also fought the administration over Bangladesh, the site of the other great famine of the Nixon years. The Nixon administration firmly backed Pakistan's central government in Islamabad as it sought to keep control of what was then called East Pakistan, a thousand miles away with India in between. Henry A. Kissinger, the President's national security adviser, used Pakistan in July 1971 as a jumping-off place for his efforts to open relations with China. India, Pakistan's enemy and a supporter of Bengali independence, had not been well regarded in Washington since John F. Kennedy was President. The administration ignored India's complaints of the burden of caring for millions of refugees fleeing Pakistan's army in 1970 and 1971. It ignored the demands of American diplomats in the region that it denounce Pakistan's brutal suppression of Bengal.

To Kennedy, U.S. backing for Pakistan was a cynical contradiction of the whole rationale for the war in Vietnam, the goal of letting the South Vietnamese people decide their own destiny. Pakistan had arrested and threatened to execute Sheikh Mujibur Rahman after he won a free election in East Pakistan in December 1970 and demanded virtual independence. Kennedy's attacks on the administration for shipping arms and not food intensified after he went to India in August 1971 to visit the refugee camps near the border. On his return he accused the administration of contributing to "slaughter and starvation." His criticism had an impact. The administration offered some money for refugee relief, and the timing suggested it was as interested in upstaging Kennedy as in feeding starving Bengalis.

The President was putting even more energy into discrediting Kennedy himself. On April 9, 1971, Nixon told Haldeman they should find some way to "cover Kennedy," emphasizing, "I'd really like to get Teddy taped." On May 28, he asked for "permanent tails and coverage on Kennedy and other Democrats." On June 23, the President listened eagerly when Kissinger told him that Kennedy was now, sexually, "a total animal." On July 27, Nixon told Haldeman, Colson, and another White House aide, Peter Flanigan, that they needed at least $2 million in "Nixon discretionary money to use if we want to hire somebody to get on Ted Kennedy or some other tactic." On September 8, Nixon discussed a new head of the Internal Revenue Service, saying, "I can only hope that we are, frankly, doing a little persecuting." He asked if Kennedy's taxes should be investigated. Ehrlichman told him that Ted's personal life was a riper target. But Ehrlichman assured the President that they already had contacts at Hyannis who would tip them off if anything happened there. On September 18, the President suggested using material on Diem's assassination to make Ted appear "ruthless, heartless, bloody." On October 14, the President chuckled when Colson proposed exploiting Chappaquiddick with placards showing Ted's picture and asking, "Would you ride in a car with this man?"

Those episodic conversations were nothing like the concentrated attention the President gave Flanigan's October 28 discovery of a *London Daily Express* cartoon, published in outrage after Kennedy proposed that British troops leave Northern Ireland. The cartoon showed Kennedy with a sign around his neck reading "Britain! Leave Ulster Alone!" A framed picture showed a car in the water, off a dock with a sign reading EDWARD KENNEDY LEFT HERE. In the cartoon, Kennedy was saying, "Gee! I don't see what's so difficult about leaving someone alone." For the next month Nixon and

Colson repeatedly discussed ways to get the cartoon circulated in the United States. On November 29, Nixon told Colson, "That's the greatest thing I've seen."

They pursued this adventure despite Nixon's judgment that Chappaquiddick had finished Kennedy as a Presidential candidate. "He will never live that down," the President told Ehrlichman on September 8. "That will be around his neck forever."

Nixon was probably right, and in 1971 Kennedy made no real moves toward running for President. But polls suggesting he could win the nomination would come along and get him thinking. He had a long off-the-record visit in April with James Wechsler, editor of the *New York Post*. He made it clear he had not ruled out running, especially if nothing changed his impression of the declared candidates as "a lot of guys scurrying around" with none of Robert's impact, and, except for McGovern, with none of his own commitment to end the war immediately. In late May he talked to Apple of the *Times* and said he thought he could win the nomination if he tried, pointing out the states where he would be strong and the ones that would be tougher. He said he would like to be President because that is "where the power is." The interview kept the door open, even though he said he would not run—not because he feared losing to Nixon, though beating him would be hard, but because he wanted "breathing time" to look after his and Robert's children and gain more experience. Now, he said, "It feels wrong in my gut."

CHAPTER 15

Investigations by and of Nixon

An Irish woman came up to the Senator as he was walking in a park in London in September 1971 and asked if he was Ted Kennedy. He said yes, and she then demanded to know why, as an Irish-American, Kennedy could make speeches about the National Guard shootings at Kent State but had nothing to say about Northern Ireland, where British troops were enforcing rule by the Protestant majority. They were locking up Catholics without trial but standing by when Protestant paramilitary groups attacked Catholics, she told the Senator. Kennedy, traveling in Europe to see various national health insurance plans, thought she had a point and considered going to take a look. He sent Jim King off to visit refugee camps just across the border from Northern Ireland. King found families who had fled the violence and the constant tear gas of Derry. He rejoined Kennedy in Sweden, but warned that the security did not look good enough for the Senator to visit himself.

So instead Ted made his first speech on Irish issues, the speech that produced the cartoon Nixon and Colson drooled over. "Ulster is becoming Britain's Vietnam," Kennedy told the Senate on October 20, 1971. He demanded that British troops leave and arbitrary arrests stop. He said London was fighting another colonial war: "The tragedy of Ulster is yet another chapter in the unfolding larger tragedy of the Empire—it is India and Palestine and Cyprus and Africa once again. It is the struggle of men everywhere for the basic rights of freedom and self-determination." He called for unifying Ireland—removing the northern counties from Great Britain and making them part of the Republic of Ireland. "Without a firm commitment to troop withdrawal and unification," he said, "there can be no peace in Northern Ireland. The killing will go on, and the intolerable mounting violence will continue."

The British were furious. Lord Cromer, their ambassador in Washington, issued a statement saying the troops were there only to prevent killings by extremists. The next day in London, sixty Conservative MPs offered a resolution stating: "This house deeply resents the quite unwarranted incursion of Sen. Kennedy into the domestic affairs of the United Kingdom but is confident that it does not reflect the views of responsible American leaders." In Belfast, the prime minister of Northern Ireland, Brian Faulkner, called Kennedy's comments "deplorable." Kennedy hit back, and said the British must have a "guilty conscience" to respond so angrily. Then Britain's prime minister, Edward Heath, dismissed Kennedy's arguments as an "ignorant outburst."

"The troubles," as they were called, had been going on since 1966, when police broke up Catholic civil rights marches that demanded housing and jobs, and were accompanied by singing "We Shall Overcome." The Protestant local authorities resisted efforts by the Labour Party government in London to make concessions. Troops were sent to maintain order in August 1969, and succeeded for a time. But things turned bloodier after the Conservative Party took power in Britain in June 1970. The Irish Republican Army's Provisional, or violent, wing grew stronger, as did Protestant paramilitary forces. Imposing internment without trial in the summer of 1971, accompanied by torture of prisoners, did little to weaken the IRA's leadership and much to build Irish tolerance for its violence.

Ted had been to Ireland several times, though never to the North. His first trip was in 1938 when his grampa, Honey Fitz, visited Rose and his grandchildren at the embassy in London and led them on a pilgrimage to Limerick, where the Fitzgeralds came from. After President Kennedy's triumphal visit in June 1963, Ted went over to Jack's house at Hyannis Port to see the films of the television specials on the trip for three nights in a row, when all other members of the family had tired of them. And the ecstatic crowds for Ted's own visit to Limerick in 1964 closed the city down. The Kennedys also kept in touch with Irish visitors to Boston, often inviting them down to the Cape, where in the summer Irish college students usually were earning money working in the Kennedy compound.

Back in Washington from his health insurance trip, Kennedy joined Senator Abe Ribicoff of Connecticut and Representative Hugh Carey of New York to offer a resolution calling on the British to withdraw their troops. They urged unification of the nation partitioned in 1920. The resolution condemned British tactics: "The random midnight roundup of suspects on the night of August 9 this year—the knock on the door, the violent entry, the arrest in the dark of night—rank as yet another flagrant example of the repression of the Ulster minority."

Kennedy was hardly bothered by the British response. When a Catholic bar in Belfast was bombed in December and fifteen died, he wrote the *Times* of London to ask, "How many more men, women and children must die?" British MPs and the tabloid *Daily Mirror* demanded to know how the survivor of Chappaquiddick dared to lecture them. But when British paratroopers killed fourteen Catholic civil rights marchers in Derry on January 30, 1972, a day that quickly became known as "Bloody Sunday," Kennedy compared their action to a massacre of three hundred Vietnamese civilians by American troops in 1968, saying this was "Britain's My Lai."

The focus a Kennedy could bring to Northern Ireland did not anger Dublin publicly as it did London and Belfast, although Prime Minister Jack Lynch disagreed with him in a *Washington Post* interview. But Dublin's *Irish Times* welcomed his intervention as proving "one thing: Northern Ireland is no 'domestic U.K.' problem. It is now a problem on the world stage."

Irish diplomats, who had no influence with the Nixon administration, welcomed the attention he directed to the issue but were concerned that his speech translated into a simplistic "Brits Out" case that would lead nowhere. After Bloody Sunday, the diplomats began trying to educate friendly politicians like Kennedy on the situation, especially on the importance of rejecting violence, and of dropping demands for unification, which sounded romantic in the States but was utterly unrealistic in Ireland. The Irish were late in this effort, and for two or three years the only alternative to British policies heard in the United States was the IRA's.

That fall Kennedy found someone else to learn from. On his way to Germany for a meeting about NATO, he telephoned John Hume, a schoolteacher in Derry who had led civil rights efforts and was by then a leader of the newly formed Social Democratic and Labour Party. When he said who was calling, Hume thought someone was playing a joke on him. Kennedy said, "I need to know what's really going on in Northern Ireland and I am told that you are the person I should talk to." Convinced by now of his caller's identity, Hume agreed to meet with him in Bonn, where the Irish ambassador had them to dinner on November 21, 1972. Hume recalled later that the point he made to Kennedy was that "at that stage our objective was to get equality of treatment and civil rights for everybody in Northern Ireland. And our strategy was, having got that, that we begin the process of reconciliation and breaking down the bias between the two sections of our people, because reconciliation can only be based on equality." For decades after, Kennedy never took a step on Ireland without consulting Hume and twice nominated him for the Nobel Peace Prize in the nineties, a prize Hume shared after the Good Friday agreement in 1998.

TED TOOK A stand on Northern Ireland, though none in his family ever had before. But in those same months he was taking up two of Robert's unfinished causes—hunger in America and the plight of the American Indian. Robert Kennedy had been shaken by the sight of malnutrition in Mississippi in 1967, sitting with a child with a bulging belly and telling Charles Evers, "My God, I didn't know this kind of thing existed. How can a country like this allow it?" Robert's passion and his influence with reporters nourished the nation's awakening to the startling fact that many Americans did not have enough to eat.

One response had been an experiment with federal support for programs to feed the elderly, either with meals delivered to their homes or in group settings. The program was patterned on volunteer programs in many cities, and began experimentally in 1968 under the Johnson administration. By 1970, Democrats like Representative Claude Pepper of Florida wanted to make the program permanent and national in scope. The Nixon administration opposed the Pepper bill in 1970, and its budget plan in 1971 called for ending the eighteen existing pilot programs.

Senator Charles H. Percy, an Illinois Republican, persuaded the administration not to cut off the experimental programs and leave their elderly patrons hungry. In March he joined in when Kennedy introduced legislation for permanent federal subsidies for these programs. Senator Eagleton, who headed a subcommittee on the aging for the Labor Committee, held hearings in June, and Kennedy ridiculed the administration for trying to save $1.7 million on the food program while it sought $2.8 million for aeronautical exhibits at Dulles International Airport. He quoted Arnold Toynbee as saying a society can be measured "by the respect and care given its elderly citizens."

Kennedy continued, "At this time, the United States would be judged harshly by that criteria. The elderly in America are the Nation's most neglected minority group. They have low incomes, inadequate kitchen facilities, and a budget in which food is the only element that can be cut."

One of his constituents, Ella Reason of Roxbury, was even more compelling. She told the subcommittee about feeding Roxbury's elderly at three centers: "Eighty percent of the elderly in Roxbury are poorly fed. . . . My office receives calls daily asking us for meals in areas that we cannot afford to service with our limited operation. The seniors ask, 'Is it worth having a longer life if we slowly starve to death?' "

On June 14, the administration finally sent witnesses to testify on the $300 million bill, under instructions to do anything to appease the elderly short of endorsing Kennedy's bill. They argued that an integrated program would be more help to the elderly and that the Kennedy bill would feed only fifty to sixty thousand. They urged Congress to wait for the President to make a comprehensive welfare proposal. Those arguments had no political appeal at all, not in contrast to the vision politicians had of greeting groups of elderly people, all gathered together at lunchtime, and available for a little low-profile campaigning. Indeed, while the program was known as "Meals on Wheels," in its early years only a small percentage of its meals were delivered to the homebound elderly.

When the bill came to the Senate floor on November 30, no one spoke against it, and it passed 89–0. In the House, Representative Gerald R. Ford of Michigan, the minority leader, blocked action that year, but in February 1972, the bill passed there easily. By the time the Senate adopted the conference report on March 7, and Kennedy said it would help "one of the nation's most oppressed minorities," the Nixon administration had given up its opposition.

But it took well over a year, until August 1973, before the new, national program began. In the interim, Nixon vetoed bills to fund it because they included prohibitions on bombing Cambodia—provisions Kennedy voted for.

He also took time to follow through on Robert's commitments to Indians. Ted had succeeded him as chairman of the Special Subcommittee on Indian Education—the Alaska trip was under its authority—which concluded in 1969, "Our national policies for educating American Indians are a failure of major proportions." In February 1971, he introduced legislation to provide help for Indians, especially in nonreservation schools, with the requirement that school districts claiming the money work with Indians to design the new programs. As Mondale, a leading cosponsor, observed, the bill "puts Indian education into the hands not of Indian experts, but of expert Indians."

On a 57–0 vote the Senate passed a slightly narrower version of the bill on October 8, 1971. This legislation dropped the idea of one national board of Indian education, intended to take control of reservation schooling from the traditionally unresponsive Bureau of Indian Affairs. Division among Indians prompted this change. But the bill did authorize $390 million over three years, and Kennedy said that with passage of the bill, "the Senate can be proud of this step toward giving Indians a controlling voice in the education of their children." The House did not act on this bill itself, but it became law in 1972 when it was included as one title of a higher-education bill, and it has been widely credited with significant improvements in the education of American Indians since then.

Other than finally speaking for it, Kennedy played no part in one of the major bipartisan movements of the time, the proposal to add the Equal Rights Amendment to the Constitution, reading, "Equality of rights under the law shall not be denied or abridged by the United States or by any state on account of sex." Birch Bayh of Indiana managed the resolution, and on March 22, 1972, the Senate sent it to the states for ratification, on a massive 84–8 vote. The House had adopted it the year before.

On March 21, Kennedy told the Senate he had changed his mind and now supported the proposal because efforts to guarantee women's equality through other routes, like the Supreme Court and the Equal Employment Opportunity Commission, had turned out to be "blind alleys." An amendment was necessary, he said, because "sex discrimination remains the general rule in most states. It continues to force women into roles below their intellectual and physical capabilities."

On June 7, 1971, Kennedy accused Nixon of delaying serious Vietnam peace efforts to coordinate them with his reelection campaign. "The only possible excuse for continuing the discredited policy of Vietnamizing the war, now and in the months ahead, seems to be the President's intention to play his last great card for peace at a time closer to November 1972, when the chances will be greater that the action will benefit the coming presidential campaign," he said. Bob Dole, the freshman Senator from Kansas who was chairman of the Republican National Committee, replied that Kennedy was indulging in "the meanest and most offensive sort of political distortion" and in "the height of irresponsibility and blind personal ambition." Even Hubert Humphrey chided Kennedy, invoking the old-fashioned claim of being "the titular head of the Democratic Party" as its most recent Presidential nominee. Nixon was not "playing politics," Humphrey said. "I believe the President does want peace."

Bangladesh was another continuing sore point. After Pakistan—stupidly—attacked India along the border of West Pakistan, India invaded East Pakistan on December 4, 1971. Two days later, India recognized a provisional government in Bangladesh. The United States ineffectually favored Pakistan, and Kennedy accused Nixon of forcing India and Bangladesh to seek support from the Soviet Union because of his overtures to China and his admiration for General Yahya Khan. But after India routed Pakistan's army in twelve days, Yahya and his military government resigned. Zulfikar Ali Bhutto, his civilian successor, released Sheikh Mujibur Rahman. He returned to Dacca on January 10, 1972, and became President of the new nation. Five weeks later, Kennedy visited with Joan and his nephew Joe, getting a grand welcome. He demanded that the United States extend diplomatic recognition. It did so in April, two months after Nixon had returned from his historic visit to China.

Democrats could do little but stand back and congratulate Nixon over his China trip. But their own Presidential contest was largely fought over Vietnam, and under a new set of nominating rules that worked to favor candidates who backed the most unconditional approach to withdrawal by drawing passionate opponents of the war to primaries and caucuses. Kennedy, like most analysts, misread the process. Impressed by a string of endorsements, he thought Muskie would be nominated. But the new rules, written by a commission headed by McGovern, were designed to make sure delegates were elected by real people, not picked by party bosses. The people who took the trouble to vote were the ones who were excited about issues—in 1972, the war—and not bosses worried about who could win and help elect their own candidates for lesser offices. McGovern understood the system best. He scared Muskie in New Hampshire on March 7 and beat him in Wisconsin on April 4. Back at the White House, Nixon decided Kennedy was now the obvious candidate. He assumed that Democratic leaders would not let McGovern prevail.

Some Democratic leaders thought so, too, and Kennedy gave at least a little thought to the prospect of a deadlocked convention that might ask him to run. When Muskie dropped out in late April, Mayor Daley invited Kennedy to speak at his annual party dinner the next month in Chicago and said he was beginning to hear talk of a draft. The AFL-CIO's head, George

Meany, who had no use at all for McGovern, asked Kennedy in May to consider accepting a draft, and thought he did not reject the idea.

The hope that Kennedy would run loomed over the McGovern candidacy. Old Washington hands like Larry O'Brien and Robert S. Strauss, a fund-raiser from Texas, speculated over lunchtime martinis whether he would jump in. But Kennedy, who admired McGovern for his opposition to the war, was moving to help him. Robert Kennedy's children started campaigning for McGovern, an improbable development without Ted's approval. And Ted told Anthony Lewis of *The New York Times,* a friend of Robert's, that he favored McGovern and would endorse him if he thought it would matter, a message Lewis conveyed in his column.

Kennedy also saw a way to use the campaign to advance national health insurance in Congress. He had been holding hearings, drumming up interest. But the Labor Committee had no jurisdiction over bills that involved taxes to pay for health insurance. The Finance Committee did, but its chairman, Russell Long, was interested only in insurance that covered medical catastrophes. So, at the urging of Lee Goldman, who was heading the health subcommittee's staff, Kennedy called Representative Wilbur Mills of Arkansas, who had been chairman of the House Ways and Means Committee for fourteen years. Ways and Means governed taxes in the House, and Mills was even more powerful than Long. As Kennedy knew, Mills was interested in the Vice Presidential nomination, and a public connection to Kennedy would help with McGovern's delegates. Mills jumped at Kennedy's proposal that they work together and appear jointly at the Democratic Platform Committee hearings. In St. Louis, on June 17, they announced, "The Federal Government should establish a system of compulsory national health insurance which covers all Americans with a standard, comprehensive set of basic health insurance benefits." With a bow to Long, they said the package should be "supplemented by protection against catastrophic costs."

Meanwhile McGovern was facing an unexpected late challenge. He had a comfortable lead over the other delegates and was just short of a majority when labor and his other foes got together in the Democratic convention's credentials committee and voted to take away nearly half the 271 delegates he had won in California's winner-take-all primary on June 6. McGovern's side retaliated by voting to oust the 59 elected delegates from Illinois who were loyal to Mayor Daley, replacing them with a delegation almost devoid of white ethnics.

The last-ditch opposition was known as the ABM movement—"anybody but McGovern"—and it was not based just on his opposition to the war. Traditional Democrats were uneasy at the variety of odd liberal movements that coalesced behind the idealistic Senator with a reedy voice. There is no evidence that Kennedy did anything to encourage them, but he was the only possible beneficiary. McGovern's delegates would not have stood for the nomination's going to Humphrey or George Wallace or any of the others they had beaten in the field. Nor could the opposition have united behind any of them.

Once the convention met in Miami Beach on July 10, McGovern overcame the California credentials challenges and won the nomination. But for several weeks he had been distracted from choosing a running mate. Moreover, McGovern had the unshakable idea that Kennedy's rejection would melt away once McGovern was the nominee, that he would accept second place, just as Lyndon Johnson had agreed to run with John Kennedy.

Ted tried to make it clear he would not run for Vice President. On June 13 he told reporters he would not accept a draft. A week later, a bid came from Ted's old Army tentmate Matty Troy, who was now the Democratic leader in Queens County, New York, and a McGovern backer. When McGovern told Troy, "I'd jump out a window to get him" to run for Vice President, Troy announced a press conference at New York's City Hall, with broad hints that he would get Kennedy to run. He fooled the *Times*'s Apple, who wrote that Troy "would be unlikely to propose a running mate for the South Dakotan unless both men had indicated their approval." Early the next morning, Paul Kirk, now Kennedy's political aide, telephoned to chew Troy out, and then put Kennedy on the line to ask, "Are you crazy?" The press conference was canceled.

Still, McGovern was not persuaded. The night he was nominated he talked to Kennedy at Squaw Island and asked again. He said that campaign polls showed Kennedy would add votes to the ticket, and argued that, win or lose, being willing to run for Vice President would help him put Chappaquiddick behind him. Kennedy, at McGovern's urging, promised to sleep on it. But on the convention's last morning, Thursday, July 13, he called McGovern back to say, "I just can't do it."

McGovern and his staff started scrambling. The polls showed McGovern way behind, and the Vice Presidential nomination did not seem like a great honor. Muskie and Humphrey had no interest. Governor Reubin Askew of Florida turned them down. The McGovern camp showed no interest in another new Southern governor who did want the nomination, Jimmy Carter of Georgia. They focused instead on Boston's Kevin White, who seemed, as a big-city Catholic mayor, a plausible balance to McGovern's prairie background.

But Kennedy, when asked what he thought, was decidedly unenthusiastic. He suggested Mills instead. McGovern and his aides dismissed that idea as unacceptable to their delegates. Mills had no antiwar credentials. That was followed by calls from two prominent Massachusetts delegates, Professor John Kenneth Galbraith from Harvard and Father Robert Drinan, an antiwar freshman Congressman from Newton. They both said the Massachusetts delegation would walk out if McGovern chose White, a strong early supporter of Muskie. Then Kennedy called again, and left McGovern with the impression that he would not campaign for the ticket if White was on it. Kennedy himself never made his reason clear to McGovern, but it seemed obvious that he was being territorial. To paper over Kennedy's veto, the shift was blamed publicly on Galbraith and Drinan. Kennedy's personal role was not revealed until December by Pierre Salinger in *Life* magazine. When the article was published, Kennedy went to see White and told him he hadn't done it, a position he still maintains. For a long time after that White kept a picture on his desk of the moment in 1961 when he swore Kennedy in as a prosecutor.

Rushed by a deadline for filing nominating papers, McGovern settled on Senator Eagleton, who was eager to run. But it took a long time to get him nominated. The newly empowered Democrats, a movement, not a party, voted for everyone they could think of—from Frances "Cissy" Farenthold, the choice of the women's caucus, to Cesar Chavez, to Roger Mudd, a CBS News correspondent. Eagleton won, but it was about 2 A.M. in Miami, after 11 P.M. in California, and the television audience had dwindled.

Kennedy spoke next, brilliantly. He had been practicing while the Vice Presidential proceedings dragged on.

He brought the crowd to its feet again and again. He compared McGovern in turn to Jefferson, Jackson, Wilson, Roosevelt, Truman, Kennedy, and Johnson. He cited their credos and their accomplishments. Jefferson, he said, "attacked and unseated a government based on secrecy, repression and misuse of the machinery of justice—and so will George McGovern."

"It was John Kennedy," he said, "who summoned every citizen to ask what he could do for his country—and so will George McGovern."

"It was Lyndon Johnson," he said of the absent former President whose widened war he and McGovern opposed but whose civil rights efforts they honored, "who promised that we shall overcome, and beginning next January, under the leadership of President McGovern, we shall overcome." His praise of the former President was unique at that antiwar assembly; Johnson's picture was even omitted from a tribute to former Democratic Presidents.

Ted offered an appeal to unity, apparently with Daley in mind. "No one in this hall," he said, "has earned the right to fault or doubt those who have worked a lifetime in the cause of human dignity and worth, which they have made the cause of the Democratic Party." McGovern followed, promising to end the war. He said he would have all troops and prisoners home within ninety days of his inauguration, and then "never again will we shed the precious young blood of this nation to prop up a cruel and corrupt dictatorship ten thousand miles from our shores."

The convention adjourned well after 3 A.M., but there was little respite for McGovern. The next day there was a messy fight over who would be chairman of the Democratic National Committee. Then the McGovern campaign learned that Eagleton had been hospitalized three times for depression, twice receiving electric shock treatments. McGovern first backed him "1,000 percent," then wavered, then decided he had to go, the first nominated candidate dropped from a national ticket since Charles W. Fairbanks in 1916. But Fairbanks, the Republican Vice Presidential candidate, had at least died before he was replaced.

Again McGovern turned to Kennedy. He did not refuse immediately, asking if there was some way a Vice President could serve simultaneously as secretary of state. But after a day or two, he said no.

Again, Ted objected to McGovern's final choice, Sargent Shriver, as he had four years before when Humphrey considered his brother-in-law. But McGovern ignored his objections. On August 8, the Democratic National Committee met in Washington and nominated Shriver.

The McGovern–Shriver campaign never really made peace with labor or Democratic regulars. McGovern's campaign message was a hodgepodge of a guaranteed income, tax reform, and savings in defense spending, with an idea picked up for a day or an event and then forgotten. There was one continuing theme: ending the war. He would recall a grim moment as the pilot of a crippled World War II bomber, saying he told the frightened crew of his B-24, "Resume your stations, we're going to bring this plane home." Then he would tell his audience, "Resume your stations, we're going to bring America home." But even the war lacked the impact McGovern expected. There were only 27,000 American troops left in Vietnam, and since midsummer, North Vietnam had been seriously negotiating in Paris for a cease-fire. South Vietnam balked, but on October 26, twelve days before Election Day, Kissinger told a press conference, "Peace is at hand."

Kennedy campaigned with McGovern a few times, bringing a lift to the campaign plane, exciting the crowds. He told Minnesotans that McGovern and Nixon replicated the choices of 1960 between "the comfortable and the concerned." In Chicago, he urged Daley's precinct captains on. In Jersey City, on Halloween, he said that at the White House that night, "there's a fellow dressed up like the President. The doorbell rings and he goes to answer and there are the people from ITT and the Dairy Lobby and the Wheat Lobby and the National Rifle Association and do you know what he's going to say? He's going to say 'come on in boys, because it's all treat tonight.' "

He campaigned for other Democrats, too. In Vermont, he raised $11,000 for an underdog candidate for governor, Tom Salmon, enough to get him on television. In West Virginia, John D. Rockefeller IV didn't need money for his race for governor, but he needed the certification of concern for common folk that Kennedy, recalling his brother in 1960, provided as he drew hundreds to each of eight stops on a rainy day in October.

Salmon won. Rockefeller lost. McGovern and Shriver were crushed, carrying only Massachusetts and the District of Columbia.

Though Kennedy never fulfilled Nixon's fears by running against him, they fought repeatedly that year. In the winter, Kennedy took a leading role in an investigation of ITT, which contributed generously to a fund for the Republican Convention in San Diego after having an antitrust investigation dropped. The issue came up in confirmation hearings for Richard Kleindienst to be attorney general. The Democrats neither quite proved the connection nor stopped Kleindienst. But they worried Nixon. Once, after ITT lobbyist Dita Beard broke down under questioning in a hospital, Nixon suggested to Colson that Kennedy receive a few telegrams saying, "Isn't one woman on your hands enough?"

Nixon's almost fanatical desire to get inside information on Kennedy was a reason to assign Secret Service protection to him after George Wallace was shot in May. After only a few weeks, Kennedy had that protection halted. But his mother was worried, and called Nixon to tell him. Then, over Labor Day weekend, the *New Hampshire Sunday News*, a right-wing paper with a history of hostility to the Kennedys, reported that the Senator and John Tunney had been sailing in Penobscot Bay with "lovely" women "not their wives," including Amanda Burden, wife of a New York City councilman. Mrs. Burden denied being on the boat. (The reported affair was also the occasion, Joan Kennedy said years later, for a pious warning from Rose Kennedy, who had managed to overlook her own husband's affairs. She told Joan, "My dear, you can't believe any of these things you are reading. Women chase after politicians.")

When the report appeared in his daily news summary on Thursday, September 7, Nixon talked with Haldeman and Ehrlichman and asked if they had anyone in the Secret Service they could rely on to spy on Kennedy. Ehrlichman said, "We got several." Haldeman said, "Give orders to the detail that they are never, at any hour of the day or night, to let him out of their sight." That evening Nixon checked with Haldeman and Alexander Butterfield and was assured that the Secret Service would put forty men on Kennedy. Mrs. Burden's name came up, and Nixon said with relish, "We might just get lucky and catch this son of a bitch. Ruin him for '76." The next day the Secret Service began covering Kennedy. Butterfield gave Haldeman an urgent message that Kennedy was about to leave on a trip to Florida. Haldeman promptly talked to Robert Newbrand, a retired agent who now sometimes worked in Haldeman's office, and he became part of the detail. In 1997, Butterfield told the *Washington Post*'s George Lardner that it was his impression the agents never found anything juicy.

BUT IF NIXON'S probes were fruitless, Kennedy's weren't.

On June 17, 1972, at 2:30 A.M., five men were arrested with bugging equipment and cameras inside the Democratic National Committee's headquarters in the Watergate office building. They were working for the Nixon reelection campaign. Even though Ronald L. Ziegler, the White House press secretary, brushed it off as a "third-rate burglary," the President and his top aides were quickly involved in an effort to cover up their connections to the group. When Congress returned from the conventions in the fall, Representative Wright Patman, chairman of the House Banking Committee, tried to investigate. But the administration pressured committee Republicans and some vulnerable Democrats to block him.

Kennedy and Jim Flug, now serving as chief counsel to the Judiciary Subcommittee on Administrative Practice and Procedure, faced fewer inhibitions. At first, they just kicked up enough dust to remind the administration that someone was watching, someone more independent than Henry Petersen, the assistant attorney general in charge of the criminal division, who told Nixon about grand jury proceedings. Flug explained years later, "We wanted them to know that whatever they were doing, they couldn't hide it. They couldn't cover it up." They asked another assistant attorney general, Robert C. Mardian, to appear at a hearing on wiretapping. He refused. Another time, the FBI was testing the bugs found at Democratic headquarters. Flug borrowed equipment from the Federal Communications Commission to monitor what the FBI was doing.

When the Republicans thwarted Patman on October 3, Kennedy urged Sam Ervin, who headed a committee on wiretapping, to investigate. Ervin in turn urged Kennedy to do it, and on October 12, Kennedy wrote the subcommittee's members telling them the staff was now working full-time on the case. They avoided the burglary itself, because a trial was pending, but pursued reports in the *Washington Post* that the break-in was connected to other political dirty tricks designed to stir feuds between Democratic Presidential hopefuls. So Kennedy and Flug started pursuing the man identified as the ringleader, Donald Segretti, and found evidence of his hiring saboteurs to plant stink bombs at Muskie events and disrupt his Florida primary operation, and to concoct phony ads, letters, and picketing to spread dissension among Democrats. Through telephone records, they connected him to Dwight Chapin, a White House aide, and E. Howard Hunt, who was under indictment in the Watergate burglary.

They also investigated the man who paid Segretti, Herbert W. Kalmbach, a California lawyer and longtime Nixon fund-raiser. Flug and Carmine Bellino, a veteran investigator, issued subpoenas for his bank and telephone records. With the subpoenas as authority, banks and phone companies readily turned over information that showed the connection, including a series of bank accounts established for just these transactions. When Kalmbach refused to be interviewed, they subpoenaed him to testify before the subcommittee. That attention may have made Kalmbach reluctant to obey when the White House asked him to raise more hush money for the Watergate burglars, who were still awaiting trial.

Kennedy wrote Eastland on January 13, 1973, that their probe "strongly indicates that a wide range of espionage and sabotage activities did occur during the recent Presidential campaign . . . , that at least one key participant [Segretti] was in repeated contact with the White House . . . , that at least part of the financing was arranged through a key Republican fund-raiser who is a close associate of President Nixon's, and that neither the federal criminal investigation nor the White House administrative inquiry included any substantial investigation of the alleged sabotage and espionage operations apart from those surrounding the Watergate episode itself."

Meanwhile Mike Mansfield, the majority leader, was angered at the dirty-tricks stories. On October 29, 1972, while campaigning in Helena, Montana, he told reporters there was going to be a Senate investigation: "I think the American people want to know the truth." In November he wrote to Eastland, as chairman of the Judiciary Committee, and Ervin, as chairman of Government Operations, saying one of them should conduct an investigation that would "lay bare all the facts on the Watergate affair and other insidious campaign practices." He then decided that a select committee would be better, one including no Democrats with Presidential ambitions. He asked Kennedy to put his investigation aside. Kennedy, after two efforts to schedule closed hearings, agreed. "It was very obvious to me that I was not going to be the one that was going to be able to investigate Richard Nixon. I could see that was a nonstarter," he said years later. The Senate voted, 77–0, on February 7, 1973, to create a select committee headed by Ervin. Flug's material was turned over to that committee.

CHAPTER 16

A Crisis at Home

The war had not ended before the 1972 election. South Vietnam had refused to agree to any cease-fire that did not require North Vietnamese troops to leave. North Vietnam would not consider that. After the election, Nixon tried to pressure Saigon to accept the inevitable. And to validate a secret promise to President Thieu that the United States would continue military aid and even resume bombing if North Vietnam attacked, Nixon ordered massive bombing of North Vietnam, including Hanoi.

There was neither warning nor explanation of the "Christmas bombing," as the attacks beginning December 18 were quickly labeled. It startled Washington and the world. Pope Paul VI called it "an object of daily grief." William O. Saxbe, a gruff Republican Senator from Ohio, said Nixon appeared to have "left his senses." In *The New York Times,* James Reston called it "war by tantrum." And Kennedy, who a week earlier had spoken of an "olive branch" toward Nixon, had an op-ed piece in the same day's *Times* asking, "How can any American be proud of the face that our country is presenting to the world during this holiday month of 1972?"

Nixon halted the bombing on December 30, after Hanoi agreed to return to the negotiations in Paris. But on January 4, 1973, Kennedy led an effort in the Senate Democratic Caucus to put his party on record to end the war, immediately. He told the closed-door meeting that he had heard for years that the time for peace was not just now. It never seemed timely to end the war, he complained, and so here were Democrats dithering about whether to take a stand. He said the administration had denied obvious evidence about the severity of the bomb damage. He said it was incongruous that the administration claimed to know how many bicycles came down the Ho Chi Minh Trail but could not say whether the Bach Mai Hospital in Hanoi had been hit.

Kennedy won on a 36–12 vote in the caucus. His resolution declared it was party policy that "the U.S. has more than fulfilled any obligations it ever had to South Vietnam and the Thieu regime" and that no more money be appropriated "for U.S. military combat operations in or over Indochina, and that such operations be terminated immediately, subject only to arrangements necessary to insure the safe withdrawal of American troops and the return of American POWs and accounting for the missing in action." Two days later, House Democrats adopted similar language on a 154–75 vote.

The resolutions underlined the warnings Nixon was sending to Thieu that the cease-fire on the table was the best deal he could get and that Congress was increasingly likely to end American involvement, with or without Thieu's consent. Kissinger returned to Paris, and on January 9, after a couple of cosmetic changes in the October deal, he reported to Nixon that they had an agreement. But Thieu delayed and deprived Nixon of the chance to boast of peace in his second inaugural. On January 20, Kennedy said, "Our first priority is still as it was four years ago—to end the war." But the pact was announced on January 22 and signed on January 27, and the first U.S. prisoners were freed on February 12.

LATER THAT YEAR, Kennedy tangled with the administration's support for the generals who overthrew Salvador Allende's leftist government in Chile on September 11, 1973. Kennedy had become fascinated with Chile's democratic history when he and Joan went there on a belated honeymoon in 1959, and the role of an O'Higgins in Chile's independence stirred his Irish pride. The administration had poured millions of dollars into efforts to defeat Allende in 1970 and tried to stimulate a coup then. After Allende took power, it sought to squeeze Chile through international financial institutions. There is no evidence of U.S. government involvement in the 1973 coup, but the administration quickly recognized the new government despite the murders of Allende and thousands of others.

Less than three weeks later, Kennedy convened a hearing with witnesses who had escaped and who described mass executions in the national stadium. While Chile and the American government said matters were calmer, Connie Sue McDuffee, a translator, testified, "Every night—I think the curfew is still eight o'clock—the shooting starts, the explosions and bombs can be heard." Pat Garrett-Schesch, detained with her husband at the national stadium, told of four hundred to five hundred executions by automatic weapons. She said they hoped to be freed, and that "we felt we had a moral responsibility upon our release to be able to report accurately. So we made a point to count." Jack B. Kubisch, assistant secretary of state for inter-American affairs, said, "Our Embassy has come back saying that based on all of the staff's sources and contacts with official representatives, the press and others they have been unable to find any evidence of mass executions."

Kennedy demanded that the United States admit refugees from Chile. He opposed the administration's aid plans, saying, "We should be in no hurry to provide general economic assistance to a regime which has come to power through a violent military coup—especially after years of denying such bilateral assistance and impeding multi-lateral assistance to a democratically-elected government. I do not believe that military aid ought to go to Chile whatsoever." He did not persuade the administration, but later that year won Congressional adoption of a resolution calling on the military government to permit the International Committee of the Red Cross access to prisoners.

WHEN MANSFIELD EXCLUDED Kennedy from the Watergate investigation as a potential Presidential candidate, he was perfectly right. Soon after the 1972 election, Kennedy began talking to friends about running in 1976, although he was concerned about whether he would have to abandon his Senate seat to run for President, since that term was up in 1976, too. He also wondered if he would have to trim some of his more liberal positions. He told Milton Gwirtzman in December that several Democrats, including Gene Wyman, a prominent Humphrey backer from California, had each offered to raise $1 million for him. "I won't have the problem Bobby and Jack had" with spending the family's money, he said.

In the spring, the national polls were encouraging. Gallup showed him beating either Vice President Agnew or John B. Connally, the former Democratic governor of Texas turned Republican and President Nixon's preferred successor. There was a buzz in Washington. The *Times*'s Johnny Apple wrote in May, "Mr. Kennedy has given his close friends the impression in recent days, as he never did in 1972, that he believes his moment to strike for the summit has come."

The most dramatic step he took toward running was to fly to Decatur, Alabama, with a planeload of reporters for a July 4 event called "Spirit of America Day" honoring George Wallace. Kennedy had developed some feeling for Wallace after he had been crippled by an assassin in 1972. He borrowed from the Alabamian's themes of state's rights and high taxes. And he told the audience, "Let no one think I come to lecture you on racial justice, which has proven to be as deeply embedded and resistant in the cities of the North as in the counties of the South. . . . We are no more entitled to oppress a man for his color than to shoot a man for his belief."

The trip was widely regarded by political writers as a shrewd, healing step that would advance his chance for the White House. But he also got a taste of what a campaign would be like in return for a blast he leveled at Nixon in Decatur. Kennedy said, "Those who are proven to have abused the people's power, shall forfeit their right to exercise that power. Those who usurped authority shall be forced back within the barriers of the Constitution." A week later, Barry Goldwater told the national convention of Young Republicans, "Until all the facts involving the Chappaquiddick tragedy are made known, the American people can do without moralizing from the Massachusetts Democrat." He said Kennedy was the "last person in the country to lecture us." Kennedy worried about Chappaquiddick's impact, telling Gwirtzman he did not want to take the personal risk of running if Chappaquiddick was sure to defeat him in the end anyway.

He plunged ahead on health care, shaping a piece of legislation that changed the face of delivery of medicine in the nation. This was a bill that encouraged the development of health maintenance organizations, which in time led to the economies, efficiencies, and irritations of managed health care. The concept of health maintenance organizations, in which a multispecialty group of doctors practice together for annual fees from patients, had been around since the twenties. But many local medical associations fought the movement bitterly. Membership could lead to a denial of hospital privileges, or even, in Texas, loss of a doctor's license. The legislation had initially been proposed in 1971 by the Nixon administration, urging efforts to get 80 percent of the population enrolled in HMOs.

Under pressure from the doctors' lobbies, the administration backed off. Kennedy pushed ahead with a $5.2 billion bill to subsidize start-up costs for new HMOs. The Senate passed it in 1972, but the House never took it up. In 1973, Kennedy cut the money back to $805 million, and the Senate passed his bill on May 17, although the administration threatened to veto it. The House came in at a much smaller level, $240 million over three years, and won the money battle in conference, for $375 million over five years. That was enough, said Representative Paul Rogers of Florida, chairman of the House conferees, to develop a hundred new HMOs. But Kennedy and the Senate prevailed on two other issues. One, noted at the time, prevented states from blocking HMOs or penalizing doctors for joining. Another, more important in the long run, required that employers who offered health insurance had to make HMOs an option for their workers. That provision guaranteed access to the market and was essential for the future development of HMOs. Nixon signed the bill on December 29, 1973.

By then the Nixon administration was moving on national health insurance, too. Despite Kennedy's intense opposition, Caspar W. Weinberger, who had been director of the Office of Management and Budget, was confirmed to succeed Richardson as secretary of health, education, and welfare. When Weinberger took over, he set his staff to work on a new health insurance proposal for the administration. He thought that it was foolish to deny, as doctors' lobbyists did, that some Americans lacked decent health care, but equally mistaken to create a new government agency to run it, as he said the Kennedy–labor proposal for a system would do. His solution was to make employers bear most of the burden by requiring them to insure their workers, but at the same time keep health care in private hands.

The outlines of Weinberger's plan, which would require that all Americans have health insurance coverage, began to leak out to the public in late summer, with Kennedy's help. He kept in touch through Stuart Altman, deputy assistant secretary for health, and his complaints were chiefly about the slow pace, not the content, of the administration's effort.

That fall there was a sense in the capital and throughout the health care system that national health insurance was coming. In October, *Medical Economics* magazine surveyed 110 experts on medicine and found that three-fourths of them thought national health insurance would be a fact by 1978. Their forecasts differed on what would be covered and how it would be paid for, but did not fundamentally question the assertion by Dr. John H. Knowles, Kennedy's old sparring partner on hospitals in Vietnam and now head of the Rockefeller Institute, that "the private sector of medicine has defaulted in its responsibility, and people will look to government for solutions." Kennedy told the magazine he was "certain" that national health insurance would be a reality by 1978. Altman reflected twenty years later, when another national health insurance plan was in the air, "There was a feeling in Washington that there was no question that we were going to have national health insurance." Just as it did two decades later, overconfidence worked against success.

ALTHOUGH KENNEDY'S PROMINENCE kept him out of a direct role in the Watergate investigation, he remained involved at the edges. In early February, when the Senate considered the resolution setting up the Ervin Committee, he worked closely with Senators Ervin and Javits to fend off Republican attempts to limit its powers.

He took a more important role early in May. As the scandal unraveled, Nixon fired Haldeman and Ehrlichman; John Dean, the White House counsel, who was turning on him; and Richard Kleindienst, the attorney general, who had been about to quit to get away from the mess. He chose Elliot Richardson, secretary of defense since the beginning of the year, to succeed Kleindienst. Announcing the decision on April 30, Nixon said Richardson could appoint a special prosecutor to handle the Watergate case. Nixon only had in mind a supervisor for the United States attorney, Earl Silbert, someone who could defend the failure to indict higher-ups. Richardson's idea was a bit broader, an assistant attorney general to take charge of the whole case. But when he went before the Senate Judiciary Committee for his confirmation hearings, he found even the chairman, James Eastland, focusing on the sort of truly independent prosecutor who had investigated the Teapot Dome scandals in the twenties.

By mid-May, Richardson concluded that he would not be confirmed without agreeing to a very independent prosecutor, and he fixed on Archibald Cox, a Harvard Law School professor he had known a bit for many years. Cox had served as solicitor general in the Kennedy administration, arguing the government's cases before the Supreme Court. Now Kennedy and Flug got heavily involved, telling Cox and Richardson that the charter proposed for Cox did not promise enough independence. With Richardson and his aides, they rewrote some key sections to enhance the counsel's ability to deal directly with Congress and to get the money he needed for his inquiry. Cox appeared with Richardson before the committee, and Robert Byrd asked him if he would follow the trail all the way to the Oval Office.

Cox replied, "Wherever that trail may lead." Richardson was confirmed and sworn in on May 25. So was Cox, with Kennedy present.

Kennedy and Flug thought the charter was airtight in protecting Cox, who could be fired only for "extraordinary improprieties." They thought that the only way Nixon could get rid of Cox was to fire Richardson, and that, needing Richardson as a symbol of integrity, he would not do that.

They were wrong. On Friday, October 19, fighting to keep Cox from getting tape recordings of White House conversations which the courts had said he was entitled to, Nixon directed Cox to agree to a plan he had already rejected. Under the plan, Senator John B. Stennis of Mississippi, in ill health after being shot, would get to listen to the tapes once and verify White House transcripts, which were all Cox would get.

An even more intolerable part of the order was that Cox would be prohibited from seeking tapes of any more conversations. Cox refused to agree and issued a statement saying, "In my judgement the President is refusing to comply with the Court's decree." On Saturday afternoon he held a news conference to explain why. Alexander M. Haig, who had succeeded Haldeman as chief of staff, called Richardson and ordered him to fire Cox. Richardson refused and resigned. Haig called William D. Ruckelshaus, the deputy attorney general, with the same order. Ruckelshaus gave the same answer. But he was told he could not quit; he was fired. Finally Robert Bork, who as solicitor general occupied the same number-three position in the Justice Department that Cox had once held, agreed to fire him—an act a federal court later held was illegal. Haig sent FBI agents to seal Cox's offices. The press called the event the "Saturday Night Massacre."

Kennedy called the firing "a reckless act of desperation by a President who is afraid of the Supreme Court, who has no respect for law and no regard for men of conscience," and, in what sounded like a call for impeachment, he said, "The burden is now on Congress to nullify this historic insult to the rule of law." That night he held a meeting at his house about what to do next. He and Phil Hart reached Eastland, who was in Turkey, and he agreed to call hearings on the firings for the next week. On Sunday afternoon, Kennedy got staff members started on researching how the Senate would conduct a trial if the House voted to impeach Nixon, information he turned over to the Rules Committee the next summer when the House seemed ready.

In the face of a furious public response—a cacophony of car horns responding to the slogan "Honk If You Favor Impeachment" sometimes kept

Nixon from sleeping—the President backed down. On Tuesday his lawyers told federal judge John J. Sirica that he would turn over the tapes.

LESS DRAMATICALLY, WATERGATE led Kennedy into a major legislative effort on election law reform, using the fund-raising scandals unearthed by the Watergate investigations as an argument for using government (or taxpayers') money to pay for federal election campaigns. He derived no advantage from committee jurisdiction; he served on neither the Rules nor the Finance Committee, which handled the subject. His prominence as a potential President was, if anything, a handicap. But his strategic sense, and determination, made the issue his.

The idea of avoiding corruption by using public funds had been around since Theodore Roosevelt proposed it in 1907, and Congress had been dealing with it, ineffectually, since 1966. In 1973 taxpayers could begin earmarking $1 of their income taxes for spending on the 1976 Presidential campaign. But there was no guarantee that the money would ever be spent; it would take a regular appropriation bill, subject to a veto by President Nixon.

Kennedy had begun supporting the idea in 1971, when its appeal to a debt-ridden Democratic Party was overwhelming. He returned to it in May 1973, telling the American Society of Newspaper Editors that Watergate offered an opportunity to "ring down the curtain on the insidious role of big campaign contributions." He was not the first to propose legislation; Phil Hart had been. Kennedy joined a group of Senators in testifying on various campaign law changes on June 6, saying Congress now had a chance to "dry up the torrents of influence money that pour into campaigns for federal office." But the most important words at that hearing were not his own. He listened as the witness before him, Hugh Scott, the Republican leader in the Senate, testified that he had changed his mind and now believed that public financing was necessary.

Kennedy saw the opportunity for transforming the idea from a Democratic Party bailout into one with real bipartisan support. Until then the only Republicans who supported public financing were the party's most liberal Senators. Scott was the first Republican leader since Theodore Roosevelt to back the idea. So Kennedy went to Scott and asked if they could work together.

Scott agreed, and on July 23, when the Senate was about to take up a bill that would set contribution limits and establish an independent commission to enforce election laws, they announced their proposal to prohibit private contributions in all federal elections, for the Presidency, for the Senate, and for the House, to be paid for by taxpayers who checked off a $1 contribution on their income tax returns. Scott told the Senate, "The taxpayers themselves should be condemned if they are not willing to put up a buck for honesty." This time Kennedy said, "At a single stroke, by enacting a program of public financing for federal elections, we can shut off the underground rivers of private money that pollute politics at every level of the federal Government." Their preemptive move annoyed some of their reformer allies, including the newly powerful citizens' group Common Cause. But it focused attention on the issue and got thirty-eight votes in defeat, when fifty-three Senators, many swayed by the argument that no committee had really looked into this plan, voted to kill it.

That vote was held against the backdrop of the revelation that Nixon had taped his White House conversations and was refusing to respond to subpoenas from Cox and the Ervin Committee. Things got steadily worse for Nixon as the year wore on. There was the firing of Cox, then the discovery of an eighteen-minute gap in one key tape, and revelations that Nixon had paid hardly any income taxes and used government money to fix up his homes in Key Biscayne, Florida, and San Clemente, California, with everything from ice-cube makers to petunia beds. Little of this, except for guilty pleas from corporations to illegal fund-raising, bore on the question of how to finance campaigns, but it increased the pressure to do something.

All fall, Fred Wertheimer of Common Cause shuttled back and forth between various Senate offices to forge a coalition behind a single bill. On November 13, Scott and Kennedy announced it with seven other backers, the authors of the other reform measures. Their new plan, which they discussed at a hearing held by the Finance Committee two days later, had four parts. One provided guaranteed full funding for general election for President, not dependent on an appropriations bill, and also federal funds to match private contributions for the primary campaign for President (a Mondale idea). A second provided optional public financing for general elections for the House and Senate. The third dealt with the mechanics of the $1 checkoff. The fourth allowed deductions and tax credits for small political contributions.

But the timing was as important as any of the substantive details. Neither the House nor President Nixon had shown much interest in campaign legislation, so the Senators planned to attach it to a must-pass bill to raise the legal limit on the national debt. The existing temporary limit would expire on December 1.

On November 27, Kennedy offered the campaign finance amendments. He said, "We would have a different America today if the political power of campaign contributors were measured by their votes and voices instead of by their pocketbooks." He argued, "I doubt if any money could be better spent in the national interest than the money that would be spent for public financing." But James B. Allen, a very conservative Alabama Democrat, argued, "If a man seeks the Presidential nomination, it is not up to the American people to pay his way." Wallace Bennett, a Utah Republican, complained about tying the measure to the debt limit: "This is legislation by blackmail. This is legislation with a gun to our head."

Then Bennett made a tactical mistake, moving to kill the whole package instead of first attacking the section with the weakest support, the Congressional part. "I think it's better this way, Fritz," Kennedy said to Mondale when Bennett offered his motion. He was right. Bennett was soundly beaten, 56–39, and then each of the sections was approved individually and the amended debt-limit bill was sent back to the House.

But the House would not be stampeded into a vote on public financing for Congressional candidates. Senators, with more experience of real contests, were more willing to take the risk of having subsidized opponents than were representatives, whose reelection rates were much higher. So on November 29 the House, instead of asking for a conference on the differences between the two bills, simply sent the debt measure back to the Senate. But thanks to intense lobbying by labor and various citizens' groups, House leaders had agreed to pass the measure if the Senate would only remove the Congressional provisions.

Kennedy and other proponents were willing to settle for that much. But on Friday, November 30, the last day the old debt limit was in force, Allen launched a filibuster to block them. Mansfield called a rare Sunday session of the Senate in an effort to overcome Allen. But he did not have the two-thirds vote he needed to break a filibuster on Sunday, December 2, or on Monday either. Now the pressure of time turned against the public financing advocates. The government was now running outside its lawful debt limit, and the Treasury had stopped selling savings bonds. Mansfield asked Howard Cannon of Nevada, the chairman of the Rules Committee, for help.

Cannon agreed to report a public financing bill out early in 1974, and the campaign finance sections were stricken from the debt measure. Success had been deferred again, but the reformers had come closer.

FOR KENNEDY TO play any role at all, let alone the leading role, in the campaign finance debate of November 1973 was heroic. For the fifth time in ten years, something sudden and awful and personal was happening. His son Teddy had cancer in his right leg.

On Tuesday, November 13, the day he and Scott announced their amendment, Ted was told the leg would have to be amputated. On Thursday, November 15, when they testified before the Finance Committee, a Sydney, Australia, newspaper broke the story despite a request that it be kept private until Teddy was told. And on Friday, November 30, when Allen launched his filibuster, Teddy came home from the hospital.

Teddy, an athletic seventh-grader at St. Albans School in Washington, was home sick on the afternoon of Tuesday, November 6. Theresa Fitzpatrick, the children's governess, noticed a reddish lump a little below Teddy's kneecap. He thought it was a bruise from playing football. He conceded, in the family manner of uncomplaining understatement, that it felt "not great." That worried Fitzpatrick. Joan was in Europe, where she was spending a lot of time that fall, "leading a life of her own," as the *Washington Post*'s gossip columnist, Maxine Cheshire, accurately put it. Fitzpatrick told the Senator when he got home.

Kennedy called S. Philip Caper, a doctor he had met in 1970 when he wanted to tour Boston City Hospital without any administrators in the way. Caper, who was then the hospital's chief resident, had joined the health subcommittee staff in 1971. He arrived at Kennedy's McLean estate in about half an hour, wearing a dinner jacket for an event he was going to and joking that it was his normal uniform for house calls. He examined the leg. He thought it was probably a bad bruise but worried that it might be a tumor. He told Kennedy to call him if the bruise did not go down in two or three days.

On Thursday, November 8, Fitzpatrick called Kennedy, who was in Boston and had planned to visit his mother in Palm Beach before returning home on Monday. He changed plans to be back the next morning. Caper told Fitzpatrick to take Teddy to Georgetown University Hospital for X-rays of the leg. Caper and Dr. George Hyatt, head of the hospital's department of orthopedics, thought the X-rays looked ominous at first. After a more thorough study, Hyatt called Caper and said he believed it was a bone tumor.

On Friday, Caper met Kennedy at National Airport and told him the diagnosis. At the hospital, Hyatt repeated it, but said he wanted to call in additional consultants, and Teddy was thoroughly X-rayed to see if the cancer had spread. He was puzzled about what was wrong, and his father did not want to tell him until he was certain. But on Saturday, although Teddy was let out of the hospital to go to a birthday party, the Senator called Joan in Europe and she arranged to fly back.

On Monday, Teddy went back to school for the day, but on Tuesday, November 12, he was given a general anesthetic and a sliver of the suspected tumor was cut out. Dr. Kent Johnson of the Armed Forces Institute of Pathology promptly declared the sliver malignant. But he believed it was a chondrosarcoma, a cancer of the ligament, which was much less deadly (70 percent survived after ten years) than osteosarcoma, or primary bone cancer (20 percent survived after five years).

Joan was in tears when she saw Teddy wheeled back from the operating room and again when they went to talk to the doctors. Hyatt told them it was cancer, but not the more menacing bone cancer. He told them the leg had to be amputated, and above the knee joint, where they feared the cancer had spread. Joan wanted to know what Teddy would be able to do after the amputation, and the doctors told her that with his spirit he could still swim, sail, and even ski.

The doctors wanted to saw the leg off as soon as possible. But Teddy's cold had gotten worse, and so the operation was scheduled for Friday. Teddy was still not told; the doctors said to wait until the day before the operation, and to make sure he did not hear of it from anyone but his parents.

That was hard to assure. Georgetown University Hospital is a very public place, and reporters had heard of Kennedy's visits and were beginning to call Dick Drayne, his press secretary, to find out what was going on. Kennedy and Drayne made a very unusual decision: they decided to tell them and ask them to hold the story back until the day of the operation. Everyone agreed.

Kennedy had planned to tell his son on Thursday, November 15, but when his cold lingered, the operation was put off one more day. He had his son's radio and television set removed, telling Teddy he was sending them to children who could not afford them. A stream of family visitors came through. President Nixon called. Then an Australian paper carried the news, and United Press International called Drayne to say it now was obliged to move the story. Drayne persuaded the news agency to carry a message with the story, quoting the family as asking that it not be printed or broadcast yet, and undertook a new round of calls of his own to ask reporters the same thing. Everyone agreed again.

Finally, on Friday, Kennedy told his son that he had cancer. Teddy asked if that meant he would die. His parents assured him it would not, but then he asked what was going to happen. They told him his leg would have to come off. Everyone cried, but Joan and Ted reassured him that with an artificial leg he would still be able to go camping and sailing and skiing. Teddy tried to be brave and matter-of-fact, telling his mother to stop crying.

The surgery began about eight-thirty Saturday morning and went smoothly. When the doctors came out of the operating room just after ten, they told Joan and Ted that the operation had been uneventful and Teddy, still unconscious, was doing well.

Ted hurried off to Holy Trinity Church, a few blocks away. His oldest niece, Kathleen, was to be married to David Townsend at 11 A.M. When she heard of Teddy's cancer, she had suggested postponing her wedding. Kennedy had refused; he had promised to give Kathleen away, in Robert's place. He did his best not to let Teddy's cancer dim Kathleen's day. As they waited to walk down the aisle, he spoke of the happiness of his own wedding and told her how proud Robert would be of her. When the guests at Holy Trinity heard the Senator was there, they were relieved at the signal that the operation had gone well, and the service ended with the congregation singing "When Irish Eyes Are Smiling." Ted returned to the hospital immediately with Kara.

The next day, Kennedy pushed his son to start using the artificial leg even if it hurt. He brought people in to see Teddy continually. One day the whole Washington Redskins offensive line dropped in. The Redskins' coach, George Allen, brought a football autographed by the team. George Wallace sent dolls of Confederate and Union soldiers. Thousands of letters and hundreds of gifts came. Teddy had most of the gifts sent to Washington's children's hospital after he left Georgetown on November 30.

Ted's idea was to keep his son busy and entertained so he would not dwell on his lost leg. Teddy was afraid to tell his father he was too tired, so Joan did it for him. Years later she said, "My whole marriage I was put in the position of being the spoilsport."

But the prognosis turned grim. It was not just ligament cancer. Pathology after the surgery showed that the cancer was mixed, with the much more deadly bone cancer cells turning up too. On a January plane ride from Miami to Washington, Kennedy told a science reporter, Stuart Auerbach of the *Washington Post,* that he worried about whether he had done enough for his son. He had visited various cancer research centers and was convening a meeting of experts at his home to talk about what should be done. Other doctors were telephoning all over Europe to see if there were any experimental treatments there worth considering.

The four-hour meeting in McLean brought experts from San Francisco, New York, and Boston to describe their experimental work to prevent recurrence of cancer. All of them were long shots. While radiation was established as a cancer treatment, chemotherapy was not yet widely used. But Dr. Edward Frei III, of Children's Hospital, Boston, had been successful in treating twenty-one children suffering from bone cancer with massive doses of methotrexate, a drug that killed cancer cells especially but other fast-dividing cells as well. The effects on noncancerous cells were minimized by following the methotrexate with doses of a vitamin called citrovorum factor, given by injection every six hours. The doctors decided on this treatment, and a two-year regime began on Friday, February 1, 1974. Every three weeks, Teddy, Dr. Frei's twenty-second patient, was accompanied by his father or mother and flew to Boston. For six hours the boy would lie still, with methotrexate dripping into his arm. It left him feeling sick and nauseated. Then came the citrovorum injections. The Senator learned to give them so they could return to Washington on Sunday nights and get Teddy to school on Monday.

In March, the family went to Vail, Colorado. With a ski attached to his artificial leg and his parents cringing at every fall, Teddy learned to ski again. Physical courage was a family emblem. Eight years earlier his father had shown it, skiing for the first time after he broke his back and choosing the steep, almost sheer bowls off to the left of the lift at Sun Valley.

CHAPTER 17

No Campaign in 1976

The aftereffects of Teddy's surgery—fear of recurrence, the burden of the chemotherapy itself, and the impact on Joan—turned Kennedy away from a likely Presidential candidacy in 1976. Rivals like Henry M. Jackson and Hubert H. Humphrey described him as a certain nominee if he ran, and Mayor Kevin White even chided him for hurting the party by being indecisive.

But Kennedy kept discouraging the idea. In Everett, Massachusetts, in January 1974, he said, "I have no real interest in running for the Presidency." In February, one *New York Times* reporter, Chris Lydon, reported that "the word is being passed" that he was running. Early the morning that story appeared, Kennedy called another *Times* reporter, Johnny Apple, to say he was angry over the story and that it was "extremely unlikely" he would run. Apple then wrote that Kennedy "has made it clear to his close associates that only some dramatic change in the condition of his son would persuade him to change his mind." In early April, Kennedy told a Hamburg newspaper his family did not want him to run because it feared assassination. Later that month he told a reporter in New York he would not run, then backed off a bit that evening, promising an announcement in 1975. And in June, Bob Healy wrote in the *Globe,* "You come away from a session with Sen. Edward

Kennedy these days convinced that he will not run for President in 1976."

It wasn't just Teddy. While Joan steadfastly shared some of the chemotherapy weekends with the boy, the strain of her son's illness got her drinking more and more. She was in and out of sanitariums. She was arrested for drunk driving.

Beyond those problems, Chappaquiddick was still out there, spurring intense anger even when other Democrats went out on the stump. Still, in May, when Kennedy told a public television program that if he had to decide that day he would just run for reelection to the Senate, he conceded he had not put the thought of the White House behind him. Asked if "deep down" he would still like to be President, he answered, "Yes."

But Kennedy's only intense involvement with Presidential politics that year was through campaign finance legislation. Senator Cannon, as he had promised, brought a bill out of the Rules Committee in February, and the Senate took it up in late March. This version provided public financing for all federal elections, primary and general, Presidential, House and Senate, as well as repassing the contribution limits and independent commission the House had ignored since the Senate passed them the previous July. Debate began on March 26, and Kennedy said, "At a single stroke, we can drive the money lenders out of the temple of politics. We can end the corrosive and corrupting influence of private money in public life. Once and for all, we can take elections off the auction block, and make elected officials what they ought to be—servants of the people instead of slaves to a special few."

He continued, "We get what we pay for. As a result we have the best political system that money can buy, a system that has now become the worst national scandal in our history, a disgrace to every basic principle on which our nation stands."

James Allen of Alabama filibustered again, and many Republicans fought the bill. Howard Baker of Tennessee tried, and failed, to eliminate a section that made the legality of political action committees clear. Bob Dole of Kansas argued that the cure was irrelevant to the disease of Watergate. "Public financing is simply not a solution for human stupidity, individual criminality or personal greed."

But, as Kennedy said when the bill was taken up, there was an "increasing probability that the President will be impeached" as the campaign scandals around Nixon continued to deepen. Votes eventually switched. The filibuster was broken on April 9, on the second try, and on April 11, after thirteen days of debate and fifty-one roll calls, the bill was passed. The vote was 53–26. While Kennedy called that moment "one of the finest hours in this or any other Congress," the House prospects were still uncertain. Indeed, the House only began holding hearings, occasionally, on March 26, the day the Senate took up the issue for the third time in a year.

Few years of Kennedy's career have been so busy as 1974. In the spring he also fought, unsuccessfully, to eliminate the oil depletion allowance. And he worked with Ed Muskie to make the 1966 Freedom of Information Act more effective by allowing judges to decide if a document the government wanted kept secret was properly classified. While both Houses passed the bill in May, a conference report did not get to the White House until October. By then, the President was Gerald Ford, who vetoed the bill. Kennedy led the effort to override the veto on November 21, and won by a 65–27 vote. The legislation worked, and more documents were made public, though it still often takes several years.

But the most dramatic example of 1974's hectic pace came between the two filibuster votes on the election bill. Kennedy went to Germany for the weekend. He met Chancellor Willy Brandt in Bonn, and went on to Berlin on April 8 to congratulate Germany on a quarter century of democracy and Brandt for setting an example to other Western nations in seeking better relations with the Soviet Union. He said that détente "has taken us beyond the cold war. Leaders in both East and West now accept joint responsibility for preventing mankind's final and cataclysmic war."

This was not a casual trip, but part of a carefully planned effort to assert a voice in foreign affairs. In September 1973, before Teddy's cancer clouded any Presidential race, Kennedy had hired a full-time foreign policy aide, Robert E. Hunter, who was then the editor of *Foreign Policy* magazine. Because his brother had been President, Kennedy had access to foreign leaders without having to spend a committee slot on membership in the Foreign Relations Committee. But Hunter was hired to help him make his approach to foreign affairs more organized and less episodic.

The first major effort was a December 6, 1973, speech on energy. Protesting against American support of Israel in the war which Egypt and Syria had launched on Yom Kippur, Arab nations had raised oil prices from $3.01 to $5.12 per barrel and cut production by 5 percent. Nixon responded by proposing that the United States make itself independent of foreign oil sources by 1980. Kennedy disagreed, arguing that the long-run solution was for Europe, Japan, and the United States to work together and avoid cut-throat competition—and to work with the oil states to create greater stability in oil production and trade.

Soon after the campaign finance bill was passed, Kennedy traveled to the Soviet Union. The Communist leadership had pondered his request for a visit and high-level meetings. It concluded from a study of his writings and speeches that he should be welcomed as a powerful advocate of nuclear disarmament. He arrived in Moscow on April 18, 1974. Joan, Kara, and Teddy came too, and the Senator and his children had a snowball fight near the Kremlin.

He was treated as the next American President. Occasionally during the trip he was asked if he thought Nixon would survive Watergate. He told his hosts, "No." Leonid Brezhnev, the general secretary of the Communist Party, gave him the unusual courtesy of a four-hour meeting in the Kremlin and lent him his own jet for a trip to Tbilisi in Soviet Georgia, to Leningrad, and back to Moscow, sending his senior foreign policy aide, Alexei Alexandrov, along with him.

On April 19, Kennedy spoke in Moscow on arms control. He said the United States and the USSR should not test new nuclear warheads, and he endorsed a Soviet proposal to try to halt an arms race in the Indian Ocean, where the Soviet Union used Indian ports and the United States was developing an island base at Diego Garcia.

But he also clearly challenged his hosts when he asked why the USSR was "building new missiles and testing still others." He said, "Secrecy on intentions and doctrine in nuclear arms can only cause difficulties—and dangers—for everyone." That same day, at a luncheon at Spaso House, the residence of the U.S. ambassador, Walter Stoessel, Kennedy urged Soviet officials to be more flexible on Jewish emigration and to understand it as an emotional issue for a nation of immigrants. He warned that the value of détente was being questioned in the United States, and Soviet rigidity on emigration strengthened her critics on that issue and arms control as well.

On the twenty-first, Kennedy was permitted to make a speech at Moscow State University, though few students were allowed in and awful simultaneous translation made him hard to follow. He answered complaints that American criticisms of Soviet oppression amounted to interference in her internal affairs. He said, "A central principle of developing relations between countries is frankness." He said he had been critical of human rights abuses in Chile, in Vietnam, and in the United States, "and I do not believe in silence." But the striking part of his university appearance was a question-and-answer session. He started by asking the questions. This technique puzzled the audience, but the university's rector told them it was all right. Kennedy asked if they thought the Soviet Union should be spending more, less, or about the same for military purposes. Every hand was raised for existing levels, except for one who voted for more. Less sensitive questions got more mixed responses. And he drew laughter by pretending not to be able to hear a question about whether he would run for President.

Hunter cut short his speech after a couple of hostile questions. But when the *Washington Post* reported that the authorities had hustled him off the stage, his escort, Andrei A. Pavlov of the State Committee for Science and Technology, was called in by his bosses and asked how he had allowed such a discourtesy to occur. That same day Kennedy met Andrei A. Gromyko, the Soviet foreign minister, and then on the twenty-second, Brezhnev.

That meeting went ninety minutes beyond schedule. They chatted about Brezhnev's grandchildren, and how they, like Ted's mother, relished fresh fruit. They talked of the battle of Stalingrad, where Brezhnev had served, and of Joe Jr.'s death in the war. The Soviet boss warned Kennedy of the Georgians and their hospitality, telling of a late departure when his hosts there insisted he join them as jars of liquor were passed.

They talked of arms control. Two years earlier, both nations had agreed to all but abandon antiballistic missiles, and they were discussing limits on intercontinental missiles. Kennedy, picking up from his brother's success with the atmospheric test ban, pressed Brezhnev to agree to ban all nuclear tests. Interested, Brezhnev listened carefully as a draft treaty Kennedy carried with him was translated. When it was finished, Brezhnev told Kennedy, "If you were President of the United States now, I would ask you to sit over here in front of this fireplace. We would light a fire, and we would have some vodka and both of us would sign it and celebrate a great step toward halting nuclear expansion."

Toward the end of the meeting, Kennedy brought up an issue of Joan's—an exit visa for the great Soviet cellist Mstislav Rostropovich. He was then barred from performing in the Soviet Union and from traveling abroad, because he had befriended Aleksandr Solzhenitsyn, the writer who was the regime's most powerful critic. Leonard Bernstein, former music director of the New York Philharmonic, had urged Joan to bring up his case. She was thwarted when she tried to meet the cellist at the Moscow Conservatory. But the first night in Moscow, at a dinner with Soviet officials, Joan brought his case up with Yekaterina Furtseva, the minister of culture. She mentioned it to Brezhnev himself at a picture-taking ceremony as the Kremlin meeting began. When Kennedy raised the issue later, Brezhnev was evasive. First he claimed he thought Rostropovich had already left. Then Brezhnev said the cellist's training had been paid for by the Soviet people, and he should perform for all of them before appearing in the United States. But he told Kennedy he would think about it.

Rostropovich had been on a list of people Kennedy wanted to get out of the Soviet Union, a list shown his hosts in advance. The rest were mostly Jewish dissidents. When Kennedy brought their problems up with Brezhnev, the atmosphere of the meeting soured. Brezhnev insisted his country had no discrimination against Jews. Suddenly, it was time for lunch.

Later that day, the party flew to Tbilisi. The next morning, Kennedy plunged into a huge market crowd that was excited to see him. (One old man was convinced he was President Kennedy, not his younger brother.) But Jim King tried to keep them on schedule and annoyed their hosts by giving scant attention to a sixth-century monastery of great historical significance. So the atmosphere was strained when Eduard Shevardnadze, then the Communist Party boss in Georgia, arrived late for a celebratory lunch. Kennedy made up for the earlier slight by having King, the scheduling villain, leave to make room for Shevardnadze. The Georgian, who had been told of Teddy's leg amputation, offered a toast for "our children, for everything that we are doing for their future."

From Georgia they flew to Leningrad, visiting Piskaroszkoya Cemetery, which honors 400,000 dead from the World War I siege, and the Hermitage. They also toured a turbine factory with Grigory V. Romanov, the local party boss, as a host, and struggled to find Polina Epelman, who wanted the Senator's help in joining her husband in Israel. Her telephone suddenly stopped working, and Kennedy was kept from meeting her.

On the flight back to Moscow, Kennedy quarreled with Alexandrov, who told him that it was silly to jeopardize his relations with Brezhnev over dissidents. But late that night, Kennedy went to call on a group of them, gathered in the tiny walkup apartment of Alexander Lerner, a mathematician. The KGB had opposed the visit, which an embassy officer had arranged, and ordered Kennedy's driver to go home. But Grace Kennan Warnecke—daughter of former ambassador George Kennan—was along on the trip as a family friend. She told the driver in Russian his departure would cause an "international scandal." She apparently scared him more than the secret police had. He stayed on duty.

Sipping tea in Lerner's apartment, Kennedy asked the dissidents why the Soviet Union would not let them leave. They were not sure, suggesting they were kept either as bargaining chips in some larger game or because official arbitrariness might intimidate others. They called for continued protests in the United States, saying protests provided protection. Kennedy recalled years later "a sense of serenity in the room. . . . There was nobody that was looking like they were a firebrand or a revolutionary, or somebody that was selfish of any kind. You just got this sense of enormous faith. . . . They were at peace with themselves, all had suffered, but nothing could be done to them that was going to interrupt their own kind of mindset, and the fixture of their soul and their hearts as to what they believed, and what was at risk." Kennedy told them he would raise his voice for them back in the United States and only wished he had come "with a pocketful of visas." Some were given exit visas as a result of Kennedy's efforts, which got almost no publicity at the time, though the late-night meeting itself was reported.

Despite Soviet irritation over the meeting, as the Kennedy party prepared to leave Moscow on the morning of the twenty-fifth, Brezhnev sent word that Rostropovich would get his visa.

KENNEDY'S FIRST SPEECH about his trip was also the occasion of his first meeting with Governor Jimmy Carter of Georgia, punctuated by an exposure to Carter's competitiveness. Kennedy went to Atlanta on May 3 and spoke about his meeting with Brezhnev. "From my conversations in Moscow," he said, "I believe that it is possible this year to reach a workable agreement on banning all underground tests of nuclear weapons." But he criticized the Soviet Union for denying Jews the right to emigrate, and he also said if the Soviet Union really sought peace, it could improve relations with Israel.

At Carter's invitation, Kennedy stayed overnight at the Governor's Mansion. Carter had decided to run for President in 1976, though he had not announced. He expected to have to defeat Kennedy for the nomination, and he wanted to size him up. But he also had been told Kennedy might stay out. If Kennedy did not run, he could be helpful, and so Carter wanted to impress him, too. On Saturday, May 4, they were both to speak at Law Day at the University of Georgia: Kennedy at a major event unveiling a portrait of Dean Rusk and Carter at a luncheon later.

Carter had offered Kennedy his state plane, but on Saturday morning said he had to withdraw the offer, and so Kennedy and his party had to drive hard to arrive on time. Kennedy gave an adequate speech, saying, "Watergate is more a challenge than a crisis" and that "the process of impeachment is not a process to be feared" but a test of Congress that the House was already meeting with dignity.

Then Carter arrived and gave the speech of his life, challenging the comfortable lawyers and judges before him to commit themselves to "search with a degree of commitment and urgency" for ways to help the people of Georgia and not just themselves. And, noting the Nixon tape transcripts that had been released that week, showing the plotting to cover up the Watergate case, Carter said, "I believe that everyone who is in this room who is in a position of responsibility as a preserver of the law in its purest form ought to remember the oath that Thomas Jefferson and others took when they practically signed their own death warrant, writing the Declaration of Independence—to preserve justice and equity and freedom and fairness, they pledged their lives, their fortunes and their sacred honor."

Kennedy was impressed and sensed the competitive spirit in his host's speech, although, since he was not inclined to run for President, he didn't think he was competing with Carter. But one line in the speech made it clear to Carter-watchers that the governor was competing with Kennedy. He recalled talking with Kennedy "about his trip to Russia." That reminded him, Carter said, that he had read Tolstoy's *War and Peace,* all 1,415 pages, "when I was about twelve years old."

KENNEDY WAS BACK in Washington the next week for one more battle over spending on the war in Indochina. Although American troops and prisoners of war had left Vietnam, the U.S. role in the war was still continuing, with arms aid to South Vietnam and American bombing of Cambodia. Kennedy had demanded to know the legal justification for the policy. The imperious American ambassador to Vietnam, Graham Martin, had attacked him as part of an "effort to aid Hanoi by seriously crippling" South Vietnam. He cabled the State Department on March 21, "It would be the height of folly to permit Kennedy . . . the tactical advantage of an honest and detailed answer." Kissinger, by now secretary of state, did not take the advice. He wrote Kennedy that the Paris cease-fire agreements were authority for the United States to provide aid to South Vietnam "as long as it is needed."

On May 6, Kennedy led a fight in the Senate to block the Pentagon from spending another $266 million in Vietnam (beyond the $1.126 billion Congress had already approved for the year). On a 43–38 vote, he defeated Senator John Stennis of Mississippi, the highly respected chairman of the Armed Services Committee, who had been shot by a mugger the year before. Stennis argued that the extra money, available because of revised accounting of earlier aid, was part of the process of "winding down" while "carrying out our obligations." From his back-row seat, Kennedy challenged Stennis. "How long are we going to hear those arguments?" he asked. "We have been hearing them long enough." Kennedy's view prevailed in a House-Senate conference, and the $266 million was not spent.

Kennedy kept arguing all summer. On July 8, he held a hearing on refugees in Cambodia and scoffed at the argument of a "moral commitment." He asked, "What is so moral about providing vast amounts of ammunition for Indochina? What is so moral about an aid program that places a priority on fueling war and keeping a war-economy afloat, rather than helping to meet the needs of war victims?" Later that summer, on August 20, in an unsuccessful attempt to cut funding still further, Kennedy told the Senate, "It always amazes me to find that the great majority of Americans believe this war is over and ended, simply because we are not shedding American blood."

ONE SATURDAY AFTERNOON that summer, as newspapers were anticipating the next week's votes in the House Judiciary Committee on recommending Nixon's impeachment, Kennedy saved five people off Cape Cod. In his sloop *Curragh,* with Ethel Kennedy and several children along, he spotted them in a rubber raft and a leaky motorboat. David Lamkin and his wife and friends thought they were being blown out to sea. Kennedy, who said years later they were understandably alarmed but more frightened than they really needed to be, heard their cries for help and brought them aboard. It happened on July 20, 1974, five years and one day after Chappaquiddick.

THE YEAR HAD begun with an expectation of Congressional action on national health insurance. While Kennedy had reintroduced a bill to create the Canadian-style single-payer system favored by labor, he was also in touch with Caspar W. Weinberger, the secretary of health, education, and welfare. As the Nixon administration's plan developed, he offered encouragement. In

January, after leaking the unfinished plan to *The New York Times,* he praised it as "serious and carefully prepared." His major complaint was that it would provide windfall profits to the health insurance industry. He renewed both the praise and the criticism when the plan was actually sent to Congress the next month. Weinberger's plan would require employers to offer insurance to their workers, paying 75 percent of the premiums. Employees would not have to buy it. Private companies would sell the insurance and states would regulate them. Families would pay the first $450 of medical bills, and 25 percent of everything afterward up to a maximum of $1,500.

He was also working with Mills, trying to develop the bill they had agreed to offer when they spoke at the Democratic platform hearing in St. Louis in 1972. Kennedy would walk across the Capitol to Mills's office just off the House floor, and they would talk politics and Washington gossip for most of an hour, and then get down to business. On April 2, they announced their new plan, which differed from Weinberger's in several major ways. Enrollment would be compulsory. It would not be paid for through private insurance premiums, but primarily through a new payroll tax of 3 percent on employers and 1 percent on workers. A federal board, not states and insurance companies, would manage the system. Insurers would just pay claims, as they already did for Medicare. Weinberger had reservations about the large federal role, but called the proposal an "encouraging step forward."

Labor, which had provided most of the energy and money for the national health insurance effort, was furious. Max W. Fine, the executive director of the Committee of 100 for National Health Insurance, which Reuther had recruited Kennedy to join, said the committee would now try to block any legislation and wait for a "veto-proof" Congress they expected in an anti-Nixon landslide that fall. "Anything this Congress would pass that Mr. Nixon would sign wouldn't be worth having," Fine said. The National Council of Senior Citizens and many unions took similar positions. Kennedy would speak at union halls and find flyers and labor papers labeling him a "sellout." The *Progressive,* a liberal magazine published in Wisconsin, said he was "taking a dive." Kennedy tried to persuade labor that however big the next year's majorities were, they would still have to deal with Chairman Mills at Ways and Means and Chairman Long at Finance, so there was no good reason to wait.

Kennedy talked with Weinberger, hoping they could bridge their differences. The White House, despite its neuroses about Kennedy, needed a legislative success badly to turn attention from Watergate. So aides to Kennedy and Weinberger met secretly in a church basement behind the Supreme Court. Stan Jones of Kennedy's staff and Stuart Altman, a deputy assistant secretary at HEW, met several times in late spring. At one point in May when they were hopeful of success, Nixon made a speech calling for compromise. Kennedy, chosen by Democratic leaders to respond, proclaimed that "a new spirit of compromise and progress is in the air" and challenged the labor line by saying Congress felt "there has never been a better time to do the job."

Kennedy and Weinberger went to Seattle in June to try to sell their respective plans to the National Governors' Association. Georgia's Carter asked Kennedy what the chances were for action that year, and he said, "It's an uphill battle at best." Weinberger said that "1974 is one of those rare years" when action was in reach. Ronald Reagan, the governor of California, demanded, "Who has asked for national health insurance?" He said U.S. medicine was the best in the world and was already available to "virtually all Americans." Weinberger told his former boss in Sacramento there were "gaps" to be filled.

But even though Jones and Altman once thought Kennedy and the administration could agree, their efforts failed. Weinberger had already moved about as far as he could go, and Kennedy had gone out on a limb with labor and also had little room left. Mills took up the idea of working with the administration on his own. But by early summer it seemed likely that Congress would have to spend the fall on impeachment, and so with none of the major groups, especially the American Medical Association and labor, ready to compromise, the effort seemed dead.

It was revived briefly after President Nixon resigned on August 9 to avoid impeachment by the House and conviction by the Senate. President Gerald Ford, in his first speech to Congress, called for compromise and action. Mills presented something that was close to the administration's plan. But when twelve of the twenty-five Ways and Means Committee members insisted instead on the AMA's modest alternative of tax credits to help pay for private insurance, he knew he could not get it done, and he gave up. Kennedy, who had hoped Mills might somehow get something to the floor and keep the game going, called Weinberger and suggested they try again in 1975. But before they had a chance, Mills was fished out of the Reflecting Pool on the Mall with a stripper named Fannie Foxe. He admitted to having a drinking problem and resigned as chairman of Ways and Means; he was succeeded by Al Ullman of Oregon, who had no interest in health insurance.

EVEN AS THAT year's effort to enact national health insurance failed, Kennedy succeeded in making another law in the health field, but in a very different way. Instead of persuading Senators and Representatives, this time he persuaded judges.

The case of *Kennedy v. Sampson* arose out of another clash between President Nixon and Congress, this one over a pocket veto Nixon had issued at Christmastime in 1970, when Congress was out of town on a five-day recess. The bill was a $225 million program of grants to medical schools to encourage them to teach the family practice of medicine. Its huge margins of 64–1 in the Senate and 346–2 in the House were far beyond the two-thirds majorities required to override a normal veto. But Nixon did not return the bill to Capitol Hill for another vote, saying he could not because the Congress was in adjournment.

Kennedy sued in the federal district court in Washington. He contended that only a final adjournment of a Congress allowed a pocket veto. Since the President had not sent the bill back to Congress for a vote to override within ten days, he argued, it had become law.

Appearing in court for the first time since he left the district attorney's office in 1962, Kennedy argued the case himself, after careful prepping with Carey Parker and other aides. He argued that he had a right to sue, because Nixon's action had taken away his right to vote in the Senate. If a President could stretch the Framers' idea of the pocket veto this far, he said, it would logically apply whenever Congress took a holiday break. The Justice Department concentrated its argument on the point that Kennedy lacked standing to bring the case. Kennedy won in the district court on August 15, 1973, but the Justice Department appealed.

On June 1, 1974, Kennedy argued his case again in the U.S. court of appeals in Washington. This time he raised images of a banana republic, saying his vote had been nullified just as surely "as if there were armed troops surrounding the Senate, threatening to come in." The court of appeals agreed. On August 14, Judge Edward A. Tamm wrote, "No more essential interest could be asserted by a legislator." On the merits, the court decided the bill had become law, in language that strongly suggested that only a veto after the final adjournment of a Congress could escape a possible vote to override. Although the case was not appealed to the Supreme Court, it remains the rule because the court where it was decided handles virtually all such cases.

The curb on pocket vetoes, involving the issue of the separation of powers, carries a broader constitutional message. But the question of standing to sue had greater political impact. It has been used to allow members of Congress, and state legislators, too, to go to court to defend the laws they pass. But in 1997, when six members of Congress sued to attack the newly enacted line-item veto, the issue finally reached the Supreme Court. It ruled they had no right to sue, in language that suggested *Kennedy v. Sampson* had been wrongly decided. The earlier case was not mentioned, not even by Justice Stephen G. Breyer, a former Kennedy aide.

CAMPAIGN FINANCE WAS one of the last legislative items on the House schedule before an August vacation that would give the Judiciary Committee time to prepare its impeachment case on the House floor. On the afternoon of August 8, a few hours before Nixon announced to the nation that he would resign the next day, the House finally voted. First it rejected public financing for Congressional campaigns, 228–187. Then it passed the bill with Presidential campaign financing and a commission 355–48.

The House Senate conference dragged on for another two months. Kennedy, though he served on none of the committees with jurisdiction over election or tax laws, was a conferee, but he found himself fighting a House group determinedly opposed to public financing. Wayne Hays of Ohio, chairman of both the House Administration Committee and the Democratic Congressional Campaign Committee, led the House conferees in rigid opposition. Kennedy found that even old progressive allies like Frank Thompson of New Jersey and John Brademas of Indiana had little appetite for funding potential opponents, and neither did the majority leader, Tip O'Neill of Massachusetts. The House group even rejected a proposal that public financing apply only to Senate, not House, races, a result that may have relieved quite a few Senators.

On October 8, the Senate passed the conference report, 60–16. Kennedy voted for it. But he complained that the legislation ignored the sins of Congressional campaigns: "Abuse of campaign spending and private campaign financing did not stop at the other end of Pennsylvania Avenue. They dominate Congressional elections as well. If the abuses are the same for the

Presidency and Congress, the reform should also be the same. If public financing is good enough for Presidential elections, it should also be good enough for Senate and House elections."

Even so, Kennedy had played a central role in passing a major piece of legislation that produced cleaner elections. The old law would not have prevented a future hostile Congress or President—especially in tough fiscal times—from changing the rules through simple inaction, by not appropriating the money, but the new law guaranteed it. The $1,000 limit on individual contributions—though it remained stuck despite inflation, long past its time—combined with the assurance of public financing to make it impractical for Presidential candidates to reject the public money. Until Steve Forbes used his own fortune in 1996, only John B. Connally in 1980 tried to do without public funds. Neither got very far. The system worked pretty well for three Presidential elections, until the increasingly feeble Federal Election Commission and the rise of "soft money" in 1988 put corporate and union money back into politics as directly as they had ever been.

ON SEPTEMBER 8, the Sunday after Labor Day (and the day President Ford burst his own popularity bubble by pardoning Nixon), a *Globe* columnist called on Kennedy to get involved in the school busing issue that was scaring Boston. "You have the one voice that can help keep this city calm, leaving the clear ring of justice and common sense in homes and streets where people sit, uncertain," wrote Mike Barnicle, a Fitchburg native who often spoke for the working-class Irish who, in fact, rarely read the *Globe*. "To you, Senator Kennedy, they would listen."

Four days later, Boston's schools were scheduled to open, with 20,000 of the system's 93,000 pupils to be bused to other schools. A suit filed in February 1972 had charged the city with segregating its schools by assorted techniques, from gerrymandered school district lines to racially based school feeder patterns to biased assignment practices. The racist intent was documented in the Boston School Committee's minutes.

Kennedy had not been involved in the suit at all—except that in 1966, after the Morrissey debacle, he had recommended the appointment of W. Arthur Garrity to the federal district court, and Garrity, a friend of his, drew the case. Garrity, whose work in John F. Kennedy's 1958 and 1960 campaigns had got him the post of United States attorney, was a serious, careful judge. But he had tin ears about the practicalities, as opposed to the legalities, of the issue. He ruled on June 21, 1974, fifteen months after the trial ended. He had no time before school opened to devise a desegregation plan of his own, so as a first step he adopted one the state education department had been pushing. It shifted large numbers of poor Irish kids to black Roxbury High School, and large numbers of poor black children to South Boston High School, where the Irish kids had been. The opportunity to attack busing had already made a few political careers, especially that of Louise Day Hicks, who had gone from the Boston School Committee in 1962 to a close run for mayor in 1967, to the city council in 1969, to South Boston's House seat when John W. McCormack retired in 1970. She lost the House seat in 1972 to a regular Democrat, Joe Moakley, running as an independent.

On Monday, September 9, 1974, Mrs. Hicks led a crowd of several thousand opponents to City Hall for a rally she said was intended to get Kennedy and Senator Ed Brooke, a black Republican who had succeeded Leverett Saltonstall in 1967, to oppose busing because it threatened the safety of children "who are going to be sent into the high-crime areas of the city." The group called itself ROAR, an acronym for "Restore Our Alienated Rights."

ROAR had asked to meet with Kennedy that summer. The group said he had ignored its letter. His involvement over the summer had been slight, although Bob Bates, a black aide who dealt with urban issues, had been working to ease blacks' fears about sending their children into South Boston.

But Barnicle's column sparked Kennedy. On Sunday, Bates was told to get to Boston. On Monday morning, Kennedy flew to the city from the Cape and, with Joseph P. Kennedy II in tow, began a tour of high schools that would be affected. After visiting South Boston High School, he called his office in the John F. Kennedy Federal Building overlooking the City Hall plaza. Eddie Martin, now his administrative assistant after Burke had left, told him there was a group waiting to see him, so instead of going on to more high schools, he then suddenly sped downtown.

When Kennedy arrived, after Judge Garrity had been hung in effigy, there were several thousand people at the rally. The delegation had left his office. There was brief discussion in the car about what to do, and Kennedy said he wanted to talk to the crowd. The police told him he should go to his office. Jim King warned, "For Chrissake, don't. These people are wacko." The Senator turned to his nephew Joe, who understood how he thought of the working-class Irish as his core constituency, the ones who had rallied behind him after Chappaquiddick and were entitled to hear where he stood today. Ted said, "Let's go," and they walked into the crowd and up to the microphones.

For five minutes the crowd jeered at him. The taunts were as rough as a Kennedy had ever heard in Boston: "Why don't you put your one-legged son on a bus for Roxbury?" "How would you like your daughter to be raped by one of them?" "Why don't you let them shoot you like they shot your brothers?"

Then the crowd members turned their backs to him and started singing "God Bless America," and Kennedy gave up and left the platform. That seemed to enrage the crowd even more. One woman punched him in the shoulder. He was hit by a tomato. As he entered the federal office building, security guards blocked the crowd from following, but someone shattered a glass door. Kennedy, shaken, was hustled into an elevator. Barnicle broke the tension by saying, "I think you really had them, Teddy."

Up in his twenty-fourth-floor office, Kennedy held a news conference and told reporters that while he understood the crowd's emotions, it was "deplorable" not to let him speak: "I think what I would have tried to have said to them is that first of all that I have at least some kind of understanding of what their interests are in the safety of their children and the quality of education of their children, and that no one is a bigot because they have that concern, and that in my view the real issue is not a busing issue but what's at the other end of the line.

"And in the twelve years that I have been in the United States Senate I have voted for equal rights for all Americans. And I think I have had the overwhelming support of the people of Massachusetts and the people of Boston, and that we in Boston can't express one rule for Birmingham, Alabama, and another rule for Boston, Massachusetts."

Back inside City Hall, the staff of Mayor Kevin White grumbled that all Kennedy had done was to guarantee that the anti-busing rally got more attention than it would have otherwise. That night White broadcast an appeal for calm on all four of the city's television stations. "No city, or group within it, can stand in defiance of the law," the mayor said. "To those who would violate the order and peace of our city, and to those who would exploit the tensions of next week and the weeks to come, jeopardizing the young and the innocent, I promise swift and sure punishment."

When school opened on Thursday, September 12, 1974, most of the eighty affected schools with bused pupils were quiet. But at South Boston

High School, angry whites shouted "Niggers, go home" in the morning and then injured nine black students riding the buses home when they threw eggs, beer bottles, and rocks. The next week, police broke up racial brawls in school cafeterias, and the next month black students in Roxbury attacked whites.

THAT WEEKEND KENNEDY was off to the West Coast. On Friday, September 13, the band played "Hey, Look Me Over" when he appeared at a convention of the painters' union. Its president, Frank Raftery, introduced him as "the next President of the United States." Kennedy criticized the pardon of Nixon, saying it looked to many Americans like "the culmination of the Watergate coverup." He said, "I don't think the country will stand for it." He also scoffed at a poll showing him running behind President Ford, saying the result "might have been different had it been taken after the pardon." On Saturday in San Francisco, he teased an audience with a laugh line he had employed, with appropriate numbers, all over the country that year. He said his appearance had "absolutely nothing" to do with the state's 45 electoral votes or its 271 delegates to the 1976 Democratic Convention.

That audience, and the chuckling reporters, thought he was joking. But a week later he made believers of them.

On Monday, September 23, he held a news conference at the Parker House in Boston and announced he would not run for President or Vice President in 1976. Joan had flown in from a California sanitarium that tried to cure alcoholism with high doses of vitamins. He was unequivocal: "There is absolutely no circumstance or event that will alter the decision. I will not accept the nomination. I will not accept a draft. I will oppose any effort to place my name in nomination in any state or at the national convention, and I will oppose any effort to promote my candidacy in any other way."

He said he had learned from his brothers' campaigns that running for President demanded "a candidate's undivided attention and his deepest personal commitment," and that his obligations to his family and their fears made that impossible.

Answering questions, he said Chappaquiddick was not a factor, though aides undercut him by telling reporters the issue would have made things much tougher on his family had he run. Kennedy brushed off questions about future candidacies, saying, "I've seen in my own family where it never served much purpose for planning beyond the immediate future." He predicted that some other Democrat would win in 1976 and be reelected in 1980.

The decision had actually been made no later than Labor Day weekend on the Cape. That was when Kennedy told Dave Burke, by then working for the Dreyfus Corporation, that he had decided not to run. "He was not looking for advice," Burke told *Newsweek* of their conversation at a cookout. "I know cold turkey when I see it, and this decision had been in the freezer for a long time." Kennedy told Mary McGrory of the *Washington Star* that there was more involved than Teddy's leg; there was Patrick's severe asthma, too. And there was Joan's emotional state, not just her dread that he might be shot, but also a fear that he might be elected and she might not be able to handle the job of First Lady. Years later Kennedy spoke of Teddy and Joan and said, "There was so much on my mind that I wasn't going to be doing the kinds of things that you ought to be doing in terms of preparation and get moving on it."

CHAPTER 18

Unconventional Wisdom

In the mid-seventies, Kennedy was actively looking for new issues beyond those like health care, which had become his signature, and the mandates he had assumed from his brothers, from Vietnam to arms control to Indians.

None was more ambitious, politically and intellectually, than deregulation. Conservatives routinely complained that government regulation of business meant red tape and harassment. Traditional liberal thinking held that regulatory agencies protected consumers from being gouged by big business. But in several industries, regulation protected existing companies from competition and kept prices high.

After Jim Flug left the staff in 1973, Kennedy and Flug's successor, Thomas M. Susman, were looking for a more concentrated focus for the Subcommittee on Administrative Practice and Procedure. In Flug's years, "Adprac" had tackled one liberal cause after another, from the draft to Watergate. Now, the way the federal regulatory agencies functioned was an obvious subject consistent with the subcommittee's title.

In the spring of 1974, Kennedy invited Stephen G. Breyer, a Harvard Law School professor who taught administrative law, to come down to his house in McLean for dinner. They talked about regulation, and Breyer pointed to the Civil Aeronautics Board as a conspicuous example of regulators captured by the industry they were supposed to regulate: stifling competitive economic forces by keeping fares high, excluding new airlines from the industry, and limiting access to profitable routes. Inflation was a serious worry at the time, and the CAB kept approving rate increases—20 percent in 1974 alone, compared with a 12 percent increase in the Consumer Price Index. It also stalled applications by new carriers who wanted to attract new customers with low fares. Its critics' argument that lower prices could fill up planes and keep airlines profitable was not merely theory; it was working in the large, unregulated intrastate markets of Texas and, to an extent, of California.

Kennedy offered him a job, and Breyer took a sabbatical leave and went to work in August. The new Ford administration was intrigued with deregulation, and Breyer was in touch with its officials, hoping to influence administration policy. James C. Miller III, a staff economist for the Council of Economic Advisers and a specialist on the airline industry, cheerfully offered advice on structuring hearings. Another key contact was Philip Areeda, a law school faculty colleague of Breyer's and an antitrust expert who had just joined Ford's White House counsel's office. Breyer started requesting information from the government and the industry, planning to hold a series of hearings early in 1975. But in the fall of 1974 he found an earlier target that was too juicy to resist.

On Saturday, September 27, President Ford held a White House conference on inflation, saying the problem required "sacrifice and a strong common effort." A couple of miles away the Department of Transportation brought transatlantic charter airlines together to urge them to raise their fares. With just about full planes, the charter airlines could make money at lower fares than could the scheduled airlines, flying with half or more of their seats empty. The trigger for the meeting was that Pan American World Airways, on the way to its sixth consecutive year of losses, was going broke. If it went bankrupt, it would be the first major airline to do so.

Breyer went to the airline meeting, at least until the public was kicked out. At his urging, Kennedy announced a two-day set of hearings on charter fares in early October. But two powerful Senate friends of the airline industry, Senators Howard Cannon of Nevada and Warren Magnuson of Washington, the home of Boeing, protested sharply. They said the hearings alone might push Pan American over the financial brink.

Cannon was chairman of the Commerce Committee's Subcommittee on Aviation, and Magnuson was chairman of the full Commerce Committee. Besides that, Cannon was chairman of the Rules Committee, which was just then considering a request from Kennedy for money for his investigation. Kennedy agreed to postpone his hearings until November, and got his money.

Postponing the hearings gave them even more material. On October 18 the CAB had announced, without a hearing, that it would force charter rates up. On October 30 it decided to make it harder for the charter airlines to offer services to groups.

On the morning of November 7, the hearings began with complaints by charter airline operators that forcing them to raise their prices would put them out of business, leaving the market to the higher-priced scheduled carriers.

The theater was provided by Freddy Laker, the self-made British millionaire who ran cheap charter flights all over Europe and was trying to get permission to sell New York-London tickets for $125. He said the attempt to raise charter fares was a government response to complaints from big carriers: "We are now suffering from a disease which I call 'PanAmania.' Because of 'PanAmania,' people seem to have lost their senses out of a concern over what will happen to Pan American and TWA and British Airways." He said the scheduled carriers, "the high cost carriers, do not like competition. They don't want it; they don't do it among themselves."

The result, Laker said, was that fifty million American and British workers and their families were being deprived of cheap travel "because one, or two or maybe three . . . airlines, are having a bad time." Laker made this argument for customer choice: "If a man has the money and he wants to fly with a certain airline at a certain price, he should have that right and privilege. If he has not got any money, and he wants to fly on Laker Airways, he should have that privilege as well."

The news of the hearing came just before the lunch break, from Keith I. Clearwaters, deputy assistant attorney general in the antitrust division of the Justice Department. He announced that the Ford administration now opposed minimum charter rates. He said the CAB had overstepped the law in setting them, and contended it was "inappropriate" for the Department of Transportation to press the airlines to raise rates. "No justification whatever has been shown for government-sanctioned price fixing in the charter industry."

One witness looked beyond the specific charter fare issue. Wesley J. Liebeler, an economist from the Federal Trade Commission, argued that this step was only one piece of a misguided CAB effort that kept flying expensive.

With no price competition, he said, airlines only "compete in terms of scheduling frequency or in terms of how many delicacies they can pile on the sandwiches they serve you in flight." When Breyer asked what should be done, Liebeler said, "If you put more people on the plane and reduce the prices, the consumer is going to get what he wants. . . . Total deregulation is the solution that I would offer if I were asked."

The next day, November 8, Kennedy tangled with Robert D. Timm, chairman of the CAB, over why the board would not just let airlines cut prices to attract customers—and go out of business if they lost money by doing so. Timm tried to answer by saying stability in the industry was important. Kennedy asked why a Republican administration feared the free enterprise system, in the process denying travel to the "workers in plants and factories, and the elderly people that cannot afford those astronomical costs at the present time."

On January 19, 1975, the subcommittee issued a staff report on the charter fare issue. It found that the government's purpose was to prop up Pan American. It said that fears of predatory low fares by charter airlines against the bigger scheduled carriers were illogical. The CAB's sloppy procedures and lack of hearings "documented a classic example of bad agency procedures resulting in bad policy," Kennedy wrote in a preface. The report said the board's actions amounted to setting international fares, "which is illegal." The result was backward social policy; "since passengers of charter airlines are often poorer than passengers on scheduled airlines, raising charter rates to help Pan American is a classic example of government action that subsidizes the rich by taxing the poor."

The charter hearings were really just batting practice. Breyer had managed them in a way that made news, especially in *The New York Times,* where David Burnham was trying to establish a regulatory agency beat. But the big issue of deregulation itself was ahead, and Kennedy foreshadowed it in a December 16 Senate floor speech. He repeatedly stressed his subcommittee's authority to examine CAB "procedures," a word he used thirteen times in less than fifteen minutes. But he also made it clear the hearings would focus on high fares, asking why a Boston–Washington flight cost twice as much as one from Los Angeles to San Francisco; on the CAB's refusal to authorize new routes, saying only one airline was allowed to fly nonstop from Boston to Detroit; and on restricting charter flights and prohibiting price competition.

Before the new CAB hearings began in 1975, Kennedy moved into several other issues. Two days after the charter hearings ended, he was in London for a NATO meeting, then went on to Paris for lunch with President Valéry Giscard d'Estaing and to Vienna for briefings on European troop reduction talks. The trip also included meetings with Prime Minister Yitzhak Rabin of Israel and his predecessor, Golda Meir; King Hussein of Jordan; and Premier Anwar Sadat of Egypt.

Its most poignant moment came in Beersheba, where Kennedy had gone to make a speech dedicating a medical school. Jim King, following Kennedy as he shook hands, found a woman tugging at him and asking to talk to the Senator. She explained that she was Polina Epelman, the Jewish "refusenik" he had been prevented from seeing in Leningrad earlier that year. Thanks to pressure from Kennedy and Marc Ginsberg of the U.S. Embassy in Moscow, she had been allowed to leave. Now she and her husband and daughter wanted to thank him personally. Kennedy went to her tiny apartment, and he asked her how she felt. She looked out the window and said, "It isn't Russia. We are a family again." Kennedy responded with a variation on the traditional Passover plea: "The Epelman family, it's not next year in Jerusalem, it's this year in Beersheba."

The final stop on the trip was the most interesting politically. On November 19, 1974, Kennedy went to Portugal, where the forty-two-year dictatorship of Antonio Salazar and Marcello Caetano had been overthrown that spring by military officers. Political parties had been suppressed for decades, and the armed forces were newly politicized in the wake of bloody losing fights to retain Portugal's African colonies. Abroad, there was widespread, realistic fear that Communists would take over in the wake of economic unrest. The government already included one Communist. The Ford administration urged Kennedy not to go. But he had been there in the fifties, and southeastern Massachusetts had plenty of voters of Portuguese descent. He disregarded the advice.

On November 19, 1974, he met with the new President, Francisco da Costa Gomes, and with army officers and socialists. Mario Soares, the foreign minister, hosted a dinner for him. The next day Kennedy told a Lisbon news conference that he would urge the United States to provide economic aid to Portugal, but warned it would not come if Portugal turned Communist. Kennedy found the demonstrations and the uncertainty about the future exciting. On his return to Washington, he wrote Soares of his admiration for the nation's "democratic experiment," and then promptly met with Henry A. Kissinger, who was now secretary of state, urging him not to give up on democracy in Portugal yet.

Kennedy offered a similar message in the Senate, saying, "I believe that the United States should support efforts by Portuguese democrats to forge popular institutions in their country. Let it not be said that this experiment was jeopardized because we turned a blind eye to what is happening in Portugal, because we failed to demonstrate our concern for what these people are trying to do." After Kennedy got the Senate to approve $55 million in aid on December 13, Kissinger startled the State Department bureaucracy by supporting him. Kissinger also offered Portugal immediate economic aid, directing the embassy in Lisbon to tell Costa Gomes that "this demonstration of U.S. interest and confidence" was offered because "we are fully conscious of the efforts he is making to guide Portugal toward democracy."

Frank Carlucci, a career foreign service officer who spoke Portuguese, was sent to Lisbon. He and Kennedy can both claim credit for helping preserve Portugal from the dubious achievement of becoming the first European country outside the reach of the Soviet Union to go Communist.

One of Hunter's concerns had been to broaden Kennedy's approach to arms control, which had been largely focused on the hope of a complete test ban treaty. One opportunity arose in December. Joining with Mondale, who by then had also dropped out of the Presidential race, and a Republican, Charles McC. Mathias of Maryland, Kennedy worked to stiffen the Ford administration's commitment to arms control. Ford had met Brezhnev in Vladivostok in November and agreed to ceilings on bombers, missiles, and multiple-warhead missiles. That deal had been attacked by Senator Henry M. Jackson of Washington, a foe of most agreements with the Soviet Union, who said the ceilings were too high.

Instead of urging rejection of the Vladivostok pact, Kennedy, Mondale, and Mathias introduced a resolution on December 12 calling on Ford to "make every possible effort to negotiate further nuclear arms reduction measures." Kennedy complained the limits were so high that if both sides reached them, the arms race would not be slowed. Mondale scoffed, "The President said it put a cap on the arms race, but if it is a cap, it rides ten or fifteen feet above our heads."

Then, after working with the State Department, they introduced a new version of the resolution just after the 94th Congress met in 1975. The January 17 statement went further, insisting that the Vladivostok principles not be used as a basis for increasing the American missile arsenal. But the Senators settled for a promise of immediate follow-up talks, once the Vladivostok pact was spelled out, aimed at reducing the levels of arms allowed. They told reporters that a major purpose in backing the administration, despite their doubts, was to strengthen support for détente, as it was becoming increasingly unpopular under attack from Jackson and others. Kissinger welcomed their efforts. So did the Soviet Communist Party newspaper *Pravda,* which called them "sober-minded U.S. leaders."

One last postscript to the Soviet trip—more elegant than the meeting with the Epelmans—occurred on February 28, 1975. Rostropovich appeared at the Kennedy Center, as Joan had suggested to Brezhnev. The cellist performed Bach, Schumann, Brahms, Shostakovich, and Bernstein to prolonged applause and ecstatic reviews. An hour later, he swept into the Kennedy home in McLean, still toting his cello, embraced Joan, and gave her a box of long-stemmed roses. When he saw Kissinger, he hugged him, too. Kennedy took Rostropovich down a hall to meet Teddy, even though it was midnight. The Senator recalled years later, "I said, 'This is the greatest cellist in the world.' And Teddy met him. And then we walked out of the room, and Rostropovich said, with tears in his eyes, 'I can never thank you enough for helping me get out of the Soviet Union. But the best gift I could ever give is to teach your son how to play the cello.' Very, very moving. And we were never able to do it, but it was wonderful."

The new airline hearings had begun earlier that month. They had an effect even before testimony began, for the Ford administration felt compelled to settle internal differences and reach a common position before it began testifying. The Transportation Department was hesitant. John W. Snow, an assistant secretary, insisted to Areeda on January 29 that the idea being pushed by a task force from Justice, the Council of Economic Advisers, and the Council on Wage and Price Stability "to phase out CAB regulatory authority in three years is simply unrealistic. It has no chance politically and it gains us nothing to tilt at that windmill."

The Transportation Department prevailed. On February 6, 1975, the first day of new hearings, John W. Barnum, the acting secretary of transportation, announced the policy, which called for greater price flexibility and easier entry by new carriers onto busy routes, but kept the CAB to administer the new rules. He said, "Passengers find airlines competing for their patronage through elaborate cuisines, free drinks, attractive stewardesses, multicolored planes and piano bars," although many would prefer crowded, cheaper, no-frill service. He and other administration witnesses promised legislation "in the near future," and one defined that as "six weeks."

Lewis Engman, chairman of the Federal Trade Commission, followed by observing that a transcontinental passenger might be pleased to find empty seats so he could spread out, "but I wonder how pleased he would be if he were aware that he has paid not only for the seat he was sitting in, but for the seat his briefcase was sitting in, too." Economists scoffed at industry fears of competition. Alfred E. Kahn, chairman of the New York State Public Service Commission, explained, "No businessman protected from competition ever believes competition is anything but destructive. In the same way, he does not use the word 'competitor' but 'chiseler.' "

The hearings resumed in Boston on February 14. Antibusing demonstrators interrupted with shouts. One demanded, "Why are you having hearings on airlines? I have never been able to fly." "Well, that's why," Kennedy said before recessing the session to talk with them. The meeting was unsatisfactory to ROAR, which heckled him in Boston in March and shoved him and slashed his tires in North Quincy in April.

When that Valentine's Day hearing resumed, William A. Jordan, an economist from York University in Toronto, testified that regulation increased fares between 40 and 100 percent. Charles A. Murphy, executive director of the Texas Aeronautics Commission, made the point colorfully as he described how price-cutting increased business for his state's Southwest Airlines, which cut its normal $25 fare to $15 for early-morning, evening, and weekend flights. "These people are people who are coming to the airline instead of going on the bus," he said of the new customers. "It is almost like they are getting ready to take off on some of the late flights and you kind of wait for them to tie the chicken coop on top of the airplane."

Back in Washington on February 18, Kennedy saw he had already begun to change policy without introducing a bill. Richard J. O'Melia, acting chairman of the CAB, testified. He contended that there had never been a moratorium on awarding new routes. (The next week evidence showed there had been.) But whatever had been the case, O'Melia said, "as far as I am concerned, there is no route moratorium as of now." The importance of that shift was underlined by William A. Kutzke, a Transportation Department lawyer, who explained the logical connection between deregulating prices and deregulating routes: "Liberalized entry and the threat of potential entry is the regulator and enforcer which assures that pricing flexibility will not be misused."

The hearings were structured to keep hammering on fares. On February 25, Lamar Muse, head of Southwest Airlines, the unregulated Texas carrier, said he made money with the philosophy "Give as much service as you can at the lowest rate you can get away with, and the traffic will come to you and you will make more money than you will the other way." The next day Kennedy himself was pursuing Civil Aeronautics Board members, asking what they were doing to lower fares and telling them that allowing new airlines in would surely reduce prices.

While Kennedy occasionally badgered witnesses he thought evasive, his questioning throughout the hearings seemed informed and alert. He did the homework Breyer assigned. From the first day of the hearings, he could debate load factors and other rate issues with industry spokesmen. He pushed Jack Yohe, head of the CAB's newly created consumer office, to testify on how little authority and staff he had. The next week O'Melia was back to announce a doubling of Yohe's staff and say, "I know the Board wants to give him as much help as he needs to do his job." When the board's staff was accused of damaging the prospects of an Air Europe plan to fly between Tijuana and Luxembourg by denouncing it to the press, Kennedy gave the offending aide a hard time about who had authorized his efforts. No one had.

The Civil Aeronautics Board itself was perhaps the best argument against regulation. When commissioners told Senators there had been no route moratorium, they were simply not telling the truth. O'Melia said there never had been one. But Breyer turned up a 1973 memorandum from an administrative law judge referring to "informal instructions of the chairman's office in connection with an unofficial moratorium on route cases."

Even more serious was an April 21 hearing into CAB handling of an investigation of illegal campaign contributions by airlines. O'Melia, the acting chairman, clashed with Timm, the former chairman. CAB investigators testified that they had been forced by Timm to curtail their 1973 questioning of airline executives and ask no follow-up questions. Then O'Melia, who had been head of the board's enforcement division, swore that he had been handed a note from Timm in late 1973 telling him to stop the investigation. Timm swore he had written no such note. He said he had only wanted the questioning short to get quick results. Kennedy sent their testimony to the Justice Department, suggesting that the contradictions in sworn testimony amounted, at least, to an effort to obstruct his investigation. No one was prosecuted, but President Ford forced Timm to resign.

The deregulation idea had momentum. The industry's self-proclaimed guardian *Aviation Week & Space Technology,* which covered the hearings less energetically than *The New York Times* did, wrote furious editorials warning, "A dangerous fantasy has appeared on the Washington horizon that could ruin this country's air transport system." The "simplistic and theoretical approach," it said, was invented by "a cloistered gaggle of lawyers and economists," and it imperiled passengers.

President Ford appointed a new board chairman, John E. Robson, who took over in late April. He was honest and open, but timid. He got the board to move forward on hearing applications for new routes, but spent months pushing a scheme for an experiment in price competition his own staff found unworkable.

While the Ford administration had a general position favoring less regulation, it took months to translate that into legislation. Kennedy, with no legislative jurisdiction over the issue and no confidence that the CAB would move very far on its own, needed an administration vehicle to support and cheered at every sign of life.

When Ford went to New Hampshire on April 18, 1975, and promised to send up legislation encouraging more flexibility in pricing and easier entry into new routes, Kennedy pledged "enthusiastic support" for what he said would be "a highly constructive program that is likely to create meaningful regulatory reform." On June 26, the day after a White House meeting with members of Congress, where he spoke up for reform, Kennedy wrote Ford, warning that the delay in putting a bill together "suggests to some a lessening of the Administration's commitment to genuine reform, making it correspondingly more difficult for me to maintain the same momentum on the Congressional front."

A few days earlier, Kennedy had circulated a 328-page draft report, describing improper and perhaps illegal board efforts aimed at protecting "the industry at the expense of the consumer." It urged freedom to fly new routes and an end to price regulation. A month later, a CAB staff study reached similar conclusions and said, "The ability to adjust quickly to change, to innovate and to experiment by trial and error is blunted by relatively crude and ponderous regulatory controls."

Internally, the deregulators in the administration cited these reports as evidence that Kennedy's hearings had changed the political dynamic and the Transportation Department's caution was outdated. Even so, the eventual Ford proposal was a compromise, aiming for less regulation but not an end to it. More important, it was not finished until October. Even though *Aviation Week* had softened its opposition, calling for "Re-Regulation, Not De-Regulation," it was too late for a not very enthusiastic Senator Cannon to hold hearings that year.

But it was clear that Kennedy and Breyer had put the issue on the map. Robson said Kennedy gave "political legitimacy" to an issue which until then had only intellectual backing. Two even more unlikely supporters, the *Wall Street Journal* and *Fortune,* chimed in. The *Journal* said Kennedy was "crusading for less government regulation and more marketplace competition . . . with considerable impact and success." *Fortune* said that "Senator Kennedy, who conducted a penetrating series of hearings on the CAB last winter, has provided important leadership."

Moving back and forth from one issue to another, Kennedy took up an entirely different kind of arms control in 1975. In the Senate, in the Middle East itself, and in an article in *Foreign Affairs,* he pressed for a curb on sales of weapons to Persian Gulf states. United States arms sales had doubled to $8 billion from 1973 to 1974, and most of the weapons, from missiles to fighter jets, were going to a few oil-rich states on the Gulf. In the Senate on February 22, he complained of an "apparently indiscriminate administration policy of selling as much military equipment, and training as many foreign soldiers, as foreign countries will pay for." James A. Schlesinger, the secretary of defense, rejected a pause in sales bluntly, saying if the United States did not sell arms, other countries would.

But Kennedy kept pushing the idea, telling an unimpressed university audience in Teheran that "an unregulated flow of arms may in time prove to be neither in your interest, those of your neighbors, or those who depend on the gulf for oil and stability." Back in the Senate from that trip, which also included stops in Saudi Arabia and Iraq, he testified on a bill he had introduced calling for a six-month moratorium on Gulf arms sales. "We are running incredible risks of reducing rather than enhancing security in the region," he told the Foreign Relations Committee on June 18, saying the policy would "hook these nations on the heroin of modern arms."

Then, in the October issue of *Foreign Affairs,* he warned that military modernization might create an officer class that would threaten internal stability in the Gulf states and, more immediately, risk an explosion of the "traditional rivalries and conflicts, with deep roots in ideology, religion (between Sunnis and Shi'ites), ethnic and cultural differences, territorial disputes and endemic distrust of one's neighbors." He wrote, "An arms race between different Gulf states—or even a leveling-off at high levels of armament in several countries—contains built-in risks of increased political tensions or even conflict, by accident or design."

He had other scraps that year. In March he joined with Ernest F. "Fritz" Hollings of South Carolina to fight the oil depletion allowance, a bonanza that allowed oil producers to exclude 22 percent of their revenues from any taxes. This provision, which dated from 1926, had been a perennial target of liberals, but had always survived before. In February, though, a changed House, with the huge Democratic majority that followed the post-Watergate election of 1974, voted to repeal it as part of a bill to cut taxes. A few days later, on March 4, the Senate Finance Committee dropped that provision. Russell Long of Louisiana, the committee chairman who had little use for repeal anyway, and Mike Mansfield, the majority leader, said a contentious issue like this one should be put off until later to get the tax cuts passed quickly.

But the issue was not all that contentious after all. Kennedy and Hollings rallied nineteen other Senators behind them in a day, announcing their challenge on March 7. That was the day the Senate, with Mondale leading the fight, had changed the filibuster rule so only sixty (not sixty-seven) Senators were needed to end debate. On the floor, Hollings took the lead, because Kennedy missed two days of debate to attend the funeral of Aristotle Onassis. But he returned on March 20 when the Senate, 82–12, adopted a compromise that trimmed the allowance for independent producers and ended it for the major oil companies. He called the depletion allowance "the most unconscionable, intolerable tax loophole in the Internal Revenue Code." A House–Senate conference agreed to reduce it, and President Ford signed the cut into law.

In April, he won a Senate fight to keep a ban on federal funding of abortions off a bill to promote nurse training and to establish on their own various health programs that had been part of the War on Poverty, including neighborhood health centers. His abortion stand drew an angry rebuke from Daniel A. Cronin, the Roman Catholic bishop of Fall River, who said it must amount to a "weakening of your personal convictions." In March he urged an end to the embargo on trade with Cuba, prompting veterans of the Bay of Pigs to demand the return of a battle flag they had given President Kennedy. That summer, he mused to James Stevenson, a *New Yorker* writer, about how many groups he could antagonize and still be effective. Kennedy said that if he could, he might like to tackle "fewer issues, more intensively, at a slower pace."

He received one huge piece of good news. Teddy's chemotherapy was a complete success. The doctors took him off it in June 1975, six months ahead of schedule. But that news only enhanced a new round of speculation that Kennedy would change his mind and run for President. In April; Johnny Apple reported in *The New York Times* that many Democrats were "enacting a familiar scenario—waiting for Teddy." In June, *Newsweek*'s "Ready for Teddy" cover story reported that the Ford White House was sure he would be the Democratic nominee in 1976. On Sunday, July 20, the *Boston Herald-Advertiser* pointed the other way with an interview with Rose Kennedy. She said, "There is not any point in discussing politics since Teddy has taken himself out of the Presidential thing." But even so, Tip O'Neill, who had been pushing the idea for months, claimed on July 28 that Kennedy had asked him to "keep me alive" and promote his possible candidacy. Four days later he said Ted had called him to say, "Tip, I am not going to be a candidate for President of the United States."

Marty Nolan, no less sure of Ted's intentions than O'Neill was, scoffed in a column in the *Globe:* "The discussion about nominating Edward M. Kennedy for President in 1976 originates in a conspiracy of hacks. Many hacks in the Democratic Party—young, old, middle-aged—have decided that the party needs Kennedy. Many hacks in the media agree.

"The discussion seldom considers whether such an event would be good for Kennedy or his children or indeed the country. . . .

"The hacks whose coattails burned in trying to escape from the Muskie bandwagon in '72 are looking for an armored car with velvet upholstery. The Camelot Express looks less like a mirage the more they talk about it."

CHAPTER 19

Crime and Taxes

In late 1975, Kennedy began to define himself on one of the major political issues of the seventies—crime. Republicans had put Democrats on the defensive on the subject for several years, even though the only major federal entity to help states and cities cope with crime, the Law Enforcement Assistance Administration, was a creation of the Johnson years. Though most crime is under jurisdiction of the states, the issue was before the Senate because of an effort to codify the assorted federal criminal laws that had been written over two hundred years.

The basic task of recapitulating and simplifying existing law and eliminating duplication and contradiction was complicated by the efforts of conservative Senators like John McClellan of Arkansas, who headed the Senate Judiciary Subcommittee on Criminal Laws, and Roman Hruska of Nebraska. They sought to add a series of new measures easing the imposition of capital punishment, narrowing insanity defenses, and creating crimes of publishing classified information. Opponents like the American Civil Liberties Union complained bitterly, and the bill made slow progress in the Senate and none in the House.

That fall Kennedy decided he needed a specialist on his staff who knew about criminal law. Dave Burke, who was in New York, was impressed with an assistant United States attorney named Ken Feinberg whom he had seen prosecuting former attorney general John Mitchell, and recommended him for the job. Feinberg was intrigued, but doubted that he would get very far with the Senate's liberal icon. Kennedy peppered him with questions: "What do you think about preventive detention? What do you think about bail? What do you think about sentencing? What do you think about wiretapping?" After Feinberg had outlined his views, he thanked Kennedy for his time, but said he doubted they were compatible. "Don't be too sure," Kennedy replied. "I'm not sure what my views are on criminal justice."

He hired Feinberg and began developing those views. Kennedy laid them out October 20 in a Chicago speech in which he called for gun control, efforts to relieve burdens on courts, and minimum two-year imprisonment for violent crimes. He scorned "eight years of federal efforts to reduce crime by talking tough." But he also took aim at the usual liberal response by saying, "Nor can we counter law and order slogans with arguments that crime can only be controlled by demolishing city slums, ending poverty and discrimination, and providing decent health and education to all of our citizens." He said, "Let us not confuse social progress with progress in the war on crime. . . . We fool ourselves if we say 'no crime reform until society is reformed.' "

Kennedy put Feinberg to work with McClellan's and Hruska's staffs. McClellan had been Robert's boss in the late fifties and was pleased to find the younger brother joining in anticrime efforts. They made enough deals so that early in 1976 the ACLU's Washington director wrote to Kennedy that Feinberg was "out of control." Kennedy called the aide in and told him simply, "Good work." But the job was too big. They had to agree on a bill, get it through the Senate, and get the House to hold hearings and act—too tall an order for an election year.

KENNEDY'S MOST DRAMATIC Senate success of 1976 came only three days after the Congress returned to work on January 19. It had nothing to do with legislation. He was in the back of a crowded elevator in the Russell Senate Office Building when its door started closing as Carol Chealander, a young woman who worked for the Senate Republican Policy Committee, tried to get out at the last moment. Her neck was caught in the closing door. The elevator operator did nothing, apparently thinking the door would open automatically.

But it was an old elevator, of the sort Kennedy was used to in the

Bowdoin Street apartment Jack had kept and where Ted sometimes stayed. It did not spring open. Sharply, Ted told the operator to let go of the handle. The operator froze. Kennedy pushed through the passengers and knocked the operator's hand off the control lever. The door opened, and Chealander was freed. She sustained serious neck injuries, but told columnist Jack Anderson, "If Senator Kennedy hadn't kept his head, I'd be dead."

Otherwise, it was a tough, plodding year for Kennedy. Democrats saw little need to compromise with President Ford on legislation. They were willing to wait until 1977 because they expected to win the White House, especially since Ford was being challenged for the Republican nomination by Ronald Reagan.

Breyer, back at Harvard Law School, finally completed the report on airline deregulation and the failures of the Civil Aeronautics Board, and it was made public on February 21. It said, "Many of the Board's procedures fail to meet commonly accepted standards of fairness and openness," and said the CAB had failed to provide "the low-fare service that is technically feasible and that consumers desire." The report recommended allowing competition in fares and opening up new routes. Kennedy's covering letter said the study made it clear that regulation had been a failure. Ford, hoping to spur his own bill forward in an election year by avoiding lengthy new hearings, praised Kennedy's "comprehensive and thoughtful efforts."

Kennedy introduced his own bill in April, and Senator Cannon held some hearings on that measure, Ford's, and a proposal from the Civil Aeronautics Board itself. Kennedy told one hearing that the Congress should try to make air travel cheaper and available to more Americans, since only one adult in four flew at least once a year. Cannon was worried about the impact on labor, fearing that new airlines would probably be nonunion, and he worried that small cities would lose service. But the Nevada Senator was persuaded to support deregulation when John Robson, testifying for the CAB, came out in favor of it. Still, there was no urgency behind Cannon's effort that year, and Senator Magnuson, who was chairman of the full Commerce Committee, made it plain no bill would come to the floor in 1976.

Kennedy had one important success in foreign policy. The Ford administration was friendly to the Chilean military who had overthrown the Allende government. But Kennedy fought against arms sales to Chile, trying to punish the junta for its torture of its opponents. He pushed a ban on all sales—government or private—through the Senate in February, and found it gutted in a House-Senate conference on the foreign aid bill. Ford vetoed that bill over other issues, and Kennedy got the ban through again in April. This time it stuck in conference and became law, although the administration signed $9.2 million in spare parts contracts just before the bill was finally passed on June 25. Hubert Humphrey, who managed the bill, and Kennedy denounced the contracts. Kennedy said the action was an "outrageous breach of faith with the conferees and the Congress," since the conference had rejected the administration's request to continue spare parts sales. Later, the junta outraged even its supporters in September when it had Orlando Letelier, Allende's foreign minister, murdered in Washington by a car bomb. Kennedy said the administration's attitude on Chile amounted to "human rights be damned."

The Kennedy who made the most headlines in the early part of the year was Joan. After boasting in January to the *National Enquirer*, a supermarket tabloid, that she had stopped drinking (a Kennedy aide added that she was no longer seeing a psychiatrist), she spent a month at the Smithers Alcoholic Rehabilitation Center in New York City. The *National Enquirer* was enabled to set its millions of readers straight when two other patients talked about her time there.

It quoted her as telling them, "I am a lush. I've been an alcoholic for at least four or five years. I kept a bottle of vodka in my bedroom closet and drank about a bottle every night." One of her fellow patients said, "She told me that Chappaquiddick was where her alcoholic nightmare began. She said, 'I'd been drinking before, but after that time it started to get out of control.' She indicated she was completely destroyed by all the speculation" about Ted. The tabloid quoted her savagely on her sisters-in-law, said she called Rose Kennedy "a harridan," and said that when she first told Ted she was an alcoholic in 1975, he replied, "Don't be ridiculous. You just drink a little once in a while." It quoted her as saying, "After eighteen years of marriage, my own husband didn't understand what my most pressing problem was—booze." And it said that though the staff at Smithers had urged her to stay longer, she had refused because she planned to go skiing with her children. Ted, it said, never visited but picked her up when she left in early March.

This was a big enough story to get out of the supermarkets, and the *Washington Star* ran an account. That provoked Eunice Shriver to write and ask, "Do the writers or the editors of the *Star* not consider themselves human beings as well as guardians of the public's 'right to know'? Do they understand moral integrity? Do they feel appreciation for family life? Do they have compassion for Joan Kennedy's children, ages 8–16, who read the *Star* every night, including the edition in which this story appeared . . . ?" She denied the parts of the article that said Joan disliked her in-laws.

KENNEDY TOOK ON yet another intellectual challenge in 1976, tax reform. This time he brought down a Boston College Law School professor, Paul McDaniel, as a consultant to drill him on the ins and outs of minimum and maximum tax rates, artificial losses, and tax deferrals for U.S.-controlled foreign companies. The issue occupied him from March through August. Despite a very painful back, he led the forces trying to cut loopholes and tax shelters on the floor for two months starting in June. While many amendments were only narrowly defeated, his victories were rare and small, like a success in undoing a special break slipped into the bill for four California businessmen who had missed a deadline for setting up foreign trusts for their children.

Kennedy's own analysis of the reform effort, given on the floor on September 16, was largely on the mark. He said then, "All of our major efforts were defeated in the Senate Finance Committee and on the Senate floor. The few victories we achieved were largely of the finger-in-the-dike variety—an occasional success in preventing a new loophole from being added on the Senate floor, or in striking some of the worst special interest sections and other regressive provisions from the Finance Committee bill."

Even so, his attacks on tax breaks cut uncharacteristically close to the bone, angering colleagues who thought he was calling them corrupt. Bob Dole said, "You impugn the integrity of some of us and we don't like it. . . . You have said in effect that all members of the committee cannot be trusted, we are slipping in little amendments in the dark." Carl Curtis, the Nebraska Republican whom *The New York Times* had identified as the benefactor of the California businessmen, objected to demands by Kennedy and William Proxmire, a Wisconsin Democrat, that the Finance Committee keep records of the cost and sponsorship of amendments. Curtis said, "We have too much work to do here to provide demagogues with specified bills of particulars."

Stanley S. Surrey, a Harvard Law School professor, credited the reform effort Kennedy led with some success beyond what was voted on the floor. Working with Robert Brandon of Ralph Nader's Tax Reform Group, Kennedy and Carey Parker had identified the beneficiaries of some of the narrowest special-interest breaks in the legislation, and a few were actually withdrawn because of their efforts. And, Surrey argued in a review of the legislative effort, their attacks and the close votes weakened Senate conferees in negotiations with the House, which had passed a much more reformist bill. The House conferees prevailed on many issues and achieved an overall reduction in special tax breaks in the final legislation.

BY THE TIME the tax bill reached the floor, the Democratic Presidential nomination had been secured by Jimmy Carter, who, incidentally, denounced the tax code as a "disgrace to the human race." But just as he was rolling easily to the Democratic Presidential nomination largely untouched by a suspicious but feeble party establishment, Carter had no influence on the Senate's consideration of the tax bill.

Carter was a loner, disconnected from Washington-rooted Democrats. His willingness to discuss his deep religious faith and his moralizing set him apart from most politicians. When he promised to deliver "a government as good and as kind and as full of love as the American people," and never to lie, he was operating on a plane most traditional Democrats and many reporters could not relate to.

But while that made him hard for Washington to deal with, it translated into national appeal. It was little more than a year after Nixon had resigned in disgrace and Ford had pardoned him. After closely studying the tactics used four years earlier by another long shot, George McGovern, Carter defeated one of the largest fields in nomination history: Senators Birch Bayh of Indiana, Frank Church of Idaho, Fred Harris of Oklahoma, and Henry M. Jackson of Washington; Governors Edmund G. Brown, Jr., of California and George C. Wallace of Alabama; Ted's brother-in-law Sargent Shriver, for whom Kennedy provided almost no help; and Representative Morris K. Udall of Arizona. Carter was easily the best organized and most determined of the group. He profited by positioning himself as a moderate, nonracist Southerner who could beat Wallace.

Kennedy got some exposure to the Carter phenomenon as early as February. Touring Cape Cod, he found Carter the preferred Democrat in two high schools. Intrigued, he told Joe Lelyveld, a *New York Times* reporter, that if Carter won, "it would be an 'only in America thing.' "

A major element of Carter's strategy was to stay fuzzy on a lot of issues, avoiding the detailed policy debates in which Democrats delighted. His rivals attacked him for it and tried to draw him into dangerous specificity.

One example was critical to the relations between Carter and Kennedy. On April 16, Carter gave a speech about health care. It had been written in large part by two top aides to Leonard Woodcock, president of the United Automobile Workers and, as such, Reuther's successor as head of the Committee of 100 for National Health Insurance. Indeed, to a group of UAW officials, Carter said the speech had been written by Woodcock himself. The union leader said he had needed a clear declaration on national health insurance if he was going to be able to get the UAW to endorse Carter and use union machinery for him in Michigan.

Carter announced a series of "principles of a national health insurance program." They included "Coverage must be universal and mandatory," "Every citizen must be entitled to the same level of comprehensive benefits," and "Benefits should be insured by a combination of resources: employer and employee shared payroll taxes, and general tax revenues. As President, I would want to give our people the most rapid improvement in individual health care the nation can afford, accommodating first those who need it most, with the understanding that it will be a comprehensive program in the end."

There was nothing in the speech inconsistent with the proposed Health Security Act, making health insurance a government-provided social insurance benefit, which Kennedy and Representative James Corman of California were sponsoring. And Carter aides made a point of telling reporters that it was "fairly close," "close," or "almost identical" to Kennedy–Corman, as the *Washington Star-News,* the *Washington Post,* and *The New York Times* put it. But the *Atlanta Constitution,* with perhaps a better ear for Carter if not for health insurance, stressed Carter's efforts to make distinctions between his ideas and Kennedy's, though it also nailed down what it called his promise to "begin implementing a national health insurance program within the first year of his Administration if he is elected President."

Carter's health insurance stand was not really called into question for more than a year. Kennedy said nothing public about it, though he let friends know he doubted Carter's commitment. But on May 25, Kennedy joined other general critics, complaining that Carter "intentionally made his position on some issues indefinite and imprecise."

There was a weak "Anyone But Carter" movement in late May, with its hopes on Hubert Humphrey. Kennedy's comment seemed like a part of the conspiracy to Carter, because on May 20, Jim Wieghart of the *New York Daily News* talked to Kennedy and then reported that Kennedy might change his mind and run for President, or even run for Vice President on a Humphrey–Kennedy ticket. Wieghart was close to Kennedy, but Kennedy's press staff denied the story immediately and said Kennedy had told Wieghart he would not run. But the irritation lingered with Carter, and when he heard the "indefinite and imprecise" charge, he exploded. On May 27, Carter said, in front of David Nordan of the *Atlanta Journal,* that he was glad he did not have to depend on Kennedy: "I'm glad I don't have to kiss his ass," he said.

In fact, Kennedy seemed to have been trying to use the idea of his jumping in at the last minute to push Carter in a more liberal direction. At least that was how he explained it to Johnny Apple of the *Times* a week later. His lack of any real commitment to stopping Carter had been shown on the same day he talked to Wieghart. That day he told Marty Nolan of the *Globe* that denying the nomination to Carter, who by then had close to a majority of the delegates, "would be a real distortion of the expressed will of the working members of the Democratic Party."

Hamilton Jordan, Carter's campaign manager, said years later he appreciated that comment at the time. But that sentiment didn't show often. Carter himself could say of Kennedy, "He's a friend of mine and I need his help." But a few weeks later, Carter's son Jeff, not known for independent political thinking, popped off to *Women's Wear Daily,* saying Kennedy "probably has a little animosity toward us, because he sees a Carter victory as the end of the political dominance of the North, where it's been for so long, and I think he's probably right."

But beyond the regional chip on their shoulders that made the Carter camp see almost every question raised about them as an anti-Southern slur, there was an even more important obstacle to good relations between them. Carter looked down on Kennedy, morally, because of Chappaquiddick. In later years, when Carter had a real political grudge as well, he called Kennedy "a woman-killer."

But it was his charge of vagueness that led the Carter operation to offer Kennedy no important assignment in the 1976 convention. He turned down an invitation to speak for two to four minutes on health care, but sent word that he would like a more prominent role. Some people thought he wanted to nominate Carter, who chose instead Representative Peter W. Rodino of New Jersey, the chairman of the House Judiciary Committee when it recommended Nixon's impeachment.

On Tuesday, July 13, 1976, the second day of the convention, Kennedy came anyway, apparently still hoping. On the drive in from the airport, he mused about what might have been with Haynes Johnson, who by then was working for the *Washington Post*. They talked about 1968. If he had run and won, said Kennedy with a wry laugh, "I'd just be finishing up now, wouldn't I?"

At a fund-raiser for Tip O'Neill that night, Kennedy joked he had so little influence that he couldn't even get a perimeter pass to Madison Square Garden, where the convention was held. The next day he talked about health care to a ballroom audience in Carter's hotel, watched campaign ads for his own race for reelection that fall, and dropped in at the convention hall, where he got thirty seconds of applause. Carter was nominated that night. Kennedy got one lone delegate's vote for President. Jeno D. Berta of Iowa, a thirty-eight-year-old Hungarian immigrant, said, "I love him and believe he should have been the nominee."

Late that night, Carter called Kennedy from his hotel suite, with reporters present, to ask for his help in Massachusetts. On Thursday, the convention's last day, Kennedy dutifully visited the Massachusetts delegation and urged support for the ticket, praising Mondale, whom Carter had chosen for Vice President, as "a good friend of mine." Then he left New York, not staying for Carter's acceptance speech and the ritual arm-waving by leading Democrats on the platform afterward. Paul Kirk told reporters the platform invitation had come so late that it did not seem important, and a few days later Kennedy let Bob Healy know that he felt he had received shabby treatment at the convention. Later that summer he told a reporter he had confidence, but not complete confidence, in Carter. Asked if he would campaign for him, he said, "I suppose I'll be glad to help in some way." Another time he said, "I guess he's looking over his shoulder at me, although God knows why at this point."

The Carter campaign operation was worrying about bigger problems than Kennedy's feelings. Its once-huge lead slipped after Ford narrowly edged Reagan for the nomination. Carter gave a disastrous interview to *Playboy* magazine, confessing "lust in his heart" for many women. In many states his forces did not mesh well with the regular Democrats they had beaten in the primaries. Still, Kennedy exemplified the regulars they had to work with, and there were occasional overtures. First the Carter campaign sought the services of Jim King, who served as a cheerful wagonmaster for the traveling press. Then, wanting a political pro who was not allied with any local faction to run its New York State effort, it asked Kennedy for help and he got a reluctant Gerard Doherty to spend two months there running an unusually smooth New York campaign.

Kennedy and Carter finally met for a half hour in Washington on August 31, for the first time since 1974. Carter told reporters he sought advice about the problems of New England, and Kennedy paid the nominee the high compliment of saying Carter's "sincerity" recalled the sincerity of John F. Kennedy.

Although he had done badly in the Massachusetts primary, the state was not one Carter worried about losing in the fall. He made one visit, on September 30, and Kennedy gave him a stem-winder of an introduction to a packed hall at Boston College. Zeroing in on Carter's promise to support national health insurance, Kennedy said Carter had to be elected "so we can have a decent health-care program, so that no mother that hears her sick child call in the night has to make a decision whether that child is $30 sick or $40 sick because that is what it costs to go to a hospital." The roaring crowd unnerved the usually cool Presidential candidate, and when Carter got up to speak, he thanked Kennedy as "the next Preside ," stopping just before he finished the word, and correcting himself to refer to "the next Senator and still the senior U.S. Senator of Massachusetts."

The only serious issue of Kennedy's own 1976 Senate race had been settled two weeks before, when he carried the city of Boston in the Democratic primary. Busing had threatened to cost him a majority in the city, which George Wallace had carried in the Presidential primary. Campaigning, Kennedy avoided South Boston and Charlestown completely.

As Ted had served as a titular campaign manager for Jack in 1958, learning his way around Massachusetts politics, his nephew Joe filled that role for him in 1976. He worked hard, but neither renomination nor reelection had ever been in doubt, though he did have to warn Ted that the city itself might reject him.

But a smooth school opening without the racial fights of recent years, lots of television ads listing accomplishments in the Senate, and a reliance on organization (with some help from Kevin White in black areas) brought Kennedy the votes to carry the city with 55 percent over Robert Emmet Dinsmore, a lawyer who represented ROAR, and Frederick C. Langone, a six-term city councilman. Statewide, Kennedy had 74 percent.

The general election campaign was not much tougher. The Republican nominee, Michael Swing Robertson, a curtain manufacturer, ran as the anti-busing candidate. In a televised debate in New Bedford, he also blamed

Kennedy for the growth of the federal government since he arrived in the Senate in 1962. Kennedy answered that voters had some modest but appropriate expectations of government: "They want a vote for jobs, decent health care as a matter of right, not privilege. They want someone to work for the elderly and someone to provide a voice for the education of their young."

A rare lively moment in the Massachusetts campaign came when Kennedy, who usually avoided any ballot issues besides his own candidacy, backed a referendum to ban handguns. At a party for Robert J. DiGrazia, the retiring police commissioner, Kennedy boomed an endorsement, saying, "The Saturday night special and short barrel revolver are just to kill people." But the gun proposal lost. Sixty-nine percent of voters opposed it—the same percentage as voted for Kennedy. Carter, who got 55 percent in Massachusetts, squeaked through nationally with 50.1 percent of the vote.

After the election, Kennedy put together another privately printed family book, called *Words Jack Loved.* He began by recalling Jack's love of reading and his encouragement to Ted, his godson, "to read the lives of Melbourne, Charles James Fox, Edmund Burke and Daniel Webster. . . ." He wrote, "I marveled at how he could read so fast and so widely while the rest of us plodded along from page to page!"

Some passages were brief, heavily annotated, like this one under the heading "Country": "This generation of Americans has a rendezvous with destiny." He wrote that this line from Franklin Delano Roosevelt's acceptance speech in 1936 may have inspired John Kennedy to say in his own inaugural, "Each generation of Americans has been summoned to give testimony to its national loyalty. . . . Now the trumpet sounds again."

There were many selections about the Irish, including Robert E. Lee's tribute to the Irish brigade at the Civil War battle of Fredericksburg—"Never were men so brave. They ennobled their race by their splendid gallantry"—and a rhyme attributed to the soldiers:

War battered dogs are we
Gnawing a naked bone
Fighters in every war and clime
For every cause but our own.

Under "Faith," there was this passage from Luke 12:48:

For unto whomsoever much is given,
of him shall be much required;
And to whom men have committed much,
of him they will ask the more.

These favorite passages, Ted wrote in the prologue dated Christmas 1976, "carry us back to those great days when his energy, spirit and brilliance bound us together in that glorious adventure we shall never realize, and shall never forget."

CHAPTER 20

Strains with Carter, Strains on Joan

For Ted Kennedy, the four years of Jimmy Carter that began in 1977 were his most difficult and ultimately least successful time since he entered politics.

In personal terms, it could not compare to the six years from 1963 to 1969, with Jack's assassination, a plane crash, Bob's assassination, and Chappaquiddick. Most important, Teddy's cancer was behind them. But Ted's daughter Kara, a student at National Cathedral School in Washington, sometimes ran away from home. Some of the nephews for whom he tried to be a father were getting in trouble, arrested for speeding wildly and sometimes involved with drugs. Joan was an alcoholic, and while he felt guilty over that, his marriage was dead, and he asked friends who were divorced about how it felt.

His political circumstances were awkward. He was by far the most important Senator who was neither a committee chairman nor an elected leader. But 1976 had brought no legislative successes. And while he could talk realistically of the pace of the Senate, where, to quote Phil Hart, "you measure accomplishments not by climbing mountains, but by climbing molehills," he seemed restless as he sought new issues like taxes, crime, and China.

But most of all, he no longer seemed to be the most important Democrat in Washington, a status he had held with the press and the public, though not always with his colleagues, since Robert's death.

The most important Democrat was President Jimmy Carter, who had campaigned to end not just the Nixon scandals but the indifference to deficits of Ted Kennedy's kind of Democrats. At first Carter thought he had no need of help from the Northern Democrats who symbolized the city he had run against.

Even so, Carter and Kennedy had a lot more issues in common than they had in conflict. They thought alike on deregulation, on tax reform, on race, and on the importance of human rights as a marker in American foreign policy. But they had different priorities. Kennedy's prime concern was national health insurance. Carter had difficulty ranking his goals, but balancing the budget, reorganizing the federal government, and dealing with the energy shortage easily outranked health insurance.

That priority gap might have been bridged if the new President had been someone Kennedy knew—or even someone who tried to know Kennedy. But while there were occasional coincidences of purpose between them, there was never a real alliance, much less a political kinship, between the Senator who had held office in the capital for fourteen years (and whose brother the President had been there for sixteen before that) and the first President since Calvin Coolidge with no Washington experience.

Kennedy began the Carter years with a visit to the President-elect's state. On January 15, 1976, five days before the inauguration, Kennedy went to Atlanta to celebrate what would have been Martin Luther King's forty-eighth birthday. With Jack Carter, the President's eldest son, and Bert Lance, the banker he had chosen as budget director, in the audience at the Ebenezer Baptist Church, Kennedy led a ceremony to dedicate King's tomb. He said, "We commemorate the extraordinary life of the man who believed that the crises that divide Americans are not nearly as important as the qualities that unite us." Without King's efforts, he said, "blacks in this state could not have gone to the polls to put Andy Young in the Congress or to put Jimmy Carter in the White House." He concluded by comparing Carter to the martyred civil rights leader: "It is time for America to return to the principle that set the freedom movement into orbit. And it is fitting that the rededication take place here in Georgia. Georgia gave America Dr. Martin Luther King, Jr., and Georgia has given America our new President—Jimmy Carter."

Except for the one issue that mattered most to him, national health insurance, where he was more a gadfly than an opponent, Kennedy was a consistent ally of Carter's in 1977. He voted with him more than three-fourths of the time and rarely contributed a vote to an embarrassing defeat. Kennedy fought, unsuccessfully, for Carter's tax reform proposals, worked with him on Ireland and China, and supported the energy plan which was Carter's chief concern. Hamilton Jordan, by now installed as Carter's chief White House aide, recalled years later that at the outset, Kennedy "was very helpful and supportive of the President."

The tax issue came first. Carter had initially sought a $50 individual tax rebate to put money in circulation and spur the economy along with some modest business breaks. But in April, facing growing opposition, he suddenly dropped the request for the rebate, saying the economy was improving. He asked that the business breaks—an increase in the investment tax credit and a new credit for hiring new workers—go too, leaving basically a spare bill dealing with the standard deduction.

Russell Long's Senate Finance Committee disagreed. Long argued that the new breaks were necessary to increase investment. Kennedy and Dale Bumpers of Arkansas fought to delete the provisions. Kennedy said, "Simply put, it is not fair to kill the rebate for millions of average taxpayers and families, while enhancing the lucrative tax breaks already available for business. In the present posture of the bill, the Senate would be smashing the Easter eggs we had promised the average citizen, but offering a new golden egg to business." But they were overwhelmingly beaten, 74–20, on April 20.

On Ireland, Kennedy and Carter helped each other, though the issue meant far more to Ted than to the President. The violence in Northern Ireland was increasing on both sides, and when the IRA set off deadly explosions in England, British authorities knowingly convicted innocent men and women. Carter had worn a button proclaiming "Get Britain Out of Ireland" in the 1976 St. Patrick's Day parade in New York. Then, the week before his election, he proclaimed the Democratic Party's commitment to Irish unity and said the United States should get involved.

That not only angered the British and the State Department, but to John Hume in Derry and the Irish government in Dublin it seemed unrealistic. They thought Carter recklessly encouraged the Irish Republican Army and its American supporters in a group called NORAID who provided much of the money the IRA needed for arms and explosives. Kennedy and Tip O'Neill, just installed as Speaker of the House and deeply loyal to Carter, worked to get a more moderate Presidential statement of concern issued on

St. Patrick's Day. But they failed, largely because the British opposed any statement at all.

So instead Kennedy and O'Neill—joined by New Yorkers Hugh Carey, who had become governor, and Daniel Patrick Moynihan, elected in 1976 to the Senate—issued a St. Patrick's Day statement of their own, drafted with Hume's help. They reaffirmed their commitment to righting the "underlying injustices at the heart of the Northern Ireland tragedy" but insisted that "continued violence cannot assist the achievement of such a settlement." Then, in a warning directed squarely at NORAID, they appealed to "Americans to embrace this goal of peace and to renounce any action that promotes the current violence or provides support or encouragement for organizations engaged in violence."

They quickly became known as the Four Horsemen (a nickname sportswriter Grantland Rice had given the great Notre Dame backfield of 1924) and swapped pictures of each other on horseback. But it was Kennedy and O'Neill who took the lead in pressing the administration to disregard British objections and take a stand. They urged the administration to promise $50 million in economic aid for Northern Ireland once a settlement was negotiated. The British objected.

When Carter finally issued the statement on August 30, 1977, he condemned violence and any American support of the groups that committed it, called for the establishment of a government that could "command widespread acceptance throughout both parts of the community," and promised: "In the event of a settlement the United States Government would be prepared to join with others to see how additional job-creating investment could be encouraged to the benefit of all the people of Northern Ireland."

Kennedy was delighted. He issued a statement saying that "for the first time in memory, a United States President has spoken out for the human rights of the minority in Northern Ireland," and followed it with a handwritten note telling Carter, "No other President in history has done as well by Ireland."

But even at the beginning, Kennedy could not work effectively with the Carter administration on national health insurance. He thought Carter had assured him that after the energy plan was sent to Congress, health insurance would be next on Carter's agenda. The energy plan was unveiled on April 20, but there was no sign that anything was about to happen on health, other than the appointment of an advisory committee by Joe Califano, Carter's secretary of health, education, and welfare.

Kennedy, who had pushed Califano hard about health insurance at his confirmation hearings, had him over to his house in McLean on May 2, 1977. He also had Representative Jim Corman of California, the leading House sponsor of the unions' health proposal; Woodcock, who was about to retire as President of the UAW and had a meeting with Carter the next morning; and various Kennedy and Woodcock aides, including Max Fine and Mel Glasser, who had helped write Carter's April 1976 health insurance speech. Califano argued that the administration needed time to study the costs of health insurance and to produce a sensible plan. "The issue isn't working up a new program. We already have a program we have been working on for years," Kennedy replied. The meeting got intense, and ended with Califano seeking to assure Kennedy he would move as fast as he could.

The problem was simple, but neither side really spelled it out to the other. Kennedy, Corman, and the unions, especially the UAW, regarded that 1976 speech as a commitment to their program. That was not Carter's view. "Labor is wrong. I was the only Democratic candidate not to endorse the Kennedy bill during the campaign," he told Califano and Stuart Eizenstat, the White House domestic policy adviser, later that year.

Two weeks later, Kennedy took the dispute public. He flew to Los Angeles to speak at the UAW convention on May 16, at an event honoring Woodcock's retirement. He knew that Carter would be there the next day and told thousands of delegates that while Carter was keeping most of his campaign promises, "Health reform is in danger of becoming the missing promise in the administration's plans." He said, "With all respect, I say it is time for the Carter administration to begin to carry out its commitment on national health insurance and health security." He dismissed Califano's excuse of a crowded agenda, saying, "The American people should not tolerate delay on national health by Congress simply because other reforms are already lined up bumper to bumper." He demanded that the administration set a date—in 1977—for introducing a proposal. The autoworkers were ecstatic; the next day they made Kennedy an honorary member after one delegate said, "Kennedy talked like he is one of us."

When Carter spoke, he gave himself a deadline, but a later one, saying that "we're aiming to submit legislative proposals early next year." And he got a standing ovation when he said, "I'm committed to the phasing in of a workable national health insurance system." Kennedy said he was "gratified" and "delighted to have the commitment from the President since people were beginning to think he had forgotten."

Kennedy got help from Carter's attorney general, Griffin Bell, on the criminal code. Even though the effort to pass it in 1976 had been abandoned, Feinberg kept working with Paul Summit of McClellan's staff and Emory Sneeden of Strom Thurmond's. In May, with Bell on hand to convey administration support, Kennedy and McClellan announced the introduction of a new measure. Its bill number was no longer an easy-to-remember target like the old S. 1, but the unmemorable S. 1437. Even so, liberals immediately dubbed it "son of S. 1," though it had been stripped of measures like expansion of the death penalty and restrictions on publishing classified documents. Kennedy was still concerned about a liberal backlash. Feinberg enlisted Alan Dershowitz, a liberal professor at Harvard Law School, who defended the new measure as a net gain for civil liberties. *New York Times* columnist Anthony Lewis also argued for the plan, saying that "it would drop entirely from the statute books a number of laws that have been criticized as repressive and outmoded." He warned that "perfectionists will no doubt oppose it unless it does such politically impossible things as repeal the obscenity laws."

While the criminal code bill no longer had new crimes like publishing official secrets, it did have one major new concept—greatly reduced judicial discretion on sentencing and an end to parole in the federal system. Back in 1975, Feinberg had brought experts like federal judge Marvin E. Frankel and Dean Norval Morris of the University of Chicago Law School to dinner at Kennedy's house, where they made their case for "determinate sentencing." They argued that different judges gave widely disparate sentences for the same offense, and that parole decisions were thoroughly inconsistent. One result was a justified sense of unfairness among prisoners. Another was the absence of predictability required for any deterrent effect.

The bill proposed establishing a sentencing commission, made up of judges, that would govern sentencing for each offense, considering variously whether it was a first offense, the age of the offender, and the use of violence or a weapon. For each combination it would authorize, subject to a Congressional veto, a quite narrow range of possible sentences, such as from six to seven and one-half years, instead of the much broader, existing range of possible punishments like five to fifteen years. A judge could go outside those limits, but would have to state his reasons, and if he did, his sentence could be appealed.

The bill they introduced ran to 297 pages, and even with the support of Kennedy, McClellan, and Thurmond, moving it forward was a drawn-out effort, with dozens of hours of hearings and markups first for three months in McClellan's criminal laws subcommittee and then for a month in the full Judiciary Committee. McClellan was dying, and Kennedy often sat in for him in the chair. He impressed the conservatives when he stayed put in one meeting after another and demanded that his restive, more liberal allies do the same.

The liberals were the ones with the most amendments, from efforts to decriminalize possession of marijuana (which failed) to limiting penalties for criminal contempt of court (which succeeded). Finally, on November 2, the Judiciary Committee approved the bill on a 12–2 vote, with only its most conservative member, James B. Allen of Alabama, and its most liberal member, James Abourezk of South Dakota, voting no.

Another area where Kennedy cooperated with the Carter administration in 1977 was China policy, working with aides like Richard Holbrooke and Michael Oksenberg to nudge Carter along to implement his campaign promise to normalize relations. Since the late sixties, Kennedy had been calling for normalization of relations with China. He had hoped to be the first American politician to visit Communist China. Once, in 1971, he had gone to Ottawa in pursuit of a visa from Huang Hua, then China's ambassador to Canada. But Huang, who was later foreign minister, indicated that the price of a visa was a public declaration that Taiwan was a part of China, and Kennedy backed off. Then, after Nixon's visit opened up China in 1972, Kissinger made it clear to the Chinese they should not invite any Democrats.

On August 15, 1977, after weeks of meetings with China experts in and out of the administration, Kennedy spoke in Boston to the World Affairs Council and proposed that the United States move to establish full diplomatic relations with the country that was home to a fourth of the world's people. To do so, he said, the United States had to break its official ties with Taiwan. "We must end our military presence there, our defense treaty, and our formal diplomatic relations with the island," he said. Breaking diplomatic relations would automatically end the mutual defense treaty with Taiwan, he said, and was easier than a laborious effort to terminate it with a year's notice. (That was the same view the State Department's legal advisers were taking.) Unofficial ties and trade could continue to flourish, he argued, and Taiwan was protected by a hundred miles of open sea and strong military forces against attack from the mainland. The Taiwan issue, he said, "is one for the Chinese themselves to resolve over time. The United States has, and should have, no other fundamental interest than that this resolution be a peaceful one."

The speech had been shown to the State Department in advance. It noted that Secretary of State Cyrus R. Vance would leave the next week for China, and it was widely seen as foreshadowing the approach he would take. And because it was Kennedy, it refocused many in the Carter administration on the importance of China policy.

Kennedy was not much more than a highly interested onlooker on the airline deregulation issue in 1977. He testified before Howard Cannon's Commerce Subcommittee on Aviation, and reached agreement with Cannon on a bill. The Carter administration provided the key impetus that year, sending aides around the country to persuade chambers of commerce and other local groups of the benefits of deregulation, and appointing Alfred E. Kahn to head the Civil Aeronautics Board. Kahn, who as chairman of the New York Public Service Commission had testified in Kennedy's hearings, not only backed the theory of deregulation but implemented as many of the ideas of freedom in pricing and new access to routes as the existing Civil Aeronautics Act allowed.

Still, with all the unions and most of the airlines opposed and Senators from rural states fearing loss of service, it was slow going. The full Commerce Committee took until almost the adjournment of Congress to vote out the bill on October 27, 1977. Before acting, it added a major weakening amendment, one that put the burden of proof on airlines that wanted to open new routes, while the bill originally had put the burden on opponents.

JOAN MOVED OUT in the fall of 1977. She went to Boston to live in the Beacon Street apartment that had been bought to use in Ted's 1976 campaign. There was no public announcement of a separation, and she would rejoin the family at some holidays and for special events. Early in 1979, doing yet another round of magazine interviews about how she was regaining her self-confidence, Joan told Lester David her marriage was in a state of "limbo." When David asked if Ted still loved her, she answered, "I don't know, I really don't know." When Ted was asked about "the state of your marriage" by Roger Mudd of CBS later that year, he stammered: "Well, I think that it's a—it's had some difficult times, but I think we have, we, I think have been able to make some very good progress and it's—I would say that it's, it's, it's—delighted that we're able to, to share the time and the relationship that we, that we do share."

As Joan described her departure from their home in McLean, she was not running away from Washington or Ted. She said she was taking the only step she could to get her own life under control, even if it meant moving away from her children, too. She told Joan Braden of *McCall's* early in 1978 that she could see a psychiatrist regularly and go to Alcoholics Anonymous meetings with more privacy in Boston than in Washington. "I go to AA in Boston and it's wonderful. I can walk in, and there's no big flutter whatever, and I'm just like any other person who is an alcoholic and who needs help to stay sober."

In Washington, she told Braden, she had collapsed and drunk after Teddy's hospital stay was over and he was back in school, and when she heard rumors about Ted's affairs, "I began thinking, well, maybe I'm just not attractive enough or attractive any more, and it was awfully easy to then say 'Well, after all, you know, if that's the way it is, I might as well have a drink.' " In Boston, she said, she didn't hear the rumors about Ted and other women that were common gossip in Washington.

CHAPTER 21

Beijing and Moscow

In 1977, Kennedy finally got to China.

He got his long-sought visa after Secretary Vance's August visit to Beijing had failed to produce progress toward diplomatic relations. The Carter administration stumbled by asking more of China than Ford had—it sought to maintain official United States representation in Taiwan—and then calling the Chinese "flexible."

Before going, Kennedy worked closely with the administration—which had too many other foreign policy problems to spend much political capital on China—on what to say. Zbigniew Brzezinski, Carter's national security adviser, urged him to stress three points: the administration's continued commitment to establishing diplomatic relations, the idea that the United States and China would each have to make an effort to achieve it, and the fact that Congress would have a role in any change.

On Christmas Eve 1977, the Kennedy party left the United States for China. Joan rejoined him and the children for the trip but worried what the Chinese would think about her having a separate bedroom. There were also two aides, Carey Parker and Hunter's successor, Jan Kalicki; three sisters and a brother-in-law; two Boston reporters; Professor Jerome Alan Cohen from Harvard Law School; and one niece and one nephew, Caroline and Michael Kennedy, both Harvard sophomores.

During the trip, Kennedy's itinerary required constant negotiation. He kept asking to see a university, and the Chinese kept putting him off with explanations that this one was closed or that one was having examinations, apparently fearing he would start asking and answering questions as he had in Moscow. He did that, in a small way, at a teachers college in Changsha, asking students what they thought his country and their country needed most. They wanted agriculture, industry, and education for China, but housing and military modernization for the United States—so it could check the Soviet Union. Finally, on Kennedy's last day in China, he spent two hours at a real university in Canton.

Sometimes the Chinese response was more artful. He asked to see a prison and a trial. In Shanghai, the first stop of the trip, he did get to see a prison. But the script was specially written for him. Kennedy stopped a prisoner, thinking he was making a random choice, and asked, "What are you in for?" The prisoner's answer showed that the Chinese had done their research on Chappaquiddick: "I am in for negligent homicide. The bus I was driving went over a cliff and killed nine people." When he asked to see a trial, he was told repeatedly that none were going on. Finally, in Changsha, in a metropolitan area of a million and a half where there were supposed to be no trials occurring, he was taken to a Hunanese opera, *Fifteen Strings of Cash,* about an honest magistrate in the Manchu dynasty. Kennedy, in a stage whisper intended for his hosts' ears, told Professor Cohen, "This is the only trial we're going to get to see in China."

But Kennedy also made some unusual political headway in Shanghai. He was allowed to visit Johnny Foo, who had been left behind when his parents fled China twenty years before and was seeking permission to visit them in Massachusetts, and then return to China. He told Kennedy he had tuberculosis and wanted to see his mother and father. State Department officials had told Kennedy not to waste his time on such personal cases. He kept raising them, and while the Chinese declined to discuss the issue directly, one official told Kalicki they would look at his list.

The most interesting political lesson of the trip may also have come in Shanghai, as a parable of how China was not a unitary state for all the efforts at control made in Beijing. Kennedy visited a local Communist Party boss. When Kennedy learned that he had been in power for more than twenty years, he said, "My God, that means you went through the Cultural Revolution. Party leader then and party leader now. How did you manage it?"

The Chinese official replied, with the confidence of a Richard J. Daley putting up with reform movements, "Well, I'll tell you. During the Cultural Revolution we investigated very carefully whether there was a single, whether there were any counterrevolutionaries in our district. And do you know, we didn't find a single counterrevolutionary in our district."

But Kennedy's big objective was a meeting with Teng Hsiao-ping, the vice premier who ran the government. The Chinese kept telling him that Teng was ill, and that Kennedy was only one Senator, asking what would Teng do when the other ninety-nine also asked to see him. Kennedy never thought of himself as just one Senator, and Cohen made this point every way he could think of to the Chinese, including making elaborate telephone calls home to his wife so the Chinese could overhear his "private" despair at their stupidity in ignoring Kennedy's importance.

After Kennedy reached Beijing on New Year's Eve, he went through a familiar routine of meetings, working his way up a ladder of officials. He saw Huang Hua, now minister of foreign affairs, on January 3, and Teng the next day.

The two meetings were of a piece, though Teng was probably less confrontational. Kennedy's message was that while American politicians like him wanted diplomatic relations with China and were willing to break relations and the defense treaty with Taiwan to get them, they needed some help from China. After Democrats had been blamed in the fifties for "losing" China to the Communists, they needed to be able to say China would not try to conquer Taiwan, where the Chiang Kai-shek government had fled.

"It is our problem," Kennedy told Huang. "It can and must be done soon, but it depends on the climate."

Huang replied with the formula Chinese answer. He said it was up to Americans "to act in accordance with the spirit of the Shanghai communiqué, truly to admit there is only one China, and to give up any efforts to create two Chinas, or one China, one Taiwan, in any form, to stop interfering in the internal affairs of China. The liberation of Taiwan is the continuation of the Chinese people liberation cause that has not finished yet. . . . This is China's internal affair. . . . We cannot make any commitment to any party that we are not going to use force. . . . We can wait but we cannot wait indefinitely. The longer you stall on the settlement of this question the more the political and moral debts you will owe to the Chinese people."

Kennedy tried again: "What I am observing is the harsh reality of the American political climate, and that is something the Foreign Minister is familiar with and knows about. As the Foreign Minister remembers, most

Senators want to get reelected first and too often prefer that to being statesmen."

The meeting with Teng, punctuated by his coughing up phlegm and launching it into a spittoon, was a little gentler. Where Huang had complained about delay, Teng said, "The question now involves whether the United States needs Taiwan now. If it needs Taiwan now, I will reiterate that we will be patient. In the end, the Chinese people will resolve the issue of Taiwan. People of my age may not be able to see the goal fulfilled."

When Kennedy, picking up from Huang's language, said, "I am opposed to a two China policy or to a one China, one Taiwan policy or any other similar policies," Teng told him that commitment was appreciated.

But when the Senator said that whether diplomatic relations came quickly or slowly depended in part on China, Teng answered, "Of course we hope we can push forward the process of normalization as fast as possible. That would make us very happy. . . . But there is one thing that I cannot agree with. Namely that normalization of relations depends on the actions of the Chinese government and the Chinese people."

Kennedy kept trying, saying Chinese statements on Taiwan would affect how the United States acted. He said hopefully, "If I were able to go back to my constituents and to the United States Senate and to say that I was completely convinced that people in Taiwan would be able to live in peace and prosperity, that would have an impact on American opinion."

"This is an old question," Teng replied. "You should break off diplomatic relations with Taiwan. You shouldn't have any contact between governments. But you can have non-governmental exchanges and trade." But he said China would not promise not to use force, for that hedged China's sovereignty: "As to how and when we liberate Taiwan, it is China's internal affair." The only opening he offered was to remind Kennedy he had told some Associated Press editors "that China would take note of particular conditions in Taiwan in settling this issue."

A twig of an olive branch was offered as the meeting ended when Ting Yuan-hung, a diplomat who headed China's American desk, told Kalicki that China would not promise the outside world not to use force, "but I think perhaps we may tell our people internally—including our people on Taiwan—they may expect a peaceful and prosperous future." David Dean, the senior American diplomat in the liaison office, cabled Secretary Vance that China might be "seeking ways of giving subtle assurance—short of an explicit agreement or negotiated understanding—that armed liberation is not in the cards." He added, "We find it worth noting that the spectre of armed liberation was not once raised in the meeting with Senator Kennedy."

Teng also sounded a familiar Chinese warning ("He *was* the party line," Kennedy recalled years later) against "appeasement" of the Soviet Union. He said war was inevitable, and "in my opinion Soviet military strength has surpassed that of the U.S." Kennedy debated that conclusion with a wealth of detail about U.S. arms and Soviet weaknesses.

The meeting ended cheerfully. Kennedy asked how long Teng had been a soldier. Teng replied fifty years. Kennedy said he had been in the army for just two years and never risen above private first class. Teng replied, "But I have no rank either. We are all equal in the Chinese Army. So, like you, I am a private too."

The Kennedy party saw the sights, too, from the Great Wall and the Forbidden City to authentic scenes like a huge field near Shaoshan where thousands of peasants with picks, shovels, and baskets were leveling a huge field for an irrigation project. Joan said that sight of "feudal backwardness" made her wonder if China had changed at all. They dined in grand restaurants on duck and mai tais and orange soda (and had hamburgers, martinis, and Tab at Leonard Woodcock's official residence as a break).

The family bought so many souvenirs, especially posters and rubbings, that it needed to buy extra luggage to get them home. Joan and Eunice bought expensive antique scrolls, and Teddy started a run among the cousins on Chinese fur hats and padded coats. The young people sometimes felt like circus attractions as Chinese who had never seen white people before surrounded them to stare. They fascinated their observers by taking Polaroid pictures and giving them the prints. It was a moving trip for them. At a banquet in Canton on January 8, the last night in China, Patrick spoke of his pleasure in making friends his own age. A Chinese host, obviously touched, said it was significant that Patrick had had a baby tooth come loose a few days before. Chairman Mao, he observed, said that when you leave something behind in China, you leave a bit of yourself with China.

In Hong Kong the next day, Kennedy held a news conference. He said he "did not expect and cannot report progress" in resolving the issue of Taiwan and its sixteen million people. But, he said, the discussions "were virtually free of ideology and polemics and were basically practical." Kennedy told the reporters he had given Chinese authorities twenty-two requests to permit Chinese to leave China. The China-watching American press in Hong Kong saw this as a first test of a policy change announced, in general terms, while the Senator was in Beijing but not communicated to him until later. Kennedy spoke of his hopes that scientific exchanges in fields from oil to earthquakes to cancer would improve relations. He reported that the Beijing Symphony, at Joan's urging, said it would like to send several musicians to Tanglewood that summer to study with the Boston Symphony.

WHEN THE SENATE came back to work on January 19, the criminal code bill was its first piece of legislative business. Kennedy spent long hours on the floor, contending with the Senate's most conservative members—freed to oppose the bill by McClellan's death in December. Kennedy managed the bill, saying it would transform the present code of "some three thousand laws without plan or structure." He repeatedly sought to assure other Senators that the Judiciary Committee had approached this measure with a determination to avoid the controversy. Birch Bayh of Indiana praised the effort from a liberal perspective, saying that "most of the rough edges have been sanded off."

Kennedy argued hard for the bill's major reform of fixed, predictable sentences, established by a sentencing commission, leaving judges little latitude and permitting appeals when they went outside the commission's guidelines. Today, he said, "in the great majority of federal criminal cases a defendant who comes up for sentencing has no way of knowing or reliably predicting whether he or she will walk out of the courtroom on probation or be locked up for a term of years that may consume the rest of his or her life, or something in between." Widely varying sentences and unpredictable parole policies were based on the hopeful but failed reliance on prison as a route to rehabilitation. "Prison should be used, I believe, to punish a particular offender who violated the law, and we should separate the issue of rehabilitation from that."

In the first important votes, on January 24, the Senate overwhelmingly backed him on the sentencing scheme by tallies of 83–9 and 82–8. The conservatives, led by James Allen of Alabama and Jesse Helms of North Carolina, managed to delay the process but never to threaten it. Allen forced separate votes on dozens of technical amendments, like changing the word "appearing" to "appearance" at one place in the bill or replacing "issued" with "used" at another. Kennedy insisted that the bill was a net gain for civil liberties; Helms observed, "Codification is not automatically reform and revision is not automatically improvement."

After eight days of floor debate, the Senate passed the bill, now 382 pages, by a 72–15 vote. Kennedy said that effort proved "that law enforcement can be a bipartisan issue, that Democrats and Republicans can work together to improve the quality of criminal justice in America."

Three months later, the Senate had a much easier time with airline deregulation, taking up the legislation and passing it in a single day. The pace was a relief because on the day before, April 19, the Senate had finished thirty-eight days of debate over treaties turning over sovereignty of the Panama Canal to Panama, acting with one vote more than the constitutionally required two-thirds majority. Senator Cannon managed the bill to an easy victory. But Kennedy played an important role.

Cannon opened debate by saying that when the issue was considered in committee, "We learned that it is not as easy to support 'free enterprise' and 'less regulation' as a lot of us make it sound in our general speeches. When powerful corporations fear leaving their protective cocoon, a cocoon which has guaranteed survival, but discouraged profit, it is surprising to me that these companies are able to change the basic philosophy of some of my most respected colleagues."

Kennedy summarized his and Cannon's hearings by saying they proved "beyond doubt that the aviation regulatory scheme devised in the depression to protect an infant industry has resulted in higher fares, less service, especially to smaller communities, fewer competitive alternatives and generally less competition than the public deserves." He cited Kahn's work at the CAB as proof that allowing airlines to choose their own fares benefited consumers, as would lower barriers to entry into the industry. "This bill," he said, "frees airlines to do what business is supposed to do—serve consumers better for less."

Kennedy took the lead in debating George McGovern, who feared that small cities like those in South Dakota would lose service under the bill. McGovern sought to limit access to new routes, warning that free competition would replicate the nineteenth-century success of the Overland Stage Company in monopolizing stagecoach traffic and jacking up rates. Kennedy, thoroughly familiar with the bill and the supporting studies, argued, "To be able to compete in price is to be able to protect the consumer." He said, "Only sixteen percent of the routes that have been granted to the airlines in this country are actually being serviced today. The rest have a lock on them."

McGovern's amendments were beaten badly, and then Kennedy succeeded in undoing the Commerce Committee's caution by passing an amendment to put the burden of proof in complaints against new service back onto those who objected, instead of on whoever favored the new service. He said the question was "whether you are going to put that particular burden on the hundreds of small communities in this country, or on an incumbent carrier who protests entry by a competitor." Senator Dale Bumpers of Arkansas backed him bluntly, saying, "If we oppose an amendment like Senator Kennedy's, we are saying that we are scared to death that competition is going to spoil free enterprise." Kennedy's amendment passed, 69–21, and the bill passed, 83–9. President Carter, calling the bill an "important step in the fight against inflation," praised him along with Senators Cannon and James Pearson, a Kansas Republican, for their leadership.

Carter needed legislative victories badly, and the Panama Canal treaty hardly counted as one with the country, because the majority of the public who favored it cared less than the bitter minority who opposed it. Carter's approval in the Gallup Poll dropped from 75 percent in March 1977 to 40 percent in mid-April 1978. Kennedy, though he denied it firmly, was reported to be talking to friends about running against Carter in 1980. A Gallup Poll conducted in early April suggested that he might have a good chance if he did. He led Carter among Democrats, 53 percent to 40 percent. Other questions in the poll, however, indicated it reflected more dissatisfaction with Carter than a Kennedy boom. *The New York Times* reported, "Never before in the 43-year history of the Gallup Poll has an incumbent President eligible for re-election stood lower in his party's esteem." That meant Carter was weaker than Truman in 1952 or Johnson in 1968.

Kennedy's alternative, his Senate career, was becoming clearer and inviting. In March, Eastland announced his retirement, and Kennedy was next in line to head the Judiciary Committee. Whatever became of the criminal code bill as the legislation played out, it had already established him with the conservatives on the Judiciary Committee as a fair leader. Judiciary Committee business was an area where he could work closely with the Carter administration, on old causes like the criminal code and new ones like encouraging other Senators and the Justice Department to propose more women to be federal judges. In alliance with the administration, he had moved on from deregulation of airlines to the trucking industry, using the antitrust subcommittee, which he had inherited from Phil Hart in 1977, to fight Cannon for jurisdiction. Most of the time, he did not seem to be spoiling for a fight with Carter. The U.S. Conference of Mayors, annoyed that Carter would not address their convention in June, invited Kennedy instead. But when he spoke on June 19, he urged them to "give President Carter a chance."

Meanwhile, national health insurance was coming to a boil again. Joseph A. Califano, secretary of health, education, and welfare, considered Kennedy's plan too costly and thought proposing legislation in 1978 would not get that bill passed and would create problems for other bills. Kennedy thought Carter was breaking his promise at the UAW meeting the year before "to submit legislative proposals early next year." In March, Califano persuaded the President to announce that he would offer "principles" in April, followed later in the year by legislation. In April, Kennedy and union leaders emerged from the White House praising a "general agreement" on those principles.

But the "general agreement" was no real agreement at all. Within the administration, Califano and Stuart Eizenstat, Carter's chief domestic policy adviser, were arguing that Carter should keep his campaign promise and propose universal health insurance. But Charles L. Schultze, chairman of the Council of Economic Advisers, fearful of inflation he believed greater federal spending would cause, urged him to go slow. So did Michael Blumenthal, the secretary of the treasury; annual inflation had reached 7.4 percent in June. Kennedy and the labor leaders were leading a coalition increasingly suspicious of Carter. One of their leading academic voices, Rashi Fein of the Harvard School of Public Health, warned Kennedy, "This is an Administration which uses words loosely, apparently without any sense of history. . . . We have been told that the President is committed to NHI, but his Cabinet and others have been told that he won't increase the budget."

It was July before the President was ready with principles in enough detail to announce them publicly. Kennedy and the unions were unhappy with what they heard. They had agreed to abandon their core idea that health care should be provided by the government as part of basic social insurance and to keeping the private health insurance industry alive with an important role. But they wanted a commitment to a single, comprehensive piece of legislation which, however slowly it was phased in, would assure health insurance coverage for all Americans. They argued that the major effort needed to pass the legislation could be mounted only once. The unions also wanted a bill out soon so they could force candidates to take a stand on it in November. At a July 28 White House meeting, Kennedy told Carter that multiple bills would lead supporters to question his seriousness. Carter replied that he wanted a single bill, but circumstances might force him to choose multiple measures. Turning to debate Larry Horowitz, Kennedy's chief health aide, he insisted, "If there are unforeseen circumstances, difficulties in administration, excessive costs, then secondary phases must be delayed. There must be flexibility." Carter said Califano would announce his principles the next morning.

But that afternoon, joined by George Meany of the AFL-CIO, Kennedy held a news conference to describe Carter's "principles" and attack them. Kennedy called Carter's plan "unacceptable." He said separate bills would amount to "built-in self-destruct buttons to halt the program in its tracks if things go wrong."

On July 29, Califano told a news conference of his own, "The President is deeply concerned that the present health care system fails to serve millions of Americans." But, he said, Carter was "equally concerned about the intolerable inflation in the health care industry." And he said that health care problems "are closely tied to two other major national priorities: the need to bring inflation in the economy as a whole under tight rein and the need to spend the Federal dollar prudently."

There were political ways out of confrontation, but they were unlikely between rivals who did not understand each other. Several Carter aides, and perhaps the President, believed Kennedy was not really committed to health insurance but was looking for an excuse to challenge Carter. Kennedy suggested to David Broder of the *Washington Post* that Carter should just have endorsed Kennedy's strategy and left it up to him and labor to get their bill passed. That was an obvious Washington way of doing business, and it might have satisfied Kennedy. But Carter, though he suffered for it, could not do business that way.

The first concrete result of Kennedy's China visit appeared that summer when Johnny Foo and his wife, Wen-fong Foo, arrived at Logan Airport. After he had hugged his father, Yao-cheng Foo, then eighty, and his mother, Lucy K. Foo, Kennedy greeted him, and then told reporters that the elder Foo had written without success to Nixon and Kissinger seeking a visit from his son. In fact it was more than a visit, for they came on immigrants' visas and stayed.

Kennedy, with an eye on the other twenty-one families, said simply, "I hope the process of normalization can continue, and then this can continue." Indeed, Foo was only the first of those on Kennedy's list to be allowed to leave. Without public announcement, the Chinese government gradually dribbled the others out, too, as steps toward full diplomatic relations progressed.

By the time of Foo's release, Kalicki was involved in a new set of negotiations with the Russians. Kennedy was being welcomed for a second visit because he and the Soviet leadership both thought it was possible to get somewhere on arms control negotiations. They had virtually stalled when Carter proposed, and Brezhnev rejected, a whole new approach to strategic arms talks early in 1977, casting aside the Vladivostok framework that Brezhnev and Ford had agreed on. Carter called for much lower limits on nuclear weapons arsenals, with the larger cuts coming from the Soviet side. Even though considerable progress was made that September, other major obstacles kept arising: Carter's outspoken and public criticism of the Soviet Union on human rights; American plans to deploy cruise missiles in Europe, just ten minutes from Moscow; Russian efforts to stir things up in East Africa; the USSR's fear that better relations between the United States and China were aimed at threatening Moscow. Then Carter gave a speech at the Naval Academy on June 7, 1978—a speech that was read widely in the United States as a confusing mixture of appeals for détente and for Cold War firmness. But in Moscow it was taken mainly as a threat.

Brezhnev was puzzled about Carter and his goals, but he thought of Kennedy as committed to nuclear arms control. For that reason, he put up with sharp (but politely private) messages from Kennedy asking for better treatment for dissidents such as Andrei Sakharov and Anatoly Shcharansky. The Russians wanted the visit because of arms control. So did Kennedy, whose constant goal was a total ban on nuclear weapons tests to complete the work of President Kennedy. But he also wanted action on dissidents and refuseniks. The Politburo had agreed to that condition for the trip, and Kalicki was negotiating with Andrei Pavlov over who would leave. Sometimes the discussions were held by long-distance phone calls in the middle of the night; calls from the United States were enough of a rarity for Pavlov to be summoned to KGB headquarters to be asked what was going on. But Kalicki also visited Moscow in person in late August.

Kennedy, accompanied by Kalicki, Larry Horowitz, and Ken Regan, a photographer friend, flew to Moscow on September 4. They were met by Pavlov. He took them in a government jet to Alma-Ata for an international health care conference, where Kennedy said the example set by the eradication of smallpox should be applied to other diseases, especially those that killed children. The speech was uneventful, but the visit was shadowed by word to Pavlov from Moscow that the meeting scheduled on Friday with Brezhnev had to be canceled because he had to go into a hospital for tests. After traveling ten thousand miles, Kennedy was furious. Pavlov appealed through his superiors at the State Committee for Science and Technology to Konstantin Chernenko, who as head of the general department of the Central Committee was more or less its manager. The argument that the Politburo's promise of a meeting should be kept persuaded Chernenko to reschedule the session for Saturday, September 9.

On their way back to Moscow, Kennedy's party had collected crates of melons and bread from Central Asia. When Kennedy's party arrived at the Kremlin, the guards were baffled, but Pavlov assured them there was nothing sinister in the boxes. When they opened them for Brezhnev and Kennedy urged that he present them to his grandchildren, the Soviet leader breathed in their aromas and seemed deeply touched, exclaiming that it took a visitor from America to bring him these wonderful products of his country.

Despite that opening note, the meeting itself was not as lively as the 1974 session, though it lasted nearly two hours. Brezhnev's speech was slurred, and Alexandrov had to cut the session short. Brezhnev began the meeting by reading a long statement—as formal as many of the letters he had exchanged with Kennedy—which put all the blame for worsened ties on the United States:

"The facts show that the United States is stepping up the arms race and adopting more and more new military programs. At the same time, the brakes are being applied to the negotiations on several important problems of arms limitation and of disarmament.

"Links and cooperation between our two countries are being phased out under all kinds of invented pretexts. Ever more new obstacles are being erected in the trade and economic sphere instead of rectifying the existing situation, which is abnormal as it is. And finally, overt attempts are being made to play the so-called 'Chinese card' and to encourage Peking's aggressive anti-Soviet line.

"Take, for example, the hullabaloo around the so-called question of 'human rights.' It has long been clear that there are no humanitarian considerations behind all this. Rather, there is a deliberate line pursued by those who want to poison the atmosphere, to undermine trust and, generally, to frustrate or, at least, to seriously impede the positive development of relations between the USSR and the USA. 'Concern about human rights' in the American version has become an unclean political game. This is a crude attempt to interfere in our internal affairs and to bring pressure to bear upon us."

Brezhnev said that while he had no objection to efforts to improve relations between China and the United States, the "anti-Soviet hidden motives" behind such efforts were unacceptable, and American backing for a "strong" China supported China's aggressive intentions.

Fundamentally, he accused Carter of weakness, of being afraid to challenge opponents of better relations. He said, "It is becoming evident that we are already coming up against not only inconsistencies and vacillations, but rather a conscious line reflecting the designs of the foes of détente which run counter to the principles of equality in our relations."

Kennedy spoke up for Carter's personal commitment to arms control. Brezhnev did not disagree, but spoke of "confusing signals" from the administration. Kennedy also tried to defend American concerns about human rights and the need for greater openness in relations between the two countries. Kennedy told Brezhnev that no arms agreement would get through the Senate unless the Soviet Union showed greater sensitivity on human rights. A weary Brezhnev was obviously drooping, and Alexandrov passed Kalicki a note saying a sudden meeting had come up so the session with Kennedy had to be ended.

Alexandrov gave Kalicki a copy of the Brezhnev statement, and Kennedy passed copies on to Carter and Secretary of State Cyrus R. Vance when he returned to Washington the next week. Marshall D. Shulman, the chief Soviet expert at the State Department, wrote to Vance that he and Carter should both read it because of "how little overlap there is between the world view reflected in them and our own. Throughout there runs an aggrieved tone about the US responsibility for the deterioration in US–Soviet relations. There appears to be no awareness of the effect of Soviet actions as contributing to that deterioration."

Kennedy had one more meeting in Moscow, after 1 A.M. the next morning. He returned to Professor Lerner's apartment to meet with Soviet dissidents, including Shcharansky's mother and brother; Sakharov, the renowned physicist who had won the Nobel Peace Prize in 1975 for his human rights campaign; and Sakharov's wife, Yelena Bonner. He told them he had discussed human rights with Brezhnev, but did not go into details. Kennedy asked their views on how he and the U.S. Senate could help the cause of the activists, and whether in voting on a strategic arms limitation agreement, he should take the human rights struggle here into account. Sakharov told him that arms limitations issues should be kept separate from human rights issues and decided only on both countries' sense of their own interest. The dissidents said they were worried about that summer's spate of trials and thought it was important to try to do something for people who had been exiled or jailed, like Shcharansky. They praised a U.S. law, sponsored by Senator Henry M. Jackson of Washington and Representative Charles A. Vanik of Ohio, which denied favorable trade status to the Soviet Union because of its limits on Jewish emigration.

Kennedy listened carefully to the personal stories of the dissidents and told two of them privately that he had news that they would be allowed to leave. One, Lev David Roitburg, broke down in tears and hugged Kalicki. The better known of the two, Boris Katz, kept his emotions in check because his hopes had been raised before. He had become a cause célèbre in the West because his infant daughter, Jessica, had a feeding problem that allowed her to digest only a formula imported from the United States, which visitors would steadily bring to her parents in Moscow. Part of Kennedy's deal with the Russians was to say nothing while he was in Moscow. So Katz walked home near dawn, unable to tell his friends, and fearful of raising the hopes of his wife, who was pregnant with their second child.

When Kennedy made the list public at a news conference in Washington on Monday, September 11, there was an even more surprising name on it than Katz's. The Soviets had decided to allow Benjamin G. Levich, a physicist working on hydrodynamics, to leave. Never before had a member of the Academy of Sciences been allowed to emigrate, and for several years he had been rejected because of his access to nuclear secrets. Pavlov said years later that he was allowed to leave only because of Kennedy's insistence, and before he departed on November 30, Levich had to swear to the KGB that he would never do anything hostile to the USSR. He kept that promise, first in Israel and then at the City University of New York.

At Kennedy's news conference, he said the promises of exit visas, along with leniency toward an American businessman accused of spying, "offer hope that the recent downward spiral of Soviet-American relations may be at an end." Robert Kaiser, a *Washington Post* reporter who had been its Moscow correspondent, saw a blunter message, calling the visas a signal that Brezhnev did not like the Carter administration's singling people out. He wrote, "Carter's public interventions on behalf of the dissidents have not produced more favorable treatment for them. Kennedy has made no public statements comparable to Carter's, and now the Soviets have apparently 'rewarded' him with these new exit visas."

The reward was delayed when *Pravda* attacked Kennedy in an editorial, but he sent a message to Katz that this was all to be expected. At the beginning of October, Katz got a postcard telling him to come to the office which handled such cases, and he was given his visas. He left on November 29, and Kennedy sent Kalicki to meet Katz and his family in Zurich and bring them to Boston (his planned destination had been New York).

Kennedy had wanted to bring Brezhnev's views—and his impressions of their meeting—immediately to Carter. But the President had been fully involved in what proved to be his greatest foreign policy success, getting Anwar Sadat of Egypt and Menachem Begin of Israel to agree to the basics of a peace treaty. For thirteen incredibly difficult days Carter had gone back and forth between the two of them at Camp David, and he had managed to get them to agree.

So Kennedy settled for reporting to the State Department and sending Carter a copy of Brezhnev's statement. He did meet the President for half an hour on September 25 and urged him to move ahead on a strategic arms limitation treaty curbing intercontinental missiles, saying that the remaining differences were "readily resolvable." He also told him the Russians were prepared to accept the United States' most important conditions concerning inspections for a comprehensive test ban treaty. In a memo he sent that evening, Kennedy wrote, "The Soviets are confused and dismayed" by statements from the Department of Energy and others in the government opposing a treaty "even on these terms, at complete variance with Administration policy."

Legislatively, Kennedy's year ended with one major failure—the criminal code bill stayed bottled up in the House Judiciary Committee—and with a far-reaching success on airline deregulation.

On September 22, the House adopted a bill providing for the abolition of the Civil Aeronautics Board by 1983, and a House-Senate conference—on which Kennedy did not sit—changed that to 1985. The Senate adopted the compromise bill, 82–4, on October 14, and the House followed the next day. And even before Carter signed the bill into law on October 24, 1978, dozens of airline representatives were lined up outside the CAB's office waiting to file requests for new routes under the new law.

CHAPTER 22

Against the Wind

"Sometimes a party must sail against the wind. We cannot afford to drift or lie at anchor. We cannot heed the call of those who say it is time to furl the sail."

With that call to arms in Memphis on December 9, 1978, Ted Kennedy transformed his quarrel with Jimmy Carter over national health insurance into a debate over the soul of the Democratic Party, a debate that continues today. And he set himself on the road to run for President.

First he defined himself, and not just in terms of commanding an embattled issue, when he proclaimed, "National health insurance is the great unfinished business on the agenda of the Democratic Party." As he neared the end of the speech to the Democrats' midterm convention, he seemed to draw strength from the cheers and applause. He discarded a dry paragraph about what expensive health care had meant to the Kennedy family. Instead, he expanded, ignoring the text, and made that thought his peroration. He spoke of Teddy's cancer, his father's stroke, and his own broken back. "We were able to get the very best in terms of health care because we were able to afford it. It would have bankrupted any average family in this nation. . . . But I want every delegate at this convention to understand that as long as I am a vote, and as long as I have a voice in the United States Senate, it's going to be for that Democratic platform plank that provides decent quality health care, North and South, East and West, for all Americans as a matter of right, not of privilege, for all."

The 2,500 Democrats and reporters in the room knew they had been present at a great event. The moderator, presiding over what was formally described as a "Workshop on Health Care," was William J. Clinton, the governor-elect of Arkansas, who had spent the morning in a motel room boning up on health care. He pounded Kennedy on the back when he returned to the speakers' table as an ovation continued. Califano, who had spoken first, congratulated Kennedy and told him he was glad he did not have to follow him. In the back of the hall some junior White House aides were screaming like the other listeners. Not Hamilton Jordan, the White House chief of staff. He turned angrily to Pat Caddell, the President's pollster, and said, "That's it. He's running."

The Saturday speech was the event of the mini-convention. Everyone talked about it that night, comparing it to Carter's tedious account of his administration's successes the day before. Newspaper articles said the speech showed how Kennedy could tug the heart of his party while Carter and Vice President Walter Mondale, who was little better on Sunday, got nowhere with cool reason. Jody Powell, Carter's press secretary, sourly observed that Kennedy's speech proved only that "demagoguery is not exclusively a Southern characteristic."

The next day an all-out administration effort defeated a resolution demanding no cuts on domestic spending. But the victory was unimpressive, with 39 percent of the party establishment delegates voting against the administration—roughly equivalent to the 42 percent who told CBS News they were not backing Carter for renomination.

Jordan's instant analysis—that the speech meant Kennedy was running against Carter—was almost certainly wrong. The classic "sail against the wind" line had been inserted only Saturday morning into a pointed but familiar text of passages such as "Every day, parents are deciding whether they can afford the twenty-five-dollar doctor office charge and the twenty-five-dollar laboratory bill when their child is sick." The addition was designed to make sure the speech got the administration's attention. It succeeded—far beyond the expectations of Kennedy and Carey Parker, who wrote it.

Mondale, an old Kennedy ally from Senate days, had a subtler view than Jordan's. He felt Kennedy was trying to make health insurance a rationale for running. Carter, back in Washington, minimized the importance of the crowd's reaction, but took the speech itself as a "tentative" declaration of candidacy.

There is no question that the event moved Kennedy toward running. The spectacular audience reaction surprised and impressed him, and underlined the message he was reading in public polls. And the Jordan and Carter views led the White House into a run of petty decisions—Jordan urged Carter that the worst thing he could do was "to behave in a way that suggests we fear a Kennedy candidacy"—that only got Kennedy's Irish up.

Three days later, on December 12, Carter held a news conference and patronized Kennedy, calling their differences "very minor." He said, "I have a unique perspective in this country as President. I can look at a much broader range of issues than does Senator Kennedy." Then he dismissed Kennedy's appeal as not earned on his own but derived from "a family within the Democratic Party which is revered because of his two brothers and the contribution of his family to our party. There is a special aura of appreciation to him that's personified because of the position of his family in our nation and in our party."

LATER THAT WEEK, Carter regained control of the political agenda, announcing on television on December 15 that the United States would break relations with Taiwan, abrogate the mutual defense treaty, and open full diplomatic relations with China with a visit the next month by Teng Hsiao-ping. The timing was chosen to capitalize on Congress being away, and only Robert Byrd, by now the majority leader, got advance word. Kennedy did not, though he had known for months that the action was coming; Vance had told him during the winter. A subsequent omission of Kennedy from the guest list for a January 29, 1979, state dinner honoring Teng was deliberate. He was invited only after Vance complained to the White House.

Carter's announcement, without prior Congressional consultation, drew a storm of conservative protest. Barry Goldwater called it "one of the most cowardly acts ever performed by a President of the United States." Many Democrats, who had been unable to prepare the ground with constituents, were worried. Some demanded that China explicitly promise not to use force against Taiwan.

Kennedy stepped in with Senator Alan Cranston of California to help. While the Carter administration was still struggling to get its implementing legislation prepared, they offered a resolution in language China would tolerate. It said that "the United States has a continuing interest in the peaceful resolution of the Taiwan issue and expects that the Taiwan issue will be settled peacefully by the Chinese themselves."

Carter thought he needed no help from Kennedy, though the resolution's language had been worked out with the State Department and the National Security Council and generally echoed what Carter had said himself. Before the Foreign Relations Committee, Kennedy answered Goldwater and his allies, "There are some who say that normalization was a reflection of American weakness. I say the opposite. Normalization is a reflection of American strength; our strength to recognize the reality of nearly one billion people controlled not by Taipei but by Peking."

Teng himself went to considerable lengths during his visit to reassure Senators and Representatives that the mainland would not attack Taiwan. That made it easier to treat the Kennedy-Cranston language as sufficient, and its language was adopted almost verbatim into the eventual legislation.

EXCEPT FOR THE Teng dinner episode, Kennedy and Carter usually managed to maintain at least the appearance of diplomatic relations in the first few months after Memphis. They shared a Kennedy Center box for the charity premiere of *Superman* in December. On February 7, Kennedy hosted Rosalynn Carter, honorary chairman of a mental health commission, before his subcommittee. He compared her to another First Lady who testified before Congress, Eleanor Roosevelt, saying she and Mrs. Carter shared the view that "all Americans are first-class citizens, that the most vulnerable among us can and should be cared for." Kennedy cited the circumstances of his sister, Rosemary, and they agreed on the need to "remove the stigma" of mental illness. But Mrs. Carter corrected him quietly when he said, "Funding for basic research has grown impressively." She said, "In the 10 years from 1967 to 1977 research funds dropped considerably—to where they can provide only half of what they did in 1967."

On March 21, Kennedy met with the President at the White House. They continued to disagree over health insurance, but Kennedy told Carter he had "tentatively" decided to support him for reelection. In May he praised the President for his policies on deregulation and on arms control. And in June, when the strategic arms talks—delayed by Russian pique over the diplomatic breakthrough with China—finally produced a treaty to limit bombers and missile launchers while allowing increases in warheads, Kennedy praised Carter and Brezhnev on the Senate floor. He also said, "I believe that history will judge the Senate harshly if we fail to fulfill the promise now at hand."

On May 23, as the head of his family, Kennedy invited Carter to Boston. He wrote: "After sixteen long years of planning and hopeful anticipation, the John F. Kennedy Library will open its doors to the public on October 20th. . . . On behalf of the Kennedy family, I would like to invite you to speak at the dedication of the Library on Saturday, October 20th at 11:00 A.M. Your participation in this ceremony would honor my brother's memory in a most meaningful way. It is our sincerest hope that you will be able to be with us on this occasion." Carter accepted.

But none of this stopped speculation that Kennedy might run against Carter for the Democratic nomination in 1980. Moreover, Kennedy frequently criticized Carter, not only for a "piecemeal approach" to health insurance, but on an area where the nation was already more engaged and irritated, a growing energy shortage that led to higher oil prices and, even worse, long lines at gas stations, where squabbles sometimes led to violence.

Kennedy spoke for his home state's interest in cheap heating oil and natural gas and became a continual critic of Carter's energy policy during the winter. He opposed decontrol of natural gas prices, which Carter favored as an incentive to increase U.S. production. On April 30, Kennedy told the American Society of Newspaper Editors the administration had been "intimidated" by the oil industry into agreeing to the decontrol of oil prices and offering only a "fig leaf" windfall profits tax. "That's just a lot of baloney," the President answered in a news conference that afternoon.

Kennedy and Carter had an unerring ability to get under each other's skin. Kennedy's formula comment on 1980—"I expect the President to run, I expect he will be renominated and re-elected, I intend to support him"—stopped just short of commitment, and sounded like a childish bit of teasing to the President's men. But Kennedy could deliver it easily, with a smile.

When Carter hit back, it was clunky. On June 11, with polls showing Democrats preferring Kennedy over Carter by margins of more than two to one, Carter told several Congressmen at a White House dinner that if Kennedy ran, he would "whip his ass." The Congressmen tried to persuade him he hadn't meant to say that, so he repeated it. Then White House aides came around to tell them to confirm it if reporters asked. They passed out their unlisted phone numbers to make sure reporters did. *The Washington*

Monthly said Carter "had the manner of a certified public accountant trying to be one of the boys."

Kennedy was plainly annoyed. After the story was printed, he told a reporter, "I think what he meant to say was that he was going to whip inflation," a dig at another Carter vulnerability that had reached 10.9 percent a year. But he also told Tom Brokaw of NBC's *Today* show, "Well, if I were to run, which I don't intend to, I would hope to win."

Another disagreement had more weight to it. As chairman of the Judiciary Committee, Kennedy had been working with Attorney General Griffin Bell to get judicial nominees confirmed quickly. Under his chairmanship, the committee investigated nominees and did not simply rely on good or bad chits from home state Senators. But one vacancy mattered more than the rest. A seat on the First Circuit Court of Appeals was open, and Kennedy strongly supported Archibald Cox, who was back at Harvard Law School. So did an eleven-member commission appointed by Carter, unanimously ranking Cox as its first choice. But Bell opposed the nomination. He explained that Cox was almost sixty-seven, and the American Bar Association considered prospective judges over sixty-four unqualified.

Bob Lipshutz, Carter's White House counsel, sharply disagreed. On May 3, he wrote Carter that Cox was a "national treasure" and "a folk hero with an extremely distinguished career," certainly "the best qualified of the candidates." Kennedy talked with Carter and insisted that Cox was utterly independent and "nobody's man around here." Carter would not budge. According to Kennedy, he said that he was not rejecting Cox because of age or the Kennedy connection, but because Cox had supported Representative Morris K. Udall of Arizona in the 1976 Democratic Presidential race. That was hard to believe. (Carter remembers it differently; he says he stuck, too rigidly, to the age question and never mentioned Udall.)

Even nastier than the public clashes was the staff atmosphere. Each side felt the other looked down on it. The Carter folks felt the Kennedy contingent considered them dumb hicks. The Kennedy people thought the Carter side considered them snooty, with nothing but memories of John F. Kennedy to go on. Caddell, who had polled for Kennedy in Massachusetts in 1976 and had friends in each camp, said in December 1980, "Both sides would always read the worst intentions into each other's actions. . . . If somebody didn't get a return phone call, I would hear from friends on the Kennedy side that it was because the Carter people were putting it to Kennedy, and any time Kennedy said anything on policy, the Carter people would erupt and say 'he's really trying to destroy us.' "

Kennedy was getting plenty of advice about running that spring, either privately from friends, who usually urged against it, or from Democrats like Henry Jackson of Washington, who encouraged him after looking at their polls and concluding that they and their party had a better chance of surviving if he and not Carter headed the Democratic ticket. He was also getting a lot of public encouragement from a draft-Kennedy movement. That was an agglomeration of liberal Congressmen, some union leaders, especially William Winpisinger, head of the machinists, and various local politicians, most notably Tim Hagan, party chairman in Cuyahoga County, Ohio, and Joanne Symons and Dudley Dudley, two New Hampshire Democrats experienced in that state's Presidential primary. The draft movement sought to spare him the charge of ambitious interloper that had dogged Robert Kennedy. Symons said, "We wanted to provide him with a way of saying that the party had called him, rather than his being represented as trying to usurp an elected Democrat's place."

The Senator or his office would occasionally issue public disavowals of the draft-Kennedy campaign, but in fact his staff encouraged some of it. When Carl Wagner, his current political aide, heard from supporters he considered serious, he told them to talk to Paul Kirk, his predecessor and now a private lawyer in Boston. Kirk would redirect them to Mark Siegel, a political consultant in Washington who had been executive director of the Democratic National Committee and had quit the Carter White House over arms sales to Saudi Arabia. Sometimes those fictions of distance were ignored. In the late summer, Wagner called Siegel directly and said he could tell potential contributors that Kennedy had decided to run. At an August fund-raiser in Hyannis for a pro-SALT campaign, a reporter overheard Kirk telling Symons and Dudley, "You're doing a great job."

As Kennedy moved toward making the race against Carter, the chasm between their ideas of how a President should lead mattered at least as much to Kennedy as any policy differences. On April 30, Kennedy told the newspaper editors the nation needed "vision for longer run" or else it would face a decade of "reaction and desperation, or drift." Jody Powell, Carter's press secretary, responded by comparing Carter to Aesop's thrifty, hardworking ant and Kennedy to a chirping grasshopper. A few days later, in an interview with Walter Mears of the Associated Press, Kennedy said that if the public were challenged, it would respond, but Carter was failing to "mobilize the American people in pursuit of the great goals we share."

Instead Carter, with his energy program stalled, retreated to Camp David for ten days of meetings with politicians, civic leaders, and ordinary Americans to talk about what was going wrong in the nation. He came down to make a nationwide television speech on July 15 to discuss "a fundamental threat to American democracy." The menace, he said, was neither economic nor military and was "nearly invisible in ordinary ways. It is a crisis of confidence. It is a crisis that strikes at the very heart and soul and spirit of our national will. We can see this crisis in the growing doubt about the meaning of our own lives and in the loss of a unity of purpose for our nation." In a thought that came straight from his outsider's campaign for the Presidency in 1976, Carter said Washington was an "island," cut off from the people and "incapable of action."

Though some skeptics like Francis X. Clines of *The New York Times* ridiculed the speech for blaming the people—Clines called it the "cross of malaise" speech and the term stuck even though Carter had not used the word "malaise"—it was generally popular. Polls recorded an immediate boost in Carter's approval ratings. That advantage was squandered three days later when he fired four cabinet members, including Califano (Kennedy thought Califano was ousted because they were friends, though Califano steadfastly held up Carter's side on health insurance). That sudden action renewed the public doubts about the President.

Kennedy had no doubts. He spent two hours on the telephone the night of the fifteenth, calling friends and complaining that Carter's speech failed to challenge the people, as a good Democratic leader should. "The essence of political leadership," he said later, "is basically challenge and response. Constantly, that's sort of the central element in terms of my makeup, and I think in terms of politics you are trying to challenge and move the process, and then you are getting response to the challenging again. In this country, I think it is an essential aspect of the soul of this society, and one of the things that's made it sort of great. And it seems to me that that is sort of what the Democrats have been about." He felt that it "violated the spirit of America," Paul Kirk explained more concisely in 1980.

Health care, not energy, was still Kennedy's central concern. The message he took from Carter's emphasis on costs was "that this was not really a serious effort, because you were only going to be able to get the kind of momentum, the national kind of sense and purpose, probably one time, on a major kind of a policy issue like that. . . . We were losing what I thought was the most powerful and important issue and opportunity of our time. I was convinced that under different leadership we would be able to get this passed." He said health insurance was the "motivating, driving force" behind his challenge.

The "malaise" speech was probably the last outside influence on Kennedy's decision to run. A bold move by Carter to accept Kennedy's terms on health insurance might have forestalled a contest, and Carter insiders argued about it. Stuart Eizenstat favored it, and Hamilton Jordan opposed it, convinced that Kennedy would run anyhow if he thought he could win.

Kennedy set up and crossed a series of barriers that summer. He decided he would take the risk of being shot. Concerned about his children's reaction to his running and their reluctance to argue with him, he asked his friend Tunney, Teddy's godfather, to talk to his oldest son about it. Tunney, never an enthusiast for the race himself, persuaded Teddy that his father was in a unique position, with a chance to become President, and the children should not deny him. Horowitz also talked to the children about how their father would be protected, and then he spoke to Carter's personal physician, Dr. William Lukash, about providing Secret Service protection.

Kennedy worried about the impact on Joan and convened a meeting of psychiatrists to discuss what effect they thought a Presidential campaign would have on her. They concluded it would not hurt, and might even help because of the structured, busy life it would provide.

There exists no calendar, tangibly or in his memory, to record when each of these barriers was passed. Most were settled in August at the Cape, and the meeting about Joan came in early September—a few days after Kennedy had met Carter on September 7 and let him know he was going to run. They lunched on broiled fish at the White House, following the publication that morning of leaked stories in *The New York Times* and the *New York Daily News* saying Joan and his mother had agreed to his running. Carter called Jordan right after the meeting to say he was "certain" Kennedy would run. The next day he sent a letter to his state coordinators, telling them to prepare for opposition.

The Secret Service went to work on September 20. When the agents arrived at his office, he gave a chilling, fearful glance to Melody Miller. She had pleaded with him not to run. She had joined his staff eleven years before after being the last worker to close down Robert Kennedy's office.

FOR THOSE ELEVEN years, the idea of running for President had been before him, as his brothers' goals lay unfulfilled. "I always thought I would run sometime," he said years later. One poll after another showed that twice as many Democrats wanted him as wanted Carter. "The political opportunity that presented itself," he said years later, "was really irresistible." Once again, as in 1962 and 1969, there was an opening.

David Burke put it more eloquently, recalling an obscure 1951 poem of Robert Frost:

> How hard it is to keep from being King
> When it's in you and in the situation.

Kennedy's summertime discussions about running were with family, friends, and staff. Some were professional politicians, but they were friends first. One enormous subject seems to have been all but left out of those summertime talks: Chappaquiddick, and how that event would affect his chances to be President.

Tip O'Neill recalled bringing it up—he called it "the moral issue"—when Kennedy came to his ornate Capitol office on September 17 to tell him that he was running. He said Kennedy replied, "Yeah, my pollsters tell me it's a three- to five-percent moral issue." The Speaker replied, "Let me tell you something. You can't measure morals. I said that's an issue you just can't measure. My advice this year, not to run."

There may not have been any way to deal with the death of Mary Jo Kopechne and keep it from disabling his Presidential candidacy. But Kennedy had not explored the issue and his friends had not forced him to. Chappaquiddick was an issue that had seemed to slip by in July, the tenth anniversary of the accident. Kennedy gave a series of uninformative interviews, as if to inoculate himself. They were overshadowed by Carter's cabinet shakeup. But Caddell's polls had made it clear this was a devastating Kennedy weakness. Despite what Kennedy told O'Neill, he did not really have a pollster, someone actually doing polls for him, though he talked occasionally with Lou Harris, who had done some work for John Kennedy and minimized the importance of Chappaquiddick as an issue.

No episode captures that hole in campaign thinking more than Kennedy's interview with Roger Mudd for the disastrous *CBS Reports* program

"Teddy." Mudd, who knew him from covering Capitol Hill and also socially, got Kennedy to agree to do two interviews, one on his family and his personal life and the other on political issues. Kennedy, who had decided to run but was not yet saying so, did the first at Squaw Island on Saturday, September 29. The Senator, who usually had aides around to prepare him before scheduled interviews, surprised Mudd by being alone. It was soon clear that he was also unprepared for tough questioning, that he had somehow expected a much easier time, something casual about life on the Cape.

Mudd asked about Chappaquiddick. As the interview was shown on television, Mudd took a nighttime drive on Chappaquiddick Island with a camera in his car. He quoted Judge Boyle as believing Kennedy had lied. When Mudd asked if he thought "that anybody will ever fully believe your explanation of Chappaquiddick," Kennedy replied, "Oh, there's, the problem is, from that night, I, I found the conduct, the behavior almost beyond belief myself. I mean that's why it's been, but I think that's the way it was. Now, I find that as I have stated that I have found the conduct that in, in that evening and in, in the, as a result of the accident of the, and the sense of loss, the sense of hope, and the, and the sense of tragedy, and the whole set of circumstances, that the behavior was inexplicable. So I find that those, those, types of questions as they apply to that, questions of my own soul, as well. But that happens to be the way it was."

Kennedy's lack of preparation was also inexplicable. Carter had made it clear he would use the issue just four days before. In a visit to New York's Queens College, Carter answered a question about leadership by saying, "We've had some crises where it required a steady hand and a careful and deliberate decision to be made. I don't think I panicked in a crisis." When the next day's newspapers treated that as a Chappaquiddick reference, Carter wrote Kennedy, saying, "I won't make a habit of this."

Kennedy did no better with another obvious question: "What's the state of your marriage?" He stammered his way to a claim that "we, I think, have been able to make some very good progress."

Those were the hard, predictable questions that this candidate should have prepared for, though they were obviously difficult and personal. He was just as bad, with aides all around, in the second interview in his office on October 12. Mudd asked a question almost any Senator can answer better than Kennedy did: "Why do you want to be President?" He answered, "Well, I'm, were I to make the announcement, and to run, the reasons that I would run is because I have a great belief in this country, that it is, there's more natural resources than any nation of the world; there's the greatest educated population in the world; greatest technology of any country in the world, and the greatest political system in the world. And yet I see at the current time that most of the industrial nations of the world are exceeding us in terms of productivity, are doing better than us in terms of meeting the problem of inflation; that they're dealing with their problems of energy and their problems of unemployment."

When political America watched the program on November 4—up against a ratings blockbuster in the first televised showing of the movie *Jaws*—it was startled. Carter aides were amazed at how inarticulate he was. Reporters who covered him, like Mudd, were used to his false starts and unfinished sentences, and Mudd said later he wanted to convey that to an audience that has been used to seeing him in short, coherent sound bites when he was prepared on a subject or giving a speech.

But no one expected to see him unable to explain why he wanted to be President. Some friends, like John Tunney and Larry Horowitz, guessed later that it reflected lingering second thoughts about running. Others said he had never really thought about the why of running, only the whether.

Another explanation, which corresponds with his presentations in speeches and more casual interviews in that period, is that he felt gagged by the fact that as the leader of the family he was about to be host to the President at the John F. Kennedy Library dedication. Ever since his brother's Presidency, Kennedy had held the office in awe and had always persuaded himself that his criticisms of various incumbents were political, not personal, and certainly not insulting. In an interview years later, he said that in those weeks he felt he had to avoid drawing sharp lines that would insult his guest. Kennedy called that inhibition "very, very, very strong." Horowitz said Kennedy told him, "This is more important to me than anything else, getting my brother's library launched. We've invited the President to our home. He's coming to our home. You don't trash people in your own home." Still, a subjunctive "If I were to run" construction need not have been insulting, if he had figured out what to say next.

THE SUMMER OF deciding had left the Kennedy side, for all the encouragement the polls offered, at a disadvantage in the mechanics of running for President. Steve Smith, Paul Kirk, and Carl Wagner did an effective job of lining up talent in September and October. Before Labor Day, overtures had been made to Peter Hart, who ranked with Caddell as the Democrats' leading pollster, and to Morris Dees, who had used direct mail brilliantly to raise money for McGovern in 1972 and had taken a major role in Carter's 1976 campaign. But neither went to work until the end of October, and Hart did not get the money to do a poll before December—a poll that if taken earlier could have helped define the Chappaquiddick problem, and more broadly define what the public knew and thought about Kennedy. So the campaign was left with two dangerous assumptions. The first was that everybody knew what Kennedy stood for. The second was that the applause, the cheers, and most of all the published newspaper polls showed that the public liked it.

The time Kennedy took to decide was a luxury he could not afford. The Carter side was not waiting for Kennedy to get moving, but it had been working on the campaign for months on Jordan's assumption that he would run.

Mondale told his old Senate friend when Kennedy told him of his decision, "I don't intend to leave voluntarily. . . . This is going to get rough. And I'm sorry about it."

Jordan persuaded several friendly Southern governors to move up their state's Presidential primaries to two weeks after New Hampshire's. He expected to lose in New Hampshire, but wanted the South as a firewall to slow Kennedy's momentum. In July, while Carter was off at Camp David worrying about the national mood, the aides responsible for liaison with states, Jack Watson and Gene Eidenberg, engineered a Democratic governors' resolution that called for Carter's renomination and reelection. It passed with twenty votes for, seven absent, and four abstentions. No one voted no. This was a telling signal of the limits of Democrats' disaffection with Carter, but it was little noted. In September the Carter camp threw manpower and money into Florida to win a bizarre October 13 election for delegates to a state convention where a straw poll would be taken later that fall.

Even more important than those events was the fact that while the Kennedy side was signing up talent, it was not really planning a campaign to show how Kennedy would solve problems that stymied Carter. The leadership argument was not enough.

Nor was Kennedy making the case himself anymore. For whatever reason, he seemed to have lost the voice he had found at Memphis, the voice of the fighting liberal in a tentative, conservative time. As 1979 proceeded and he came closer and closer to running, there was a tactical air to his speeches; he was talking more about energy and inflation. Those were two politically hot topics where Carter was weak, but Kennedy's alternatives were vague or complex, and he could not present them with the singular intensity he brought to health insurance or poverty. And when he said what he would offer was not new policies but new leadership, he played into the hands of Carter aides who said he had no reason to run but ambition.

Kennedy and Carter came together, for the last time before they were formal opponents, on Saturday, October 20, 1979. On a sunny and breezy fall day, seven thousand people gathered to see the John F. Kennedy Library dedicated on Boston's Columbia Point, overlooking a sparkling Dorchester Bay. There were awkward moments, none more vivid than when Jacqueline Kennedy Onassis winced as the President kissed her on the cheek. Kennedy looked at his shoes as his nephew Joe demanded to know who would fight the "vested interests" and the oil companies. Joe complained that the people Robert Kennedy had championed—the migrant workers, miners, tenant farmers, Indians, blacks, Chicanos, Eskimos, folks of the hollows of eastern Kentucky and West Virginia, and ordinary workers—"too often get a raw deal from those who think of themselves as their betters, their leaders."

The President rose to the moment. In one of the best speeches of his career, he told of his own sense of loss on hearing of the death of John F. Kennedy. "My President. I wept openly, for the first time in more than ten years, for the first time since the day my own father died." President Kennedy, he said, "embodied the ideals of a generation as few public figures have ever done." Carter found imperatives for his own time in President Kennedy's efforts in civil rights and nuclear disarmament, but he also made an immediate political point when he said, "The world of 1980 is as different from what it was in 1960 as the world of 1960 was from that of 1940." He said, "The carved desk in the Oval Office which I use is the same as when John F. Kennedy sat behind it, but the problems that land on that desk are quite different."

If that was not clear enough, he had begun his speech by recalling a 1962 Presidential news conference when Kennedy was asked if he enjoyed the Presidency and would recommend it to others. Carter said, "The President replied, 'Well, the answer to the first question is yes and the second is, no. I do not recommend it to others, at least for a while.' " Ted joined in the laughter, and Carter said, "As you can well see, President Kennedy's wit—and also his wisdom—is certainly as relevant today as it was then."

As the applause for the President died, an advance man whipped the Presidential seal off the lectern. Kennedy came forward and spoke of his brother, more personally, less politically, using a line he had used before. "He and I had a special bond, despite the fourteen years between us. When I was born, he asked to be my godfather. He was the best man at our wedding. He taught me to ride a bicycle, throw a forward pass, and to sail against the wind." He spoke of Jack's adventurous love for the sea: "He might have sailed with Magellan, navigating beyond the charts to the new and better world he sought." And he closed with a line he would use again, saying, "Now, in dedicating this library to Jack, we recall those years of grace, that time of hope. The spark still glows. The journey never ends. The dream shall never die."

On his way home from the ceremonies, President Carter stopped for an airport interview with Boston television reporters and praised Kennedy, saying his standing as a leader was excellent, not dependent "on his relatives or his family position, but on his own record." But he quickly jabbed at Kennedy, saying he favored "the old philosophy of pouring out new programs and new money to meet a social need," in contrast to Carter's policy of "fiscal prudence . . . holding down unnecessary spending." Carter also said he favored "much stronger defense commitments" than Kennedy did, and said he felt that his kinship to President Kennedy was as strong as the Senator's.

THE LIBRARY DEDICATION liberated Kennedy. He started making test trips, trying out themes and bringing a large press corps along. In Philadelphia, he used the word "leader" or "leadership" thirteen times in a speech, and said when Americans faced problems before, "we didn't talk of the malaise of the American spirit." He called around the country to let Democrats know he would run. He walked the corridors of the Senate, asking his Democratic colleagues to support him. More said they would than actually did.

At Georgetown University on October 24, he demanded that the administration do more for refugees in Cambodia, instead of worrying about which warring dictator, Pol Pot or Heng Samrin, held the country's United Nations seat. The issue mattered to him; on September 6, the day before he let Carter know his intentions, he had managed Senate passage of a bill to triple the number of refugees allowed into the United States annually.

Kennedy said, "Cambodia is on our conscience. . . . We cannot escape the moral consequences of our actions during the Vietnam War which helped to launch the descent into hell of that once beautiful and peaceful land." The attack prodded President Carter, just as the threat of Kennedy sometimes moved Presidents Johnson and Nixon to action; that same day, Carter announced he would send more aid; Rosalynn would go to visit refugee camps.

On Monday, October 29, Steve Smith announced the formation of an "exploratory committee." Fund-raising had begun before that. Smith said Kennedy would formally announce on November 7. Chicago's mayor, Jane Byrne, endorsed Kennedy after all but endorsing Carter. Kennedy was eager, telling reporters, "I'm tired of screwing around with this. I want to get going."

Still, there was some concern. In September, a former McGovern aide, Richard G. Stearns, had scoffed, "The Carter campaign will be like the French army in World War II, an obstacle but not a deterrent." On October 29, now installed as a delegate hunter for Kennedy and busy trying to sign up state chairmen, Stearns said, "If Kennedy doesn't win well early on, we're going to have problems."

Then on Sunday night, November 4, the Roger Mudd interview was shown. The reaction was awful.

But that was only the second-most-important campaign event that day. The most important was thousands of miles away. Iranian students, angry that their former ruler, the shah, was in the United States for medical treatment, overran the U.S. Embassy in Teheran and took about sixty Americans hostage.

CHAPTER 23

Losing

The last Kennedy brother announced his candidacy for President on Wednesday, November 7, 1979. For the eleven years, five months, and one day since Robert Kennedy's death, he had always been someone's candidate. Now he was his own.

Though the Kennedys, from his mother to his nieces and nephews and his brothers' allies, filled the hall and made it clear that this was an event of family fulfillment, Ted sought in small ways to assert his own identity. John and Robert had announced in the Senate Caucus Room, not two miles from the White House they sought. Ted chose Boston's Faneuil Hall, with its statues of Presidents John Adams and John Quincy Adams and a huge painting of Daniel Webster. He invoked not the twentieth but the eighteenth and nineteenth centuries—Boston's revolutionary history, debated in that very hall, and the hopes of immigrants like his great-grandparents, who had faced difficulties far greater than today's in "seeking a brighter destiny on a distant continent."

Where John Kennedy said in 1960 that the "hopes of the globe" focused on the Presidency, and Robert Kennedy said in 1968 that the election involved "the moral leadership of this planet," Edward Kennedy's declaration was far more national in direction, with only 122 bland words about the rest of the world. He complained about inflation and touched on policy, mentioning health care, poverty, and equal opportunity. But he concentrated on leadership:

"For many months, we have been sinking into crisis. Yet we hear no clear summons from the center. . . . Government falters. Fear spreads that our leaders have resigned themselves to defeat."

Picking up Carter's 1976 themes, he said, "Before the last election, we were told that Americans were honest, loving, good, decent and compassionate.

"Now, the people are blamed for every national ill and scolded as greedy, wasteful, and mired in malaise. Which is it? Did we change so much in these three years? Or is it because the present leadership does not understand that we are willing, even anxious, to be on the march again?

"The most important task of Presidential leadership is to release the native energy of the people. The only thing that paralyzes us today is the myth that we cannot move."

It was not an exciting speech. Campaign aides immediately explained that they had not meant it to be. The most intriguing moment of the morning came when Jim Wieghart of the *New York Daily News* asked Kennedy what role Joan would play in the campaign. There were some boos and hisses. Kennedy shushed them and turned to Joan, who was prepared for the planted question. She came to the podium and said, "I look forward to campaigning for my husband. I look forward very, very enthusiastically to my husband being a candidate and to his being the next President of the United States."

A lot of worry and preparation had gone into that answer, and indeed into her appearance at the event. Leaving her at home would have reminded people of their tenuous marriage and the rumors of his womanizing. So might her presence. Kennedy beamed proudly at her answer.

But perhaps the most telling foretaste of the campaign came when Tom Oliphant of the *Boston Globe* picked up on Kennedy's complaints about inflation and asked him what he would do differently from Carter. His feeble answer seemed a measure of how little thought Kennedy had devoted to deciding how to run for President. He seemed so intent on avoiding economic policy that he did not even remark on a recent comment by Carter's appointee as chairman of the Federal Reserve Board, Paul Volcker, who said that to curb inflation, "the standard of living of the average American has to decline." Instead all Kennedy had to say about curbing inflation was "I would devote my full energies and effort to deal with it in every respect of American public policy."

After speaking briefly to several thousand fans outside, Kennedy was off to Manchester, New Hampshire, and Portland, Maine, to talk about the price of home heating oil, and Chicago, where the embrace of Mayor Jane Byrne repelled her enemies in the Democratic organization and left a rally half empty.

His first few weeks on the road were rocky. He raced around hectically, but was sometimes flat, monotonous, and fumble-mouthed, uttering phrases like "the rising price of inflation," "What we're here to talk about is the issue of the substance," and that all-time Iowa favorite, "fam farmilies." His shouting sometimes worked in the room, but looked awful on television. And the technical, Senatorial differences he sought to establish with Carter on questions like farm policy or how to keep regional railroads alive usually failed.

He was sometimes very good, though reporters, conditioned by the Mudd interview, took a while to notice. On November 8 in Chicago, after a speech on health care to senior citizens, he found a retired factory worker who came forward to say, "I worked hard for forty-seven years, and if I ever have to go to the hospital, I'll lose everything I saved." At Grinnell College in Iowa on November 13, he impressed the traveling press corps when he said: "The President has adopted all the old Republican policies. Well, they didn't work for William McKinley, nor for Herbert Hoover or Richard Nixon. It's time to have a real Democrat in the White House again." There and again that night in Mondale's Minnesota, he said, "President Carter likes to say he tackles the tough issues. Most of the time, he misses the tackles."

Those early trips, and the perceptions of them, were handicapped by expectations. Kennedy was expected to be a great campaigner, as his brothers were believed to have been great. He also had no time for a shakedown cruise, a chance to test lines and approaches, develop a basic speech, and get comfortable with a campaign style in relative solitude. Like no Presidential candidate except Ronald Reagan or Robert Kennedy, Ted had a big press corps from his first day on the trail. Moreover, they were reporters with attitude, determined not to be fooled or flattered, as they believed their predecessors on other Kennedy campaigns had been.

But Kennedy's overwhelming campaign problem in those first weeks was Iran. First, it kept him off national television, which devoted the first several minutes of each evening news program to the hostage situation. After his announcement, the next time he was covered on an evening news broadcast was two weeks later. Meanwhile the reporting of Carter standing firm but rejecting military action (although the seizure of an embassy is nothing less than an act of war) started to regain him the credentials of leadership he had lost. Early on, Kennedy tried a couple of times to blame the administration for the mess, saying it should have been forewarned by a previous embassy takeover and the resentment over the shah's presence. "Either you can protect the Americans there, or they shouldn't be there," he told his Nashville audience. But the public was rallying around its President in a time of international crisis, and it did not want to hear how he had helped cause it, so Kennedy dropped it.

But on December 3, Kennedy gave the administration an opening, and paid for it. In a San Francisco television interview, he said the shah "ran one of the most violent regimes in the history of mankind—in the form of terrorism and the basic and fundamental violations of human rights, in the most cruel circumstances, to his own people." Kennedy also said the shah stole "umpteen billions of dollars from his country." The White House, newspaper editorial pages, and even George Bush jumped on Kennedy for giving aid and comfort to the Iranians. Kennedy and his campaign made it worse by taking several hours to respond while they tried to figure out what to say about his untimely, impolitic, but fundamentally accurate statement. By the time Peter Hart finally conducted a national poll for the campaign a few days later, 54 percent of the public agreed with the statement "I feel that Edward Kennedy has hurt America by speaking out against the former Shah of Iran."

At least by December the campaign finally had Peter Hart as a pollster, but it still did not have a message, nor a someone to make television commercials to sell it. When it came to vetting a possible ad man, no one seemed to be in control. Instead, sisters and friends joined in critiquing presentations, and ended up preferring the soft work of Charles Guggenheim, who had done moving documentaries about Kennedy's brothers and most of George McGovern's spots. He did not have enough edge for 1980.

Kennedy put Steve Smith in charge because he knew him and trusted him like a brother. He had not given Smith enough time to prepare a campaign. Smith had hired well, if perhaps too lavishly in salaries and overlapping authority. He had hired people like Carl Wagner and Richard Stearns who knew how much the rules had changed since 1968, Smith's last campaign, and Morris Dees, who understood how the Federal Election Campaign Act Kennedy had championed made it harder to run for President. Still,

Smith was not a firm manager, and he did not make sure that the campaign on the plane could rely on informed, consistent advice from headquarters in Washington. Kennedy, known for the best staff in the Senate, had not created—or seen to Smith's creating—a campaign operation worthy of a President.

The staff did handle mechanics. The advance work of turning out crowds went very well—8,600 in Nashville for his second day on the road. And after Kennedy had said—on the night of Carter's success in Florida—that Iowa would be the first real test between him and the President, his campaign had no choice but to pour enormous resources into that state. Iowa's precinct caucuses on January 20 would be the first step in choosing delegates for the national convention in New York in August. Carter, who had won headlines with a plurality there in 1976, had more than twenty paid staff working the state for weeks, but by the end of November, Iowa reporters thought Kennedy had caught up. In a state where the task was to get supporters to turn out, perhaps for hours, on a cold January night, organization was essential. Mondale warned Carter in mid-December that he might lose Iowa. The Vice President tried to get Woodcock, who was now ambassador to China, to go out and campaign in Iowa, but Woodcock refused to work against his union's choice and his own old friend.

The best staff in the country might not have figured out how to overcome Chappaquiddick. The problem was not that Kennedy would not answer lingering questions, or that the right questions had not been asked. As one Kennedy adviser told Martin Schram of the *Washington Post,* "They've all been asked and all been answered. It's that people don't like the answers."

In the first few weeks of the campaign, Chappaquiddick was not an issue thrust in his face by hecklers, though reporters found it was on voters' minds. Carol Kelderman, an Oskaloosa, Iowa, teacher, told John Walcott of *Newsweek,* "He's hung onto the lie for so many years—that makes him seem like a megalomaniac, like Nixon. He should just admit that he got drunk and screwed up." Campaign aides, like Patti Saris, a junior member of the Judiciary Committee staff who took an unpaid leave to work plant gates in Keokuk, found "people just couldn't get over it."

But from the campaign's first days, the press kept reminding the voters by bringing up Chappaquiddick on its own. There were stinging columns from usually friendly quarters. Tom Wicker of *The New York Times* wrote that Kennedy "cannot or will not yet explain what happened at Chappaquiddick or rectify the numerous inconsistencies in his 10-year-old account of the matter." Jimmy Breslin in the *New York Daily News* was rougher. Seizing on Kennedy's references to "the conduct," Breslin wrote, "What was required here was an old-fashioned Catholic confession: 'Bless me, Father, for I have sinned . . .' And when that confession is made, you talk about yourself, first person, as the sinner. You don't say the sins were committed by some guy standing on the side someplace; you were there and you did it, so tell what you did and how you feel. Do not try these little shifts and evasions to give the impression that perhaps it wasn't you who did it in the first place." Even a *Globe* editorial said: "Chappaquiddick was not just an auto accident. Many Americans suspect, not without reason, that Kennedy's handling of its aftermath is another case of a politician stonewalling. And they wonder whether Kennedy would lie to the American people in a more public crisis."

Kennedy sometimes did better at confronting the subject than he had with Mudd. On NBC's *Meet the Press* on November 19, he said he was sure there would be no "new information that is going to challenge my testimony. . . . If there was ever going to be any new information that was going to be different or challenge the sworn testimony that I gave, there would be absolutely no reason that I should consider either, one, remaining in public life, let alone run for the Presidency of the United States."

There was a logic to that, and to his attempt to broaden the definition of his character by continuing, with rare self-justification, "The fact of the matter is, I have been impacted over the course of my life by a series of crises, by a series of tragedies. I lost my brothers under the most trying and tragic circumstance. I have also faced the illness and sickness of a child that has been impacted by cancer. I have had other tragedies in my life, and I have responded to those challenges by one, acting responsibly, and two, by the continuing commitment to public service." He said he would not be running for President "unless I was completely satisfied that I could deal with any of the pressures that would come."

But the public was not persuaded. A *Des Moines Register* poll found in December that 36 percent of Iowa Democrats and 52 percent of all Iowans said they did not believe his account. Hart's poll found 35 percent agreeing with the statement "The accident at Chappaquiddick revealed basic problems in Edward Kennedy's character which still will affect his performance as a public official." Thirty-six percent disagreed, with the rest uncertain.

And Chappaquiddick played an important role in the hesitant response of politically active feminists to his candidacy. Feminist Democrats were of two minds about Kennedy when he ran. Iris Mitgang, head of the National Women's Political Caucus, conveyed them both when she said in September, "He is a known womanizer, or has been, and the relationship in his marriage gives me reason for pause." She said, "When you're a public figure, your personal life becomes public property." But she also praised his support for paying for abortions under Medicaid, a position he took quickly after *Roe v. Wade* upheld abortion in 1973, and his stand on women's health issues and on picking women to be federal judges. Suzannah Lessard, writing in the December *Washington Monthly,* was more critical: "Kennedy's womanizing is widely known. . . . The idea evidently is lunch and a dalliance, over and out, on with the pressing schedule. . . . Certainly it suggests an old-fashioned, male chauvinist, exploitative view of women as primarily objects of pleasure. It gives me the creeps."

He made an effort. Susan Estrich, a young lawyer who moved from the Judiciary Committee staff to the campaign, pushed to get Carter's failure on the Equal Rights Amendment issue into his speeches, and found him more comfortable with women's issues than most of the men who were close to him. They worried about his being photographed with any woman he was not related to and worried about raising any subject that could lead anyone to think of Chappaquiddick.

It was not that many women's groups were satisfied with Carter, who had not fought to get the last three states needed to ratify the Equal Rights Amendment. The board of the National Organization for Women, for example, voted on December 10 not to endorse Carter, even if he beat Kennedy and ran against Reagan. But for a campaign that needed an outpouring of support to oust a determined incumbent, this was still an area of hesitancy.

THE OVERALL BAD news continued to roll in. Kennedy's strongest rationale for running had been that he could beat Reagan or Gerald Ford or whomever the Republicans nominated, and Carter could not. By mid-December the polls stopped saying that, and indeed gave Carter a better chance. Jerry Brown, the governor of California, was running a lonely campaign, overshadowed by Carter and Kennedy.

Over the holidays, Kennedy and his top aides gathered at Palm Beach to discuss how to overcome the focus on Iran, which had taken away their argument that Carter was no leader. They debated whether to move openly to the left on policy questions, but in the end Kennedy decided against any change, hoping the staggering economy would take hold as a dominant issue.

Then another foreign policy disaster gave Carter still more help. On Christmas Day, Soviet troops invaded Afghanistan to support a coup engineered by their allies in that poor, strategically located country—where Russia had dueled with Britain in the nineteenth century. Once more the public rallied to Carter, and his position seemed so strong that on December 28 he withdrew from a debate in Des Moines against Kennedy and Brown. Carter sent a telegram to the editor of the *Des Moines Register,* whose invitation to debate on January 7 he had accepted even before Kennedy announced, saying, "We are going to try to bring the Iranian matter to a head in the next ten days or so," and "I don't think there is any way I can leave Washington." The *Register* proposed a debate in Washington instead. Carter refused.

Carter's political aides saw an advantage in freezing Iowa with him ahead and making sure the economy did not supplant Iran as an issue. Others worried that he might lock himself in to never campaigning until the hostages were freed. Carter wrote a pious intended-for-publication note to Jody Powell, his press secretary, saying, "I cannot break away from my duties here, which are extraordinary now and ones which only I can fulfill. We will just have to take the adverse political consequences & make the best of it. Right now both Iran and Afghanistan look bad, & will need my constant attention."

"Constant" was a fib. Carter made twenty or more telephone calls a day to Iowa, frequently talking to improvised campaign meetings assembled around a speakerphone.

Iowa has a squeaky-clean, high-road image in American politics, but the stakes were too big in these caucuses for much restraint. John Pope of Americus, a Carter friend who led a "peanut brigade" of Georgia volunteers, used this message about the opponent: "Ted Kennedy leads a cheater's life—cheats on his wife, cheats in college." He told a reporter, "I find a lot of folks receptive."

The last weeks of the Iowa campaign took a toll on two Kennedy friendships. Carter had reacted to the invasion of Afghanistan by ordering an embargo on grain sales to the Soviet Union, although he had pledged in Iowa in 1976 that he would never use an embargo. Kennedy attacked, arguing the embargo would hurt Iowa farmers but not the Russians, who could buy grain elsewhere. On January 10, just as the *Des Moines Register*'s Iowa poll found most Iowans (though not farmers) supporting the embargo, Mondale accused Kennedy of pursuing "the politics of the moment" in opposition. Supporting the embargo, said the Vice President, who had urged Carter not to do it, was "the patriotic route to take." When a reporter asked if he was calling Kennedy "unpatriotic," Mondale replied only, "I've said what I've said." Kennedy shot back, "I don't think I or the members of my family need a lecture from Mr. Mondale or anyone else about patriotism." Then, on Saturday, January 12, Kennedy upstaged Mondale in Waterloo at a joint appearance; Mondale was there in place of Carter, and Kennedy said he might have used him as a stand-in himself, because Mondale was on record for national health insurance, against decontrol of oil prices, and against grain embargoes. Then he presented him with a New England Patriots football shirt.

A few days later, sensing defeat, Kennedy came by the Senate office of his old friend John Culver and urged him to abandon his public neutrality and come out to Iowa for the last weekend. Culver, facing a tight campaign of his own for reelection to the Senate, told Kennedy he did not think it would help, promised to sleep on it, and then told him flatly on Friday, January 18, that he would not come along. Some Kennedy aides said Culver promised to come and did not show up at the airport, and the friendship dimmed for a while.

Instead of Culver, Kennedy's main ally on that last trip to airport news conferences was Joan, who fielded tough questions about Ted and other women. In Sioux City, responding to another burst of Chappaquiddick stories in the *Washington Star* and *Reader's Digest,* she said: "I believe my husband's story which he told me right after the incident. It seems to me that I have heard my husband's story and various versions of the story by other people for over a decade. I believe that when my husband testified, under oath, all that was made public ten years ago. I don't think anything new has come in the last ten years."

ON MONDAY NIGHT, January 21, 1980, Kennedy took a pounding. Carter got 59 percent of the precinct delegates to Kennedy's 31 percent. The turnout was huge, over 100,000. The advertising and intense personal campaigning had turned a pleasant political folkway—folks sitting around living rooms or firehouses debating and then choosing sides among Presidential candidates—into the functional equivalent of a Presidential primary.

Kennedy and his top campaign staff met at McLean that week to decide what to do next. The dimension of the Iowa defeat turned off contributions. Money was a severe problem. Most staff salaries, some a then-lavish $50,000 a year, were stopped. The lease on a chartered United 727 was canceled in favor of scheduled airline flights. There was some internal talk of dropping out. But on Thursday, January 24, Kennedy decided to keep going, and he went by campaign headquarters to cheer up the troops. He signed some autographs, told a reporter "No!" when asked if he would quit, and after mangling a telephone call that came in from Puerto Rico while visiting the forlorn Southern delegates' desk, quipped, "Now you see how much trouble I had with Spanish."

The plan to rescue the campaign was to give a hard-hitting speech that would define his differences with Carter and his reasons for running. He stayed in Washington, canceling a New England campaign trip to work on it, and gave it at Georgetown University on January 28.

He began by attacking Carter on his current strength, foreign policy, for declaring the Persian Gulf to be an area of vital American interest, and for contending that the invasion of Afghanistan was the "gravest threat to peace since World War II." "Is it a graver threat than the Berlin blockade, the Korean War, the Soviet march into Hungary and Czechoslovakia, the Berlin Wall, the Cuban Missile Crisis or Vietnam?" he asked.

He ridiculed Carter for saying he was "surprised" by the invasion, saying the administration had ignored the warning signals of troop concentrations on the border. "Exaggerated dangers and empty symbols will not resolve a foreign crisis. It is less than a year since the Vienna summit, when President Carter kissed President Brezhnev on the cheek. We cannot afford a foreign policy based on the pangs of unrequited love."

But his central complaint was not about foreign policy itself, but the chilling effect it had on the Presidential campaign. "The political process has been held hostage here at home as surely as our diplomats abroad," he said.

He said his campaign would speak for those Americans whom Carter had ignored in his State of the Union address a few days earlier. "Let me tell you what we did not hear from the President last week: Inflation will continue. Unemployment will go up. Energy prices will rise to even higher levels."

He called for gasoline rationing to cope with reduced oil supplies. To combat inflation, he proposed a six-month freeze on prices and wages, but also on profits, dividends, interest rates, and rent. As he came to the end of the speech, having tried to establish Carter as an incompetent and himself as a reliable leader who could face immediate problems, he turned to a vision of a Kennedy Presidency:

"An America where the many who are handicapped, the minority who are not white and the majority who are women will not suffer from injustice, where the Equal Rights amendment will be ratified and where equal pay and opportunity will become a reality rather than a worn and fading hope.

"An America where average-income workers will not pay more taxes than many millionaires, and where a few corporations will not stifle competition in our economy. . . .

"An America where the state of a person's health will not be determined by the amount of a person's wealth. . . .

"As I said a year ago, sometimes a party must sail against the wind. Now is such a time. We cannot wait for a full and fair wind, or we will risk losing the voyage that is America. A New England poet once wrote: 'Should the storm come, we shall keep the rudder true.' Whatever comes in the voting of this year, or in the voyage of America through all the years ahead, let us resolve to keep the rudder true."

In a brief preamble to the speech, shown on television in New England that night, Kennedy spoke of Chappaquiddick. He said he knew some people would never believe him or forgive him, but he had told the truth and carried the burden of sorrow in knowing that "nothing is more difficult than when you know you alone are responsible for the loss of a young life and all that it means."

Kennedy's campaigning improved dramatically in the next few weeks. It had a purpose to it. But he did not win anyhow, except in Massachusetts. Maine's caucuses and the New Hampshire primary went to Carter, not by great margins but in what both sides had expected to be Kennedy country.

As the losses mounted, more of the old hands disappeared from headquarters. Steve Smith even took a vacation in Majorca. A trimmer staff seemed to work better in making decisions, and even the suddenly Spartan travel arrangements—flying commercial or in chartered puddle-jumpers, staying in tired hotels—helped create the "plucky Teddy" stories that had to serve as ersatz victories. "Uncomplaining" was a constant theme, though reporters saw him grimace when his back gave out. Even the *Times*'s William Safire, hardly a fan, conceded that "if he continues to build character this publicly, his day will come."

In some ways the campaign was hardest on the younger Kennedys. Kara, twenty, was leafleting in New York City when a passerby told her, "You know, your father killed a young woman your age." Teddy, on the road all the way through, recalls elliptically, "It's pretty rough. People say mean things." Maria Shriver, who quit her first job as a television producer to campaign for her uncle, speaks angrily of voters who wouldn't try to "find out what he was as a human being." But it was sometimes a great adventure, and perhaps for no one more than for twelve-year-old Patrick. He had a lot of time on the plane with his father, playing "crazy eights," or discussing what worked in the Senator's last speech, which Patrick followed with a duplicate set of cue cards and marked up with stars to show what went best. But there were scary moments for him, too. On at least two occasions when his asthma flared up, the campaign plane was diverted so he could be treated on the ground. And every night when Patrick was not on the road, Ted called his son to reassure him that another day of campaigning had gone safely.

Illinois, whose primary came on March 18, was the worst of all. Kathleen Kennedy Townsend, who traveled the country nursing her infant daughter, Megan, recalls Chicago as "awful," because the family's memories of Mayor Daley and its vision of the city as a working-class Democratic stronghold that would naturally back Ted were so far off the 1980 mark.

Both sides had seen Illinois as a pivotal state, the place where they could finish off the opposition if they came to it with a winning streak—which Carter did. Illinois was especially important because its rules were different. In most states delegates had to be divided in proportion to the vote, but in Illinois the rules permitted a narrow majority to be translated into a landslide. Carter's campaign was well organized, bolstered with the sort of patronage and federal aid Illinois pols could appreciate. And it was solid downstate, where Kennedy supporters were rare.

Kennedy's Illinois effort had almost no money, having virtually shut down after Iowa. Jane Byrne's endorsement proved a handicap, not an asset, because she was profoundly unpopular with the organization precinct captains who turned out the vote on Primary Day. (One diehard antagonist, Jeremiah Joyce, arranged to interrupt her speech to about a thousand captains with a recorded rendition of Mayor Daley proudly asserting that the Chicago organization supported the Democratic Party, not any challengers.)

But the most important factor was Chappaquiddick and what it symbolized in conservative Catholic parishes. The issue bit more deeply in Chicago than anywhere else Kennedy had been. Grade-school straw polls, going for Carter two to one, were one early measure that delighted the Carter campaign. Not that anything was left to chance. Carter's television commercials played on doubts about Kennedy—about Chappaquiddick and about his marriage. One showed the President in the White House, with the message "You never find yourself wondering if he is telling you the truth." Another, with Rosalynn and Amy, had Carter talking about how a "good family life" was essential to "being a good President."

St. Patrick's Day fell the day before the primary, and it put the disaster of the Kennedy campaign on public display. He had never been as exposed to the public, and the Secret Service had him wear a bulletproof vest. Joan had heard there were assassination threats, and she worried that either Ted or one of the children might be shot. The Kennedys started out marching together in a chill rain with Mayor Byrne. But as the mayor was greeted with boos, the Senator began to fall behind, half a block at times. He zigzagged from one side of State Street to the other, plunging into the crowd, shaking hands, but finding that some of the boos were for him, too. "When are you going to learn to drive?" called one heckler. "Where's Mary Jo?" a drunk shouted into his face. The rain had turned to snow, and a youth with his beard dyed green tossed some firecrackers out into State Street, and Kennedy's legs buckled from fear as Joan grabbed for him and the Secret Service agents clustered around. And then he pulled himself together and marched some more, looking for hands to shake, until the parade reached its end.

The next day Carter crushed Kennedy, winning a nonbinding preference vote 65 percent to 30. Delegates were elected independently of the statewide vote, and Carter's organizational strength won him 155 out of the 169 at stake. Carter's national lead in delegates was now 615 to 192, and his aides started telling reporters that the way the rules worked in the remaining states, Kennedy could not win—without getting an impossible 60 percent in all the remaining states. They were right. Stearns, the chief delegate hunter for the Kennedy campaign, said after the election that Carter's huge victory in Illinois "gave him an insurmountable lead that Kennedy under no conceivable set of circumstances would overtake before the convention."

The next primary was New York's. Kennedy never stopped insisting that he was in the race to the end. On *Meet the Press* on Sunday, March 23, he said he believed "that the soul of the Democratic Party is being tested. . . . I believe in the things which we're putting forward, and I'm going to continue to press them."

But the money was drying up. As he hurried around the state, his bad back plainly bothering him, he was either flying commercial or in very small planes—while the press corps trailed in smaller ones, called the "paper tigers." The crowds were friendly, but not huge. There was little money for phone banks or other organizational help, although several unions—especially the teachers, communications workers, and autoworkers—pitched in, with lawful efforts to stir their members and with not quite legal cash to keep unpaid Kennedy volunteers fed. He had a few Congressmen and local officials behind him. But Mayor Ed Koch of New York made television commercials hinting at Chappaquiddick. Worse yet, two leading Democrats who had encouraged him to run in August stayed on the sidelines, Senator Daniel Patrick Moynihan and Governor Hugh Carey. The governor's neutrality was especially galling; not only had they worked together on Irish issues, but in 1972, when the anti-McGovern backlash threatened Carey's House seat, Kennedy had braved thrown eggs and tomatoes to campaign for him in Bay Ridge.

Though the television budget was slim—$200,000 to Carter's $750,000—Kennedy did have two new, effective television commercials. Charles Guggenheim, the family's favorite, had been replaced as ad man by David Sawyer in New York. Carroll O'Connor, who had starred as Archie Bunker in the hugely popular *All in the Family* television program about a right-wing blue-collar worker from Queens, appeared in them. In one, O'Connor redirected the old Democratic device of running against Republicans, saying Carter "may be the most Republican President since Hoover and he may give us a depression that will make Hoover's look like prosperity." Apparently aiming to blunt the character attacks, he said of Kennedy, "I believe in him, friends, in every way."

Along with support for wage and price controls, opposition to a higher gasoline tax, and criticism of Carter's proposed budget cuts as a threat to New York City, Kennedy did have one new issue by New York. On March 1, the United States had abandoned past abstentions and voted for a United Nations Security Council resolution on March 1 that rebuked Israel for establishing settlements in Jerusalem and on the West Bank. The message was a deliberate one of impatience with the government of Menachem Begin, and Carter had conferred with Donald McHenry, the ambassador to the UN, and with Secretary Vance before the vote was cast. But two days later, after a blizzard of criticism from Jewish groups, Israel, and Kennedy, Carter issued a statement saying the aye vote was cast because of a "communications failure." McHenry responded by saying he voted "according to their instructions."

The issue simmered. Kennedy called the vote a "betrayal of Israel." On March 19, Robert S. Strauss, Carter's campaign chairman, blamed the State Department, saying, "I think there are some goddamn Arabists over there and they ought to be fired." But Secretary Vance defended the administration's position in Congressional hearings on March 20 and 21, saying the settlements were illegal and Israel had been ignoring quiet efforts to urge restraint. On *Meet the Press,* Kennedy assailed the administration for dictating to Israel. The next day, when Mondale mentioned the President's name at a Young Israel banquet, the crowd booed.

The New York primary was also the occasion for a more imaginative Kennedy venture in foreign policy. For several weeks he had been hearing from Andrei Pavlov that the Soviet leadership was looking for a way out of Afghanistan and wanted his help. Of the small group that set foreign policy in the Kremlin, Pavlov was closest to Yuri V. Andropov, the head of the KGB, who had been slow to support the invasion. Pavlov passed his message through John Tunney, who had business interests in Moscow, including marketing the bear chosen as the symbol of the 1980 Olympics.

Pavlov said years later that the message he passed, in good faith, was that Moscow, trusting Kennedy more than Carter, would be willing to pull out at his request. He said Kennedy told him that if the Soviet Union did pull out, he would work to get the SALT treaty ratified. After Kalicki conferred with Marshall Shulman, Secretary Vance's chief Soviet adviser, Kennedy worked out some proposals. On March 20 at Columbia University, he urged a "fresh political effort to restore Afghanistan to genuine independence and nonalignment." He said he believed Brezhnev "would now respond positively." He called for immediate Soviet withdrawals of 20,000 of its 100,000 troops, and a promise to get them all out by the end of the year in exchange for a guarantee that Afghanistan would have a coalition government and be nonaligned under UN supervision.

But, as Kennedy had pretty much expected, no response came from Moscow.

In the final days, the public polls in New York were grim for Kennedy. A Lou Harris survey in the *Daily News,* published on Friday the twenty-first, said he was trailing Carter, 61 to 34 percent. Caddell's polls for Carter were very different, showing Carter behind by 9 points on Sunday and "in a free fall." But the Kennedy camp did not know that. When another Harris poll published Monday showed Carter ahead 56 percent to 36, they braced for another humiliating defeat. Kennedy's top staff, his sisters, and Steve Smith met to persuade him to drop out. Under an assumed name, Kirk rented a room at the Parker House, Boston's political hotel, for a Wednesday-morning withdrawal statement. Steve Smith told Shrum to draft it.

"First of all, let me say, I love New York anyway," Shrum wrote. "We would have preferred victory, but we can live with our loss," he continued. "I have withdrawn from the campaign," he wrote, but "I have not withdrawn from my commitment to speak for those who have no voice, to stand for those who are weak or exploited, to strive for those who are left out or left behind."

That speech was not needed. Instead Joan sang for reporters, "I like New York in March, how about August." Kennedy twitted two of his early, vanished cheerleaders, Moynihan and Carey. "They gave me very good advice," Kennedy told reporters. "They said I could carry New York and, by God, I did." Kennedy won a landslide that approached Carter's triumph in Illinois, 59 percent to 41. But New York's delegates were split much more closely, with Kennedy only taking 164 to Carter's 118.

One obvious factor in his victory was Jewish unhappiness with Carter. Jews cast a quarter of the primary votes, and 78 percent of them voted for Kennedy. Abe Transky, a retired captain at the Latin Quarter, explained his vote: "Kennedy, because he's with the Jews." Another Kennedy voter in heavily Jewish Midwood, an election worker at P.S. 153, said she had decided three weeks earlier, after "what Carter did at the United Nations."

But there was more to Kennedy's victory than Jewish unhappiness. He also won that night, 47 to 41 percent, in Connecticut, where Jews made up a very small slice of the electorate. These were states where Carter had always seemed a little strange, and Kennedy's liberal values echoed those of Democratic voters. They voted for a candidate who said, "The poor and the cities may be out of political fashion this year, but they still have a rightful claim on public policy," and "Someone has to speak for them. Someone has to stand against the temporary tides of reaction—against a politics and a Presidency that may see injustice, but then quietly looks away. . . ."

There was another difference—the Catholics. In New York and Connecticut the Catholic voters did not seem to hold Chappaquiddick against Kennedy as they had in Illinois. While 64 percent of Illinois Catholics voted against him, just 47 percent of New York Catholics did.

EVEN THOUGH KENNEDY told Tom Oliphant of the *Boston Globe* that he did not share the sudden high hopes of his followers and knew the vote was anti-Carter rather than pro-Kennedy, his victory in New York led a suddenly worried Carter campaign to take two steps that helped immediately but cost dearly in the long run.

The first came the next week. The schedule now required only a one-week attention span between primaries. That week was Wisconsin's. Kennedy did not have the resources for a major campaign there, but he started gaining anyway. Carter, seeing some progress in negotiations with Iran over the hostages, lost no time in using the issue. The President briefed editors and television anchors over the weekend, and the Wisconsin papers played up his hopes and helped his polling numbers. Then, after an encouraging message arrived at 5 A.M. on primary day, Carter called a brief news conference. At 7:13 A.M., as the polls opened, he called reporters into the Oval Office, an event the usually friendly *Washington Post* called "more an exercise in domestic politics than international diplomacy." Carter announced that the President of Iran, Abolhassan Bani-Sadr, had assured him that the hostages would be taken from the students and held by the government—a move regarded as a first step to their release.

Carter did not promise their prompt release, but he implied it. A reporter asked, "Do you know when they will be actually released and brought home?" Carter answered, "I presume that we will know more about that as the circumstances develop. We do not know the exact time scheduled at this point."

As it turned out, the exact time was more than nine months away; Bani-Sadr backed down. But, as Hamilton Jordan noted in his memoir, *Crisis,* "At least there was good news from Wisconsin: we defeated Kennedy handily." There is no way of knowing how many votes the wishful announcement on the network morning news shows brought Carter. A lead which Caddell had measured at 15 percentage points and falling before the weekend mushroomed to 26 on Election Day. Kennedy's aides pointed out that the victory came on April Fool's Day.

In retrospect there seems no reason to accuse Carter of more than taking political advantage of an exaggerated hope, blearily measured when the cables came into the White House Situation Room before dawn. But that press conference worked against Carter in the fall. The Ronald Reagan campaign, persuaded that Carter would manipulate anything to win, spread the idea that he was planning a last-minute release as an October surprise, and their drumming on the charge kept the hostage crisis in the public eye.

His other post–New York move hurt Carter long before autumn. Despite all their mathematical optimism about delegate counts, Carter's people were shaken by the New York defeat. Carter was now the issue, and, as Caddell told him in a blunt memo a couple of months later, "by and large the American people do not like Jimmy Carter." If the rest of the primary season focused on Carter, they feared, a string of losses might enable him to be beaten, if not by Kennedy (nor Jerry Brown, whose oddball challenge never went anywhere), then by some late entrant.

Facing a bad defeat in Pennsylvania, where some Carter aides argued for giving up, Caddell and Jerry Rafshoon unleashed a set of television commercials intended to turn the campaign directly on Kennedy's character. They shifted from the oblique messages of praise for Carter's steadiness and family to gritty man-on-the-street interviews knocking Kennedy. "I don't believe him," said one Pennsylvanian. "I don't think he could deal with a crisis," said another. "I don't like his policy on weakening our defenses," said a third. "I don't trust him," said a fourth. "You're taking a chance with Kennedy," said a fifth.

The word "Chappaquiddick" was never mentioned—on the air. It was edited out of many of the interviews. Rafshoon said the purpose was to "remind people that they don't like Kennedy." Caddell acknowledged the risk of a wound that would never heal because Kennedy and his people would "never forgive us." He compared it to "dropping a nuclear bomb on your position," a desperate step but necessary so as not to lose in Pennsylvania.

Kennedy made his attacks in person. Steadily gathering the support of more unions, he charged the President would "abandon the Democratic Party's commitment to working people." He kept insisting he could try harder and do better than Carter, whom he called "a pale carbon copy of Reagan," seeking reelection "on the basis that no one can do the job." On April 22, Kennedy won the Pennsylvania primary, but by just 4,840 votes out of 1,613,223 cast. The attack ads had done their job.

In May, Kennedy lost eleven of the twelve primaries he contested against Carter. He won only in the District of Columbia, but lost in states he had hoped to win, like Indiana, Maryland, Nebraska, and Oregon. Carter eventually abandoned his Rose Garden strategy. After a failed hostage rescue mission in Iran, he announced that the situation was now "manageable" and he could campaign again. In fact, Carter campaigned personally only once, in Columbus, Ohio, on May 29, where he made the obviously foolish claim that "we're turning the tide on the economy." The next day the leading economic indicators showed their sharpest drop in history, and a few days later unemployment figures spurted to the highest level in Carter's term. The boast proved false so quickly that it cost Carter 5 to 7 percentage points in California, New Jersey, and Ohio, Caddell said.

Those were the three big states left. Along with five smaller ones, they held primaries on June 3. In California, Kennedy used O'Connor/Bunker to urge voters to back him if only to keep the contest from ending. In Ohio, Carter brought back the attack ads he had used in April in Pennsylvania and in May in Indiana and wherever else they seemed necessary. Kennedy responded with a few of his own, with ordinary Americans saying of Carter, "I think the job is a little too big for him," and "I think the nation is embarrassed by the leadership we have." It is not clear how often the Kennedy ads were actually shown. National headquarters paid to have them made and then shipped them to state campaigns, which had to find the money to air them. In California, manager Ron Brown made up for a nearly empty treasury by flying around the state and holding airport news conferences at which he gave away tapes so the spots of the "new Kennedy ad campaign" would at least be shown on news programs.

Kennedy flew back and forth across the nation, filling an enclosed four-story shopping mall in Cleveland, or holding a Senate hearing in Helena, Montana. He blasted Carter from coast to coast for "the worst Presidential leadership in half a century" and "the worst recession since the great depression." He was concentrating more on unemployment than inflation now, charging Carter was happy to throw people out of work to hold down prices.

He constantly demanded that Carter debate him. On May 29, Kennedy even volunteered to release his delegates to vote for whomever they chose if Carter would debate him sometime before the August convention. Though Strauss brushed that suggestion off as potentially divisive, it was in fact a clear offer—opposed by Kennedy's more dogged aides like Stearns—to withdraw from the race. The Carter camp plainly missed the signal, perhaps because it was overshadowed by persistent Kennedy threats of a major rules fight at the convention to free Carter's delegates.

As the primary season ended, the Kennedy campaign was in key ways more comfortable with constituencies that had been difficult early. Kennedy promised to put a woman on the Supreme Court, insisted that the government help provide child care for working women, and argued that equal pay was required not just for equal work, but for jobs of comparable worth to society. That would mean raising pay in female-dominated professions like teaching and nursing to the level of male work like driving a truck.

Even more striking was something he did in May that no Presidential candidate had ever done. He held a fund-raising event with Los Angeles's gay and lesbian community. Earlier in the campaign he had promised, if elected, to issue an executive order barring bias against homosexuals in federal hiring, a move that got him support in the District of Columbia primary, his only May victory. And he had sent aides like Estrich or relatives like Robert F. Kennedy, Jr., to gay political meetings.

The $25-a-ticket event was held in a huge Spanish mansion in the Hollywood Hills, the home of Clyde Cairns and John Carlson, on May 24. Kennedy told the crowd he would issue the executive order on hiring and said, "I want to say very clearly that I stand for the rights of gays and the rights of lesbians as a Senator and I will do so as President of the United States." (He also said, "The members of the gay and lesbian community of Southern California know all about Jimmy Carter's refusal to debate.") At one point Carter's most prominent gay supporter, Sheldon Andelson, accused him of meeting in secret and keeping the press out. Kennedy waved toward the balcony, where some reporters watched while others mingled with the crowd. Kennedy took the same antidiscrimination message to a public rally in San Francisco the next week and said he supported legislation to protect gay rights in jobs—including schoolteachers—and in immigration policy.

As May ended, the Carter campaign was trying to figure out how to end the war with Kennedy and his supporters. Richard Moe, Mondale's chief of staff, was in charge. He had been meeting with Kirk, and he counseled patience, urging the Carter campaign to avoid public pressure or private jabs that would "force them to dig in their heels," as he put it in a May 29 memo to Jordan, which Carter read and praised. Moe urged that Carter call Kennedy on June 4, "congratulate him on a hard-fought campaign, and indicate he'd like the Senator to come in to see him when he has rested. I have reason to believe this kind of call, if accompanied by some gracious public comments, will be well received by Kennedy and will lay the groundwork for our ultimate goal, party unity." In the margin, Carter marked the proposal "OK." But when Moe warned that publicity would lead peace efforts to be discounted as self-serving, Carter wrote "?"

Carter started trying to conciliate Kennedy on Saturday, May 31, saying in a television interview, "My feeling toward Senator Kennedy is one of respect, one of personal admiration for mounting a very forceful and a very determined campaign." He said Kennedy was a "loyal Democrat" and promised Kennedy "concessions in every direction" on the platform. But Kennedy, campaigning in California, replied that what he wanted was a two-man debate on the economy. To Carter aides, it was a desperate Kennedy ploy; but to some of the reporters who had lived with him on the road for months, Oliphant and Susan Spencer of CBS News, it was a desperate plea for a dignified way out.

Kennedy won big on Tuesday night, carrying New Jersey and California soundly, while losing Ohio, and taking three of the five smaller states, for an overall, one-day delegate advantage of 372 to 321. He spoke of the "first night of the rest of the campaign" but barely mentioned winning the nomination. In defeat, Carter had still won enough delegates to pass the barrier of 1,666, a majority of the delegates who would vote in New York. Kennedy insisted the results meant the voters wanted a debate, and renewed his promise to release his delegates if he got one. "I will insist that the great issues be debated. And after that debate, I will release my delegates to vote at the convention in accordance with their conscience."

If Carter was looking for conciliation—and his diary for June 3 sounds sympathetic to Kennedy over the defeats his rival never expected—Kennedy was not. He ignored two of Carter's telephone calls, a completely uncharacteristic personal discourtesy to a President. But on June 4 he finally called back and they arranged to meet the next day at the White House. Beforehand, when Kennedy met with aides in McLean, Kirk's was a lonely voice arguing that the time had come to end the campaign. Kennedy, feeling that he and the party had been shortchanged, meant to press, again, for a debate.

The June 5 meeting went badly. Only the two of them were present for the fifty-minute session. There are no neutral notes. But neither's immediate recollections were amiable. Carter wrote in his diary that Kennedy "took about an hour to fumble around" and "seems to be obsessed with the idea of a personal debate."

Kennedy's recollections, dictated immediately to Bob Shrum, his speechwriter, are more detailed. According to them, Kennedy began by saying, "I'm realistic about arithmetic." While he told reporters later, "I'm planning to be the nominee," there is no indication in Shrum's notes that he told Carter that. Kennedy immediately demanded a debate, a subject to which he kept returning.

Each complained about the other's campaign. The Shrum notes say Carter said, "You had a lot of hard words to say about me. I know campaign rhetoric, but I've never attacked you personally." Kennedy broke in to say, "Those ads of yours didn't handle me with kid gloves." Kennedy reminded Carter of his earlier willingness to debate. Carter did not respond.

Carter, disregarding Moe's advice, asked Kennedy if he would pledge to support the party's nominee. When Kennedy avoided the question, Carter said, "I've been very hurt that you refused to support the nominee." Then he asked Kennedy whether, if there was a debate, Kennedy would commit himself "now" to supporting the nominee. Kennedy said that would still depend on Carter's approach to economic issues. (Carter's diary agrees on this point.) Carter said debate on the platform would suffice to settle party policy. Kennedy, remembering health insurance, told him, "I'm most interested in substance—not what's in the platform." Carter underlined Kennedy's point when he told reporters later that he would have to "reserve judgment" on how much of the platform he would support. They did agree to stay in touch, through Kirk and Moe, to avoid misunderstandings.

When Kennedy came out to meet reporters, he said a debate "appears unlikely." And when Carter talked to them a few minutes later, he poured cold water on the idea just as he had with Kennedy. But it is clear, both from Kennedy's notes and Carter's diary, that Carter never flatly refused to debate.

Later that day, Strauss—perhaps cued by Paul Kirk—called the President and urged him to debate Kennedy. Strauss's reasoning was that Carter would do well, and even if he did not, he would get the matter behind him. Carter, who plainly regarded Kennedy as a fat rich kid too accustomed to getting his way, said he didn't like the idea but would think about it.

The next day, Friday, June 6, Carter had lunch with his campaign team in the Cabinet Room. Rosalynn was there. So were Hamilton Jordan, Jody Powell, Jerry Rafshoon, Pat Caddell, and Vice President Mondale. Carter startled them by saying he had decided to debate Kennedy the next week before the platform committee. Carter's aides were delighted. They thought Carter could handle Kennedy, and some thought that Kennedy would keep his word and release his delegates, breaking the logjam. They also thought Carter was very rusty as a campaigner—as he had showed the previous week in Ohio—and needed the exposure in a tough situation. Only Mondale argued against it. He said debating Kennedy would be degrading to the Presidency, and may have feared elevating his obvious rival in 1984.

As the hour-long lunch ended, Carter said they would meet again Sunday to decide finally. Carter went off to Camp David for a quiet weekend and some trout fishing.

On Sunday night, Mondale made his argument again, even more strongly. Then Charles Kirbo, Carter's chief adviser from Georgia, telephoned the President. When Carter left the room, the remaining Georgians realized that none of them had bothered to get in touch with Kirbo. When Carter returned, they saw how bad a mistake that was. Carter said Kirbo thought the idea was terrible, and he would not debate.

With Kennedy's last chance for a preconvention exit strategy with dignity blocked, the Democrats entered a summer of bickering. There were rare interruptions. On July 1, Carter signed Kennedy's major legislative success of the year, a bill to deregulate the trucking industry. (His refugee bill was enacted, too, but he was off campaigning when it passed.) Carter invited Kennedy to the White House ceremony, and they appeared together in public for the first time all year. Carter praised Kennedy's persistence on the bill, and then invited him to speak. Kennedy said the law would cut prices and dampen "the fires of inflation."

But sniping was more the order of the summer. The Carter White House got the nation's mayors to revoke an invitation to Kennedy to speak, saying the President would not appear on the same day as Kennedy. The Senator's forces thwarted an effort by Carter to keep Mayor Bill Green of Philadelphia, a Kennedy supporter and the state's ranking Democratic officeholder, from heading the Pennsylvania delegation to the Democratic National Convention. Kennedy's people filed eighteen minority reports on the platform, giving them the power to tie up the convention for at least a day.

Kennedy posed for pictures with Representative John B. Anderson, the maverick Republican from Illinois who was running for President as an independent, and Anderson said he might reconsider his own candidacy. And as the Kennedy forces spread their demand for an "open convention," with delegates free not to vote for the candidate they had supported when they were elected, various other Democrats like Ed Muskie (who had become secretary of state when Vance quit in April after the failed hostage rescue mission), Senator Henry Jackson of Washington, and Governor Carey started dropping hints of their availability in the party's hour of need. Kennedy asked Senator Robert C. Byrd, his onetime foe, if Byrd would be his Vice Presidential nominee—a preconvention gambit Ronald Reagan used to gain attention in 1976 with Senator Richard S. Schweiker. Byrd declined, though he was impressed that Kennedy thought him worthy of serving as President if Kennedy died. But he did come out for an "open convention."

One important problem did develop for Carter in the summer. On July 14, as the Republicans opened their convention in Detroit and prepared to nominate Reagan, the Justice Department announced that Billy Carter, the President's brother, had registered as a foreign agent. Libya had lent him $220,000 to try to get its oil accepted in the U.S. market. For weeks, the White House dribbled out bits of information, followed by clarifications, in a way that made it seem as if it had something to hide.

That led to serious slippage in support by Carter delegates for the Carter position on the rule. The Kennedy camp was vigorously arguing for an "open convention" in the recent antiboss tradition of the party. When the Carter people made a "hard count" of delegates they were certain would vote with them on the rule, their majority had slipped perilously, to about one hundred. Carter delegates were telling his campaign of their doubts, but not telling Kennedy aides, who thought they had picked up perhaps half a dozen votes. Carter, just as he had before Iowa and New Hampshire, worked the phones. If a campaign aide detected a wavering delegate, he could get the President of the United States to telephone him, and return a card with notes of the conversation the next morning.

Carter effectively stopped the slide himself, not on the phone but with a commanding appearance on television. The week before the convention he told the nation that his brother's business deals had not affected American policy in any way, and that using him to seek the hostages' release might have been "bad judgment" but nothing more. The importance of that broadcast was demonstrated by a *New York Times*/CBS News Poll that was half-finished when Carter spoke. Among those interviewed before he spoke, Democrats divided evenly 43 percent to 43 between Kennedy and Carter. Among those interviewed after he was finished, the Carter margin was a robust 57 to 32 percent. The *Times* published the poll Sunday, August 10, as delegates arrived in New York City for the convention; if the arriving Democrats had read a headline proclaiming the President no better than even with Kennedy, Carter's support might have fragmented.

But they did not read any such news, and the convention's first piece of serious business, on Monday the eleventh, was to adopt the rules that had been proposed months before, binding delegates to vote for one ballot for whomever they had been elected to support. It passed comfortably, 1,936 to 1,390.

THE DEMOCRATIC PARTY then saw Kennedy at his best, and at his worst.

On Monday night he called Carter to congratulate him, and then announced that he was withdrawing from the race. On Tuesday he gave a great speech.

He began with a self-deprecating touch saved from his unused April withdrawal speech—"Well, things worked out a little differently than I thought, but let me tell you, I still love New York"—and moved quickly into classical oratory, when he told the delegates that he came "not to argue as a candidate but to affirm a cause . . . , the cause of the common man and the common woman."

He argued two of the year's political causes. First, dealing implicitly with the disputes over the Democratic platform's economic plank, he said, "Let us pledge that we will never misuse unemployment, high interest rates and human misery as false weapons against inflation. . . . We cannot let the great purposes of the Democratic Party become the bygone passages of history."

But he turned with more zest to the Republicans, saying, "The 1980 Republican convention was awash with crocodile tears for our economic distress, but it is by their long record and not their recent words that you shall know them. The same Republicans who are talking about the crisis of unemployment have nominated a man who once said—and I quote—'Unemployment insurance is a prepaid vacation plan for freeloaders.' "

Kennedy, at ease with the speech he had rehearsed, played speechwriter Shrum's contrasting cadences elegantly:

"The commitment I seek is not to outworn values but to old values that will never wear out. Programs may sometimes become obsolete, but the idea of fairness always endures. Circumstances may change, but the work of compassion must continue. It is surely correct that we cannot solve problems by throwing money at them; but it is also correct that we dare not throw our national problems onto a scrap heap of inattention and indifference. The poor may be out of political fashion, but they are not without human needs. The middle class may be angry, but they have not lost the dream that all Americans can advance together."

From that allusion to Martin Luther King's dream, the speech turned to draw on the rhetoric of Franklin Roosevelt, which Shrum constantly studied:

"We recognize that each generation of Americans has a rendezvous with a different destiny," Kennedy said. "The answers of one generation become the questions of the next generation. But there is a guiding star in the American firmament. It is as old as the revolutionary belief that all people are created equal—and as clear as the contemporary condition of Liberty City and the South Bronx. Again and again Democratic leaders have followed that star—and they have given new meaning to the old values of liberty and justice for all."

He recalled the campaign: "There were hard hours on our journey. Often we sailed against the wind, but always we kept our rudder true. There were so many of you who stayed the course and shared our hope. You gave your help, but even more, you gave your hearts. Because of you, this has been a happy campaign. . . . I have listened to young workers out of work, to students without the tuition for college, and to families without the chance to own a home. . . . Yet I have also sensed a yearning for new hope among the people in every state where I have been. I felt it in their handshakes, I saw it in their faces. I shall never forget the mothers who carried children to our rallies."

For a moment he paused to congratulate President Carter, and predict that a Democratic Party reunited on Democratic principles would win in November.

Then he rose to an emotional conclusion which offered nothing to Carter:

"Someday, long after this convention, long after the signs come down and the bands stop playing, may it be said of our campaign that we kept the faith. May it be said of our party in 1980 that we found our faith again.

"May it be said of us, both in dark passages and in bright days, in the words of Tennyson that my brothers quoted and loved—and that have special meaning for me now:

> " 'I am part of all that I have met. . . .
> Tho' much is taken, much abides. . . .

That which we are, we are—
One equal temper of heroic heart . . . strong in will
To strive, to seek, to find and not to yield.'

"For me, a few hours ago, this campaign came to an end. For all those whose cares have been our concern, the work goes on, the cause endures, the hope still lives, and the dream shall never die."

Madison Square Garden went wild, with a demonstration that lasted more than half an hour.

But Kennedy could not go out on a note of triumph. Two months of bitter trench warfare could not be erased in a few hours. That morning he had been angered to read in *The New York Times* that Hamilton Jordan had gracelessly sought his support with faint respect, saying, "It will be easier with him. We could do it without him, but it will be easier with him. He doesn't matter so much himself, but his people do."

When he gave the speech, an aide who feared TelePrompTer sabotage stood by with a typed copy for an emergency. On his own side, when Kirk negotiated a series of platform compromises highly favorable to Kennedy, he encountered accusations of surrender and demands to keep the convention going for weeks with time-consuming roll calls. Carter's formal nomination was delayed when Harold Ickes of Kennedy's staff invoked a rule barring balloting until six hours had elapsed since Presidential nominees revealed what parts of the platform they planned to ignore. When Kennedy demanded that the Massachusetts delegates move to make Carter's nomination unanimous on Wednesday night, they reluctantly complied, but did not wave their standard as other delegations did.

On Thursday night, after Carter's acceptance speech, Kennedy did come to Madison Square Garden. But he arrived late because of traffic—and because the Carter victory demonstration fizzled out ahead of schedule. When he came to the podium, he shook hands with Carter, spoke with Rosalynn, and milled about among dozens of Democrats, looking as if he did not know what to do.

He did. He had practiced, with Shrum playing Carter, the stereotyped convention unity picture of victor and vanquished, arms raised together. But on the podium, he never did it. He waved to the crowd, he shook hands with Carter and others, he chatted, but he never raised Carter's arm. Kennedy was there for two minutes and sixteen seconds, but on television it seemed to go on much longer, a tableau of Democratic division.

KENNEDY MADE A few appearances for Carter that fall, and cut an indifferent television spot endorsing him. Carter sought his help, asking former aides to give money to help retire Kennedy's campaign debt. Jordan did likewise, telling campaign staffers that the $1,000 he gave was like cutting off an arm, and he wanted to see the rest of them bleed the way he did.

Carter lost, badly, to Reagan. Many of his people felt it was Kennedy's fault. But Kennedy felt Carter would lose if he stayed away from Democrats' issues: "The support in the country wasn't going to be there." Carter and Jordan blamed Kennedy's long challenge for the trouble they had locking up Jewish and labor voters—and their consequent need to focus in the fall on their base, not the political center.

Kennedy or not, Carter would have needed a big lead to overcome the last week of the campaign, when he debated Reagan childishly and the challenger came across not as scary or reckless but as confident, relaxed, and Presidential.

That was followed by several days of awful economic news—unemployment, inflation, high interest rates—and ultimately by a final weekend of Iranian posturing over the hostages that only reminded the American public that Election Day, November 4, was the anniversary of their capture. In the last couple of days, the race stopped being close, as Carter and Anderson voters turned to Reagan, or decided not to vote, and Reagan won, with 52 percent and 489 electoral votes. Carter had 42 percent and only 49 electoral votes, winning only in the District of Columbia, Georgia, Hawaii, Maryland, Minnesota, and Rhode Island. Massachusetts, which had last supported a Republican in 1956, went for Reagan.

But to blame Carter's defeat—or Kennedy's, for that matter—on timing, or strategy, or the press, or bad news or bad luck, is to miss the point of 1980. If only Reagan had won, there might have been smaller reasons, or just his own masterful campaigning. But voters gave Republicans the Senate, too, dumping veteran Democrats like Gaylord Nelson, Birch Bayh, Frank Church, George McGovern, Warren Magnuson, and John Culver. In the House, Republicans gained thirty-three seats, ousting twenty-seven incumbents, including Al Ullman, chairman of the Ways and Means Committee, and John Brademas, the majority whip. Democrats retained control, but were frightened. Democrats also lost statehouses; Kennedy was the first to call to console the defeated freshman governor of Arkansas, Bill Clinton, telling him there would be another day.

The nation's impatience with the state of the country, the gas shortages, the recession, the spurting interest rates, and the humiliation of the hostage-taking may not have moved the United States decisively to the right, as some claimed. But they left it ready to give Ronald Reagan's conservatism a chance.

CHAPTER 24

In the Minority, Leading

The 1980 campaign changed Kennedy, but not the way it changed many Democrats. They sought to accommodate a conservative mood. He became prouder and more determined. He cheerfully used the word "liberal" to the Massachusetts chapter of Americans for Democratic Action: "Let us resolve that we will not run out on those who have always depended on our party as their best hope and their last help. Let us resolve that we will not run away from our commitments as Democrats, as progressives, and as liberals."

Two political changes came from the campaign itself, which he called later "an education like no other." Kennedy became a champion of women, and of gays and lesbians.

As Kennedy met women's groups around the country, his support for women's issues changed from abstraction into serious concern. He called it an "intense experience" to hear from women about their lives and see their opportunities and barriers. Nancy Korman, a Massachusetts friend since the mid-seventies, said the campaign changed him: "He's a person that really has epiphanies."

By the end of his campaign, he was an advocate not just of the Equal Rights Amendment but of the concept of comparable worth—an idea that would raise women's wages dramatically. And personally, after his campaign experience with Susan Estrich, he had women on his Senate office staff in more than secretarial roles.

If Kennedy's commitment on women's issues sometimes produced snickers, his involvement with the nascent political movement for homosexual rights certainly raised no eyebrows. "Look at it this way," he told aides. "It's the upside of my downside. No one will think I have a self-interest." Before the campaign ended, he was comfortable with gay leaders and their issues. At the convention, he offered to support their goal of a platform plank promising the President would issue the executive order against hiring discrimination. Fearing a high-visibility defeat, though, gay leaders declined.

Still, the most profound impact of his campaign—and then Reagan's victory—on Kennedy the politician was how it changed his priorities, his attitude toward economic issues and the federal government's duty to look out for the poor. "Campaigns are an educational experience," he said years later, in which he saw the "kind of human conditions that cried for attention."

November's results changed Kennedy's position in the Senate, too. Like all but three of his Democratic colleagues, he had never served in the minority. The Republican victory cost him not just the agenda-setting power of a chairmanship, but also the dozens of jobs a chairman controls. His hope of bringing many campaign aides onto his Senate staff was blocked, although he did make Shrum his press secretary.

But the musical chairs game following the change in party control gave him an opportunity to act on his new focus. His eighteen years in the Senate did leave him in a position to choose between being the ranking minority member (or Democratic leader) on the Judiciary Committee, or on Labor and Public Welfare, where until now he had been content to dominate health issues. The shifts began with the vacancy created on Appropriations with Magnuson's defeat. Harrison Williams of New Jersey then moved from Labor to Finance, opening up the leading position there for Kennedy.

Kennedy chose the Labor Committee, telling reporters on December 2 that it would be the "cutting edge" of Democratic concerns. Eleven years later he told John Aloysius Farrell of the *Boston Globe,* "I think that was really one of the more important decisions that I have made, career-wise. The reason was that I felt that it was going to be in areas of human need for the average family that were going to be under greatest assault in the Reagan Administration. And unfortunately, it worked out that way. And I thought it was most important to spend energy and effort in at least holding the forces of reaction and retreat to as little gain as possible."

There were two other reasons to choose Labor. It was important to the AFL-CIO, still a major player in Democratic Presidential politics. And it was the only committee left in the Senate which a Democrat could lead. Not formally, for the chairman would be Orrin Hatch of Utah, a very conservative Republican who considered Kennedy "one of the major dangers to the country." But in practice, Kennedy knew he could often have control, with seven Democratic votes, the regular support of his friend Republican Lowell Weicker of Connecticut, and sometimes the help of Republican Robert Stafford of Vermont. With Stafford, they would have a 9–7 majority over Hatch and the other Republicans. Without him, they could still force a deadlock.

But that was for 1981 and the Republican 97th Congress. There was still the last month of 1980 and the Democratic 96th Congress. Before the election, Democrats had hoped to get a lot done. Now, with Republicans about to set the agenda in Washington, there was no reason for them to allow last-minute passage of the Democratic bills.

Byrd and Kennedy decided to try one measure that enjoyed some Republican support—a fair housing bill that would provide enforcement mechanisms largely absent from the 1968 housing act passed in the wake of the murder of Martin Luther King, Jr. That law had proclaimed the right to rent or buy housing regardless of race. But it gave the federal government the right to go to court only when it alleged a widespread pattern or practice of discrimination. The new bill allowed federal suits whenever the Department of Housing and Urban Development thought them necessary.

On December 2, Kennedy and Byrd broke a filibuster against even debating the bill and entered a frantic week of negotiations. They added provisions like trial by jury to the House version of the bill, but critics like Hatch wanted more. In particular, he took off from a Supreme Court interpretation of the Voting Rights Act issued that spring to contend that the government should have to prove, affirmatively, an "intent" to discriminate. That would probably have barred the easiest sort of evidence in housing cases, sending black couples and white couples of similar background to buy or rent the same housing and finding the whites were accepted and the blacks rejected.

As a vote to end a filibuster against the bill itself neared on December 9, 1980, Kennedy called Hatch's demand "a drastic weakening of current law and a severe setback to the cause of civil rights. It would plunge the government and private plaintiffs with a morass of endless litigation over the motives of defendants. The damage caused by the adoption of an intent test would far outweigh the significant benefits achieved in the enforcement provisions of the bill. The result would be a clear retreat on civil rights that none of us should be prepared to make." Byrd agreed, telling the Senate, "If we really are serious about wanting to work out a bill that will deal with the problems that confront millions of Americans in the field of the securing of decent housing, whether it is by rental or by purchase, then we will vote to invoke cloture. . . . The time is now."

But Howard Baker, the incoming majority leader, who had played a central role on the 1968 bill, assured the Senate that he would see to it that a housing bill came up in 1981, meaning that it was not now or never. He joined with thirty other Republicans to block cloture. Although forty-five Democrats and nine Republicans voted to end debate, that was six votes short of the sixty needed.

Kennedy had one victory in the lame-duck session. At a late-August meeting with Carter, he made two personal conditions for campaign support. One was that the federal government rehire Eddie Martin, who had left his job as New England regional administrator for the Department of Housing and Urban Development to help in Kennedy's campaign—doing everything from stuffing ballot boxes at a Maine straw poll to working as traveling press secretary. Carter agreed. The second was that Steve Breyer, who had been serving as chief counsel of the Judiciary Committee, get the appeals court seat that Cox had been denied. Carter agreed, so long as the selection commission—whose membership had been changed since the Cox dispute—supported him.

Breyer had no problem with the commission, or with the Judiciary Committee, where he had worked closely with Republican Senators and their aides for more than a year. He breakfasted every morning with Emory Sneeden, the top aide to Strom Thurmond, the committee's ranking Republican. Together they had worked out a pattern of screening nominees for judgeships which pleased the committee, though it sometimes annoyed Senators pushing marginal nominees. The committee's 17–0 vote to approve Breyer's nomination was a uniquely bipartisan action in the contentious postelection session. Republicans were determined that none of Carter's other judicial nominees be confirmed, but they liked the way Kennedy and Breyer had handled the Judiciary Committee.

The support for Breyer outraged Senator Bob Morgan, a North Carolina Democrat who had had one nominee of his rejected by the committee on ethics grounds and another stalled at session's end. Morgan, himself defeated in November, attacked both the cozy procedure and Breyer's lack of trial experience. But Thurmond argued, "He has wisdom, knowledge, integrity and of such great importance, he possesses judicial temperament." A 68–24 cloture vote forced the nomination to a decision, and Breyer was then confirmed on December 9, 80–10.

THE 1980 CAMPAIGN led to one other decisive change in Kennedy's life.

On January 21, 1981, he and Joan announced that they were getting divorced. Joan's account is that doing the campaign and staying sober left her feeling better about herself than she had in years. "I guess I felt like I could go it alone," she said in 1998. Ted said, "We sort of decided to go our own ways." Because of Joan's drinking and Ted's philandering, it had not been a strong marriage since the mid-sixties. Joan had been perhaps more hopeful than Ted that things would get better. But Ted had not been eager to divorce, for political and religious reasons, and because of his mother.

The Kennedys issued a statement saying, "With regrets, yet with respect and consideration for each other, we have agreed to terminate our marriage. We have reached this decision together, with the understanding of our children and after pastoral counseling." The Boston archdiocese weighed in sympathetically: "Few American families today are free from the unfortunate experience of divorce. No one should make a rash judgment about a family tragedy which is surely marked by personal pain."

IN THE NEW Congress, Kennedy's expectation of being on the cutting edge was quickly met. On February 18, 1981, President Reagan proposed consolidating eighty-eight federal programs into seven block grants to states and localities. Reagan, making traditional Republican arguments, said this would "reduce wasteful administrative overhead" and "give local governments and States more flexibility and control." He said, "We know of course that the categorical grant programs burden local and State governments with a mass of Federal regulations and Federal paperwork." Reagan proposed cutting federal spending 15 percent.

Kennedy led the Democratic attack, saying Reagan's budget, which also called for substantial income tax cuts, "would take the most from average families and from Americans who have the least, while preserving special privilege programs, such as $4.6 billion in federal subsidies for the oil and gas companies." As to the argument of federal inefficiency, he argued that the states spent three to five times as much as the federal government did on administrative expenses.

The basic argument against block grants was that states would eliminate or greatly reduce programs with weaker political constituencies. As Kennedy said in defense of a program guaranteeing subsidized meals for children in institutions: "Some say the States should provide for these children, but let us look at the record. The States have had their chance. Until 1975 the States were responsible, and the overwhelming evidence, collected through days of hearings, is that these children were not well fed. Many were living in substandard conditions and were fed poor quality food. Nothing indicates that the States are prepared to do a better job today."

With the help of Weicker and Stafford in the Labor Committee and the firm stands of the House Education and Commerce Committees, still under Democratic control, the plan to put most federal aid on health and education into block grants was checked, though spending was cut. Many programs, from big ones like elementary and secondary education to smaller ones like health centers for migrant workers, were kept alive separately, and even the block grants that were enacted often required states to keep up spending levels on some components. One key example: if states chose to join a primary-care block-grant program, they were required to go on funding all community health centers, the fruits of Kennedy's first health care initiative back in 1965. The terms were so unattractive that only two states tried the grants. The law was eventually repealed in 1986.

While Kennedy blistered Reagan's domestic policies as cruel toward the poor, he got on quite well personally with the President. The relationship began on November 19, 1980. Reagan was in Washington and Kennedy got Senator Paul Laxalt of Nevada, a close friend of Reagan's, to arrange an appointment. They talked for half an hour, and Kennedy came out to say they had talked about arms control and budget cuts. Kennedy said he shared Reagan's desire to cut spending, a position he had not been known to take until the election two weeks earlier, or very often thereafter.

One of the rare issues that brought Kennedy and Reagan together politically was Ireland. Before the election, Irish ambassador Sean Donlon met

Reagan, who told him that despite two trips to Ireland, he had been unable to find evidence of his Irish ancestry. The Irish found it, and the genealogical records were presented when Reagan visited the Irish Embassy on St. Patrick's Day 1981. Kennedy and Tip O'Neill reminded Reagan that he could not be Irish "one day a year."

Because of Reagan's influence with the British prime minister, Margaret Thatcher, they worked on him over Northern Ireland. On June 1, for example, Kennedy came to see the President to warn that the deaths of IRA prisoners like Bobby Sands on hunger strikes—demanding special status as political prisoners—were only increasing American contributions to the terrorists, and to urge Reagan to ask Thatcher to seek a solution in improving prison conditions. Reagan's notes said Kennedy acknowledged that prisoners convicted of ordinary crimes could not claim political status and agreed "with me that the problem is one of extremists on both sides, who are minorities but the moderate majority on both sides is silenced by fear of retaliation."

Later that week Reagan endeared himself to the Kennedys with a Rose Garden ceremony at which he presented Ethel Kennedy with a Congressional Gold Medal honoring her murdered husband. More than one hundred Kennedys and Kennedy friends, from Caroline to Averell Harriman, were guests. Reagan said, "The facts of Robert Kennedy's public career stand alone. He roused the comfortable. He exposed the corrupt, remembered the forgotten, inspired his countrymen, and renewed and enriched the American conscience. Those of us who had our philosophical disagreements with him always appreciated his wit and his personal grace." Ted responded for the family, thanking Reagan for the example that "the common love of our country transcends all party identification and all partisan difference." He said his brother, who had debated Reagan in 1967, "said that Ronald Reagan was the toughest debater he ever faced." Noting that thirteen years before, to the hour, his brother had lain dying in Los Angeles, Kennedy said his family and friends "felt a special sense of relief this year, Mr. President, at your own recovery from the attack against you," the shooting of the President in March by John Hinckley.

That day Reagan also created one of the enduring myths that embittered Kennedys against Jimmy Carter. As he made the presentation, he said, "Mrs. Kennedy, this medal has been waiting patiently to be presented." Later, as guests sipped iced tea, he told a few of them that President Carter had kept the medal in his desk drawer. Carter, hearing the story for the first time in 1998, said, "I know I didn't withhold the gold medal." Several Carter aides insist that if they had known of it, even in the busiest days of the campaign, they would have had a ceremony. Not only had Carter, from a distance, admired Robert F. Kennedy, but a presentation would have pleased the Kennedys, something that by then they were trying desperately to do.

In fact the medal, authorized in 1978, was delivered to the White House in the fall of 1980, and while there seem to be no records about it at either the U.S. Mint or the Carter Library, the Carter White House was never known for efficiency. The story of Carter's vindictive nondelivery seems to be just one of those movielike images which Reagan imagined and then believed with utter certainty—like the Chicago welfare queen or how the Navy was integrated in World War II in gratitude to a Filipino mess steward who shot down Japanese planes at Pearl Harbor.

KENNEDY'S OTHER MAIN focus in the first part of 1981 was on making sure of his reelection to the Senate in 1982. He worried after what had happened to several Senate colleagues after losing Presidential races. George McGovern and Ed Muskie had tough runs for reelection after their Presidential candidacies; Frank Church and Birch Bayh had run for President in 1976 and lost Senate races in 1980. Jack Leslie from his 1980 campaign staff went to Boston to start organizing.

Massachusetts Republicans were not much to fear. But Leslie was concerned about the National Conservative Political Action Committee (NCPAC), a right-wing organization that had helped defeat Church, Bayh, McGovern, Culver, and other Democrats with negative ads much sharper in tone than any that candidates had yet used against their opponents. NCPAC announced that Kennedy was one of its targets. But the Kennedy organization scared Massachusetts stations out of running NCPAC's television spots, telling managers that while a candidate's own ads had to run uncensored, a station was responsible for any others it broadcast, and might find its license threatened. Leslie and James Roosevelt, the campaign's lawyer, warned every television and radio station in the state and some on the borders. Gilbert Lefkowich, manager of Springfield's WGGB, put it gently, saying they "expressed a concern about what the spot might try to do." WGGB did not carry the spots. Neither did anyone else.

Leslie also started raising money for the 1982 campaign, one of three simultaneous fund-raising efforts, along with a drive to pay off the 1980 Presidential race and start up a political action committee, the Fund for a

Democratic Majority. The PAC provided money for political travels for Kennedy and cash for other Democrats—and helped keep in touch with all the people who had supported Kennedy in 1980, in case he needed them again.

The political action committee was just one of the tools Kennedy used to put himself in a position to run for President again—if he decided to. He also traveled widely, speaking in an effort to define his party. In St. Paul with Mondale on April 12, he told Democrats, "The answer to the problems of the Democratic Party is not for us to pretend that somehow we are similar to Republicans." He said this was not a time to sit back and wait for better times, because "the children who will be born retarded because of cuts in protein supplements for their mothers do not have time to wait for a better political opportunity."

"As Democrats," he told the Communications Workers in Boston on July 6, "we must be something other than warmed over Republicans." Kennedy, one of only a few Senate Democrats to vote against the budget bill and Reagan's tax cuts, insisted the administration was practicing "scorched-earth economics" that ran from eliminating the preference for the blind to run newsstands in federal buildings, to virtual abolition of the National Science Foundation, to deep cuts in job training, unemployment compensation, housing, nutrition, student loans, elementary education, and cancer research. He attacked Reagan's tax cut proposals as part of "the tried and untrue philosophy which holds that if government takes care of the rich, the rich will take care of the poor."

He wondered about reclaiming his place in the Senate itself, another worry that seemed odd after just a few months. Sometimes he used gimmicks. Larry Horowitz became chief of staff in 1981 after Rick Burke, who had been in the job for two years, left because of a nervous breakdown. Frustrated that Kennedy could no longer call hearings that would summon Senators, witnesses, and television cameras—that was now the prerogative of Republican chairmen—Horowitz and Kennedy came up with the idea of holding imitation hearings, called "forums," to discuss issues from a Democratic perspective. "I'm a Senator. I can get a goddamn room," he told Horowitz. Republicans sometimes showed up because Kennedy could still attract cameras.

The real way to power in the Senate was legislating, making laws or preventing laws. As he told the *Globe* in February, "Your effectiveness in the Senate to a great extent is dependent upon the work in the committee and work on the floor and your willingness to stake out positions and to follow through on them. And if you're prepared to make that effort—in the committee and on the floor—you're a factor and a force to be dealt with."

Opportunities to prevent Reagan and the Republicans from making laws were rare. But they could be limited to paper victories. Chile was an example. During 1981 the Reagan administration lifted a 1979 Carter prohibition on U.S. government financing of exports to Chile, invited Chile's navy to resume participation in joint exercises, and stopped voting against international loans to Chile. Those sanctions had been imposed in frustration at Chile's lack of cooperation in prosecuting or extraditing the murderers of Orlando Letelier in 1976. But the junta's major goal was a resumption of military sales, especially spare parts for its aging planes and ships. The ban had been imposed before the slaying of Letelier. Senator Jesse Helms, the extremely conservative North Carolina Republican, led the way, and seemed to have a Senate majority behind him. He argued on October 22 that the 1976 Kennedy amendment should be lifted because it was not only "punitive," but was ineffective in improving human rights conditions in Chile.

When Kennedy raised the Letelier murder, Helms brushed it off, saying the exile had been a collaborator with Cuba: "Murder is murder, but let us keep it in perspective. . . . He who lives by the sword shall die by the sword."

Kennedy gave as good as he got in debate, insisting, "A change in U.S. policy will be a clear and unmistakable message to governments and political leaders throughout the world—that the United States no longer places major importance on protecting human rights."

But the issue was not settled in public, either in debate or when Kennedy's motion to kill the Helms amendment was defeated on the Senate floor, 57–30. The question had really been decided behind the scenes, as Kennedy persuaded Senator Charles H. Percy of Illinois, the chairman of the Foreign Relations Committee, to undercut the Helms move with a "perfecting amendment."

Percy refined the Helms proposal by allowing military aid only if President Reagan certified "that the government of Chile has made significant progress in complying with internationally recognized principles of human rights." Kennedy and Percy then entered a prearranged colloquy to define the human rights on which Chilean progress was required. They agreed it meant "important matters such as protection against internal and external exile, freedom of speech, freedom of press, and the right to organize and operate political parties and free labor unions." That set of conditions was unlikely to be discerned even by a sympathetic Reagan administration, and Percy promised immediate hearings if Reagan moved. Kennedy had worked the Senate to thwart the friends of Chile's junta.

But the central issues of foreign policy, especially of relations with the Soviet Union, were not places where Kennedy could have much influence on a President who utterly distrusted Soviet leaders and believed the United States had fallen behind in strategic arms. Reagan was at work trying to build up U.S. nuclear weapons and Kennedy was among many Americans who feared a renewed nuclear arms race would get out of control.

Their vehicle was a "nuclear freeze" movement demanding that both sides stop adding to their arsenals. The movement was real, out in the country, and Kennedy joined it. Working with Senator Mark O. Hatfield of Oregon, a Republican who had seen the devastation of Hiroshima as a soldier in World War II, Kennedy offered a Congressional resolution in March 1982, calling for a freeze of nuclear weapons at existing levels and then a reduction. They dismayed some of the grassroots supporters by offering only a nonbinding sense-of-Congress resolution, the most the Congressional sponsors felt they had a chance to pass.

The administration scoffed, saying the Soviet Union had superiority in some areas, such as Europe, and that Soviet reductions had to precede any freeze. Kennedy responded by saying there were already enough warheads to annihilate the human race. He said Reagan was arguing that "we have to build more nuclear bombs today, arms today, in order to reduce the number of such bombs tomorrow," a policy he ridiculed as "voodoo arms control."

Nuclear freeze rallies attracted huge crowds, 100,000 in New York City, and spread to Europe, where they worried the Reagan administration by stirring opposition to the deployment of new cruise missiles. But legislatively, the Reagan White House never found the freeze much more than an "annoyance," as James A. Baker III, the chief of staff, put it. The resolution was beaten narrowly in the House and never got to the Senate floor.

Kennedy argues that the freeze movement "had an impact in terms of the American people's understanding of what mutually assured destruction was really all about. . . . I think we were able to have some impact in people's thinking that they could not really be winners in terms of the arms race." He thinks it changed the atmosphere.

The freeze effort clearly made a difference in two other ways. Richard Wirthlin's polling showed widespread support for the idea and fear that Reagan's commitment to new weapons risked nuclear war. Many in Reagan's administration thought negotiating with the Soviet Union was pointless, because the Kremlin would lie or cheat. But once Kennedy introduced his resolution with Hatfield on March 10, the administration decided to develop an arms control policy, announcing that decision two days later.

It mattered to Kennedy in another way, too. He started learning the intricacies of nuclear arms policy. He also began sending Horowitz on trips to the Soviet Union, where he discussed arms control and the release of refuseniks. Brezhnev died in November 1982, and then the goal became arranging a visit with his successor, Yuri V. Andropov. But he, too, was ill, and Horowitz found himself an occasional consultant on kidney disease, passing on information from the National Institutes of Health about treatment. Before and after all these trips, Horowitz reported to Jack Matlock, the Soviet specialist on the staff of the National Security Council, who in turn told a narrow circle of administration officials about them.

KENNEDY'S MOST IMPORTANT accomplishment in the 97th Congress was the reenactment of the first major bill he had helped shape, the Voting Rights Act. The 1965 law changed the political face of the South in two dramatic ways: its blacks now voted (heavily Democratic), and its whites voted Republican. The partisan change menaced the act itself. Reagan, who had opposed it and every other federal civil rights law, had won 61 percent of the votes of white Southerners. It was his best region. His attorney generals, especially William French Smith, and Bradford Reynolds, assistant attorney general for civil rights, showed no outrage at discrimination at the ballot box.

One critical section of the law was to expire on August 6, 1982. It required states with histories of discrimination at the polls to clear any changes in voting laws—from registration to voting hours to boundaries of legislative districts—with the Justice Department in Washington. Senator Strom Thurmond, the Dixiecrat candidate for President in 1948 and now chairman of the Judiciary Committee, insisted the preclearance provisions were unfair to the South and would have to go.

Working with the Leadership Conference on Civil Rights and its new director, Ralph Neas, a white Republican from Massachusetts who had succeeded the veteran Clarence Mitchell, Kennedy joined with others in both houses to introduce an extension of the act on April 7, 1981. He said, "It was sixteen years ago that Lyndon Johnson came before us and challenged us to join with him in the historic pledge that 'We shall overcome.' This is not the time to retreat or surrender." He found that signing up cosponsors was difficult. With considerable effort, he and Senator Charles McC. Mathias, a Maryland Republican, got about thirty instead of the fifty Kennedy had expected. "It just came to me," Kennedy said years later, "with excruciating clarity, that this was going to be a monumental kind of a task."

The bill's provisions reflected a bold decision. The Leadership Conference chose to seek not just an extension of the preclearance rules but a reversal of the Supreme Court's "intent" ruling, which made discrimination extremely difficult to prove. Civil rights lobbyists had seen Orrin Hatch try to apply it to housing law in 1980 and feared he and others would try to impose the stiff test to apply it to civil rights laws across the board. Since the Voting Rights Act was the most popular of all civil rights laws, they decided this was the place to fight.

For most of 1981, the action was in the House, where pro-civil-rights Democrats were still in power. Working with Representative Don Edwards of California, chairman of a Judiciary subcommittee, the civil rights groups arranged exhaustive hearings, lasting eighteen days and hearing 120 witnesses testify to continuing discrimination—brief or inconvenient registration hours, fifty-mile trips to registrar's offices, threats to registration drives, public rather than secret voting, and outright intimidation of voters. That got the attention of House Republicans, especially Henry J. Hyde of Illinois. Hyde eventually backed away from the bill, but when the House passed the bill, 389–24, on October 5, the measure had the air of inevitability its backers thought would win over Reagan.

It did, briefly. Reagan decided the politically smart thing to do was to say he would sign the House bill. Reagan, who thought of himself as a foe of discrimination, would bow to the inevitable and signal to minorities, as one White House aide told the *Los Angeles Times*, "that he cares about them." But after the paper reported the leaked decision, Attorney General Smith, a social friend of the Reagans, saw the story and talked the President out of it. Reagan issued a murky statement, saying he was committed to the right to vote but opposed an "untested 'effects' standard" and wanted the bailout provisions eased.

It was time for the Senate. Hatch, chairman of the Judiciary Subcommittee on the Constitution, announced that he would fight to require that intent to discriminate be proven before bias could be determined.

Kennedy and Mathias worked together to reassert the inevitability of renewal. Helped by the Leadership Conference, in ten days they lined up fifty-nine other Senators to join with them in introducing the House-passed bill in the Senate on December 16, 1981. It differed only slightly from the bill they had introduced in April. They were quite ready to accept the minor easing of the provisions for states to escape preclearance, and the bill's assertion, in reinstating the results test, that failure of minorities to win local offices would not in itself prove discrimination. The breadth of support they had lined up was striking—including eight Republican committee chairmen and Democrats from Arkansas, Florida, Louisiana, South Carolina, and Texas. So was the depth. As Mathias said on December 16, the sixty-one Senators were one more than the "magic number" needed to break a filibuster. Neas said they had regained momentum, with a bombshell announcement recalling that Thurmond was dumbfounded when a reporter asked him about their bill. "What are you talking about?" he asked.

Then the administration blundered, discrediting its arguments on almost any civil rights issue. On January 8, at Smith's urging, the administration decided to grant tax-exempt status to private schools and colleges that practiced racial discrimination. In the firestorm of criticism, the only voice of praise was Thurmond's, who said the decision "puts an end to a decade of trampling on religious and private rights by the Internal Revenue Service." Ultimately, the White House shifted back to the policy that dated from Nixon. But the administration had labeled itself racist. Many moderate Republican Senators wanted nothing to do with its policies on civil rights.

That mistake was fresh when Hatch opened hearings on January 27, 1982, and warned that a results test would "establish the concept of 'proportional representation by race' as the standard by which courts evaluate electoral and voting decisions." Attorney General Smith called for a simple ten-year renewal of the existing law, a renewal which would leave the Supreme Court's intent standard intact. Kennedy clashed with Smith, telling him, "You appear before this subcommittee when there really is a very significant crisis of confidence in this administration in its commitment to the millions of people of this country—the majority who are women and the minority whose skins are not white." He said the administration "has attempted to give encouragement to those committed to the concept of segregated schools," and complained of the firing of the head of the Civil Rights Commission and lackluster enforcement of the Voting Rights Act. Smith objected that Kennedy was being political and said, "The President does not have a discriminatory bone in his body." When the room broke out in laughter, Hatch angrily threatened to clear it.

As Edwards had done in the House, Hatch built a hearing record to support his case. He heard many witnesses who argued that whatever the bill said about other factors, courts operating under an effects test would find it simplest to require proportional representation. If a city was 40 percent black, they argued, courts would uphold its city council districts only if they produced a city council 40 percent black. That argument worried uncommitted Senators, and even some of the cosponsors.

After Hatch reported the bill he liked out of his subcommittee on March 24, Kennedy and Mathias had to worry about the full Judiciary Committee. It had eighteen members, and they had only nine sure votes. Seven Senators seemed clearly opposed, with Bob Dole of Kansas and Howell Heflin of Alabama on the fence. Getting Heflin's vote did not seem impossible, but Dole's seemed a better bet to Kennedy and the Leadership Conference. The Leadership Conference, and Burt Wides of Kennedy's staff, started generating pressure in Kansas, with letters to the editor and questions when Dole or his wife, Elizabeth, appeared in the state.

Editorials put more pressure on Dole. For example, the *Wichita Eagle-Beacon* wrote on April 25 that it would be "wrong at a time of serious social, political and economic flux to give any segment of the nation the mistaken impression that the basic freedoms no longer apply to all Americans. It's to be hoped that the Senate, and in particular Sen. Dole, will underscore that with their support for the House version of the Voting Rights extension."

Dole was looking for a way to support the bill. Sheila Bair of his staff was working with Wides and Mike Klipper of Mathias's staff. In substance, Dole's price was not very high. Politically, though, he wanted the public credit for saving the bill, and he pretty much got it.

On substance, the House bill had said, "The fact that members of a minority group have not been elected in numbers equal to the group's proportion of the population, shall not, in and of itself, constitute a violation." The ultimate bill hardly differed. The so-called Dole compromise said that the "totality of circumstances" had to be weighed, and the "extent to which members of a protected class have been elected to office in the State or political subdivision is one 'circumstance' which may be considered, provided that nothing in this provision establishes a right to have members of a protected class elected in numbers equal to their proportion in the population."

At a May 3 news conference, Dole called it a "compromise," a term readily accepted by the press. Kennedy avoided the word, but said the "Dole-Mathias-Kennedy agreement" should resolve "concerns raised in recent months by unwarranted scare tactics." Joe Rauh, the old-time Democratic liberal and a key figure in the civil rights coalition, scoffed to Mary McGrory of the *Washington Post,* "It was no compromise at all. We got everything we wanted."

The biggest things they got—and the reason Kennedy and Mathias swallowed their pride—were Dole's seal of approval, which brought several wavering Republicans along to assure a veto-proof majority, and Dole's use of that fact to win, finally, Reagan's backing. Dole had gone to the White House before the press conference to make the situation clear to the President, who then issued a statement promising his "heartfelt support." The next day the Judiciary Committee approved the bill 17–1. Even Thurmond and Hatch reluctantly voted for it, though Jesse Helms's North Carolina sidekick, John East, called it a "slap in the face" to the South, voted no, and threatened to filibuster.

Helms renewed the threat of talking "until the cows come home" when the Senate took up the bill June 9. Kennedy replied, "Once again, we shall overcome. We may be briefly delayed by diehard efforts to cripple the act. But we shall not be deterred by threats of filibuster." Hatch scoffed at the Dole compromise as meaningless and attacked the results test, saying that, because of it, "racial gerrymandering and block voting will become normal circumstances." Kennedy said Hatch's intent test would be inadequate:

"It will often be too difficult to prove discrimination because defendants will be able to offer manufactured evidence of an alternative non-racial explanation for the challenged procedure. But most important, the intent test asks the wrong question. If a minority citizen is denied equal opportunity to participate in the political process, then that inequity should be corrected, regardless of what may or may not have been in someone's mind one hundred years ago."

Little more than a week later the Senate passed the bill, 85–8. Mathias said, "We said seventeen years ago that the Voting Rights Act would transform this country. Today we have evidence of that." Kennedy hailed the bill's backers outside the Senate, like the Kansans who leaned on Dole. He said, "To all who worked so hard in Congress and around the country to help us reach this destination, I say this victory is yours. Your efforts made it possible. Without you, the Senate could not have overcome."

Reagan signed the bill on June 29. Kennedy told reporters, "It took some months to get the Administration aboard." Just the day before, he told the NAACP convention in Boston that this was "the most anti-civil rights administration in the history of this land."

On voting rights, Kennedy had worked to enlist Republicans. But sometimes Republicans needed him. Senator Dan Quayle, elected in 1980 over Birch Bayh in Indiana, sought out Kennedy in 1981 to help get a job training bill through first a Labor subcommittee and then the full committee. Quayle wanted to show he was a serious Senator. Kennedy, who knew that getting a bill signed by Reagan required compromise, gave in to Quayle by giving business and governors much more control of the program than they enjoyed in the expiring Comprehensive Employment and Training Act, and by abandoning CETA's "public service" jobs with local governments, a favorite of mayors and organized labor. Quayle, who knew he needed Democratic supporters to get anything through the committee, backed a Democratic summer jobs program and insisted on continuing the previous year's spending levels of $3.9 billion, when the Reagan budget asked for only $1.5 billion for job training.

Kennedy accepted defeats in the full committee, including a flat prohibition on wages for job trainees, acknowledging tacitly what Quayle had to do to pacify the administration. Under pressure from liberals, he took no significant role in a House–Senate conference, where Quayle largely prevailed. As the recession deepened, the White House and other Republicans saw more and more virtue in the bill. Reagan signed it on prime time on October 12, without any of the Congressional authors present. Kennedy, who regularly praised Quayle as the bill moved forward, attacked Reagan for claiming credit, saying, "The Administration endorsed it only after facing the prospect of ten percent unemployment and the continued deterioration of the economy." He added, "The record will show that this important job legislation was delayed for months in both the Senate and the House because of White House intransigence and repeated veto threats." The program proved more effective in helping adults than youth, and at all levels was probably more successful than the highly politicized CETA.

KENNEDY HIMSELF WAS a major reason for the administration's hostility to the Job Training Partnership Act, because, like most politicians and reporters, the White House believed that Kennedy would run for President in

1984. When Horowitz took over as administrative assistant in 1981, Kennedy told him that if things went well, then he would be interested in running again. Horowitz recalled that he "therefore instructed me to do everything until the time he decided not to go as if he was going. Everything."

The first thing they did was to try to get a grip on the rules for 1984. The Democrats, as they had done after the 1968, 1972, and 1976 conventions, put together a commission to write new rules for the next one. Governor Jim Hunt of North Carolina was its chairman, and from the start he made it clear he wanted Democratic members of Congress to be automatic delegates. Mark Siegel, who had led draft-Kennedy efforts in 1979, took charge of the Kennedy forces, who were in general agreement with labor and Mondale's supporters about the need to avoid unknown long shots. The rules were adjusted in various ways that favored well-known, well-financed candidates over the George McGoverns or Jimmy Carters who stirred few warm memories among most party leaders. The Kennedy people lost only one minor skirmish—a rule that allowed the Iowa caucuses, where Mondale thought he would win, to precede the New Hampshire primary by eight days in which Mondale might build some momentum. Otherwise the two camps usually agreed, and the Kennedy forces won a symbolic victory when the rule binding delegates on the first ballot was repealed.

Kennedy dutifully went on the road with other potential candidates to raise money for the Democratic National Committee. At the Waldorf-Astoria on February 1, 1982, he quipped that the joint appearance with Mondale, Senators Gary Hart of Colorado and John Glenn of Ohio, and Governor John Y. Brown, Jr., of Kentucky was the "first time I've ever heard of a show trying out in New York before opening in Iowa and New Hampshire."

Horowitz began recruiting. Bill Carrick, a seasoned campaigner from the 1980 race, was hired on the Senate staff. Ranny Cooper, director of the Women's Campaign Fund, promised to join after Election Day. And while they and others were clearly told that Kennedy had not decided whether to run or not, their appearances were taken by politicians and reporters as evidence that he would.

As they had in 1974 and 1978, the Democrats held a party conference in 1982—a reformist legacy from the hopes of some liberals of developing coherent party policy. But, like its predecessors, this one was a political cattle show where prospective Presidents preened for fellow Democrats.

The Kennedy operation featured a vast lobster-and-clam bash for delegates and reporters. City officials also threw a reception for him. Kennedy met with political reporters from around the country for ten-minute interviews. His best line came when he told Carol Horner of the *Philadelphia Inquirer* that one of his central mistakes in 1980 was "worrying about whether to run or not to run instead of what to do after I ran. That's an important mistake and I won't do it again."

But his big moment came at the conference's closing session. The other potential Presidents had spoken Friday, and Mondale had been the most successful, interrupted twenty-seven times for applause and seven times for cheers. He blamed Reagan for the recession and for opposition to arms control and seized an issue discovered when examinations of the 1980 vote showed Reagan much weaker among women than men, saying that "in Reagan's America, there is apparently no room for women."

Mondale had won the first speaking slot on a coin toss, and Kennedy's people then demanded, and got, a separate time to speak on Sunday, June 27. But Charles T. Manatt, the Democratic national chairman, had only hired a band for Friday (Saturday was spent in issue workshops), and all he had for Kennedy Sunday was an accordionist. No matter. Jim King, helping out on this event, wangled dozens of extra floor passes for Mummers band players.

The overriding message in Kennedy's speech was that Democrats, if they stuck to their convictions, had a future. "If we do not stand up for the hungry," he asked, "if we do not speak up for those who work with their hands, if we do not fight on for the desperate millions of our inner cities, then who will?"

Besides attacking Reagan over the stumbling economy ("Ronald Reagan must love poor people, because he is creating so many more of them"), he made three key points.

He began by promising to enact the Equal Rights Amendment, which would die, three states short of the thirty-eight needed to ratify it, on June 30. "We do not worry at the fading of three more days," he said. "We are ready to spend three more years or three more decades or three more generations. As we said with civil rights, so we say with equal rights: We shall overcome one day."

He won a standing ovation (among fifty-seven interruptions for applause and thirty-one for cheers) when he attacked the Reagan administration for abandoning Carter's commitment to human rights. "I had my differences with the past administration," he said. "But on the vital issue of human rights, Ronald Reagan is wrong—and Jimmy Carter was right."

And he demanded a nuclear freeze, saying, "A freeze on the mounting total of nuclear megadeath can bring us back from the brink of humanity's third and last world war. . . . There is no such thing as a limited nuclear war. . . . Let us resolve that this atomic age shall not be succeeded by a second stone age."

"Only a few months ago," he closed, "Democrats were scorned and told that our day was done. But now we know and all America knows that for us as Democrats, and for those who have always looked to us for help and hope, the dawn is near, our hearts are bright, our cause is right, and our day is coming again."

He spoke with the aid of a TelePrompTer. He delivered the speech eloquently. He did not shout, and even spoke softly at times. Manatt tried and failed to halt him when he kept going past the promised thirty minutes. But it was Kennedy's day.

The recruiting continued in the summer and fall. John Law, a skilled Iowa caucus operative, joined up. Bob Farmer, a top fund-raiser, agreed to work for Kennedy. And Pat Caddell, Carter's pollster but also Kennedy's in his 1976 campaign, was hired for 1982, ready to work in 1984.

Caddell was needed in 1982 because even that year's reelection campaign was thought of as part of preparing for 1984. It was not because there was anything to fear from Ray Shamie, a millionaire businessman whom the Republicans nominated when no one else would take on Kennedy. Kathleen Kennedy Townsend, serving a political apprenticeship as campaign manager as her brother Joe had six years earlier, assembled a field organization and did a lot of speaking, finding a lot of support for Kennedy over the nuclear freeze issue, which was on the Massachusetts ballot as a referendum.

A reelection campaign by a Presidential contender is often devoted to rolling up a huge margin, to alert out-of-state politicians to the hopeful's vast popularity. That had been the point of John F. Kennedy's Senate race in 1958, the one his brother Ted was nominally managing. But the point of Ted's 1982 campaign was to see what out-of-state voters—especially those in New Hampshire, Iowa, and Illinois—thought or could be encouraged to think about him.

The shorthand for Chappaquiddick in 1980 had been "the character issue." So the 1982 campaign made a determined and imaginative effort to see if Kennedy's character could be redefined in the minds of voters. Michael Kaye, a Los Angeles media consultant with relatively little political experience, was hired to make commercials that never confronted Chappaquiddick directly but sought to put Kennedy's character in a different, broader perspective.

Four of the advertisements were unusual for a campaign, running five minutes each.

In one, after pictures of Teddy leaning on his father, Kathleen said, "My wedding was the same day that young Teddy had his leg amputated." When she heard of the cancer, "I called up and said, 'Well, we could cancel the wedding.' He said, 'No, don't do that at all. We'll work it out.' "

Luella Hennessey Donovan, the family's nurse, told how he slept in a chair by Teddy's bedside every night during his hospitalization: "His father wanted him to know that he was with him, all the way and one hundred percent, twenty-four hours a day."

The Reverend Bruce Young, a priest from Woburn, told of Kennedy's tears for another boy, when he visited a family whose eleven-year-old son was at home dying of cancer caused by toxic waste. "They talked with him about the tragedy that had inflicted his family, and it was like two kindred souls trying to help each other. And he did not want to leave. And he had every reason to leave. He had three hundred voters back at a hall that he was keeping waiting. He didn't go back for the votes. He stayed with the kid."

Frank Manning, an eighty-three-year-old lobbyist for senior citizens, offered a summary: "I think he's very special in his dedication as a man who could, let us say, sail the blue waters of the Mediterranean and forget all of the problems of society and very often he would have had much justification for doing so. Nevertheless, he's never taken his hand away from the plow and he's stayed with it. No matter how great the momentary tragedy he managed to pull himself together."

Manning also provided the only connection to what politicians usually mean when they referred to Kennedy's "character issue." He said, "He's not a plaster saint. He's not without faults. But we wouldn't want a plaster saint. . . . We want an average human being who has feelings and likes people and who is interested in their welfare."

Caddell devised an elaborate survey to see whether the five-minute ads moved voters in southern New Hampshire, where people watched Boston television. The ads were also shown to focus groups in Iowa and Illinois.

In all, this effort cost close to $1 million. What it seemed to show was that while the ads helped Kennedy's image a bit, they were no magic bullet, certainly nothing that would keep other candidates from using Chappaquiddick as Carter had.

But in Massachusetts itself, it hardly mattered. Kennedy won with 1,247,084, or 60.8 percent of the total vote. And Michael S. Dukakis, whom he had endorsed in the Democratic primary over a much more conservative Democrat, Governor Edward King (a Carter backer in 1980), won almost as easily.

Kennedy's Senate campaign had begun formally with lines that teased the choice he faced. He joked in his May 22 acceptance speech in Springfield that "I finally get to come before a great convention and accept its nomination." But then he went on to say: "My pledge to you and my fellow citizens is that same one I have given before and always tried to keep—to be an advocate for the average man and woman, a voice for the voiceless, a Senator for those who suffer and are weak." At the end, Caddell found him not very interested in his findings about Michael Kaye's spots. That, and the teaser he had Shrum strike from his victory statement, were hints that he would not decide to consume the next two years campaigning for President. But those hints were private. Most politicians and reporters—though not Walter Mondale or Bob Healy—were sure he would run.

The decision was sealed the day after Thanksgiving, but it had been foreshadowed in the Senate. With the Voting Rights Act renewal, and especially the tactical deference to Dole, it was clear that for him there was a satisfying career without the White House, that the Senate could be a sufficient end in itself.

Part Three

CHAPTER 25

Little Mud and Central America

Ted Kennedy's recommitment to the Senate began the next day. Other senior Democrats assigned him to the Armed Services Committee. Under party rules, lesser figures would have had to give up another committee assignment to take that vacancy. When Howard Metzenbaum, a cantankerous liberal from Ohio, asked why Kennedy should get special dispensation, Tom Eagleton of Missouri explained, "Oh, Kennedy is Kennedy." So the Democrats put Kennedy in position to deal with arms control and military budget issues—at the table in the Senate, not just in public speeches.

Two weeks later, he assumed for a brief moment the sort of Senate leadership role that had seemed forever lost when Byrd won the whip's job from him in 1971. On December 16, 1982, the Senate was wrangling over pay. A deal had been struck with the House. Representatives would get a 15 percent pay increase, from $60,662.50 to $69,800. Senators, afraid to vote themselves a raise openly, would merely remove a recently imposed ceiling on their outside speaking fees. The measure also provided 15 percent increases for judges and 32,000 senior government officials, including Congressional aides.

Some Senators balked at voting for any pay raises at all (although not at taking the speaking fees). James Exon of Nebraska led the assault. Senator Ted Stevens of Alaska, the Republican whip, who was frequently short-tempered but now also suffered from a cold and laryngitis, pleaded for a "return to some sanity." But he wasn't getting anywhere.

Kennedy strode across the floor to the Republican side of the Senate to get Stevens's eye and asked him to yield the floor. Standing at the majority leader's desk—Howard Baker was not present—Kennedy commanded the attention of the Senate:

"What is very much at risk here is the respect of the American people for this institution. A few days ago, the Senate voted to remove any ceiling on outside earnings from honorariums and other activities. That represents a substantial and undeserved back-door pay increase—far more than the House of Representatives voted for itself in the form of a direct pay increase.

"It is sheer hypocrisy for this body now, after we have only just removed the limit on our own earnings from outside speeches, to refuse to respond to the members of the House of Representatives and other Federal employees, who do not have access to all those handsome honorariums. When the country understands the hypocrisy and the inconsistency of the Senate's position, we are going to be the laughing stock of the Nation."

The Senate quieted down. Baker and Byrd arrived and worked out a parliamentary device to kill Exon's plan, and the pay raises went through.

SHORTLY AFTER THE new Congress convened, Kennedy took his seat on the Armed Services Committee and joined a suddenly bipartisan challenge to President Reagan's defense budget. While proposing cuts in most domestic programs, Reagan had asked for a 14 percent increase in defense spending, from $208.9 billion to $238.6 billion. After two senior members, Republican John W. Warner of Virginia and Democrat Sam Nunn of Georgia, had fenced with Caspar W. Weinberger, the secretary of defense, at a February 1, 1983, hearing, Kennedy's turn came.

He demanded to know if Weinberger felt the United States had sufficient military power to defend itself. Weinberger, arguing that two years had not been long enough to rebuild defenses he said had been neglected in the seventies, would not take the bait. "We have very great strength," he said. "But if you take . . . a wholesale nuclear surprise attack on the United States by the Soviet Union, I don't believe we would have the sufficient degree of retaliatory power. . . . We do not have enough, in my opinion, to deter an attack against the United States or its allies or other threats, if that should come in a particular form at a particular time using particular weapons."

Kennedy tried to get him to say how much more would be needed before a suitable level of deterrence was reached. Weinberger said he did not know. Kennedy attacked: "If you don't know it, how are the world and the members of Congress and the American people going to know it? How do you know when you have enough? If you don't know if you have enough today, how are you going to know when you're going to have enough next year or the year after?"

Domestic issues were a more frequent battleground. In January, unemployment was at 10.4 percent, and Kennedy argued that Reagan's cuts were heartless: "We see factories closed and farms foreclosed. We see homes lost and children left without a chance to learn. In America we now see the shameful inequality of the hungry lining up at soup kitchens, and the homeless sleeping outside in the cold. And every time one of those people dies on a sidewalk or a grate, a little bit of America dies, too."

Kennedy's role as the Senate's leading voice on civil rights was displayed again in 1983. After fifteen years of efforts by Representative John Conyers of Michigan, in August 2 the House overwhelmingly passed a bill to declare Martin Luther King's birthday a federal holiday. Kennedy had testified at a House hearing on June 7 that the honor was especially relevant as a way to reassure American blacks that the Reagan administration's indifference was not shared by Congress. When one Republican said it might be more appropriate, and cheaper, just to designate some day in King's memory, Kennedy answered that Congress should "not relegate the contributions of Martin Luther King to Joggers Day. We have got a Joggers Day. As a member of the Judiciary Committee, we must have two hundred different pieces of legislation on what is Joggers Day, what is Walking Day, Non-Smoking Day, all of those kinds of little days that we designate, and they pass the Congress. Everybody puts out their press release on it."

Many Senate Republicans saw the holiday as a way to show they had not given up on black voters. Kennedy and Mathias, who introduced the legislation in the Senate, worked with Baker to get it passed quickly. Senator Jesse Helms of North Carolina blocked immediate action before the Senate went on vacation in August and forced a delay when Baker tried again on October 3. Helms called Dr. King a Marxist, with Communist advisers, though conceding that "there is no record that Dr. King ever joined the Communist Party." He pointed out that both President Kennedy and Attorney General Kennedy had urged King to exclude one central adviser, Stanley Levison, because of his Communist ties.

Kennedy was angry enough to attack Helms personally. He said the charges of Communism "were raised first and most vigorously by the arch-segregationists bent on retaining the rule of racism. It is their heirs in the last-ditch stand against equal justice who seek to divert us today." He said, "Martin Luther King dedicated his life and gave his life to complete the unfinished business of the American Revolution and the Civil War. . . . He is one of the true giants of American history and he richly deserves the extraordinary honor we confer on him today."

Helms and Kennedy were at each other again when the Senate returned to the issue on October 18. Helms said that when Kennedy objected to his charges about King's Communist ties, "his argument is not with the Senator from North Carolina. His argument is with his dead brother who was President and with his dead brother who was Attorney General," who had authorized FBI wiretaps of King.

His voice choking, Kennedy answered, "I am appalled at the attempt of some to misappropriate the name of my brother Robert, and misuse it as part of this smear campaign. Those who never cared for him in life now invoke his name when he can no longer speak for himself. . . . If Robert Kennedy were alive today, he would be the first person to say it was wrong to wiretap Martin Luther King. If Robert Kennedy were alive today, he would be among the first to stand and speak for this holiday in honor of Martin Luther King—whom he regarded as the greatest prophet of our time and one of the greatest Americans of all time." Later, Kennedy acknowledged that his brother had authorized wiretaps on Dr. King and warned him of his associates in the civil rights movement, but contended it was "for the good of that effort."

Some of the bitterness had drained out of the debate by October 19. There were arguments about the cost of a holiday, or whether other Americans deserved the tribute more. The burden of hitting back at Helms and John East, the other North Carolina Republican, was taken up by Bill Bradley of New Jersey, who said, "They speak for a past that the vast majority of Americans have overcome . . . , playing up to Old Jim Crow."

The bill passed 78–22, and President Reagan, who had shown no previous interest, told a news conference that he saw the symbolism of the vote and would sign the bill. But then, when asked if he agreed with Helms about King and Communists, he quipped, "We'll know in about thirty-five years, won't we?" Reagan was referring to the year 2027, when the FBI wiretaps would be made public. Kennedy called the statement "unworthy" and said that then, when "we will in fact have a clearer idea of the respective contributions of both President Reagan and Dr. King, I am confident that the more we know, the more we will respect Dr. King."

Nearly three weeks before, on the day Helms blocked consideration of the King holiday, Kennedy spent the evening in Lynchburg, Virginia, as the guest of a very different sort of Protestant minister, the Reverend Jerry Falwell of the Thomas Road Baptist Church. Evangelical Christians had deserted Carter for Reagan in 1980, and Falwell's Moral Majority was one of their best-known organizations. A computer error had caused Kennedy to receive a membership card and an invitation to fight "ultraliberals such as Ted Kennedy." After that got into the papers, Cal Thomas, an aide to Falwell, wrote Kennedy a lighthearted note inviting him to visit the church's Liberty Baptist College. Kennedy wrote back that he would—and would like to speak, too. Thomas told reporters that Falwell was startled by the idea, but decided that having Kennedy speak would show that the Moral Majority was not intolerant.

After dinner at Falwell's house, Kennedy's speech to an audience of five thousand, including four thousand students who were required to attend, was a sober appeal for religious freedom and respect for differing views.

"The separation of church and state can sometimes be frustrating for women and men of deep religious faith," he said. "They may be tempted to misuse government in order to impose a value which they cannot persuade others to accept. But once we succumb to that temptation, we step onto a slippery slope where everyone's freedom is at risk. Those who favor censorship should recall that one of the first books ever burned was the first English translation of the Bible. . . . At some future day, the torch can be turned against any other book or any other belief. Let us never forget: Today's Moral Majority could become tomorrow's persecuted minority."

He warned that the name "Moral Majority" offended some because it "seems to imply that only one set of public policies is moral," but he said liberals and evangelicals were equally at fault for intolerant name-calling. He said Falwell was not a "warmonger" because he opposed the nuclear freeze any more than supporters of abortion rights were "murderers." He drew applause when he criticized Harvard students for hissing and heckling Falwell earlier that year. "I hope for an America," he said, "where neither fundamentalist nor humanist will be a dirty word, but a fair description of the different ways in which people of good will look at life and into their own souls."

As Kennedy left, Falwell told a Kennedy aide, Jack Leslie, that he could see why many people wanted him to be President. He soon dropped the personal attacks on Kennedy from his fund-raising materials.

ALTHOUGH HE AND his family visited the grave at Arlington on the anniversaries of John F. Kennedy's birth and death, Ted had avoided any sort of public commemorations of the anniversary of Dallas. But, freed by his withdrawal of any suspicion that he was exploiting Jack's death for Presidential politics, Kennedy used the twentieth anniversary of the assassination to try to remind the nation of his brothers' commitment to values that seemed lost on the Reagan administration.

He began with an article in *Parade,* the national Sunday newspaper supplement, continued with a set of forums across the country, made a moving eulogy at Holy Trinity, the church where Jack worshiped before his inauguration, and concluded with one last forum in impoverished eastern Kentucky.

The November 13 article began with personal reminiscences from his childhood—pillow fights, sailing, reckless dives from great heights, and Jack's slogan as they walked the Hyannis Port beach: "On a clear day, you can see Ireland." Ted wrote, "We would skip a stone for good luck toward the island of our Kennedy and Fitzgerald ancestors."

But his real message was his judgment of what John Kennedy would have been proudest of, and before he got to disarmament or civil rights, Ted began:

"He believed that one test of a society was how it treated the elderly, the sick and the poor. So he fought for Medicare, job training and nutrition programs, and he was planning the War on Poverty when he died. He was shocked at the deprivation he had seen in West Virginia, the hopelessness etched onto hundreds of faces; he never forgot it, and he would not let America ignore it. That was the heart of his challenge to ask what we can do for our country—and for each other."

Economists were now saying that the recession had ended a year earlier—although unemployment was still 10.4 percent. So at his first forum in San Francisco on November 18, he asked: "I want to know why, if things are so much better, St. Anthony's is still feeding two thousand people every day. . . . Why are there a thousand homeless and hungry youngsters here every night? Why is Loaves and Fishes in Richmond serving twenty-six hundred more meals this month than it was six months ago? And why do six thousand requests a month come in to the food bank in Santa Cruz?"

He continued, "This cause is not new to me or my family. When Senator John F. Kennedy traveled to West Virginia in 1960, he saw the scorched earth of poverty and hunger—and he vowed that America would do better. The food stamp program was born of his caring and concern. My brother Robert Kennedy traveled to the Mississippi Delta in 1967 and to rural Kentucky in 1968, and was shocked by the sight of children whose bellies were bloated by hunger and malnutrition. The passion of his outrage touched the Congress; it sparked the creation of the Select Committee on Nutrition and led to the enactment of the landmark program we call WIC."

In San Francisco, and then in Minneapolis, Detroit, and Pittsburgh, he made the case against the administration's cuts in the budgets for food stamps and the Women and Infant Children feeding program. Though Kennedy never personalized it, President Reagan had long made a stump-speech joke out of steak or whiskey he said was bought with food stamps.

On the flight to Minneapolis that night, Kennedy told Tom Oliphant of the *Globe* that poverty and hunger mattered to his brothers, and this was a moment when he could use their memory to catch national attention because "their conviction is mine, the belief that people are decent, and that when they see there is still suffering and injustice, they will not tolerate it."

In all four cities, he heard from suffering local people and those who tried to care for them. He carried the gospel to a Detroit church on Sunday the twentieth. The Reverend Charles Nicks of St. James Missionary Baptist Church introduced him: "There is one man who we have loved, supported and we have prayed for him down through the years. Others may have talked one way and done another, but this man has been a champion for our causes and we thank God for him. . . . I sure wish he was President."

Kennedy chose Matthew 25 for text: "For I was hungry and you fed me, thirsty and you gave me drink . . . and inasmuch as you have done it unto one of the least of my brethren, you have done it unto Me." He spoke of food stamps and WIC, saying, "Now those programs have been slashed even as the number of the needy grows. We must take our stand: we must witness at the inhumanity of hunger; and with the commitment and intensity of my brother, Robert Kennedy, we must say: 'This is unacceptable.' "

On Monday the twenty-first, after a forum in Pittsburgh, he flew home to Hyannis Port to be with his mother and Jacqueline Kennedy Onassis on the morning of the twenty-second, the anniversary itself. Then he returned to Washington and Holy Trinity.

With the Reagans in a front pew, Ted recalled John F. Kennedy as someone who "had courage which permitted him to bear sharp pain almost every day and which made him respond to the pain of others. He rejected the cold affliction of indifference and the comfortable erosion of concern." He spoke of how Robert had worked with him: "The two of them were one in purpose and vitality in their capacity to dream dreams and renew our vision." And of Robert, he said, "To the end he lived out the meaning of our brother's unfinished life."

Of the late President, he asked, "How then shall we sum up this man, who had every gift but the gift of years?" He answered:

"He was an heir to wealth who felt the anguish of the poor.

"He was an orator of excellence who spoke for the voiceless.

"He was a son of Harvard who reached out to the sons and daughters of Appalachia.

"He was a man of special grace who had a special care for the retarded and the handicapped.

"He was a hero of war who fought for peace."

As he came to the end, Ted said, "For him on this day twenty years ago, the journey came to an end. But for us here and others everywhere, there are promises to keep and miles to go before we sleep."

The next morning Ted went four hundred miles to Appalachia, to eastern Kentucky, where he joined Representative Carl Perkins, who had been along fifteen years before when Robert ran for President.

At a parking lot in Jenkins, he was met by Marge Davis and Ellena Reynolds, who remembered they had given Robert pull candy, walnut fudge, and peanut brittle when he visited in 1968. He told reporters, "Our principal objective is to focus on the problem of hunger. Particularly around Thanksgiving Day, when many Americans will be enjoying turkey with their families, it is important for us to remember that there are many who can't participate and will be hungry."

He went through the hamlets of Trace Branch, Big Branch, Little Mud, and Grethel. In Trace Branch, Margaret Hall told him of her children going without heat to have money for food. At Big Branch, Glena Damron said food stamp allotments were tight; "It gets pretty short down near the end" of the month, she said. At the Mud Creek Clinic in Grethel, where a sign proclaimed "Welcome Senator (President) Kennedy," Parris Deal complained of his medical bills: "Right now, if I don't have it, I can't pay it."

The trip ended with one last forum, in Fleming-Neon, where his staffers tried to be secretive as they gave witnesses just-purchased Thanksgiving turkeys. Kennedy said that if anyone wondered what the difference was between the sixties and the eighties, the answer was "clear and simple: in the sixties, we had a President who said it's wrong to have hungry children."

A thoughtful columnist, Murray Kempton of *Newsday,* wrote on the trip, and observed that if Kennedy was forever known for Chappaquiddick, "in the arrogance of our conviction that we would have done better than he did in a single case, we exempt ourselves from any duty to pay attention to the many cases where he shows himself better than us."

He wrote, "He is the first Kennedy to be a loser in politics, and he gives every sign of not anticipating a second chance. He makes his witness now, not as a candidate, but as a kind of steward; he travels to call attention not to himself but to the needs of others. . . . Since no tactic can avail him any longer, we have to assume that only principle carried him to Little Mud. His generation of the Kennedys can never command again; it endures in him only to oppose, the most elevated of all political functions. If he lives wherever ghosts may live, John F. Kennedy, the grandest of successes, must be surprised and proud to have a brother who could bring such a victory out of failure."

Kennedy kept returning to the nuclear issue. On October 31, 1983, he forced a Senate vote on the freeze, losing 58–40. He said that while arms negotiators bargained slowly, "their governments have developed and deployed new missiles with more warheads, greater accuracy, and shorter warning time. Relentlessly, we have reduced the narrow span of minutes in which the fate of humanity can be decided by human beings."

He and Hatfield used receipts from a 1982 book on the freeze to stage a December 8, 1983, forum in Washington where top Soviet scientists joined Americans in warning that nuclear war would lead to a nuclear winter in which dust and soot would block out the sun. "A nuclear war of any scope would mean either the disappearance of mankind, or its degradation to a level below the prehistoric one," said Dr. Vladimir Alexandrov, chief of the department of climate models of the Soviet Academy of Sciences. Carl Sagan of Cornell agreed. Kennedy complained that the Reagan administration appeared to believe that nuclear war was "winnable and survivable." Sergei Kapitsa, a leading Soviet physicist, said that nuclear weapons no longer served as tacit hostages for peace because it was clear the weapons were "suicidal" for mankind. Kennedy stayed with the issue in 1984, too, though legislatively the freeze was going nowhere. As Congress adjourned on October 5, he forced another vote, losing 55–42.

TED AND JOAN'S divorce was generally smooth. They agreed to joint custody of the children. Joan kept the apartment in Boston and Ted the house in McLean, which was pretty empty with the older children in college and Patrick in school at Fessenden to be near his mother. But they argued over the Squaw Island house. He couldn't understand why she wanted it, telling her she could have the money to buy something fancier in Osterville, with "her Republican friends." After all, he said, "This is Kennedy territory." She disagreed bluntly, telling him, "The Bennetts were here in 1901." From an upstairs window she pointed to where her grandfather had built brick houses for himself and two brothers, and where her father had gone fishing in Hall's Creek as a boy. Ted gave in. By now Rose had suffered a stroke and was bedridden. Early in 1984 she gave Ted the family home a mile away. It was the house where he had grown up, with all its memories of his brothers and sisters.

His mother was at her other house in Palm Beach a couple of months later when another tragic memory was added to the family history. David Anthony Kennedy, Robert and Ethel's fourth child, died on April 25 of a drug overdose. He and several cousins were on an Easter visit to see Rose.

David was found alone in his room at the Brazilian Court Hotel, about five miles from the family house. He was twenty-eight, and had been deeply troubled ever since Robert's death. He had seen the shooting, alone, on television, on the same day his father had rescued him from a sharp undertow in the ocean off Malibu. He had been treated for heroin addiction and alcoholism since 1974. He had been expelled from Harvard, was readmitted, and dropped out again. Ted issued a statement, saying, "It is a very difficult time for all of the members of our family, including David's mother, Ethel, and all his brothers and sisters who tried so hard to help him in recent years. All of us loved him very much. With trust in God, we all pray that David has finally found the peace he did not find in life."

WHILE OTHER DEMOCRATIC Senators—John Glenn, Alan Cranston, Gary Hart, and Fritz Hollings—spent the early part of 1984 running for President, Kennedy was making the Senate debate an issue it had been avoiding. Central America was the Reagan administration's only active military battleground. In Nicaragua, it was backing a rebellion against Daniel Ortega's leftist Sandinista government, which had overthrown Anastasio Somoza's U.S.-backed regime in 1979. In El Salvador, it was backing the government against a leftist rebellion.

Several factors pushed Kennedy into the issue. It mattered in Massachusetts, both in liberal intellectual circles and because Catholic orders like the Maryknolls based in the state sent nuns to Central America. Tip O'Neill, by then Speaker O'Neill, and Eddie Boland, then chairman of the House Intelligence Committee, were involved, and Boland had pushed through legislation in 1982 to prohibit the use of U.S. funds to overthrow the Nicaraguan government. Ted's friend Chris Dodd, elected to the Senate from Connecticut in 1980, had served in the Peace Corps in the Caribbean and was sharply critical of American policy. Finally, Kennedy had just hired Greg Craig as a new national security aide to replace Jan Kalicki. Craig, a lawyer with Williams and Connolly who had gotten Reagan assassin John Hinckley off on an insanity plea, saw Central America as a critical area where Kennedy could make a difference and spent some time traveling there on his own before going to work at the Senate in January.

Reagan had sought to build support for his policy by appointing a commission, headed by former secretary of state Henry A. Kissinger, to examine the situation in Central America. It reported on January 11, 1984, and endorsed the administration's policy, backing so-called "covert" action against the Sandinistas and more military aid for the government of El Salvador. Over Kissinger's objections, though, a majority of the commission said aid to El Salvador should be conditioned on improvements in human rights, especially a curb on the right-wing death squads it said had killed thousands of civilians.

Four days later, Kennedy denounced the report in a long op-ed piece in the *Washington Post*. He said it repeated the classic United States mistake of relying on "armed force" to support "the forces of reaction and repression in the region." And military aid to El Salvador, he said, pretended that "the police, national guard and military have not been the principal instruments of repression in El Salvador for almost 100 years." The opponents of the

Sandinistas, he insisted, were led by disciples of the Somoza dictatorship. He said, "The Administration has already hinted that it will look the other way when it comes to the report's recommendations on human rights. Congress should look the other way when it comes to the endorsement of the President's secret war in Nicaragua."

Kennedy forced the issue on the Senate in March, first blocking any effort to take up Reagan's request for emergency military aid for El Salvador's government and Nicaragua's rebels until after El Salvador's March 25 election (the centrist José Napoléon Duarte led, but faced a May runoff). Kennedy's intervention puzzled and outraged some Senators. Richard Lugar of Indiana, a senior Republican on the Foreign Relations Committee, demanded that Craig tell him what it was that "riled up Kennedy" and got him involved on Foreign Relations turf. Craig replied that the issue mattered in Massachusetts and Kennedy wanted the Senate to face it.

Kennedy offered a series of amendments to cut or delay the aid to El Salvador or condition it on peace negotiations or trials of murderers of American labor advisers, or to assure that aid was denied if Major Roberto D'Aubuisson—whom he accused of ordering the murder of Archbishop Oscar Romero, a defender of his nation's poor—became President. His overriding complaints, though, were against the Reagan administration for ignoring diplomacy and poverty:

"This Administration seems to think that the answer to every tough problem in the world is more guns, more bullets, more soldiers.

"There is a blindness that seems to afflict our policy makers here almost as much as it distorted the politics of El Salvador. That blindness is the assumption that if someone is working to help the poor, is tending to the ill, is bringing food to the hungry, then that person is, for some reason, a Soviet surrogate and must be stopped at all costs."

All his amendments about El Salvador were defeated, as was his effort to block $31 million in aid to the Contras.

The administration's first justification of aid to the Contras was that it was an effort to block the export of arms to guerrillas in El Salvador. On March 28, President Reagan told *The New York Times* that the aid was also meant to force the Sandinistas to accept the Contra "freedom fighters" as part of their government and to hold elections. Kennedy insisted that the only reason was to oust the Ortega government. The Contras made no secret of their intention to overthrow him. They had no interest in guerrillas in El Salvador.

Kennedy called American policy "a case of see no evil, hear no evil, speak no evil while doing evil. We are giving covert aid and the people we are giving it to state publicly their intention to overthrow the legitimate government of Nicaragua." Kennedy lost again, 61–30, on April 4, after the Senate chose to credit a letter from Reagan to Baker saying, "The United States does not seek to destabilize or overthrow the Government of Nicaragua."

But he had helped stoke a fire with his attacks on the Contras. On the evening of April 4, as Kennedy was losing, Barry Goldwater was denouncing Senators who tried to meddle as the President protected American security. Joe Biden of Delaware, a member of the Intelligence Committee, was sitting at his desk reading a classified memo. It confirmed Sandinista charges that the Central Intelligence Agency had mined Nicaraguan harbors. The memo was shown to Goldwater, who was furious and read it on the Senate floor. It provoked no reaction there, and was deleted from the *Congressional Record* by Goldwater's staff. But David Rogers of the *Wall Street Journal* heard it and probed other sources in Congress, and in the Friday, April 6, *Journal* he described the mining operation as based on a ship owned by the CIA and carried out by Salvadoreans and others, but not by Nicaraguans.

The story boiled in other papers the next day. When the Senate came back to work Monday, Goldwater wrote to William J. Casey, director of central intelligence. He told Casey he was "pissed off" at not having been told of Reagan's authorization of the mining and the CIA's role, and he called the mining "an act of war. I don't see how we are going to explain it."

Kennedy, whose work on the Armed Services Committee had brought him Goldwater's respect, promptly introduced a nonbinding resolution as an amendment to a tax bill, saying, "It is the sense of Congress no funds heretofore or hereafter appropriated shall be obligated or expended for the purpose of planning, executing or supporting the mining of the ports or territorial waters of Nicaragua." Kennedy said the mining, which had caused millions of dollars in damage to foreign ships and killed at least two Nicaraguan fishermen, violated international law just at the moment the administration was withdrawing from the jurisdiction of the World Court in The Hague.

The vote came the next day. Kennedy closed the debate by saying it was the duty of Congress to face the issue and stop deferring to the administration: "We know the evasions, the rationalizations, the fabrications, for we have heard them from this administration until they have become as tattered as they are untrue. We have no excuse for continued inaction." He won, 84–12; the margin was that great because the White House decided not to fight and because Casey had angered Senators at a closed briefing, saying they were ignorant because they had not asked the right questions.

After the vote, Kennedy exulted, "The Senate took a first step to halt President Reagan's secret war in Nicaragua." The House followed two days later on a 281–111 vote. Even though Reagan said he could tolerate the resolution because it was "not binding," the mining was stopped.

Even before the resolution against mining the harbors, Kennedy was pleased with his role in the Central America debate. He told the *Globe* as the bill was passed that he had succeeded in making the Senate accountable. He said, "I basically canceled everything else to devote myself to this debate. It's a chance to get your teeth into a substantive issue. If you are in this business, you want to have an effect on policy."

Kennedy stayed with the issue, meeting with Duarte, and proposing more amendments to bar the use of American troops in El Salvador. But when Duarte won the May 6 runoff and pledged to eliminate the right-wing death squads, the Congress rallied behind him, and aid to El Salvador was never a big issue again.

Nicaragua was another matter. The administration suffered further embarrassment in October when the Associated Press discovered a guerrilla manual prepared with CIA help that gave advice on how to "neutralize" Sandinista officials; many in Congress thought it violated Reagan's 1981 executive order barring any U.S. role in assassinations. Nineteen eighty-three marked the end of U.S. government military aid to the Contras. The House refused to approve any money for that purpose, and the administration and the Senate backed down, though Ortega's untimely visit to the Soviet Union the next spring helped pass a measure of "humanitarian aid."

But the U.S. money was no longer essential. White House aides like Robert McFarlane, the national security adviser, and Lieutenant Colonel Oliver North were getting it from Saudi Arabia.

ON ONE MAJOR domestic issue, Kennedy was in step with the Reagan administration. That was rewriting the federal criminal code. The Senate had been trying to do it since 1977, but the House had repeatedly balked. Reagan joined the effort in March 1983, supporting the general lines of past Senate action and adding some much tougher drug sentences and a weakening of the insanity defense, which John Hinckley had used. Kennedy worked with Thurmond, Biden, and Paul Laxalt, Reagan's closest friend in the Senate, and the Judiciary Committee approved a bill with only one dissent on July 21.

Kennedy identified the two major innovations of the bill as its effort to establish uniform sentencing in federal courts and its provision to allow judges to deny bail to an accused they considered a threat to the community. At a May 23 hearing he assailed the disparities of existing sentencing as lacking "rhyme or reason. . . . Judges are free today to give violent criminals a slap on the wrist of probation or to lock them up and throw away the key." While that provision, which included a sentencing commission to set allowable ranges for punishments, was resisted by judges who thought it infringed on their discretion, Kennedy's other key provision outraged civil libertarians, who labeled the bail provisions as "preventive detention." In practice, however, judges already routinely set high bail for people they thought menaced their neighbors.

The Senate passed the bill on February 2, 1984, on a 91–1 vote, and the Republicans turned up the heat on the Democratic House. Reagan told a February 22 press conference it was time for the House "to stop dragging its feet and to act promptly." The House did no such thing. Between May and September, it passed fourteen separate crime measures, including forfeiture of assets for drug dealing, but stayed away from dealing with sentencing, bail, and insanity. Then in late September Republicans won a 243–166 vote to attach the Senate bill to a must-pass stopgap spending bill. After that defeat, and with elections coming up soon, Democrats agreed to bring up their own comprehensive bill, which passed 406–16.

The House conferees then fought all night, winning some changes in the drug provisions, but a united Senate group of Kennedy, Biden, Thurmond, and Hatch held out against any changes in the bail and sentencing provisions. After both Houses passed the final measure on October 4, Kennedy said the bill should not be claimed by Democrats or Republicans, the House or the Senate, or any one attorney general or President. Kennedy, who had a stronger claim to credit than anyone else, said, "We could not have done it without each other."

Ted played no role in the first six months of the 1984 Presidential campaign. Mondale sought his endorsement after winning the Iowa caucuses and was annoyed when Kennedy held back. But Kennedy stayed neutral, seeing no reason to upset either Jesse Jackson, the civil rights leader, or Gary

Hart, Mondale's other determined opponent. He did a fund-raiser to help Fritz Hollings pay off the debt from his feeble campaign. He chuckled when politicians as diverse as Goldwater and Willie Brown, the speaker of the California Assembly, predicted he would emerge from a deadlock as the nominee. "Nobody's talking to me about it," he said.

But at the end of the primaries, when Mondale had the delegates, Kennedy moved in to settle a quarrel. Mondale was in trouble with the Federal Election Commission for dubious fund-raising for some of his delegates, and Hart was threatening to challenge them at the convention. Hart and Jackson also wanted to change the rules for 1988, to eliminate the advantages for well-known candidates which Mondale and Kennedy aides had jointly structured for 1984. Kennedy called Hart on Friday, June 22, and said he was "absolutely correct about the rules." But, Ted added, the race was over, it was time for Democrats to get together, and he was going to Minnesota to endorse Mondale. They talked again on Sunday morning before Kennedy left, and again that night when Kennedy called from Mondale's living room and put the two of them on the line. Hart dropped his threat to challenge delegates, and Mondale supported Hart's rules changes.

On June 25 in St. Paul, Kennedy endorsed Mondale, saying, "He has been there for two decades at the center of the long struggle in our generation for civil rights and social justice." At the Democratic Convention in San Francisco in July, Kennedy roused the crowd with attacks on Reagan and introduced Mondale as the next President: "And then for us as Democrats, for America and all the world, the dream we share will live again."

Ted campaigned hard for Mondale all fall—insisting the polls were wrong in a race that seemed hopeless. In Boston at a November 2 rally, Kennedy complained about Reagan's quoting his brother two days before. "The incumbent President may come to Boston and speak in a rally in the shadow of the JFK Building," he said. "I only wish he also stood in the clear light of the principles in which John F. Kennedy believed—principles of fairness and justice and progress toward peace." But with the economy strong and Reagan popular, Mondale ended up carrying only the District of Columbia and Minnesota.

Kennedy's most interesting political speech in 1984 was another commentary on the role of religion in public affairs. But this time he was not trying to guide born-again Evangelical Christians, but leaders of his own Catholic faith, some of whom were openly supporting Reagan's campaign. On September 10 he went to New York in defense of two Catholic Democrats, Representative Geraldine Ferraro of Queens, whom Mondale chose as his candidate for Vice President, and Governor Mario Cuomo. They were under attack by Archbishop John O'Connor for not working to ban abortion. The archbishop had said a Catholic "in good conscience cannot vote for a candidate who explicitly supports abortion." Ferraro and Cuomo, like Kennedy, said they opposed abortion but regarded it as a matter for a woman to decide for herself; to many of their critics, that amounted to supporting abortion. Kennedy answered O'Connor in New York on September 10:

"Archbishop O'Connor surely has every constitutional right, and according to his faith a religious duty, to speak against abortion. And just as surely Geraldine Ferraro and Mario Cuomo are equally right that faithful Catholics, serving in public office, can agree with his morality without seeking to impose it across the board. . . ."

Kennedy insisted that not every moral command could be "written into law, that Catholics in America should seek to make birth control illegal; that Orthodox Jews should seek to ban business on the Sabbath; that fundamentalists should try to forbid the teaching of evolution in public schools."

He said, "We cannot be a tolerant country if churches bless some candidates as God's candidates—and brand others as ungodly or immoral."

He concluded by recalling John F. Kennedy's words to the Protestant ministers of Houston, two days short of twenty-four years before: "I believe in an America that is officially neither Catholic, nor Protestant, nor Jewish, where no religious body seeks to impose its will on the general populace and where religious liberty is so indivisible that an act against one church is treated as an attack against all."

A THIRD COMMITTEE was one mark of recognition for Kennedy's status as a Senate elder. A more painful measure was the frequency with which he now eulogized colleagues who had been there when he arrived in 1963. Senator Henry M. Jackson, who had pushed to bring Kennedy to join him on the Armed Services Committee and had praised him to conservative Democrats despite their differences on arms control, died at seventy-one on September 2, 1983. Kennedy spoke at services for him both at the National Cathedral in Washington and at the First Presbyterian Church in Everett, Washington, where he said, "His legacy lives in the lesson he taught us that national defense means not only a strong military, but a just society. He was an advocate of modern weaponry and greater security. But he never believed that he could or should pay for a missile or a bomber by taking food from a hungry child or hope from a jobless worker. He was a champion of labor even when unions were said to be out of style; he was a tribune of civil liberty even at the McCarthy era. Some said he lacked charisma; in reality, he had a charisma of conscience."

Jacob K. Javits, defeated for the Republican nomination in 1980 by Alfonse M. D'Amato, was only ailing, not dead, when Kennedy spoke for him on October 17 at the State University of New York at Stony Brook. He called him "the great negotiator," with a "genius for extracting the maximum common ground from passionately opposing points of view, . . . a Senator of inexhaustible indignation in the face of intolerance and injustice. The great civil rights and voting rights acts are achievements of Jacob Javits which, in a few brief years, have done more to fulfill the promise of equality than any other action in the century since the Civil War. Jacob Javits has been a scholar in the mold of Louis Brandeis; a legislator with the persuasive power of Lyndon Johnson; and a prophet of justice who has lived in the spirit of Abraham Lincoln. He is, in short, a Senator's Senator—and that is the way we all remember him."

Frank Church, a friend since the 1960 trip to Africa, who had lost in the Republican landslide of 1984, died the next spring at fifty-nine. Ted had been at his bedside often as he lay dying of cancer at his home in Bethesda, saying he was missed in the Senate. At the National Cathedral on April 10, 1984, he said, "Frank Church was among the first lonely voices of dissent raised in the wilderness of warmaking in Vietnam. He was a Senator from Idaho, but he was also a voice for those half a world away who had no votes to cast against the bombing and devastation of their own land. . . . He loved this country so much that he took the lead in exposing the abuses of the CIA and the FBI. . . . His own sense of patriotism would not let him declare for the Presidency in 1976 until that investigation was done—and so he lost the nomination he very well might have won."

CHAPTER 26

A Pilgrimage to South Africa

Just as he had stepped forward to fill a gap in Congress on civil rights in 1981, as 1984 drew to a close Kennedy began to make South Africa his issue. Countries around the world were growing more and more impatient with the apartheid regime, under which Africans could not vote, own land, or move about the country freely. But while American universities, unions, and some cities pushed to unload stocks in companies that did business there and hundreds were demonstrating and being arrested at the South African Embassy in Washington, the Reagan administration was singularly supportive of the government. Reagan had lifted arms embargoes imposed by Carter, and the State Department called its policy "constructive engagement," arguing that economic and diplomatic ties encouraged South Africa to reform its policies. It was a blind policy, for South Africa had no intention of changing its racist reality.

Kennedy planned a forum on South Africa for October 5. But he canceled it because he was fighting for a nuclear freeze amendment. Greg Craig took the South African speakers, the Reverends Desmond Tutu and Alan Boesak, to watch the losing effort. Later at lunch, Kennedy asked them what he could do to help. Tutu said Kennedy had to come, because "the world will not pay attention until someone like you comes to South Africa and brings the cameras and the spotlights with you." Once before a Kennedy had brought attention to South Africa. Robert Kennedy had paid a meteoric visit in 1966. He gave what Arthur M. Schlesinger, Jr., called his greatest speech. He told thousands at the University of Cape Town: "Each time a man stands up for an ideal, or acts to improve the lot of others, or strikes out against injustice, he sends a tiny ripple of hope, and crossing each other from a million different centers of energy and daring, those ripples build a current which can sweep down the mightiest walls of oppression and injustice."

Ted, usually eager to pick up his brothers' causes, was reluctant this time. There was no way his trip could have the impact of Bobby's. But the South Africans pressed him, and he agreed, saying he would come in early December. A few days later, when Tutu won the Nobel Peace Prize which he would accept in Oslo in December, the trip was shifted to January.

But first, in dramatic contrast to the skiing trips of past holidays, Ted, Kara, and Teddy spent Christmas in Ethiopia and the Sudan, bearing witness to the devastating famine that had killed 300,000 Ethiopians. As they arrived on December 19, Kara asked, "Will it be as bad as the pictures?" When she saw thousands of people huddled without shelter, she said, "It's going to be worse." They toured feeding stations. At Bati, Ted ladled milk into children's cups. Teddy, twenty-three, fitted needles onto syringes for measles shots. At Mekele, Kara, twenty-four, weighed children so that camp officials could be sure their parents weren't taking their food. Everywhere they went, relief officials told them there was not enough food, but they were struck by the dignity of the refugees waiting patiently in line. One evening when Teddy tried to say grace for a supper of chicken, lentils, and tef, the Ethiopian bread, he choked out, "I've never been so thankful for food. . . . I'm so glad my family is here."

Occasionally, a camp seemed to have the situation more or less under control. "We can feed everyone, because the Lord will provide," said Sister Berilla at the Jijiga camp, "and if He doesn't, Mother Teresa will." After Mass on Christmas morning, the Kennedys drove to the border to watch hundreds of Ethiopians cross into the Sudan hoping to find food. At Teki el Baab, one of the biggest camps, food had run out, although for three weeks a ship with five hundred tons of grain was offshore, four hours away by truck, waiting for bureaucrats to decide where to send the food. Ted wrote later in *People* magazine, "I have never seen a more graphic demonstration of the truth that red tape kills."

The government of Sudan seemed overwhelmed by the burden. Eileen McNamara of the *Globe* wrote of Abdel Magid Bashir Elahmadi, its refugee commissioner, as "an affable man who checks the time on a silver and gold

Rolex watch as he is chauffeured between refugee camps in a white Mercedes." But she said he could not answer questions about supplies or conditions. Kennedy told McNamara, "We can't even get a clear picture of what is happening in this country. . . . The response seems to be that we know some food is coming sometime but we don't know when and we don't know from whom. Evidently, what we're doing is not enough."

On the day after Christmas, they left El Obeid, a camp where the refugees were not starving, and drove by Land Rover to Khartoum. Kara and Teddy went to Kenya for a safari, but their father returned to Washington. He escorted John Kerry, a former Vietnam veteran and antiwar leader elected to the Senate in 1984, as he was sworn in on January 3, 1985. Ted also got his Christmas present from Patrick, who had stayed behind because the desert dryness was unsafe for his asthma. It was a print of a Van Gogh garden scene, inscribed "I know you planted seeds of hope for many people."

KENNEDY AND A crowd of sisters, nieces, nephews, friends, and reporters arrived in Johannesburg on Saturday, January 5. Tutu greeted them warmly at the airport, and Kennedy told reporters that while he came with an "open mind," he also held a belief in racial equality and "a deep opposition to the entire concept of apartheid." There were two immediate signs of trouble. The American ambassador, Herman Nickel, was regarded by Tutu as an apologist for apartheid and had been asked to stay away. But he came anyway. So did about a hundred demonstrators from the Azanian People's Organization, or AZAPO, a black consciousness group of followers of the murdered Steve Biko. The police gave them unusual freedom and allowed them outside the airport's VIP lounge to demonstrate against Kennedy before arresting a few of them. The group had resolved in December that visits by Kennedy and Jesse Jackson "give credibility to the settler regime and the forces of imperialism."

Then after the Kennedy party left the airport to visit Tutu's home in the huge black satellite city of Soweto, police stopped them, saying hundreds of people were throwing rocks and fighting around his house. Kennedy went ahead anyway and was greeted by hundreds of people, but they were holding candles and singing "Nkosi Sikilele Africa," the anthem of liberation. But the bus carrying reporters and cameras was turned back. American television only showed the demonstrators at the airport.

On Sunday the sixth, Kennedy toured Soweto, dismayed by a barracks-like hostel for black workers who left their families in their rural homelands to labor in the white economy. Afterward, he climbed up on a mound of dirt and told reporters, "This camp is one of the most distressing and despairing visits that I have made to any facility in my lifetime. Here individuals are caught between trying to provide for their families or living with their families. . . . That's alien to every kind of tradition in the Judeo-Christian ethic and I find it appalling here today." He confronted another face of apartheid Monday when he visited Mathopestad, a rural tribal community which the government called a "black spot." It had good farmland bought decades before, but was threatened with a forced evacuation. When John Mathope, the chief, told him his tribe feared being forced to leave, Kennedy called the evacuation "an inhumane policy." That morning Kennedy had met with the highest-ranking government official who would see him, Foreign Minister Roelof F. "Pik" Botha. Botha told Kennedy that removals were made only for "hygienic or health reasons."

The eighty-minute meeting with Botha in Pretoria had been testy. Kennedy told Botha that Congressional action against South Africa was likely if the government did not at least promise action in three areas: giving blacks a right to vote, treating blacks as citizens, and halting black spot removals and forced resettlements. Botha told Kennedy he did not understand the history and culture of South Africa, said that country was freer than other African countries, and contended that apartheid no longer existed. When Kennedy said he thought the situation was more polarized than it had been when Robert visited in 1966, Botha disagreed and told him not to believe Tutu and Boesak. In a follow-up letter, he told Kennedy that objective commentators know "South Africa is a beacon of hope and progress in an otherwise dark continent."

Kennedy renewed his request to see Nelson Mandela, the head of the African National Congress, who had been imprisoned since 1962. Botha tempted Kennedy with his reply, saying a meeting just might be allowed if Kennedy would tell the world after the visit that he had urged Mandela to renounce violence—which the ANC was using increasingly after the utter failure of civil disobedience. Kennedy told Craig that he thought about it, and decided that while he could honestly oppose violence, the public relations coup of a meeting should not be bought by weakening whatever leverage the issue gave Mandela.

Aside from his comments at Soweto and Mathopestad, Kennedy had made no speeches since his arrival. But on Wednesday, January 8, he addressed a meeting of business groups in Johannesburg. Leaders of those groups had dinner with Kennedy the night before and then issued a statement opposing disinvestment and boycotts, saying economic growth was a catalyst for social change. Kennedy challenged them, saying that if business could force political change, it could "save the nation." For himself, he said he would take a position on divestiture once he returned to Washington. He warned, quoting John F. Kennedy in 1962, "Those who make peaceful revolution impossible will make violent revolution inevitable."

But the speech that got the attention that afternoon was not Kennedy's but Ambassador Nickel's. He introduced Kennedy, and attacked him. In a speech he had cleared either with the White House or Assistant Secretary of State Chester Crocker (Nickel offered both versions, and the State Department's Freedom of Information staff can find no documents), Nickel sneered at Kennedy's "whirlwind tour." He attacked disinvestment, saying it would hurt minorities. He said the Reagan administration, like its predecessors, "has an interest in the stability of Southern Africa." While South Africa should know "there is no constituency for racism and apartheid in respectable American politics," he said, the Reagan administration was "confident we are on the right track" toward peaceful change. "Let us remember," he said, "that the success of policy is not measured in decibels." Nickel proudly reported to the State Department that Botha was "impressed and grateful" for his speech. Though surprised by the attack, Kennedy stuck to his text, except to add, "The Ambassador and I have different views, but I believe mine are closer to the American people."

Nickel's attack was covered enthusiastically in the South African press, both Afrikaans and English. Throughout the trip, Kennedy had been attacked both in newspapers and on state-run television. The basic argument was that his visit was designed to win black support so he could run for President in 1988, and it was embellished by references to Chappaquiddick and cheating at Harvard. Before he arrived, *Die Vaderland* had told its Afrikaner readers his trip could be explained by the fact that "the Negro vote is an important factor in an American Presidential election, particularly for a Democratic candidate."

On January 9 he went first to a desolate resettlement township called Onverwacht, where there were no doctors at the clinic and no answers to questions about infant mortality. He went on to the remote hamlet of Brandfort, where Winnie Mandela, Nelson's wife, had been exiled and "banned," a process under which she was prohibited from meeting with more than one person at a time and her words could not be quoted in print. Ted said later she had called constructive engagement "a shoulder to the wheel of apartheid." Kathleen Kennedy Townsend presented her with the 1985 Robert F. Kennedy Human Rights Award, a bust of her father. On the tenth, Kennedy met in Durban with Chief Gatscha Buthulezi, the leader of six million Zulus. Buthulezi attacked Tutu and Boesak in a private meeting, then brought Kennedy outside to meet several thousand of his followers and began the attacks over again.

Later on the tenth, they reached Cape Town, the South African city least enthusiastic about apartheid. The press actually welcomed him. He visited the huge Crossroads squatters camp, where thousands of blacks feared removal to a government-built town called Khayelitsha, twenty-eight miles away. Addressing an enthusiastic crowd, he told of asking people why they did not want to move to what was described to him by the government as "beautiful homes overlooking the sea." He said some told him it was too far from work. Others said they just didn't believe the government. And he quoted one woman as asking, "If it is so nice why don't the Whites go and live in Khayelitsha?" He told the crowd, "I will never forget you. I will never stop fighting apartheid." They sang "Nkosi Sikelele Africa" in response.

The only organized support Kennedy's visit got was from the United Democratic Front, a newly formed umbrella organization of various groups that spoke, more or less, for the nonviolent wing of the African National Congress. Even the UDF showed some confusion, largely because Boesak had not checked with any of its other leaders before inviting Kennedy. An appearance in Cape Town Friday night, the eleventh, was relabeled a "mass meeting" rather than a UDF rally for that reason. But before the meeting was held, Kennedy solidified the group's support with a brief, illegal protest outside Pollsmoor Prison, where Mandela was held. He was joined by two of the UDF leaders, Popo Molefe and Patrick "Terror" Lekota, who were briefly out of prison themselves. Kennedy said, "Behind these walls are men that are deeply committed to the cause of freedom in this land." Mandela knew of the protest, and years later he said he drew strength and hope from it as evidence that "we had millions behind us."

That evening he spoke to a few thousand at the "mass meeting" in a sports stadium. Boesak introduced Kennedy, saying he had been invited so that the world could see nonwhites "wanted freedom here—and now." Kennedy told them he was there "to signal that the vast majority of my fellow citizens oppose apartheid—and we will not be complacent about it and we will not accept endless delay and empty excuses for it." He said, "Those in South Africa who comfort themselves that Ambassador Nickel speaks for the American people have some surprising and painful days ahead of them."

On Saturday the twelfth, he visited Windhoek, the capital of South-West Africa, a former German colony which the League of Nations had turned over to South Africa. The United Nations was pressing South Africa to free the protectorate, or Namibia, as it was known to native inhabitants. He met with church leaders and officials of SWAPO, the South-West Africa People's Organization, and from the roof of a car promised to pursue freedom for Namibia and an end to apartheid.

The trip ended on a disappointing note. On Sunday the thirteenth, perhaps a hundred AZAPO demonstrators in a crowd of three thousand at the Regina Mundi Cathedral in Soweto insisted they would not let Kennedy be heard, and Tutu feared violence in the cathedral. Police equipped with fans to blow tear gas were waiting not far away, hoping for an occasion to intervene. Tutu said sadly, "The system knows how to turn us against ourselves," and the speech was canceled.

Bob Shrum had written a speech that rivaled the one Robert Kennedy had given nineteen years before. It was issued as a departure statement, getting little attention. But it is worth quoting. At times Kennedy spoke to America, saying, "To any who turn their heads, who pretend that they do not see, I reply: Let them read, as I have, the repressive words of South Africa's statutes. Like an earlier generation that saw the construction of the Berlin Wall, let them look at the clear barrier of discrimination against Black people here. The barrier is not always built of bricks or mortar or barbed wire, but it is starkly visible everywhere. And if anyone doubts that it is wrong, let them come to Soweto. Let them go to Onverwacht; let them go to Crossroads; let them go to Brandfort—let them ask Winnie Mandela."

At times he spoke to South African whites who contended that they were making progress toward democracy by allowing nonwhites, other than blacks, to cast watered-down votes: "You cannot justify making most non-whites into political non-persons by making some non-whites into second class citizens. . . . Apartheid in constitutional clothes will never mend the fabric of a society where most of the inhabitants are treated as legal strangers in their own land." He offered a different measure: "Here is a truer test of progress: No longer should Black South Africans be exiled or assigned to false and barren homelands; it is time to say to all South Africans, Black and white, this land is your land—it belongs to you—and you belong to this land." And to the claim that South Africa was a bulwark against Communism, he scoffed, "We cannot oppose tyranny by imitating its tactics."

But where Robert Kennedy had sought to engage young white South Africans to get them to end oppression, throughout his trip Edward Kennedy was addressing blacks. "Let us resolve, again and again, that one form of racism shall not be swept away only to be succeeded by another." Violating South African law again, he quoted Mandela from his 1964 trial: "I have fought against White domination, and I have fought against Black domination. I have cherished the ideal of a democratic and free society in which all persons live together in harmony and with equal opportunities." He recalled Lincoln's second inaugural as a bloody war ended:

"It must not happen here. In 1985, God has given it to the people of this country to decide, to put aside, unrequited toil, the lash and the nightstick, the sword and the bomb and the bullet. The fondest hope, the most fervent prayer we can have for your wondrous land is that you shall stay the mighty scourge of conflict among each other. Peace can be kept only through the triumph of justice; and the truest justice can come only through the works of peace. The dawn can come to South Africa—if all its people, Black and white, who read the same Bible and pray to the same God will see each other at last as all members of the same human family."

He flew to Zambia for a brief meeting with Oliver Tambo, the exiled acting head of the African National Congress. Tambo, whose words the South African press were forbidden to quote, told him that when the organization was outlawed in the sixties, "we decided then that non-violence had run its course." Kennedy told him, "I want to make it clear that I deplore violence—my family has been touched by that and I feel very strongly against it. Having said that, I remember that President Kennedy said those who make peaceful revolution impossible make violent revolution inevitable."

The trip had no clear immediate impact within South Africa, except perhaps to unite most whites in scorn of the Senator, and to make black divisions obvious. Liberal whites shared the view of the hard-line Afrikaners that this was merely pre-Presidential politicking. Ambassador Nickel thought the visit reduced U.S. influence by provoking anger at "meddling," though he conceded that the visit showed the oppression of their system to South Africans who would listen. Its effect was stronger abroad. ABC News and NBC News sent television crews with Kennedy, and they showed the face of apartheid on the evening news. Tutu credits that greater attention with the international movement toward sanctions; in 1995 he said, "We know that we would not be where we are, certainly not as quickly as it happened, had it not been for sanctions."

But the greatest mistake the South African press made was not its interpretation of Presidential politics, but its view of the Senate. Most newspapers, English and Afrikaans, seriously misled their readers, dismissing the trip because Kennedy was "not even a very important member of the United States Congress," as the Johannesburg *Sunday Times* put it.

BEFORE HE MOVED into the South Africa issue at home, Kennedy turned to the family business: politics. The political career he was worrying about was Teddy's. On the flight home from South Africa, he tried to persuade Teddy to run for Congress. This was to be Tip O'Neill's last term, and though Teddy was only twenty-three then, he would reach the constitutional minimum of twenty-five before Election Day 1986. Tip's seat was the one John F. Kennedy had held, Cambridge and Somerville and Boston's Charlestown. Teddy wasn't sure politics was the life he wanted. Ted said it was important to seize opportunities, saying his best chance for the Presidency had come and gone when he was asked to run in 1968. Ted said it was now or never and at least Teddy should move into the district and look things over.

In April, Teddy did "try the idea on," as he put it years later, by moving to Somerville, and Bob Farmer arranged some breakfasts with potential contributors in July. But he was plainly lukewarm about running. It showed when the Senator's advisers like pollster Pat Caddell or friends like Nancy Korman saw him in political settings. Teddy talked to his mother, and worried about disappointing his father. But he told him he wanted to do environmental studies in graduate school, and chose Yale. Once Teddy stepped aside, his cousin Joe moved in. Uncle Ted said, "That seat obviously has special meaning to our family. I'm proud of Joe, and I know his father would be proud of him, too." Joe easily won the 1986 primary and held the seat for twelve years.

The passing generation of the Senate also occupied him. Russell Long, with whom he had fought on some tax issues for fifteen years but with whom he also shared concerns about the poor, announced on February 25 that he had decided to retire at sixty-eight. Long was concerned that his health was not up to another campaign, especially one likely to be as mean and negative as Senate campaigns were becoming in the eighties. Minutes after Long's announcement to reporters, Kennedy, breathless, arrived at his outer office demanding to see him. They met privately for twenty minutes, but Kennedy could not change Long's mind. Later that year, Representative Barbara Mikulski of Maryland came to see him. She had been scheduled to nominate him in 1980, after first meeting him as a social worker in 1965, testifying before the Labor Committee on her work in Baltimore to get the elderly to apply for the benefits the new Medicare and Medicaid programs offered. She thought Senator Mathias, a Republican but a frequent Kennedy ally, would not run again. He was highly popular in Maryland, but Senate Republicans disliked his liberal views and had denied him the chairmanship of the Judiciary Committee. Kennedy told her he thought Mathias would run again, but told her about the rigors of a Senate campaign. He suggested she get in touch with Shrum, by now off his staff and working as a consultant, to be ready if Mathias stepped aside. She did, and on September 27, Mathias announced he would not run. That day Kennedy was one of two or three Senators who came personally to tell Mathias they were sorry he was leaving.

KENNEDY WAS ALSO testing new political themes and, perhaps, directions. On March 29, at a conference on his brother's Presidency at Hofstra University on Long Island, he reflected on Reagan's landslide reelection and what Democrats should do next. He said they needed to dispel the idea "that all the Democrats have to give is more programs, at higher cost, for lesser returns." He said, "The answer is simply not more dollars and more spending." Instead, Democrats "must do more with less. . . . As we create new programs, we must regard them as replacements, not additions. . . . For who can deny that some traditional programs have failed—and that others, precisely because they have succeeded, can thankfully be set aside. The mere existence of a program is no excuse for its perpetuation—whether it is a welfare plan or a weapons system." And he said public housing, welfare, and public service jobs were failed, sometimes counterproductive, efforts.

But the object of economies and discarding outmoded programs, he said, had to be securing the resources and ideas to "fulfill lasting ideals of liberty, of caring and mutual concern. . . . We must not free ourselves from the past in order to run from responsibility for the poor and the homeless, the hungry who must be fed, the weak who must be strengthened, and the hurt who must be healed."

The patriarch, Joseph P. Kennedy, gathered with his family in Hyannis Port about 1937. *Standing, left to right:* Joe Jr., John, Rose, Jean, Patricia. *Seated, left to right:* Bobby, Ted, Joseph P., Eunice, Rosemary, Kathleen. © *Boston Globe*

Ted's nanny helps him with his box camera as he and Jean wait for the changing of the guard at Buckingham Palace on March 18, 1938. NYT Pictures

Ted, Jack, and Robert relax at Palm Beach in the winter of 1960. ***Look*** **magazine/John F. Kennedy Library**

In April 1960, Jack got laryngitis during the West Virginia primary, so Ted was summoned to speak for him. Here Ted's older brother seems amused. ***Look*** **magazine/John F. Kennedy Library**

Ted's campaign duties were sometimes strenuous. He stayed on a bronco for five long seconds in Miles City, Montana, on August 27, 1960. **© George Larson**

Ted smiles early in his August 27, 1962, debate with Attorney General Edward J. McCormack. "If your name were Edward Moore," McCormack said later, "your candidacy would be a joke," and Ted felt like slugging him. Erwin D. Canham, editor of the *Christian Science Monitor,* moderated the debate, and the candidates were questioned by, *left to right,* Leo Egan of WHDH-TV, A. A. Michaelson of the *Berkshire Eagle,* C. Edward Holland of the *Boston Record-American,* and David McNeil of WCAB.
AP/Wide World Photos

The second debate, on September 5, was quieter, but a Boston crowd watched it on television sets in the window of Ted's headquarters on Tremont Street, next door to McCormack's. © ***Boston Globe***

Kara, *left*, and Teddy walk on the beach at Squaw Island with Joan and Ted in the summer of 1963.
© ***Boston Globe***

John F. Kennedy, Jr., salutes his father's casket on November 25, 1963. The President's brothers, Ted and Robert, watch as his widow, Jacqueline, holds the hand of her daughter, Caroline.
AP/Wide World Photos

After the burial at Arlington National Cemetery, Jacqueline and Ted welcomed foreign leaders who came to the funeral. Here they greet Anastas I. Mikoyan, first deputy premier of the Soviet Union, in the Red Room at the White House. **John F. Kennedy Library**

Ted survived the crash of this Aero Commander on June 19, 1964, but his aide Ed Moss and the pilot, Ed Zimny, died. © *Boston Globe*/Charles Dixon

Ted won election to his own full term in the Senate in 1964 while still hospitalized after his plane crash. Robert, elected from New York the same day, visited him at the hospital. When a photographer told him, "Step back a little, you're casting a shadow on Ted," the younger brother said, "It's going to be the same in Washington." AP/Wide World Photos

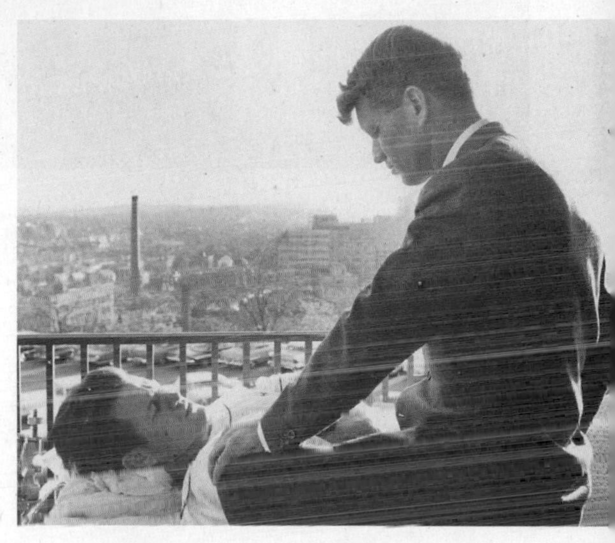

The Senators Kennedy leaving the Capitol after a vote in April 1965. George Tames/NYT Pictures

Touring a refugee camp at Thungdoc near Danang on January 5, 1968, Ted questioned Vietnamese about why they had fled their homes, how they were doing in the camp, and whether government allocations of rice or money had been provided. Corbis-Bettmann

Ted reported to an unhappy Lyndon Johnson on his January 1968 trip to Vietnam. He said, "We're not winning hearts and minds" but creating refugees who became recruits for the enemy. Yoichi R. Okamoto/LBJ Library Collection

After Robert was murdered as he won the California primary, Ted's voice broke occasionally as he delivered the eulogy at St. Patrick's Cathedral on June 8, 1968. He quoted his brother: "Some men see things as they are and say why. I dream things that never were and say why not." Edward Hausner/NYT Pictures

The astronaut John Glenn, serving as a pallbearer, presents the flag that covered Robert Kennedy's casket to Ted at Arlington. Robert's oldest son, Joseph P. Kennedy II, stands behind Ted. © *Boston Globe*

After the authorities pulled Ted's 1967 Oldsmobile 88 out of Poucha Pond on Chappaquiddick island, a crowd gathered to gape at the car in which Mary Jo Kopechne had died. Corbis-Bettmann

His neck in a brace after the accident, Ted walks to the plane that will take him to Mary Jo Kopechne's funeral on July 22, 1969. Joan is at right. © *Boston Globe*/Joe Dennehy

Reporters and photographers surround Ted as he enters the Dukes County Court House on January 5, 1970, for the inquest into Mary Jo Kopechne's death. He testified he had only two drinks that evening. Corbis-Bettmann

After being booed off the stage when he tried to speak to a crowd gathered at Boston City Hall to protest busing on September 9, 1974, Kennedy was taunted as he moved through the crowd to his office nearby. © *Boston Globe*

Ted met with dissidents on his trips to Moscow. At 1 A.M. on September 10, 1978, he went to the apartment of Alexander Lerner, a mathematician. On Ted's right is Boris Katz. Ted took him aside to tell him he would be allowed to emigrate. On Ted's left is Rick Burke, an aide. Second from right is Yelena Bonner, and her husband, Andrei Sakharov, the noted physicist, is on her right. Lerner, his head almost obscured by Sakharov's, is two places to his right. ©1979 Ken Regan/Camera 5

Before they came out fighting, Ted and President Jimmy Carter shook hands at the October 20, 1979, dedication of the John F. Kennedy Library in Boston. Carter's speech anticipated the campaign to come: "The carved desk in the Oval Office which I use is the same as when John F. Kennedy sat behind it, but the problems that land on that desk are quite different." Joan looks on, as does former Speaker of the House John W. McCormack. © ***Boston Globe*****/Bill Brett**

"What's the state of your marriage?" Roger Mudd asked Kennedy in an interview for a *CBS Reports* television program. The interview, conducted on September 29, 1979, was shown on November 2, and Kennedy's awkward replies cast a pall over his candidacy. He answered that question by saying, "We, I think, have been able to make some very good progress." **CBS Photo Archive**

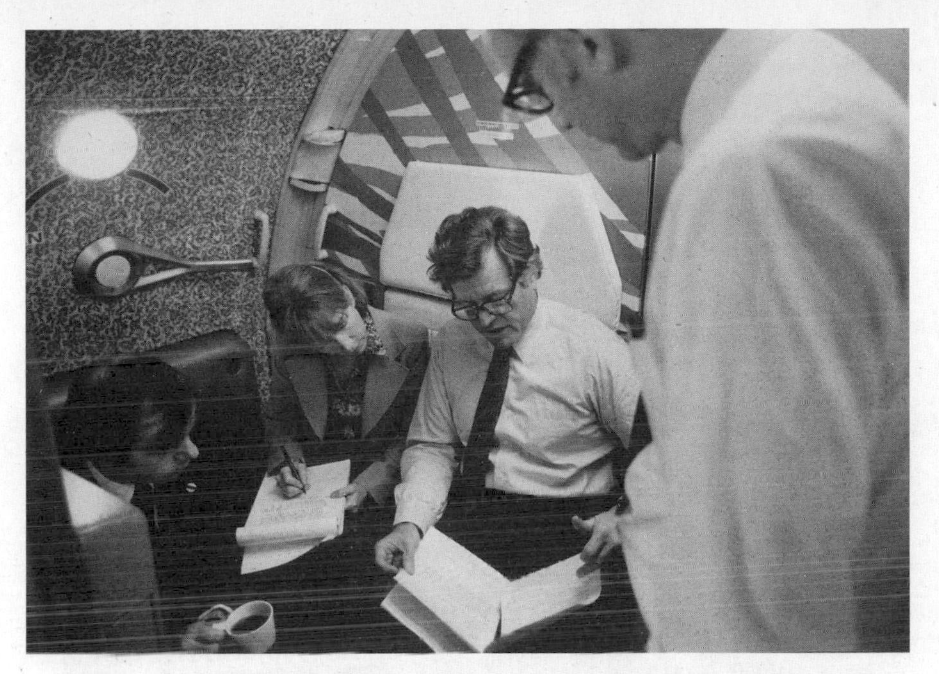

Ted goes over a speech draft on the campaign plane in 1979 with Bob Shrum, *left;* Doris Kearns Goodwin, who had joined as the campaign was beginning; and Carey Parker, who had been with him for ten years and is still with him now. **©1979 Ken Regan/Camera 5**

Kennedy lost the nomination, but his speech was the high point of the Democratic convention. On August 12, 1980, he said, "For me, a few hours ago, this campaign came to an end. For all those whose cares have been our concern, the work goes on, the cause endures, the hope still lives, and the dream shall never die." **Fred Conrad/ NYT Pictures**

The low point came two nights later. Ted did not share in the excitement on the podium after President Carter accepted renomination on August 14, 1980.

Although they were battling over the Voting Rights Act, President Ronald Reagan showed great respect for Ted and his family. On November 12, 1981, Rose Kennedy visited the White House for the first time since Jack's death and watched the President and Nancy Reagan examine a plaque hailing Reagan for understanding that "the common love of our country transcends all partisan differences." **Courtesy Ronald Reagan Library**

The villagers of Mathopestad, South Africa, a black community with good farmland, told Ted on January 7, 1985, of their fears of being deported so that white farmers could take their land. Corbis-Reuters

Ted visited Poland in 1986, and on May 24 Lech Walesa, the head of Solidarity, the union movement that was challenging the Communist regime, showed him around Gdansk, where the movement began. At dinner, Ted presented Walesa with a bust of President Kennedy. Greg Craig

Before the battle over Robert Bork's Supreme Court nomination was joined in the Judiciary Committee on September 15, 1987, Kennedy greeted one of Bork's chief advocates, Senator Alan Simpson of Wyoming. That same day Kennedy attacked Bork harshly, saying, "I believe your clock on civil rights stopped in 1964." In the amiable prehearing moments, Bork and Senators Joe Biden of Delaware, the committee chairman, and Howard Metzenbaum of Ohio watch. Jose Lopez/NYT Pictures

Vicki Reggie Kennedy and Ted walked through a marathon set of press appearances over Labor Day weekend in 1992. Here they pose on the boardwalk leading from the house at Hyannis Port to the beach. © *Boston Globe*/John Tlumacki

Senator Chris Dodd of Connecticut talks to Ted during the January 12, 1993, confirmation hearing for prospective Secretary of Education Richard Riley. Other Senators, *from left:* Paul Wellstone of Minnesota, Howard Metzenbaum of Ohio, Claiborne Pell of Rhode Island, Nancy Kassebaum of Kansas. U.S. Senate Historical Office

President Bill Clinton talks with Ted at Northeastern University's commencement on June 19, 1993. Clinton received an honorary doctorate of public service, joining Kennedy, who is wearing the regalia of his honorary doctorate of public administration, conferred in 1965. © *Boston Globe*/Mark Wilson

The Senate Labor Committee began considering health insurance legislation on May 18, 1994. Here Kennedy is talking to Senator Kassebaum and her aide Susan Hattan, as his aides Ron Weich and Nick Littlefield listen. Partially obscured by Kennedy's head is another aide, David Nexon. Senator Pell is at left. AP/Wide World Photos

In his second debate with Mitt Romney, in Holyoke on October 27, 1994, Ted and the challenger for his Senate seat clashed over crime. Romney complained that Kennedy opposed mandatory sentences for criminals using guns, and Kennedy replied by asking where Romney stood on banning cheap handguns. "I won't yield to anyone about guns in our society," Ted said. "I know enough about it." © *Boston Globe*

The head of the family with assorted nieces, nephews, grandnieces, and grandnephews at an outing on a pirate ship in Hyannis in August 1997. His son Teddy is wearing sunglasses and daughter Kara is on the left. **Kara Kennedy Collection**

The speech also included a tribute to Reagan that explained how Kennedy, who had never disagreed more harshly with a President's policies than he had with Reagan's in his first term, got along with the President and respected him. It was a conference on the Presidency, and Kennedy said that whatever one thought of his policies, "Ronald Reagan has restored the Presidency as a vigorous, purposeful instrument of national leadership."

Three months later, on June 24, Reagan came to Kennedy's home in McLean (after John and Caroline Kennedy visited him in the White House to ask his help) to raise money for the John F. Kennedy Library. Kennedy paid him the ultimate compliment of putting praise in his late brother's mouth, saying what he knew John F. Kennedy would think: "On issues I suspect the two of you would not have always agreed. But I know he would have admired the strength of your commitment and your capacity to move the nation." Reagan's response suggested why he, in turn, got on with the late President's youngest brother. Speaking of John Kennedy in terms that applied to Ted, too, Reagan said, "When the battle's over, and the ground has cooled—well, it's then that you see the opposing general's valor. He would have understood. He was fiercely, happily partisan, and his political fights were tough—no quarter asked, none given. But he gave as good as he got, and you could see that he loved the battle."

KENNEDY WAS DISMAYED that most reporting of the Hofstra speech saw it as an effort to adjust his philosophy to Presidential campaign imperatives. But he was prepared to try solutions which most liberal Democrats doubted. In June, he voted for an unsuccessful effort to give President Reagan the power to veto pieces of spending bills, not just entire bills. And he voted yes on legislation designed to force automatic spending reductions if annual spending bills exceeded receipts—a measure named for Senators Warren Rudman of New Hampshire and Phil Gramm of Texas, Republicans, and Fritz Hollings of South Carolina, a Democrat. When the Senate passed it in October, Kennedy explained, "We can no longer do nothing. We can no longer afford a deadlock, in which the Congress stands its ground and the President refuses to meet us halfway." Public service unions were furious with his vote, and hardly mollified when he said as the law was enacted in December that "I am convinced that the only way to protect the important Democratic programs that I care deeply about is to bring the budget under control," and that the Gramm-Rudman-Hollings measure would thwart President Reagan's use of the federal deficit as "the excuse for a wholesale assault on federal programs, whether they are working or not."

The second Reagan administration had little it was really seeking to accomplish. The major initiative of the 99th Congress was a reform of the tax code to reduce loopholes and tax preferences, an effort Congress pushed and Reagan tolerated. Kennedy was a cosponsor of Senator Bill Bradley's reform effort, but despite his previous interest, Kennedy was not really a player because this time the tax-writing committees, Finance and Ways and Means, were pushing for reform.

But Kennedy was deeply involved in three Congressional efforts that did not involve spending: civil rights, South Africa, and health insurance.

The civil rights issue was an effort to undo a 1984 Supreme Court decision. In what was known as the Grove City case, the Court held that when Congress barred discrimination against women in Title IX of the 1972 Higher Education Act, it did not mean that if a college received any federal aid, all of its programs had to be free of discrimination. Agreeing with Reagan's Justice Department, which had changed the policy of the Nixon, Ford, and Carter administrations, it said only those programs which directly received federal money were barred from discriminating at risk of losing their federal aid. The most immediate impact was on women athletes. Until the Reagan administration and the Supreme Court interpreted the statute anew, they had seen institutions add women's sports that they had previously ignored.

The effort to reverse the decision began in 1984, but foundered over the issue of whether new legislation should apply the nondiscrimination policy to any institution receiving federal aid or just to colleges and universities, as some Republicans wanted. With Orrin Hatch leading a filibuster, Bob Packwood, the Oregon Republican who had championed women's issues, gave up the fight on October 2, 1984, while Kennedy complained that President Reagan had offered no help at all. "I quite frankly thought these battles had been fought in the 1960s," Kennedy said.

Kennedy renewed the fight in 1985. The tactical issue he and his allies faced was how to reinstate past interpretations of the law while avoiding the charge that they were trying to expand its reach. As one aide on the issue put it, "It became very difficult to figure out how to change it in order to get it back to where it was. . . . There really wasn't a lot in the record in terms of administrative enforcement as to why things had been done a certain way." The effort to overturn the decision did not even reach the floor in 1985.

But there was one active civil rights battle that summer. Bradford Reynolds, the assistant attorney general in charge of civil rights, was nominated to be promoted to associate attorney general, the third-ranking position in the Justice Department. Civil rights groups were furious and organized in opposition.

Kennedy set the tone of the hearings. When they began on June 4, he said, "Mr. Reynolds has been the architect of most, if not all, of the Administration's retreat on civil rights. Under his leadership, the Civil Rights Division has exploited numerous opportunities to roll back the past three decades of progress on civil rights." Kennedy blamed Reynolds for the Bob Jones case, for opposing voluntary school desegregation efforts, for ignoring the constitutional rights of the mentally retarded, for overruling aides to approve voting law changes that courts later threw out, for fighting legislation to undo the Grove City decision, and for trying to reopen more than fifty settled agreements on affirmative action. He concluded, "Mr. Reynolds has done enough damage to civil rights in his current level at the Department of Justice and does not deserve to be promoted."

Biden harried Reynolds on a Louisiana redistricting case, on which his decision had been overruled by a federal court. Reynolds had argued that the gerrymandering that produced a duck-shaped district was intended to help a white Republican Congressman, Bob Livingston, not to dilute black votes—although that was just how to help Livingston. Kennedy focused on the Bob Jones case and then attacked him for approving the moving of a Selma, Alabama, polling place from a black neighborhood to the courthouse nearly a mile away. When black voting decreased, Justice Department aides said that was a clear discriminatory effect of the move. Reynolds disagreed, and his memo said that was a "fanciful conclusion." Arlen Specter, Pennsylvania Republican, asked why Reynolds was retreating on affirmative action. The next day Mathias accused him of doing too little against housing discrimination.

Reynolds did not make a strong showing in those two days, and dozens of outside groups presented statements against him. Then it turned out he had not told the committee the truth in testimony when he said he met with opponents of voting plans he approved in Louisiana and Mississippi. "My recollection may have failed me," he said when the hearings were reopened on June 18 to probe those discrepancies. Next, Specter said that Reynolds had misled the committee three years before, testifying one way about the Voting Rights Act while writing internal department memos in the other direction. Kennedy said Reynolds had given misleading testimony about the Bob Jones case. Even a nationwide radio address by Reagan, defending Reynolds as "a tireless fighter against discrimination," hardly helped the nominee. With the Democrats solidly opposed, and Mathias and Specter joining them, the nomination was killed on a 10–8 vote on June 27.

SOON AFTER HE returned from South Africa, Kennedy began discussing legislation, meeting with other Senators and with outsiders like Randall Robinson, head of an organization known as TransAfrica, who had organized the continuing demonstrations and arrests outside the South African Embassy. Kennedy had heard the arguments that blacks would suffer most from sanctions, but he was unimpressed because of his trip, on which he said he met, "at every point along this journey, this burning, burning desire for the sanctions."

The measure the Senators agreed on had four provisions. It prohibited new loans by U.S. banks to the government of South Africa, new investment in South Africa by U.S. companies, the sale of computers to the South African government, and the importation of Krugerrands, a gold coin prized by collectors and people who wanted to possess the precious metal, which U.S. law forbade private ownership of except as jewelry or coins. Alan Cranston of California, the Democratic whip and the sponsor of South African legislation in 1984, hosted the final meeting in the Capitol office that had once been Kennedy's. After they agreed on what would be in the bill, Kennedy said, "Alan, you've been out there, a leader on this, and you're up for reelection. Why don't you take the lead on this piece of legislation?" But Cranston refused, explaining that a major contributor was a gold coin fancier who was furious at the idea of banning Krugerrands.

Kennedy and Weicker introduced the bill on March 7. Ted told a press conference, "We cannot continue policies that actually encourage Americans to invest in racism or profit from apartheid. . . . There will be stronger steps to come if South Africa continues its repressive ways." He ridiculed the Reagan administration's claim that its quiet diplomacy was working, noting that in February, 18 blacks had been killed and 234 more injured when the government forced them out of the Crossroads squatter camp, and the leaders of the United Democratic Front, including Lekota and Molefe, had been arrested and charged with high treason. He said South Africa's "acts of repression speak louder than its rhetoric of change."

The State Department continued to oppose sanctions and contend that South Africa's reforms were real. But on March 21, Secretary of State George Shultz told a House hearing, "The system of apartheid is totally repugnant to me, the Administration and the President." Many Republicans also deplored apartheid and the United States' feeble policy. A freshman, Mitch McConnell of Kentucky, called U.S. policy "wishful thinking." But Reagan was their party's overwhelmingly reelected President. No others were as ready to desert him as Weicker was. In particular, Richard Lugar of Indiana, chairman of the Foreign Relations Committee, appreciated that the Kennedy-Weicker bill did not require American businesses to leave South Africa, but opposed immediate sanctions and regretted, rather than attacked, the administration view that no legislation was necessary. On March 27, Lugar brought a bill out of his committee which warned that sanctions would come in two years if "significant progress" was not made. It also offered scholarships for black South Africans.

Kennedy seized on Shultz's March 21 statement and the killings of nineteen more blacks the same day. On April 3, he got the Senate to pass a resolution condemning apartheid and the killings and asking Shultz for a detailed report. It passed, 89–4. No one spoke against it. Kennedy told the Senate, "On the question of apartheid, the United States of America should speak with one voice."

Finding that one voice proved very difficult. Although forty-one of the fifty-two Republicans had voted for the Kennedy resolution, he had made it easy by endorsing Shultz's comments. Having Kennedy out front made many Republicans even less anxious to cross Reagan, if they could find a reason not to. Lugar, hoping to stay in control of the issue, promised hearings and action on legislation that went beyond his March bill. Meanwhile Kennedy and his staff met weekly with church, union, and civil rights groups supporting sanctions and shared information on lobbying efforts.

Kennedy and Weicker testified at the Banking Committee, which shared jurisdiction over their bill, on April 16. Kennedy argued that the United States was seen by South African blacks as a supporter of apartheid and was "despised" for it. He did not argue that his bill would bring the white regime down, but he offered a long-range strategic argument for it: "South Africa will be free some day and, make no mistake about it, those in the government when it is free are going to ask whether the United States was the last country to go down with apartheid. And it certainly appears to blacks in South Africa today that this is the case." Later that day at the National Press Club, Shultz praised South Africa for the progress it was making and scorned sanctions as "well intentioned" but counterproductive.

Before Lugar and the Foreign Relations Committee on April 24, Kennedy renewed his arguments that recent South African policy changes such as integrating national sports teams and allowing mixed marriages were merely window dressing when eight million people had been sent to barren rural homelands and stripped of their citizenship in nine years. He dismissed as "hogwash" the idea that a growing economy would make apartheid go away, saying the South African economy had grown but it had not helped blacks. Then Cranston posed a thoughtful question, asking Kennedy what he would be doing if he were young and black and South African today.

Kennedy prefaced his answer by saying that if he were a young black South African, he would be separated from his family and poorly educated, facing rotten job opportunities and the likelihood of an early death. Then he said that among South African blacks, he had heard on his trip of an increasing sense of a need for "self-defense."

"They said if you do not defend yourself against apartheid, you fail to provide opportunities for your child to survive, you fail to defend your family from being physically uprooted and torn out of your home. They are moving to the idea of defending themselves against the different aspects of apartheid which are crushing to these people. I interpret 'defending' to be a sign of movement away from a non-violent position."

He told the committee, "The question before us is whether we are going to be the last country to go down with apartheid."

When the committee voted on June 4, the Kennedy-Weicker bill was defeated on a 9–7 party-line vote. But Chris Dodd then cited a bill backed by two Republicans, McConnell and Bill Roth of Delaware, and he pushed through two of the sanctions common to their bill and the Kennedy-Weicker bill: the ban on bank loans to the government and the ban on computer sales. He also got the Roth-McConnell provision for a prohibition on nuclear cooperation with South Africa adopted. Lugar voted no on Dodd's proposal, but supported the bill when the committee voted 16–1 to adopt it. It included a Lugar measure to require American companies in South Africa to follow the "Sullivan principles," proposed by Philadelphia minister Leon Sullivan, and provide nondiscriminatory employment and living conditions for white and black employees.

The next day the House adopted a stronger bill, with an immediate ban on the sale of Krugerrands and on new investment in South Africa by U.S. companies. On the Senate floor, Lugar argued against any amendments that would toughen the Senate bill. Kennedy withheld a Krugerrand amendment in exchange for the dropping of a conservative move to prohibit states and cities from imposing sanctions of their own and for Jesse Helms's withdrawing a filibuster threat.

Lugar said the bill would "distance this country from the evil of apartheid," while Barry Goldwater said it was a "blight against the United States to take this action against an ally, a friend in every war." Kennedy said the United States was implicated in apartheid and the bill recognized and attacked that involvement: "Every time a U.S. bank extends credit to the Government of South Africa, we are financing apartheid. Every time we sell a computer to the Government of South Africa, we are aiding in the administration and enforcement of apartheid. Every time an American citizen buys a South African gold coin—the Krugerrand—we are funding apartheid."

The bill passed on July 11, 80–12, but Lugar delayed in sending it to conference with the House. He urged House leaders just to pass the Senate bill, to avoid the risk of a Helms filibuster and a Reagan veto. Meanwhile, violence in South Africa expanded, arrests at the embassy in Washington reached 2,900 in seven months, including twenty-two members of Congress, and France banned new investments. On July 25, Kennedy broke Senate protocol by moving that the Senate agree to a conference with the House. Dole said he had not known that conferees had not been appointed, but insisted it was his prerogative, as majority leader, to move to have conferees appointed. He said he would do so, and Kennedy withdrew his motion.

The conferees did not meet until Wednesday, July 31. The House dropped its prohibition on new investment. The Senate agreed to a ban on Krugerrands, which the President could waive if South Africa made progress on ending apartheid. That came only after an emotional appeal from Representative Parren J. Mitchell, a black Democrat from Baltimore, who said, "You know me slightly. I'm a very proud man. I don't ask for anything unless I can't avoid it. I am pleading with you to take that one quantum step."

The agreement came too late for final action before Congress left for its summer vacation on Friday, August 2. The House passed the compromise bill that day, 380–48. Seven conservative Republicans, including a fourth-term Georgian, Newt Gingrich, urged Reagan not to veto the bill and said they would work to override him if he did. But Helms blocked an immediate vote in the Senate, and the issue dragged on until September, although at least eighty Senators were ready to pass the bill.

Over the next five weeks the administration negotiated with South Africa, and with Lugar, hoping to see some change in apartheid. President P. W. Botha's August 15 speech to a party meeting in Durban was supposed to be the occasion, but he offered only the vaguest hints of change, far less than his foreign minister had predicted to the White House. While the administration was plainly disappointed, Reagan himself seemed blasé, and on August 24 in a radio interview he said American pressure had led to South African reforms and "they have eliminated the segregation that we once had in our own country—the type of thing where hotels and restaurants and places of entertainment and so forth were segregated." He was largely mistaken, and on September 7, Lugar urged Shultz to get the President to sign the bill so the country could speak with one voice. If he did not, Lugar said, his veto could be easily overridden.

Instead, on September 9, the day the Senate returned from vacation to vote on ending Helms's filibuster, Reagan issued an executive order that imposed most of the provisions of the conference report. He barred computer sales to the South African government and prohibited most nuclear cooperation. He prohibited bank loans to the government, except for "certain loans which improve economic opportunities, or educational, housing and health facilities that are open and accessible to South Africans of all races." He left out a provision requiring new sanctions in eighteen months if South Africa made no progress toward ending apartheid, and he promised only consultations with the General Agreement on Tariffs and Trade over banning Krugerrands.

Dole and Lugar persuaded most Republicans to accept the executive order and not vote to end the filibuster. Dole asked unanimous consent to put aside the cloture vote, but Kennedy objected. Dole and Lugar won the vote; supporters of the bill fell seven votes short of ending the filibuster on a 53–34 vote. When Kennedy said there would be more votes, Dole shot back, "The focus has shifted now from South Africa to playing politics. Can we somehow embarrass the President of the United States who, though belatedly, has gotten on board?"

Kennedy said the executive order was too weak, and could be revoked. Tip O'Neill said, "This package will play in Pretoria." But in South Africa it was criticized from both sides. President Botha in South Africa called it "harmful" and negative, while Bishop Tutu said it was "not even a flea bite" and called Reagan "a racist, pure and simple." The next day, September 10, Kennedy said on television, on the *CBS Morning News,* "The Republican Party is at a crossroads on this issue. It must decide whether to be the party of Lincoln or the party of apartheid." The *Washington Post* called that comment reflective of a "frightening self-absorption." On the eleventh, the battle in the Senate widened. Weicker said Reagan's move "was done grudgingly." John Glenn of Ohio said Reagan, after all, had "credited the Botha regime with having eliminated segregation in South Africa." And Dole, defending the President again, said, "I am willing to stack my civil rights record along with those of anyone else in this chamber. . . . Unfortunately, this is no longer an issue of what is good for South Africa. It has become a raw political issue. South Africa is secondary."

On the next vote, with members of the Congressional Black Caucus using their privilege as Representatives to stand on the Senate floor as silent lobbyists, Reagan's opponents gained ground. But they lost again, with fifty-seven votes to end debate and forty-one opposed. One Republican, Bob Stafford of Vermont, accused Kennedy and the Democrats of having no higher purpose than embarrassing the President and switched his vote to oppose cloture. Weicker and Kennedy said the Republicans had started the fight.

The events of the next day were even more spectacularly bitter. Dole, angry that the Democrats were forcing the Senate to stay on South Africa with repeated votes, took the Senate's copy of the conference report from the desk and gave it to Lugar, who took it to his hideaway office in the Capitol. Under Senate rules, the bill's absence meant the Senate could no longer consider it. Dole said there were precedents for such action. But when Byrd, the Senate's ablest parliamentarian, asked him to cite them, Dole said, "We are still doing that research." The Senate broke into laughter, but Byrd was angry and Kennedy was furious, calling the move "trickery" and an "abuse of Senate rules." Kennedy was right; nothing comparable had ever been done, and the Republicans tacitly conceded the point when they quietly returned the bill two weeks later. But by then the steam had gone out of the Democratic effort to adopt the conference report and force a veto fight. Lugar, however, tired of administration delays over the Kruggerrand, threatened to bring the bill up for passage; the White House bowed and banned the South African coin.

KENNEDY'S MOVE ON health insurance was much narrower than his earlier efforts in that area and attracted none of the headlines of his previous battles. But it succeeded.

Representative Pete Stark, a Northern California Democrat who had supported Kennedy in 1980, had become chairman of the Health Subcommittee of the House Committee on Ways and Means, which, like the Senate Finance Committee, had much broader health jurisdiction than the Labor Committee did. Kennedy and Stark developed a proposal designed to cope with the effects of lingering waves of job layoffs and the fact that most Americans got their health insurance through their jobs and lost it when they were out of work.

Kennedy and Stark held a press conference on June 20 to propose that group insurance plans be required to allow a laid-off worker to keep his insurance, by paying the employer's share as well as his own, for eighteen months. Widows and divorcées and their children would also be allowed to keep that insurance for up to five years, if they could pay for it. They also called for legislation to prohibit hospitals from "dumping" indigent or uninsured patients or refusing them emergency treatment.

Stark, as a subcommittee chairman, could force the issue onto its agenda. But he only got the provision for widows and divorcees through Ways and Means; he left the laid-off workers for the Senate.

Kennedy introduced his legislation on September 9 and called it "a fairly small first step" toward dealing with the growing number of Americans without health insurance. He got it through the Labor Committee on September 27 with skillful legislative sleight-of-hand. The action was on the committee's piece of the budget reconciliation bill, which was designed to cut the deficit. This had become an opportunity to jam all sorts of unrelated legislation through with little or no debate, because the 1974 Budget Act prohibited filibusters. Hatch, whose control of the committee was tenuous, wanted it to act without a meeting and to adopt a bill with only deficit-related matters. But once other Senators made it clear they, too, had pet provisions that were the price of their going along without a meeting, Kennedy's staff got the health insurance section added, along with the antidumping measure. Approving a bill without a meeting required unanimous agreement of all sixteen committee members. But it would have been ungracious for a Senator to object after everything seemed to have been worked out. No one did, although Charles Grassley of Iowa, who was tracked down on his tractor by a state trooper, complained bitterly.

When the measure came to the Senate floor, it passed without objection. Business and the insurance industry only woke up to the issue by the time it went to conference in December. While laid-off workers would have to pay 102 percent of the combined personal and employer premium to keep their group insurance benefits—difficult for many of the unemployed to manage—that was still far less than an individual policy would have cost, especially because sicker workers were more likely to sign up. Senator Dan Quayle of Indiana tried to sink it, but other Senate Republican conferees like Senators David Durenberger of Minnesota and John Heinz of Pennsylvania had backed a similar plan in the Finance Committee and paid no attention.

The overall reconciliation bill was held up for months over other issues. When the Consolidated Omnibus Budget Reconciliation Act of 1985 was signed by President Reagan on April 7, 1986, the new program for laid-off workers had a name. The two million or so people who use its provisions each year to keep their health insurance between jobs call it COBRA, the acronym for the whole budget bill.

KENNEDY HAD ONE other important legislative success in 1985. Most of the fights on the defense authorization bill had been over weapons systems, but Kennedy and Representative Patricia Schroeder, a Colorado Democrat, worked together to improve the life of military families. The initiative was Schroeder's; she pushed in the House Armed Services Committee for a variety of provisions that would make it easier for military wives to get government jobs, increase child care programs, get the military to pay attention to the children who moved with their parents, and reduce the costs that servicemen had to pay when they were transferred from one base to another (civilians who were transferred had higher moving allowances). Schroeder said the traditional military response to families was "If we wanted you to have a family, we would have requisitioned one."

Seeking a Senate ally, Schroeder reached out to Kennedy, another untraditional member of an Armed Services Committee. He agreed readily. In a May 16 statement to the Senate, he said that as he traveled from one base to another, "everywhere I go, I find family issues on the top of the agenda—housing, permanent change of station, day care, spouse employment, education for the children—the list goes on and on." He argued, "The readiness and morale of our troops is critically dependent on the well-being of their family members, an issue which deserves as much attention as any of the more traditional components of military preparedness." His version of the plan was adopted as part of the defense bill on June 5. Most of it survived a House–Senate conference in August. Kennedy was also a successful voice for putting a 3 percent military pay raise into effect on October 1, 1985, rather than January 1, 1986, which the House budget planners preferred. He argued that military pay lagged more than 10 percent behind civilian pay for comparable jobs. He told the Senate that troops "want—every now and then—some sign that the country is grateful to them for what they do, and that the American people understand and appreciate the sacrifices involved in the performance of their duties. For us to delay this much-needed and much-deserved pay increase would send just the opposite signal."

KENNEDY WAS ALSO involved with Ireland again in 1985. In 1984 the Irish government sought to strengthen John Hume's position vis-à-vis the Irish Republican Army with an all-party forum which offered up three proposals: a federation, a united Ireland, or joint British–Irish sovereignty over the North. But British prime minister Margaret Thatcher, who had just escaped an IRA bombing designed to kill her and her cabinet, refused to discuss it, saying of each proposal, "That is out." At the urging of Irish diplomats, Kennedy joined with Tip O'Neill and Bill Clark, secretary of the interior and one of Reagan's oldest California friends, to persuade Reagan to urge her to reconsider.

Reagan and Thatcher had a singular chemistry. They helped each other; Reagan's attitudes on South Africa were strongly affected by her views. This time Thatcher took Reagan's advice and negotiated with Irish prime minister Garret FitzGerald. The result was the 1985 Anglo-Irish Agreement, which clearly acknowledged the right of the Irish government to offer proposals on human rights and the police in Northern Ireland. On November 15, Kennedy hailed the agreement as "a treaty that can work, if the extremists will put aside their prejudices and give it a chance." They did not, as Northern Protestants were infuriated by what they considered a sellout by London. But the Thatcher government's acknowledging a role for the Republic of Ireland in the affairs of the North was a precedent that mattered greatly to all future peace talks.

THAT TOTAL COMMITMENT to the Senate which had developed since 1982 was formalized on December 19 when Kennedy announced in a five-minute television broadcast that he would not run for President in 1988.

Some of his aides had hoped he would run, and most reporters thought it likely. Shrum talked it up, and Horowitz, always a doubter, had a plan for raising money and traveling. Kennedy himself told the *Globe* he wanted to learn if the people now "want an activist government or a passive government. Are they going to be ready for redressing some of the needs of the country . . . or whether it's going to be an atmosphere and a climate of 'let's just muddle through things.' "

But Kennedy had really done almost nothing to get ready for a race. There were family concerns, too, not so much about his own children, but about the personal problems of some of his nephews. Years later, he said family opposition at Thanksgiving was critical.

Perhaps his own problems were the worst. Even during a year of vast legislative accomplishment like 1985, he seemed remarkably unhappy, chasing women and drinking heavily. That year two different Capitol Hill restaurants experienced episodes with Kennedy and another unmarried Senator, Chris Dodd. At La Colline he and Dodd, drunk, smashed each other's autographed pictures. At La Brasserie, while their dates went to the ladies' room, Dodd held a waitress on his lap while Kennedy rubbed up against her.

In the broadcast on the nineteenth and the next day in a press conference, Kennedy himself insisted the main reason for announcing he would not run and for announcing it so early was that he was tired of having everything he did, from Johannesburg to the Senate floor, analyzed in terms of an assumed candidacy.

"The prospect is that the fog surrounding my political plans will cloud the far more fundamental challenge to put aside complacency, and the appeals to narrow interests to care about one another, even the least among us," was how he put it in the broadcast.

The decision was made quickly and without debate. Horowitz, who had never expected him to run, was told in early December. Michael Kaye, who had made the 1982 commercials, and his crew came from California. Shrum, by then working on his own as a political consultant, was summoned to the house in Hyannis Port on Wednesday, December 18. He sat down in the living room. Kennedy brought him a martini. Shrum recalled the conversation this way:

Shrum said, "I know, you're not going to run for President."

Kennedy replied, "That's right and I don't want to argue."

Shrum pressed on, "I'm only going to say one thing. You get to run against George Bush and that really is your best chance."

Kennedy ended it by saying, "I know that and I don't want to run."

Shrum wrote a statement. Other aides were summoned and told of the decision in the Boston office. One, Ranny Cooper, was so shaken that she drove into a curb and punctured a tire.

The statement he broadcast sounded like a calm, satisfied statement about his life in American politics. He said, "I will run for reelection to the Senate. I know that this decision means that I may never be President. But the pursuit of the Presidency is not my life. Public service is. . . . The thing that matters most, the greatest difference we can make, is to speak out, to stand up to lead and to move this nation forward. For me at this time the right place is in the Senate."

He said much the same, more awkwardly, at the next day's obligatory press conference: "It would certainly appear that the chance to gain the Presidency has been removed from my future."

He said he had no regrets, and was eager to set off with his sons the next day on another tour about hunger, freed of political suspicions. Still, the taping of his statement had been unusually awkward, with one retake after another. Horowitz said Kennedy never did quite get it right.

The next day he was driving around slippery back roads in northeast Missouri, talking to hard-hit farmers. He had stayed the night with Bill Shoop, a bankrupt hog farmer. Kennedy told reporters who wanted to ask about politics that these farmers had helped save Ethiopia and the Sudan the previous winter and "I think it's about time we in the United States respond to the crisis that exists right here."

CHAPTER 27

An Advocate for Arms Control

The never-to-be-President Kennedy began 1986 with a South American trip, visiting four countries which had recently shaken off military regimes—Brazil, Uruguay, Argentina, and Peru—and one where admirals and generals still ruled—Chile.

Aside from a rousing reception in Patagonia, where Kennedy toasted Carlos Menem, a follower of Juan Perón not long released from prison, as "the next President of Argentina," the stops in the restored democracies were uneventful. He spoke of the need to negotiate easier repayment terms for their burdensome debts.

It was never clear why Chile gave him a visa, especially since the foreign minister who announced the visa called Kennedy "permanently unfriendly" and said therefore no government official would see him. The United States ambassador, Harry Barnes, said years later he thought the government of Admiral Augusto Pinochet felt it would be most irregular to keep a prominent U.S. Senator out, especially when Chile was taking the first halting steps toward a return to democracy. Pinochet had lifted a "state of siege" in June 1985 and allowed opposition newspapers to publish, though under strict censorship, and he retained powers of arbitrary arrest. Beatings by the military continued, though at a much lower rate than in the first years after the coup. All opposition parties except the Communists had agreed in August

1985 on a patient timetable for elections that would leave the military regime in power until 1989.

Before Kennedy arrived, full-page newspaper ads in Chile asked menacingly, "Who would dare shake hands with the No. 1 enemy of Chile?" Before they left Argentina, Kennedy and Greg Craig were warned by the CIA station chief in Buenos Aires that the Chilean military had organized demonstrations, rented buses, and provided signs for protesters. Kennedy told reporters about it, saying intelligence agents were already spray-painting slogans against him on walls.

Still, they were not prepared for the scene at the Santiago airport when they arrived on the morning of January 15, 1986. Demonstrators had blocked the road to the city, carrying signs saying "Death to Kennedy" and showing pictures of Mary Jo Kopechne. Some wore life preservers marked "Chappaquiddick." They threw eggs and stones at opposition leaders who arrived to meet Kennedy. Police stood by and joked with the demonstrators, and then told Kennedy he would have to stay at the airport and cancel his meetings in town. Kennedy told Craig to tell the reporters who was responsible for the demonstrations, and Craig went out and said the government had "organized and planned" them. Kennedy himself delivered a brief arrival statement in which he said he was not an enemy of Chile, as his foes charged, but "I am an enemy of torture, kidnapping, murder, arbitrary arrest."

Angry protests by Kennedy and Barnes to the foreign ministry produced two helicopters, and after ninety minutes at the airport the party flew to the hospital which had been their first scheduled stop. But Kennedy's car was pelted with eggs when he left, and he canceled other stops and held meetings instead at a downtown social club. Church leaders came. So did politicians and mothers with pictures of their children who had been "disappeared" by the military. After demonstrators left in late afternoon, a friendly crowd gathered. Kennedy and his aides looked at a collection of flags on a desk and tried to decide which was Chile's. Kennedy guessed right and waved it from a balcony. Later he went out and spoke to a few hundred people from a park bench. He said, "I have come here from Brazil, Uruguay and Argentina where the people have said 'no' to military juntas. Will Chile be free one day, too?" The crowd replied, *"Chile libre!"*

That evening Kennedy spoke to a group of about eight hundred opposition Chileans, packed in an old auditorium. He was impressed by the courage of people who came to hear him, because "everyone who went in there had their picture taken and was marked off." He praised his hosts:

"You have spoken out on behalf of those condemned to silence. You have stood up for those bowed under the weight of tyranny and torture. You have said that the disappeared must not be forgotten. You have said that the exiles must be permitted to come home—not in fear but in freedom."

Kennedy said, "I have come to a country which has proven anew that the fire of freedom cannot be extinguished—even when the darkness descends, when dictators rule and law is lost, the flame still warms and moves millions of individual indomitable hearts. The spark still passes from soul to soul, connecting one person with another, across vast expanses of space and time, with each for a few moments or miles carrying and passing on freedom's torch—so that one day the light finally shines out again across the land."

He recalled Chile's patriots and its poets, and visits he and Robert Kennedy had made in a different time. And he challenged the military and its arguments that national security justified oppression:

"In fact, the practices of torture, kidnapping, murder and mass disappearances are the poisoned roots on which subversion best feeds and flourishes. The dictators of the East or the West, from the far left or the far right, offer only the dim security of the slave—or the final security of the grave.

"There can be no true national security without respect for individual human dignity. There can be no true defense if a nation is torn and tortured and split asunder. . . .

"Our earliest experience in the United States led our first patriots to the conclusion that there should be absolute civilian control over military institutions. The armed forces have no larger right or claim to speak for the nation than any other group. The military can and should have an honorable role. But like anyone else, the military must be prepared to subordinate its weapons to the will of the people, expressed through their elected leaders and the rule of law."

When Kennedy finished speaking, a lone man began to sing. Hesitantly, a few more added their voices, and then the whole audience joined. They were singing the banned "Internationale," known in Chile as Allende's theme song, although an outright Communist anthem elsewhere.

THE DAY KENNEDY left Chile, January 16, Horowitz arrived in Moscow to settle arrangements for a trip on which Kennedy would meet with Mikhail S. Gorbachev, who had been in power as general secretary less than a year. It would be Kennedy's first visit since 1978. After Leonid Brezhnev died in

November 1982, Kennedy had hoped to arrange meetings with his successors, Yuri Andropov and Konstantin Chernenko. But each died before Horowitz could schedule a Kennedy visit: Andropov in February 1984 and Chernenko in March 1985.

Kennedy was seen by the Kremlin as a leading fighter for nuclear disarmament and as a representative of a family that personally understood the costs of war. His steady criticism of the Reagan administration for inactivity on arms control recertified his credentials to the Soviets. Yet Kennedy was continually in touch with that administration before and after any contacts with Soviet officials. Before the 1986 trip, Kennedy and Horowitz met at the White House with President Reagan; Admiral John Poindexter, head of the National Security Council; and Jack Matlock, the NSC's Soviet specialist. They went over arms control issues, and then Reagan urged Kennedy to take on another mission, too, to find out if Gorbachev had been baptized. Reagan, who had never met a Soviet leader in his life until a brief meeting with Gorbachev in Geneva in November 1985, had been struck by seeing Gorbachev quoted as using the word "God." He told Kennedy that if the new Soviet boss was a believer, then there were new hopes for better relations.

Kennedy had set three conditions for a visit in 1986: the release of Anatoly Shcharansky or Andrei Sakharov, a personal appearance on television, and a meeting with dissidents. Shcharansky had been imprisoned for his Jewish activism since 1977. After attacking the invasion of Afghanistan, Sakharov had been exiled to Gorky in 1980. When Horowitz flew to Helsinki to prepare the meeting with Andrei Pavlov in early October 1985, Pavlov's intense criticism of Sakharov made it clear that if anyone was freed, it would be Shcharansky.

Then in early January, Horowitz went to Moscow to make sure that everything was ready. Pavlov promised that nine families would be released and hinted that Gorbachev would deliver some "ground-breaking" message on arms control in Europe by telling him that progress could be made there even if the Americans made no concessions on their plans for a space-based missile defense system. The Soviet Union considered the U.S. effort to develop a space-based missile defense that would shield the United States against Soviet ICBMs to be the main obstacle to disarmament. Soviet leaders saw that effort, formally known as the Strategic Defense Initiative or SDI, but commonly mocked as "Star Wars" after the science fiction movie, as an outrageous American effort to gain the ability to launch a nuclear first strike and be safe from retaliation. Reagan saw it as something very different, a development that would make nuclear missiles obsolete. He talked of sharing the results with the USSR so that neither country would have to fear an attack.

Pavlov also told Horowitz that Shcharansky would be released soon, but that the decision could not be discussed until it happened, when Kennedy would be free to claim credit for it. Horowitz told Pavlov that Kennedy was not interested in claiming public credit. When Horowitz returned to Washington, he brought Shcharansky's wife, Avital, to Kennedy's office. A Russian Jew who had been given an exit visa the day after her marriage in 1974, she had been traveling the West for years, campaigning for her husband's freedom. Kennedy, on the speakerphone from Palm Beach, told her the good news.

Kennedy arrived in Moscow on February 4 and was sped to a luxurious guesthouse in the Lenin Hills. Walking outside that night in the zero-degree cold to avoid listening devices, he asked Horowitz if he was sure the Soviet leaders would follow through on Shcharansky, and told him to tell Pavlov he wanted it confirmed, to him, and to make it clear he did not care who got the credit or who announced it.

The next morning they met in the U.S. Embassy's soundproof secure room with Ambassador Arthur Hartman. He gave them a dim forecast. While the Soviet position at the time was vague, Hartman said he did not think Gorbachev would "delink" intercontinental missile defense from intermediate missiles in Europe. He said Gorbachev was quick and was bringing in young aides, but was basically just another old-line Communist politician. He said Kennedy's meeting that day with Foreign Minister Eduard Shevardnadze would probably produce little.

But the meeting was productive. Shevardnadze began by recalling their meeting in Georgia in 1974. Then he told Kennedy that he knew of his concern for Shcharansky, who would be released after Kennedy's visit. When Kennedy asked about whether various Soviet citizens whose release he had sought would be allowed to emigrate, Shevardnadze said, "You will discuss that with Gorbachev tomorrow. And you will be pleased."

Then they talked about SDI. Kennedy's comments illustrated the sort of honest broker role he sought to play. In the United States, he was a severe critic of SDI as a waste of money on a technological long shot that served only to threaten relations with the Soviet Union. To Shevardnadze, he reiterated his personal opposition, but said, "Gee, you can't believe what these guys are coming up with and they're going to do it. They're going to spend all this money and they are going to do it. I wouldn't do it. I don't think it's advisable at the time. But, they're going to do it."

Then Shevardnadze, almost casually, told Kennedy that SDI presented no obstacle to negotiations on intermediate missiles in Europe. Kennedy almost missed it, but interrupted as Shevardnadze went on to another point, to ask if he was confirming that there was no linkage between the two issues. Shevardnadze said he was, and Kennedy said he wanted to hear it from Gorbachev, so the foreign minister told him to ask Gorbachev the next day. The meeting lasted two and a half hours.

That afternoon, Kennedy spoke at the Academy of Sciences, praising the Soviet Union for its high regard for scientists, and saying that "the indispensable value of science is its ability to speak truth to power." He also talked of his admiration for Sakharov, "an eminent Nobel Laureate, the first Soviet citizen to receive the Nobel Peace Prize, a member of your academy." A. D. Alexandrov, the academy's president, interrupted to say that Sakharov had worked on the hydrogen bomb and was a Hero of the Soviet Union—one of the standard explanations for keeping him in the USSR. Kennedy just said he was glad to hear that they recognized his contribution to their country.

That afternoon Kennedy went to see Gorbachev, in the same office where he had met Brezhnev in 1978. While photographers took pictures, they spoke about the disaster of the explosion of the American space shuttle *Challenger*, whose fiery crash had delayed Kennedy's visit by a week. When they left, Gorbachev began by talking of the long-standing connections between the Kennedy family and Soviet leaders. He said he wanted them to continue, and wanted to receive private messages from Kennedy. He told Kennedy that his November meeting with Reagan had been utterly without substance, and there would not be another meeting until Reagan responded to his disarmament proposals. Then he made it clear that Pavlov and Shevardnadze had been speaking for him on linkage, saying there were "no preconditions of any kind and no linkage of any kind" between INF talks and SDI. Kennedy asked if that meant that even if there was no progress on Star Wars, there could still be an agreement on intermediate nuclear forces in Europe. Gorbachev said that was what he meant.

But Gorbachev was also confrontational. He said that he would not discuss intercontinental missiles while the United States persisted with SDI, and he denounced the U.S. military-industrial complex in language that could have come straight out of *Pravda*. And when Kennedy brought up human rights, Gorbachev pounded the table as he said that the United States had no business telling the Soviet Union how to treat people. There were homeless people in America but not in the Soviet Union, he said, and human rights included the right to enough to eat, the right to health care, and the right to walk safely on a dark street at night without fear of a mugging.

Kennedy was getting angry. Horowitz saw the veins in his neck start to bulge. Two days later at a Washington news conference, Kennedy said, "He came within an inch of setting me off as well. I was about to tell him I hadn't come all the way to Moscow for a lecture, when he suddenly stopped and said that of course he was prepared to respond to individual cases. It was like he had been watching my temperature go up all along."

Then Gorbachev personally confirmed to Kennedy that Shcharansky would be freed, though he should say nothing about it yet publicly, and told him that twenty-five other Jews whose cases he had raised would be allowed to leave. As the tension eased, Gorbachev said that these decisions showed that he was ready to do business on human rights with someone like Kennedy, whose approach was professional and not designed to intimidate the Soviet Union. And he turned personal, telling the Senator that in preparing for this meeting he had watched a film on President Kennedy, and would never forget John-John saluting his father's coffin. He asked how the slain President's son was doing and told Kennedy that they had to keep their communication going.

Before he left Moscow, Kennedy had another emotional meeting with dissidents at Professor Lerner's apartment. They had been moved by Kennedy's mention of Sakharov the day before, hearing of it through the grapevine, since the Soviet press had ignored it. When Kennedy was asked if something was happening about Shcharansky, he would not confirm it. He also gave a talk on Soviet television in which he invoked "my brother Jack" and his 1963 speech that led to the atmospheric test ban treaty. He quoted President Kennedy as saying, "No nation in the history of battle ever suffered more than the Soviet Union suffered in the course of the Second World War." He reminded them that President Kennedy had said of their two countries that "almost unique among the major world powers, we have never been at war with each other."

Kennedy explained Reagan's view of SDI: "an effort to render nuclear weapons obsolete—to make nuclear war impossible." While he reminded the audience of his own "firm opposition to militarizing the heavens," it was a rare moment for Reagan's idea to be put on Soviet television. And he argued for Reagan's bona fides on reducing nuclear weapons, saying, "I know that President Reagan is ready to enter into such an agreement. As he has said, 'A nuclear war can never be won—and must never be fought.' I met with President Reagan before my departure—and he emphasized the supreme importance he places on the successful outcome of the arms control negotiations."

After trying to explain Reagan to Soviet leaders, Kennedy held a Washington news conference on February 8 to explain Soviet attitudes to Americans. "I came away with a real sense of hope that a breakthrough can be achieved in the reduction of intermediate-range nuclear forces" because Gorbachev did not link this issue to SDI. He quoted Gorbachev as saying that he doubted that another summit meeting should be held unless progress had been made in some area of arms control under negotiation in Geneva. Among the areas Kennedy cited was his personal favorite, a complete test ban treaty. Kennedy told of the promised release of dissidents, except for Shcharansky; he said the case had come up but would not say anything about it. Three days later, on February 11, Shcharansky was freed, walking across a bridge into West Berlin.

At a time in U.S.-Soviet affairs when the governments were not talking much, Kennedy was an important channel. At first the White House kept his role very secret. On this trip, Admiral Poindexter instructed him not to brief the U.S. Embassy in Moscow on his meeting with Gorbachev, but to save the news for Matlock and himself. Matlock said in 1999 that he kept Secretary of State George Shultz informed from the start of Kennedy's Soviet contacts. But Shultz in his book on his years at State said he learned of them at the end of 1986, and from Kennedy.

Looking back, Matlock concluded that because of Kennedy's political differences with Reagan, it made an impact in Moscow when, on arms control or human rights, Kennedy's message was essentially the same as Reagan's. "That made clear Reagan wasn't just playing politics," he said, "and these things represented the normal political spectrum in the United States." He said that was particularly useful in preventing the Russians from thinking "things are more split than they are, and if there are political rivalries, you can play them to your advantage." Matlock said President Reagan appreciated Kennedy's role, though he assumed that the 1986 trip to Moscow meant that he was running for President after all.

Kennedy's status as a noncandidate, whatever Reagan thought, was on display on March 22 to the Gridiron Club, the white-tie association of journalistic pooh-bahs, where he twitted himself on his 1980 campaign and "my old pal, Roger Mudd. There we were, sitting on my lawn, in front of my house, on my Cape Cod. And he started asking trick questions. Like why do you want to be President?" He explained his recent vote for Gramm-Rudman-Hollings as "If I can't run the government, we shouldn't have one." He joshed all the would-be Presidents in the room, and Reagan, too, saying, "People here may be sharply divided over the Reagan Administration's policies. But they all admire Ronald Reagan for not getting involved in them." And, he said, "I'm not bothered at all by the idea of a seventy-five-year-old President. I might need it myself some day."

ON MAY 21, Kennedy attacked Reagan over hunger. Although the Labor Committee had no jurisdiction over the food stamp program, Kennedy was working quietly to encourage the Agriculture Committee to restore the cuts made at Reagan's urging in 1981 and 1982. First, he testified before the Senate Agriculture Committee that the war on hunger was being lost because "in 1981, the Reagan Administration enacted the broadest, deepest cuts in food programs in history--and then adopted economic policies that sent unemployment into double digits. The result was inevitable, foreseeable and cruel." Kennedy called for large expansions in the food stamp program and other methods of feeding the young and the elderly. But Reagan, speaking to teenagers at the White House, argued, "I don't think there is anyone going hungry in America simply by reason of denial or lack of ability to feed them." Instead, "where there is hunger," he said, "that is probably because of a lack of knowledge on the part of people as to what is available." Kennedy shot back, "In a sense, the President is correct—hunger is caused by the ignorance of those who do not see the suffering of millions of Americans. In the last five years, hunger has swept across this country more quickly than at any time since the Depression. This did not happen because millions of people suddenly became ignorant; it happened because they or their parents cannot find work and the safety net was cut away."

That same day he renewed the battle over apartheid. All winter, things were getting worse in South Africa. Arrests and detentions ran into the thousands. South Africa attacked neighboring African countries in pursuit of black terrorists. South Africa had not responded to 1985's modest U.S.

sanctions, nor to pressure from the Commonwealth. The Reagan administration's efforts were equivocal; it seemed at least as fearful of possible Communist influence in Mozambique and Angola as it was concerned about turmoil in South Africa. The Central Intelligence Agency worked with the South African military intelligence, and William Casey, head of the CIA, urged that President P. W. Botha be invited to the White House. Shultz fought off that idea, but the administration still went out of its way to praise South Africa when it could. On April 24, it called South Africa's announced repeal of the pass laws, under which blacks were denied freedom of movement throughout the country, as a "major milestone on the road away from apartheid."

On May 21, Kennedy, Weicker, and Representative William H. Gray III, a black minister from Philadelphia, introduced new sanctions legislation. They sought to ban all new investment and bank loans to South Africa and bar U.S. help for developing the country's energy resources. They also sought to halt U.S. imports of coal, uranium, and steel, ban South African airliners from landing in the United States, bar U.S. deposits in South African banks, and keep South African companies off U.S. stock exchanges. Kennedy quoted Bishop Tutu's appeal for sanctions and said, "To those who see faint signs of progress and ask us to hold off, we say, 'Read the morning headlines; talk to anyone in South Africa; open your eyes to the truth that the progress is too little and may soon be too late.' " A White House spokesman said the administration was "opposed to further sanctions."

In June, Lugar came out against additional sanctions from Congress and urged the administration to take the lead. But despite Shultz's efforts, nothing happened, except for comments by Reagan that seemed to side with the white South African government against the blacks. So the House seized the initiative, brushing aside the Kennedy–Weicker–Gray bill and passing legislation to impose a trade embargo on South Africa and require all 284 U.S. companies to leave within 180 days. Even the disinvestment bill's sponsor, Representative Ronald V. Dellums of California, said he was "shocked" that it passed.

Kennedy hailed the House action and introduced the Dellums measure as his own, but he was maneuvering to get something like his original proposal passed. One week in July was critical.

On the morning of Monday, July 21, Kennedy took on Reagan personally in an op-ed article in *The New York Times*. "Without exception," he wrote, "every time that he has been questioned about South Africa since taking office, President Reagan has defended the white minority regime. Although he routinely denounces apartheid as 'repugnant,' he consistently refused to move beyond the abstraction of apartheid to criticism of the South African government itself. Instead, the President repeatedly praises that regime for its 'reforms' and defends its crack-downs as justified by 'violent' or 'riotous' blacks. . . . Pressure by the Administration for change in South Africa will not work until the President himself is on board."

Later that morning, July 21, Kennedy and Weicker staged a forum on South Africa at which the United Nations ambassadors from Botswana and Zambia, two nations bordering South Africa, called for sanctions, saying they could tolerate short-term hardships. But the stars of the event were the heads of the Eminent Persons' Group, a committee of Commonwealth leaders assigned the previous autumn to try to work out some solution in South Africa. The group, headed by the former Australian prime minister Malcolm Fraser and the former Nigerian head of state General Olesegun Obasanjo, had given up in June, saying the South African government was not prepared to "negotiate fundamental change."

Kennedy opened the meeting, warning, "Today in South Africa, hundreds are dying and thousands are risking their freedom to end apartheid. . . . The United States must not stand silent in the face of their appeals for our help. Above all, the United States must never stand, or appear to stand, on the side of the oppressors and the enemies of a free South Africa." Fraser told the forum the only alternative to sanctions was "a guerrilla war." He said, "No one can guarantee that sanctions will work." But, he said, "it is the only remaining option left." He argued that if the United States acted, Great Britain, with much deeper economic ties, would have to follow. But if neither acted, "black leaders will decide that they have no significant political support in the West" and their eventual victory would lead to an "anti-West, pro-Soviet regime."

Kennedy also used the Commonwealth leaders' visit for quieter persuasion. The night before the forum, he had them to dinner at his house in McLean. Obasanjo was seated next to Senator Nancy Kassebaum of Kansas, chairman of a Foreign Relations subcommittee on Africa, and Fraser was next to Lugar, who had scheduled hearings for that week. They made a case for U.S. sanctions as a way to move Britain. Fraser, a firm supporter of United States policy in Vietnam when he was prime minister, emphasized his anti-Communist record to Lugar, saying, "Senator, you should not forget what my credentials are on this issue." Lugar said years later that the evening was "very helpful" as he figured out his position.

Lugar was not sure what that position would be on July 21 when he, Kassebaum, and Bob Dole went to the White House to urge the President to shift policy, and to say nothing about "constructive engagement" or "sanctions" in a speech he was to give on the twenty-second. Shultz had proposed the speech, but lost control of it to Pat Buchanan, a White House aide deeply committed to supporting the government of South Africa. When he spoke, Reagan did criticize government violence against peaceful demonstrators and say apartheid was "morally wrong." But he also described a country "in a state of transition. More and more South Africans have come to recognize that change is essential for survival. The realization has come hard and late, but the realization has finally come to Pretoria that apartheid belongs to the past. In recent years there has been a dramatic change." Reagan, as a disappointed Shultz wrote later, "did not say much about the desperation and fear of blacks but spoke eloquently about the security of whites in this country they love and have sacrificed so much to build." South Africa's foreign minister, Roelof Botha, praised Reagan's "sincere desire" to support the South African government's approach to power sharing and his criticism of the African National Congress.

On the morning of July 22, a few hours before the Reagan speech, Kennedy testified before Lugar and the Foreign Relations Committee. While he called for disinvestment, he made it clear that a lesser set of sanctions would be acceptable. But he insisted strong action was necessary. Citing Fraser, Obasanjo, and the ambassadors of Zambia and Botswana, he insisted that "the case for comprehensive sanctions is clear. Quiet diplomacy has failed. Only strong economic pressure from the international community—combined with peaceful pressure inside South Africa—can stop the descent into violence and persuade the apartheid regime to change course."

Shultz himself testified for four hours on July 23, trying to repair the damage from Reagan's speech. Where the President had denounced sanctions, Shultz said the United States might join its European allies in imposing some, but pleaded with the committee not to put us "in a straight-jacket of rigid legislation." Some Democrats harangued Shultz, especially Senator Joseph R. Biden of Delaware, who had not previously been prominent in the anti-apartheid effort. Biden sneered at U.S. policy—"if you call it a policy"—with its "lack of moral backbone." Kennedy, sitting in with the committee, also challenged the policy as "an unmitigated disaster," but emphasized that he viewed Shultz with "an enormous regard and a high degree of respect."

He was working to support Shultz and Lugar, too. When Lugar reacted to Reagan's speech by announcing that he was drafting a bill with some sharp rebuffs to South Africa, Kennedy said that even though it fell short of disinvestment, its message would be "strong and meaningful" and he could support it. But Kennedy kept on denouncing Reagan. On Saturday, July 26, Democratic leaders chose him to give the nationwide radio response to Reagan's weekly broadcast. "Apartheid is the greatest moral challenge facing the earth," he said, and the Reagan administration was failing the test. "Time and again, instead of leading the international opposition to apartheid, the administration has protected and defended the actions of the South African government. Spokesmen for the administration have become apologists for apartheid. Our greatest Republican President, Abraham Lincoln, called America the last best hope on earth. But under Ronald Reagan, America has become the last best friend of apartheid."

Kennedy also tangled with Dole. The calendar was beginning to threaten the legislation, with a Senate vacation in August and an early-October adjournment for elections. Dole had no use for apartheid and not much more for how the administration dealt with it, but as majority leader he did not like to oppose Reagan, especially on foreign policy. Lugar usually felt the same way, but believed Reagan had abandoned leadership on South Africa. Moreover, when Kennedy pressed for quick action, Dole felt his authority was being questioned. He grumbled, "One thing about being leader. I don't have to let Ted Kennedy run the place."

Lugar held an unusual Friday session of his committee to vote out a sanctions bill on August 1. He said he was doing so to put some pressure on British prime minister Margaret Thatcher, whose opposition to sanctions as "immoral" and "utterly repugnant" Reagan repeatedly cited. She was to meet that weekend with Commonwealth leaders on the issue. Lugar's committee approved, 15–2, a bill to ban new U.S. investment in South Africa, prohibit U.S. imports of uranium, coal, steel, and any products made by the government, deny U.S. landing rights to South African airlines, bar the use of U.S. banks by the South African government and companies it owned, and encourage the President to deny visas to South African officials and to sell gold reserves to hurt its economy. By Monday, Thatcher had given a little ground. Though she agreed to sanctions weaker than those of other Commonwealth nations, she did agree to bar imports of coal, iron, and steel if other European countries did the same and endorsed a "voluntary ban" on tourist promotion and on new investment for South Africa.

On August 14, Lugar and Kennedy opened the Senate debate. Lugar said his bill sought to "target sanctions" against the regime and not its victims. Kennedy said, "The true friends of America are not to be found in the government offices of Pretoria but in South African prisons." The major vote that day came on an effort by Alan Cranston to adopt the House's disinvestment provisions. Kennedy backed it, saying, "Make no mistake about it. Every dollar that is invested in South Africa is another brick in the wall of apartheid. American business has no business doing business in South Africa." But the Cranston measure was killed on a 65–33 vote.

On the fourteenth, Kennedy failed in an effort to broaden the bill to include several sanctions supported by most Commonwealth nations. Lugar killed it on a 51–48 vote. But the next day, Kennedy dropped a provision banning renewal of existing bank loans and short-term credit and won on a voice vote. The additional sanctions were bans on imports of iron, steel, and agricultural products and on exports of crude oil and refined petroleum. On other votes, the provisions on denials of visas and encouraging gold sales were dropped and a ban on textile imports was added. The bill passed 84–14, with all forty-seven Democrats voting for it, on August 15, the last day Congress met before summer vacation.

Kennedy and Lugar then set to work to persuade the House to adopt the weaker Senate bill unchanged because a conference would allow more Senate delay and the bill would get to the White House too late to allow a vote to override a veto before Congress quit for the elections. Lugar knew some House Democrats distrusted him because of his support of Reagan's order in 1985. Representative Mickey Leland of Texas, chairman of the Congressional Black Caucus, whose members had stood witness in the Senate chamber during the August votes, invited Kennedy to a caucus meeting to discuss what to do. Craig recalled Kennedy told them, " 'We have to send this bill to the White House right now. Because if we have to have a conference, we lose the chance in this session.' And they sat there, and you know, it didn't take them that long—fifteen, twenty minutes of talking. They said, 'Yes, Ted's right. We'll do it.' " Dellums, a Kennedy supporter in 1980, agreed that it was wise to drop his measure for the Senate bill. Kennedy's message was echoed in the House on September 12. William Gray said that while the House bill was better, "we do not have the opportunity to go to conference and at the same time ensure . . . the opportunity to vote on whether or not to override a Presidential veto." The House vote was 308–77.

Reagan took all ten business days the Constitution allowed before vetoing the bill late on Friday, September 26, after Congress had left for the weekend. He said the bill would damage the South African economy and hurt blacks the most. He said that "economic warfare against the people of South Africa would be destructive not only of their efforts to peacefully end apartheid, but also of the opportunity to replace it with a free society." He warned of the prospect of "Marxist tyranny" and said, "We must stay and build, not cut and run."

The House voted to override on Monday, September 29, ignoring a last-minute offer by President Reagan to issue some of the new bans in a new executive order. Speaker Tip O'Neill called the offer "a step backward," and the vote was 313–83, or forty-nine more than the two-thirds required by the Constitution.

Before the Senate could vote, Jesse Helms and Roelof F. Botha, the South African foreign minister, made a heavy-handed attempt to sway farm state Senators on October 1. In cloakroom telephone conversations arranged by Helms with farm state Senators including Edward Zorinsky of Nebraska and Charles Grassley of Iowa, Botha threatened to halt South African grain purchases if the veto was overridden and to increase purchases if it was sustained. Lugar went to the press gallery to accuse Botha of "intimidation and bribery." Kennedy went to the floor to echo his charges and say, "We should not let the bullies and the thugs of Pretoria intimidate the Senate of the United States."

The next day, near the end of the debate, Kennedy returned to his theme that the United States was being tested by the South African issue. Supporters of the veto, he said, "have no idea of the cruelty and suffering that goes on today in the tortured land. Apartheid is not just a 'codified system of segregation,' as President Reagan described it in the cool detachment of his veto message. It is much, much more—and much, much worse. It is the torture of children and the use of terror against their parents. It is the total disenfranchisement of an entire population. It is midnight arrests and disappearances. It is the forced relocation of entire villages. It is attacks on innocent neighbors with bombs and bullets and planes. It is starvation, disease and early death. It is genocide, a crime against humanity. . . .

"The time for procrastination and delay is over. Now is the time to keep the faith with Martin Luther King, Desmond Tutu, and all those who believe in a free South Africa. Now is the time for America to act against apartheid."

Bob Dole closed the debate. He insisted the President had come around. He said Congress was unable to implement policy. And he ridiculed the measure: "This is a 'feel good' vote. This is a feel good foreign policy. Do not go near the problem, just parcel out the solution."

The roll call was solemn. Civil rights leaders were in the gallery and black Representatives stood in the back of the chamber. Senators obeyed a rarely followed rule and rose from their desks to vote when their names were called. The vote was 78–21. The only votes to sustain the President were cast by Republicans. Grassley and Zorinsky, Botha's targets, voted to override. It was the first override on a foreign policy issue since the War Powers Act was enacted over Nixon's veto in 1973.

THE IMPACT OF international economic sanctions played a crucial role in changing South Africa, but it took three years before Mandela was freed and serious talks began. Long before that, Kennedy solved a more personal, twenty-five-year-old foreign policy problem inherited from John F. Kennedy. He was told that there was one last surviving prisoner from the failed invasion at the Bay of Pigs in April 1961; more than twelve hundred had been ransomed in 1961 and 1962 in exchange for badly needed medicine. So Kennedy wrote to Fidel Castro to ask for the release of Colonel Ricardo Miguel B. Montero-Duque, saying that twenty-five years after the invasion both countries should now put it behind them. At the Swiss Embassy, which represented Cuban interests in Washington, Craig was told that if Montero-Duque was still being held, he must have committed some additional crime beyond being a battalion commander. But his request was sent to Havana, and after a few months, Kennedy was advised at the beginning of June that Cuba would release Montero-Duque, but only if Kennedy sent someone to get him, to make it clear that this was a personal gesture to him and not to the United States government.

So Craig took the weekly flight from Miami to Havana, and the morning he arrived, Wednesday, June 4, he was driven to the Combinadas del Este prison outside Havana. Montero-Duque had not been told he was being freed, but may have guessed it because he had been allowed to spend some time in the sun to remove his prison pallor. When Craig explained what was happening, the colonel said he did not think he could leave because "I'm not the last one." He said there was, in fact, one more Bay of Pigs veteran in the prison. He told Craig, "I don't know if I can leave. I'm an officer. This was just a soldier. I cannot leave my soldier here."

Craig promised to do everything he could to get the man released, as soon as he could, but pleaded with Montero-Duque to leave Cuba with him. Over the next few days, Craig and Kennedy urged the foreign ministry to let Ramón Conte-Hernández leave, too, but were told that was a decision Castro would have to make personally. Craig even asked Castro, at one of those two-hour late-night conversations the Cuban President favored, but got no immediate answer. On Sunday, June 8, Montero-Duque and Craig left by chartered plane for Miami. But four months later, Kennedy's office was told that Conte-Hernández could go, too. Craig and Conte-Hernández's eighty-two-year-old mother flew him to Miami. At the airport, the veteran of the hopeless invasion told reporters, "I'd love to do it again."

KENNEDY'S CHIEF DOMESTIC battle in 1986 was over the nomination of William Rehnquist to be Chief Justice of the United States, succeeding Warren Burger, who had retired. An associate justice since 1971, Rehnquist was clearly the most conservative member of the Court, and Kennedy led the effort to block his elevation.

He opposed Rehnquist's conservative judicial philosophy, although Senators were not generally comfortable with that as a standard for judging judicial nominees. So Kennedy began his fight in the Judiciary Committee on July 29 by arguing for a broad role for the Senate: "The Framers of the Constitution envisioned a major role for the Senate in the appointment of judges. It is an historical nonsense to suggest that all the Senate has to do is check the nominee's IQ, make sure he has a law degree and no arrests and rubberstamp the President's choice." He tried to establish that because Rehnquist had often been a lone dissenter, he was an extremist, too far out of the judicial mainstream to deserve promotion to Chief Justice.

The next day Kennedy and Metzenbaum pressed Rehnquist hard in questioning over two issues from his past. When he was confirmed in 1971, Rehnquist had denied that he had worked in Republican efforts to challenge black voters at polling places in Phoenix in the fifties and early sixties. This time, however, the issue was renewed when the FBI turned up witnesses who testified about his efforts. Rehnquist insisted they must be mistaken.

They also pressed him over a memorandum he had written in 1952 when he was law clerk for Justice Robert H. Jackson. The Supreme Court had heard its first round of arguments in the school desegregation cases, and Rehnquist wrote in favor of the old decision upholding "separate but equal," saying, "I think *Plessy v. Ferguson* was right and should be re-affirmed." The memo had come to light after his 1971 hearings and included a reference to being "excoriated by my liberal colleagues," which seemed to refer to other law clerks. Rehnquist insisted the memo was an effort to reflect Jackson's thinking, not his own. He said he was not sure how he would have voted if he had been a justice in 1952, but conceded that in retrospect, *Brown v. Board of Education* was correct. Kennedy was unable to generate much opposition in the Judiciary Committee, and two weeks later the committee approved Rehnquist, 13–5.

On the Senate floor on September 11, Kennedy wove those arguments together with Rehnquist's Supreme Court record on civil rights, including his eventual lone dissent upholding tax credits for segregated schools, to argue that minorities could not count on Rehnquist to uphold their rights. Looking at Rehnquist's votes on the court, Kennedy called him consistently hostile to claims of discrimination against women and legal aliens, to claims of prisoner's rights, and to claims that government was sponsoring religion. He sought to humanize those arguments by asking the Senate to "imagine what America would be like if Justice Rehnquist had been Chief Justice and his cramped and narrow view of the Constitution had prevailed in the critical years since World War II."

He answered, "The schools of America would still be segregated. Millions of citizens would be denied the right to vote under scandalous malapportionment laws. Women would be condemned to second-class status as second-class Americans. Courthouses would be closed to individual challenges against police brutality and executive abuse—closed even to the press. Government would embrace religion, and the wall of separation between church and state would be in ruins. State and local majorities would tell us what we can read, how to lead our private lives, whether to bear children, how to bring them up, what kinds of people we may become. Such a result would be a radical and unacceptable retreat from the protections Americans enjoy today, and our Constitution would be a lesser document in a lesser land."

Neither those arguments nor attacks on Rehnquist's candor before the committee stirred the Senate. The civil rights community, Rehnquist's main lobbying foe, failed to stir the country. After five days of debate, marked by a 68–31 vote to cut off a Kennedy-led filibuster, Rehnquist was confirmed on September 17 by a vote of 65–33, the most votes ever cast against a nominee to be Chief Justice.

Kennedy was also on the losing side of a major legislative effort to cope with a stream of illegal aliens. Liberals won an amnesty for up to five million illegals in the United States continuously since 1982. The price, however, was a law making it an offense for an employer to hire illegals. Though he worked closely with Alan Simpson of Wyoming to shape the law, he voted against it, warning that sanctions against employers would lead to bias against Americans of Hispanic origin. "If one thing is clear in the history of immigration laws," he said, "it is that whenever Congress enacts a measure with any potential for discrimination, the full potential is relentlessly realized and virulent discrimination results." The bill passed, 63–24.

NINETEEN EIGHTY-SIX also offered an interesting foretaste of what became Kennedy's most important collaboration with a Republican when Orrin Hatch called a September 25, 1986, oversight hearing into the record of the Mine Safety and Health Administration. Hatch, a union worker in his youth, had lost twenty-seven constituents in a coal mine fire at Wilberg, Utah, in 1984. But the revelations of the hearing came from Kennedy, and more particularly from Walter Sheridan, an investigator on his staff.

The MSHA was a singularly bad example of the Reagan administration's efforts to make federal regulations less painful for business. Sheridan's digging enabled Kennedy to show that when mine safety regulations were revised in the eighties, the recommendations of the safety agency's experts were ignored and industry's less stringent suggestions were adopted.

Kennedy lectured administration officials, telling them, "Recent evidence has been unearthed from the mines. That evidence suggests that this Nation's commitment to the enforcement of the mine safety laws has been undermined by a lax attitude and a solicitude for industry that does not befit this country's commitment to mine safety. . . . We hope to hasten the day when this country will again affirm its commitment that the coal the Nation needs will not be stained by the blood of miners and washed by the tears of their wives and children."

CHAPTER 28

Defeating Bork

Kennedy's efforts against Rehnquist led Reagan to use him as an oratorical prop when he campaigned for Republicans in October 1986. Warning Republican rallies of the risks of a Democratic-controlled Senate, he would ask, "Do you want Ted Kennedy controlling the confirmation of federal court judges?" The audience would boom back a drawn-out "No-o-o-o."

But the Reagan revolution was a spent force. To retain its 53–47 Senate majority, Republicans had to hold seats won by some of their feeblest 1980 winners, like Chic Hecht in Nevada, Paula Hawkins in Florida, and Mack Mattingly in Georgia. All three lost as the Democrats gained eight for a 55–45 majority.

Because the Supreme Court seemed likely to have more vacancies soon, some civil rights leaders urged Kennedy to take the chairmanship of the Judiciary Committee and do just what Reagan had warned he would do. Joe Biden, the Delaware Democrat who was next in line on Judiciary but was getting ready to run for President, did not want the distraction of the chairmanship and asked Kennedy to claim it. Tony Podesta, president of the liberal group People for the American Way, went to McLean to have breakfast and urge Kennedy to take Judiciary and block bad justices and judges. Kennedy told him, "Listen, I didn't get into politics to stop people from getting on the bench. I got into politics to help people." Anyhow, Kennedy felt he did not need the chairmanship to be a force on the Judiciary Committee. He promised Ralph Neas that if there was anything they needed him to do on Judiciary, he would.

Kennedy chose to stay where he was, leading the Democrats on Labor. On November 8, 1986, he said that committee's concerns "have been the issues that my family has been involved in since they came to Congress." He said its chairmanship "offers an unusual opportunity to reverse the retreat of the past six years" and to "halt the shameful recent trend of neglect for the needy in our society and those who have the least."

At that same news conference at Boston's Parker House, he said once more that he would not run for President. Four weeks later, he said he would back Governor Michael S. Dukakis of Massachusetts if he chose to run; "He is clearly qualified for any job," Kennedy told Boston's WNEC-TV. Dukakis was an earnest liberal who saw his own efforts to pass such measures as universal health care in Massachusetts as "an affirmation" of Kennedy's goals in Washington. Backing his own governor early also freed Kennedy from choosing among other hopefuls, some of them colleagues and allies, such as Biden, Senator Dale Bumpers of Arkansas, former Senator Gary Hart of Colorado, Governor Mario Cuomo of New York, and Jesse Jackson.

Personally, he delighted the supermarket tabloids and the *Boston Herald* by providing good copy with many brief romances. But he and Joan were together on holidays. On Thanksgiving morning they ran aground in Lewis Bay off Hyannis, with Ted at the helm of Teddy Jr.'s fifty-two-foot boat. Ted and Joan still shared a concern for their children. That spring, Patrick, then a senior at Andover, had undergone drug treatment at a rehabilitation facility, Spofford Hall, in New Hampshire. That fall he enrolled at Georgetown, but dropped out after two weeks.

A week before Thanksgiving, Kennedy had honored three figures from the outlawed Solidarity movement that was challenging Poland's Communist regime with the Robert F. Kennedy Award, given the year before to Winnie Mandela. But when Poland refused to promise to allow Adam Michnik and Zbigniew Bujak back in if they left to receive their awards, they were honored *in absentia,* along with a martyred priest, Jerzy Popieluszko. "Their final victory is not won," Kennedy acknowledged on November 20, but "they have gained the most important ground of all: once again, their fellow Poles are thinking and acting like free men and women." Kennedy then sought permission to go to Poland to present the awards at Christmas; after a month of negotiations in Washington and Warsaw, he was turned down. The government spokesman, Jerzy Urban, explained that the timing was bad because of the "crowded calendar of political events."

Kennedy did travel to the Middle East in December, in the wake of the Reagan administration's humiliation over disclosures that National Security Council officials had sought to swap arms to Iran for the freedom of American hostages and use profits from the deals to help finance the Contras in Nicaragua. Moderate Arab states like Jordan and Egypt were angered, seeing their efforts to help make peace with Israel go unrewarded while Iran was helped. Kennedy carried messages from the State Department to reassure them that the arms were no threat to them and only a factor in a war between Iran and Iraq. He spent three days in Israel and said he believed Jerusalem's version of events: that the deals, which involved shipment of arms from Israel to Iran, came at the request of William Casey, the CIA chief. In Amman, Kennedy called for more American money to help Jordan's efforts to develop the West Bank for its Palestinian inhabitants. In Egypt he quoted King Hussein of Jordan as saying the arms deal "undermines his own position in working toward peace in that part of the world." Then, after meeting with President Hosni Mubarak, he told reporters that America's friends should recognize that the American people, too, opposed the deal.

WHEN THE 100TH Congress convened in January 1987, Kennedy was ready. His six years in the minority had taught him the fine points of working with Republicans; his disavowals of the Presidency had made that somewhat easier. He had been busy with Carey Parker and Tom Rollins, his staff director on the Labor Committee, thinking through a legislative agenda. He had been hiring the additional staff that majority status entitled him to and working to get the new members he wanted. He offered Tom Harkin of Iowa a subcommittee chairmanship dealing with the disabled, and he joined. As a member of the Democrats' Steering and Policy Committee, Kennedy pushed to get the newly elected Barbara Mikulski of Maryland a coveted seat on the Appropriations Committee, where Harkin already sat; she, too, joined Labor.

Most important, unlike all but three of the other thirteen committee chairmen, he had been a chairman before and knew how to do it.

Kennedy began his Labor Committee leadership by announcing a broad plan. He wanted the committee to act to provide more medical research, raise the $3.35 an hour minimum wage, expand health insurance coverage, and find more effective ways to help the nation's schools. "We must seek ways to stretch education funds," he said in acknowledging that budget restraints would keep Congress from just providing more money, "so that every dollar spent is used effectively." Opening a series of hearings designed to generate enthusiasm for an ambitious program, he said on January 12: "I hope that these hearings will remind us not just of where we are, but of what we stand for. Full employment for our workers, compassion for our needy, first-class education for our children, quality health care and equal opportunity for all Americans are not a dying dream."

He was at work most quickly on five bills. Two of them were familiar civil rights measures. The first was an effort to overturn the 1984 Grove City decision that curbed federal efforts to use funding halts as a weapon against discrimination by holding that only programs or activities that directly received federal aid were covered. The Justice Department had interpreted that decision broadly, applying it not just to discrimination by gender, the bias covered in the Grove City case, but also to discrimination by race or against the elderly or the handicapped, which had been barred by other laws. The other was a renewal of the 1980 attempt to put teeth in the 1968 Fair Housing Act and to expand its coverage to prohibit discrimination against families with children and people with disabilities. The third bill was a new approach to national health insurance, a bill designed to make the provision of "minimum" health care a requirement like that of a minimum wage. Raising the minimum wage itself was a fourth priority. The fifth was an attempt to get the nation to begin providing an organized response to AIDS, a frightening epidemic of unknown origin and enormous threat.

The other power of a chairman is to call oversight hearings. In March, Kennedy and Walter Sheridan returned to the issue of mine safety. Sheridan found eight present and former mine safety inspectors willing to testify about industry influence. They testified on March 11, 1987, about having their recommendations for closing mines or criminal prosecutions ignored, or being pulled off inspections when mine operators complained. One inspector, Larry Layne, testified that the thirteen deaths at the Grundy mine in Whitwell, Tennessee, on December 8, 1981, could have been avoided if the Mine Safety and Health Administration's managers had followed through on their inspectors' findings. He said, "The mine should have been dangered off because methane into the bleeder system had been detected at least four days preceding the explosion."

Richard L. Trumka, president of the United Mine Workers, testified the next day that miners' deaths could be blamed directly on the first MSHA head under Reagan, Ford B. Ford. "Following his appointment, he called the top level managers of MSHA to Beckley, West Virginia. He advised them that MSHA needed to be more cooperative and less adversarial. He also advised them that they need not turn over every rock to find violations, nor engage in nit-picking violations. . . . The agency was locked into its 'cooperative approach'—which, better defined, meant reducing enforcement action and penalties against mine operators who violated the law rather than paying attention to business."

Trumka said there was "a scandalous and incestuous relationship between MSHA and the industry that it was supposed to regulate." Kennedy said, "Taken as a whole, the record of this agency is shameful and tragic. The specific nature of the testimony is especially compelling because we now know the identities of miners who have died because this agency has not carried out its duties. We even know the names of some of the people responsible for undermining the work of those who tried to bring safe practices to the mines."

The pressure seemed to have an effect. Some of the most notorious friends of the coal companies left the agency, and new safety regulations were finally issued.

KENNEDY HELD A hearing on the bill designed to overturn the Grove City decision on March 19. He said the decision crippled efforts to end discrimination not only against women in educational institutions, the core issue of the case itself, but, because of administration interpretations of its reach, discrimination against blacks, the handicapped, and the aged. Marcia Greenberger, managing attorney of the National Women's Law Center, listed dozens of complaints which the government had stopped investigating and explained that the decision seemed to tell the federal government not to investigate bias claims unless federal money could be "traced directly to the department, professor or activity accused of discrimination, even though the school itself receives large amounts of federal funds."

The hearings resumed April 1, and Mark R. Disler, a deputy assistant attorney general, said the administration saw no problems severe enough to extend the reach of any new law beyond sex discrimination. A human explanation of why a broader reach was needed came soon from Jerry Kicklighter of Belleville, Georgia. He had epilepsy but was teaching biology and botany at DeKalb Community College. In 1977 his contract was not renewed. And he couldn't get a job in the local school system either, because the college had fired him. He never got a reason in writing, but was told it was because of his epilepsy, which he told the hearing amounted, at most, to two 40-second seizures a week. "They were like daydreaming for a minute. I always went right back to work after one of my episodes."

Kicklighter complained to the Office of Civil Rights of the Department of Education, and, he said, "seven years later, on May 4th, 1984, OCR sent me a letter stating that, because of the Grove City decision, they did not have jurisdiction to pursue my case. The Government had established that DeKalb Community College received over a quarter of a million dollars in Federal funds for the 1976–77 school year, but they could not trace the funds directly to my job, as the Grove City decision required. . . . The investigation ended, not because I was not discriminated against, but because they could not prove the Federal money went to my department. My question is: Is this just?"

The hearings were tense, with Weicker and Hatch scrapping continually. The most explosive issue was the fear of Catholic leaders and the Right to Life movement that the legislation would require them to provide, or pay for, or at least provide insurance for, abortions. Proponents insisted the measure would do no such thing, but feared that amendments designed to preclude the problem would curb the right to an abortion.

Kennedy tried to get the full committee to act on the bill, but Republicans invoked a Senate rule that barred committee meetings while the Senate was in session. On May 6 he had had enough, and when Senator Gordon Humphrey, a New Hampshire Republican, objected to a 7 P.M. meeting, Kennedy postponed it until midnight, when the Senate had adjourned. Thirteen of the committee's sixteen members straggled in to rounds of applause from lobbyists who had waited to see if their pleas to members to provide a quorum would succeed. Hatch was furious. He threatened to talk all night. He said Kennedy was treating him "like dirt," because he was working day and night on the Iran-Contra hearings. Thurmond, eighty-four, with a new wife in her twenties and a small child, injected a rare bit of humor by complaining, "What about some of us young Senators who have young children" to go home to? But the Democrats and Weicker had the votes and were ready to work all night, so the Republicans, except for Humphrey, agreed to have a final vote by May 21.

The vote actually came on Wednesday, May 20, and the bill was approved 12–4, with the expected support of two Republicans, Weicker and

Robert H. Stafford of Vermont, and the surprise backing of a third, Thad Cochran of Mississippi. The opponents lost a series of abortion-related amendments and another that would specifically authorize firing anyone with AIDS. Ralph Neas and other civil rights leaders were overjoyed. "Now you see what happens when you have a civil rights champion in charge of the committee," said Neas.

That very afternoon, Kennedy spoke when the newest, fanciest Senate office building was formally named in memory of the Senator whose simple courage and integrity he admired most, Phil Hart, who died in 1977.

"Above all," Kennedy said, "for a generation he was a missionary for civil rights. He arrived in 1958, at the dawn of the civil rights revolution in Congress. And he guided every major rights bill through the gauntlet of filibusters in the Senate. He understood the meaning of 'We shall overcome'—and all his life, he helped America overcome."

Kennedy and Hart had desks next to each other in the back row on the Senate floor in those days before television made it unnecessary to be in the chamber to know what was going on, and Kennedy said, using a term he never used lightly, "He was like a brother." Phil Hart's name was never on any of the civil rights bills he guided to passage, and in praising Hart, Kennedy described the greatest legislative lesson he had learned from Hart: "He lived by the golden rule of Washington, that there is no limit to what you can accomplish in this city, if you are willing to give someone else the credit."

THE POLISH GOVERNMENT had changed its mind and given Kennedy a visa. The government of General Wojiech Jaruzelski found time on its political calendar because it was anxious to improve relations with the United States. Economic sanctions against Poland, imposed when martial law was declared and Solidarity outlawed in 1981, had been lifted in February after the government freed the last of its political prisoners, including Michnik and Bujak.

Kennedy arrived on May 22, and played down the award in his arrival statement, which emphasized America's admiration of the Polish people and the contribution her immigrants had brought. "You do not have to be Irish to appreciate the Polish," he said, "but it helps, because our two proud countries share . . . a role as victims in world history."

He went from the airport to meetings with Josef Cyrek, the second-ranking party official after Jaruzelski, and a group of members of the Sejm, or parliament. In these first few hours, the Polish officials renewed the idea—raised a couple of days before with Craig, who arrived early—that Kennedy might get to meet with Jaruzelski, who avoided Westerners. Kennedy was intrigued with the idea that he might serve as some sort of bridge between the regime and Solidarity. But the price was that the award presentation be minimized, not held at the United States ambassador's residence as planned, but at Michnik's apartment, perhaps, or that of Michael Kaufman, a *New York Times* correspondent with whom Craig had stayed on previous visits. The Poles also suggested that if the ceremony went ahead as planned, they might expel Kennedy. Michnik was irate, telling Kaufman he might ignore the award ceremony: "If they want to slip it to me in a men's room, the hell with them."

While Craig tried to reach Michnik, Kennedy went on to St. Stanislaw Kostka Church, where he prayed at the grave of Father Popieluszko, a Solidarity sympathizer murdered by security police in 1984. Several thousand Poles waited more than two hours, and children climbed trees to glimpse Kennedy, who laid a wreath at the grave, where stones from all over Poland were arranged to form a rosary.

The awards, busts of Robert Kennedy and checks totaling $10,000, were presented that evening as originally planned at the residence of John Davis, the American chargé d'affaires. The guests included Polish intellectuals and Solidarity organizers, and three Kennedy sisters, Ted's three children, Ethel Kennedy, and five of her children. Kennedy praised the recipients for "speaking truth to power," and he said, "In presenting these awards in my brother's name, we also honor the entire Polish nation—for their special virtues of courage, sacrifice, faith and endurance."

Mr. Bujak, a former tractor assembler who had eluded the police for five years, said, "The working people are a hard nut to crack, and today I have the impression that the nut has not been cracked. The teeth that tried to crack that nut have been broken." Mr. Michnik, an essayist and historian, said, "This is an award for the entire Solidarity, which still exists and continues its struggle, and will win." He said, "Thanks to this, we can be free people even if our country is not free, and we can build a sovereign society in a country that is not sovereign."

The next day Kennedy met with Poland's foreign minister and defense minister and with Josef Cardinal Glemp, the Catholic primate of Poland. Then in a speech to Poland's foreign policy establishment, he warned that more economic help from the United States would depend on progress in human rights. "Like it or not, your government must comprehend that the American people will not continue the process of normalization unless Poland continues to move on to a new openness. We will not do any business as usual with a regime that treats repression as the usual condition of daily life."

On Sunday, May 24, the party flew to Gdansk, where Lech Walesa had started the Solidarity trade union movement in 1980. Walesa climbed into their bus beside Kennedy and led them first to a monument to workers—slain in food riots in 1970—at the Lenin Shipyard, where Solidarity began. They went on to St. Brygida's Church for a mass in honor of John and Robert Kennedy. Before the service, Kennedy turned that stop into a rally. His staff handed out thousands of pictures of the Kennedys. He pinned a Solidarity button on his lapel and told the crowd, *"Jestem Polakiem,"* Polish for "I am a Pole," as John F. Kennedy had told Germans, *"Ich bin ein Berliner."* The crowd cheered wildly, and Kennedy praised Solidarity for fighting "against tyranny, repression and for human rights." Walesa took the microphone and boomed, "Solidarity will never die."

Inside the church, Kennedy spoke of his brothers, "men who had every gift but length of years." He said:

"They were heirs to wealth who felt the anguish of the poor. They were orators of eloquence who spoke for the voiceless. They were sons of Harvard who reached out and challenged all the youth of the world. They were men of special grace who had a special care for others. John and Robert Kennedy had a faith strong enough for any fate. . . . The unfinished quality of their lives symbolizes the unfinished tasks that lie ahead of all of us. As the torch is passed to each succeeding generation, I believe those who seek peace and justice, those who work for freedom in America as well as in Poland, will say of John and Robert Kennedy, 'They never left us and they never will.' "

There was a strong family note to the visit, which echoed a triumphal tour Robert Kennedy made in 1964. At a farewell dinner in Warsaw, Michnik pleased them all by saying he planned to use part of his $20,000 to begin publishing Robert Kennedy's works in Polish. At that dinner, the Kennedy family competed in offering toasts to their Polish companions, and Patrick observed that while they had hoped to see Jaruzelski, that was no loss because they had seen "the real representatives of the Polish people." Ted took him aside, told him he had done well, but cautioned that in a country like Poland, it was always best just to emphasize the positive. But the family also observed some tutoring of the Senator. One of Robert and Ethel's daughters,

Kerry Kennedy Cuomo, recalled the 1987 visit years later and said that when some of the party visited Poland's holiest Catholic shrine, that of the Black Madonna of Czestochowa, she noted two previous signatures in the visitors' book. The first was a solemn tribute from 1956 signed "Sincerly, Edward Kennedy." It had been annotated by her father in 1964; Bob wrote, "That's not how to spell it, Teddy."

Kennedy left Poland the next day after a visit to the Nazi death camp at Auschwitz. In a departure statement at Warsaw's airport, he offered Jaruzelski blunt advice: "I would urge the government to meet and talk with Solidarity. The sooner the dialogue begins, the better for all Poland."

His next stop was in Rome, where on May 27 he had an audience with Pope John Paul II, who was about to make his third papal visit to his native Poland. Kennedy told reporters that the Pope was "not optimistic" about progress in Poland because he felt its leaders were unwilling to share power. But the papal words that mattered most to Kennedy were a blessing on him as the head of one of the "great Catholic families of America." So generous were the Pope's words, Kennedy told Craig afterward, that for a few minutes at least he ranked as an even better Catholic than his mother.

FOR ALL THE fights he had with Hatch over civil rights, they were the closest of allies on many health measures, including AIDS. Hatch held the first Senate hearing on the epidemic in 1986, where he hugged an AIDS victim from Utah to try to dispel the fear that touching spread the disease. Together they had no patience for right-wing arguments that AIDS was punishment for sin, and agreed with the surgeon general, Dr. C. Everett Koop, that AIDS had to be looked at as a public health menace. "We had to get it into what was good in terms of public health, what was scientifically responsible, rather than what was ideological," Kennedy said years later.

Kennedy followed with several hearings in 1987, trying to prod the Reagan administration into seeking more money to deal with AIDS. On May 15, for example, Kennedy pressed administration witnesses on the need to teach the public how the disease was most often spread—through sex without a condom or used hypodermic needles—and to do it through paid television advertising. James O. Mason, director of the Centers for Disease Control in Atlanta, conceded that the message "must go during prime time." Stephen Joseph, New York City's commissioner of public health, insisted that the message had to be "explicit" to be effective.

With Hatch's backing, the Labor Committee approved a bill on June 17 to authorize spending $635 million on education, care, and treatment in the year beginning that October. The bill, backed by a 15–0 vote, authorized unlimited research funds. Within the $635 million was a specific $100 million to hire nearly seven hundred new AIDS researchers in federal agencies, and another $100 million to encourage home and community care for AIDS patients and $115 million (twice as much as in the previous year) for a national effort to develop educational materials.

THE OTHER CIVIL rights bill, designed to put teeth into the 1968 Fair Housing Act, did not move as swiftly as the Grove City measure. Starting on March 31, four hearings were held in the Judiciary Committee's subcommittee on constitutional rights on Kennedy's bill, cosponsored by Arlen Specter, a Pennsylvania Republican, that provided for federal administrative law judges to handle disputes and award damages in individual cases. Samuel R. Pierce, secretary of housing and urban development, said the administration did not like that idea and was working on a proposal of its own involving arbitrators. He also said the administration did not think discrimination based on family status should be covered in a federal law. Althea T. L. Simmons, director of the Washington bureau of the NAACP, said, "Fair housing is NAACP's top priority in the 100th Congress." The subcommittee tired of waiting for the administration to settle on its plan and approved the bill 4–1 on June 23, with only Hatch voting against it.

Neither the Grove City measure nor the housing bill went further in 1987, because something else came up that changed every civil rights advocate's priorities.

On Friday, June 26, the last day of the Supreme Court's term, Justice Lewis F. Powell announced his retirement. Nominated by Nixon in 1971 after the Haynsworth and Carswell defeats, Powell had emerged in the eighties as a critical swing vote, holding an increasingly conservative court in favor of abortion and affirmative action. Across the nation, civil rights leaders were alarmed; conservatives were eager. The obvious nominee for a President who had made it clear he wanted to reshape the court was Judge Robert H. Bork of the United States Court of Appeals for the District of Columbia Circuit, the most important appellate court below the Supreme Court. As a Yale law professor and as a judge, Bork had spoken out strongly against one major

Supreme Court decision after another, from *Shelley v. Kraemer,* a ruling outlawing racial covenants, to *Brown v. Board of Education,* to *Harper v. Board of Elections,* banning the poll tax, to the one-man, one-vote ruling of *Reynolds v. Simms,* to *Griswold v. Connecticut,* which threw out an anti-birth-control statute, to *Roe v. Wade,* legalizing abortion. In all those cases, he attacked the court's reasoning, not the result, but he never seemed to find a contemporary conservative decision to attack.

Bork had been passed over before, and he and some conservatives feared he would be again. But Reagan wanted him. There was never any serious debate at the White House.

Jeff Blattner, who had come to work for Kennedy on the Judiciary Committee that winter hoping to be involved in a Supreme Court confirmation battle, immediately began collecting Bork's writings and commentaries on them. He talked to Laurence Tribe of Harvard, author of one of the leading scholarly works on constitutional law, and other liberals. He assembled material for a speech, and Carey Parker drafted it.

Just five days after Powell's retirement, Reagan announced his choice. He brought Bork to the White House press briefing room to say: "Judge Bork, widely regarded as the most prominent and intellectually powerful advocate of judicial restraint, shares my view that judges' personal preferences and values should not be part of their constitutional interpretations."

At the Senate that first day of July, Kennedy got the word and called Archibald Cox to tell him he was going to lead the fight against Bork. He asked Cox to help. Cox declined; he did not want the argument over Bork to seem like a personal vendetta. But he said nothing to discourage Kennedy from making the fight.

Kennedy went over to the Senate chamber, and after waiting through a quorum call, delivered the most important and most controversial floor speech of his career. In just over three minutes, he called on the Senate to reject Bork's nomination and served notice that there would be a fight to the finish.

The speech made two points. The first, central to Kennedy though not to the ultimate confirmation battle, was that "The man who fired Archibald Cox does not deserve to sit on the Supreme Court of the United States." Recalling the Saturday Night Massacre, he praised Elliot Richardson and William Ruckelshaus, who "refused to do Richard Nixon's dirty work and obey his order to fire Special Prosecutor Archibald Cox. The deed devolved on Solicitor General Robert Bork, who executed the unconscionable assignment that has become one of the darkest chapters for the rule of law in American history."

The firing may have been the key reason for Kennedy's intensity, but the portion of the speech that was most remembered, and most criticized, came next:

"Mr. Bork should also be rejected by the Senate because he stands for an extremist view of the Constitution and the role of the Supreme Court that would have placed him outside the mainstream of American constitutional jurisprudence in the 1960s, let alone the 1980s. He opposed the public accommodations provisions of the Civil Rights Act of 1964 and the one-man, one-vote decision of the Supreme Court the same year. He has said that the First Amendment applies only to political speech, not literature or works of art or scientific expression.

"Under the two pressures of academic rejection and the prospect of Senate rejection, Mr. Bork subsequently retracted the most Neanderthal of these views on civil rights and the First Amendment, but his mind-set is no less ominous today.

"Robert Bork's America is a land in which women would be forced into back alley abortions, blacks would sit at segregated lunch counters, rogue police could break down citizens' doors in midnight raids, and schoolchildren could not be taught about evolution, writers and artists could be censored at the whim of government, and the doors of the federal courts would be shut on the fingers of millions of citizens for whom the judiciary is—and is often the only—protector of the individual rights that are at the heart of our democracy.

"America is a better and freer nation than Robert Bork thinks. Yet in the current delicate balance of the Supreme Court, his rigid ideology will tip the scales of justice against the kind of country America is and ought to be."

Kennedy intended the speech as a call to arms for opponents, and as a way to freeze Senators in place, discouraging their natural inclination to announce support of a Presidential nominee. It was also meant to make sure that Joe Biden took a clear stand in opposition. Biden did within a few days.

Kennedy defended the speech then, and does so now, as the strong language needed in desperate times. There were Bork articles or speeches that tied him to all the horribles of "Robert Bork's America," although some of the connections, especially to "rogue police" and "evolution," take a lot of explaining. On race, Kennedy dismissed Bork's recantations of such views as that banning segregation at lunch counters was a concept of "unsurpassed ugliness." He saw the retractions as tactical and insincere.

The speech's phrasing left it unclear whether Kennedy was describing what he thought America would be like in 1987 if Bork's views had prevailed when past decisions were made (as the Rehnquist speech of 1986 had done), or what it would be like in the future with Bork on the Court. As a look backward, the speech probably falls just inside the line between fair comment and demagoguery. Looking forward, it falls outside. Bork objected, with cause, that he had never said he wanted to reverse all the decisions he disagreed with, and indeed he sometimes said wrong decisions should be left alone if reversal would cause pointless disruption. Only on abortion is it clear that Kennedy had it right; analysts at the time were sure Bork would vote to narrow, if not overrule, *Roe v. Wade,* and his vote would be decisive, and nothing in Bork's increasingly conservative writings over the years raises any doubt. Nor was Bork—himself given to terms like "illegitimate," "unprincipled," "incoherent," "utterly specious," "radical," "thoroughly perverse," "frivolous," and "nutty" to describe judgments he despised—any innocent when it came to hyperbolic attacks. Even so, the Bork of Kennedy's speech was a wild-eyed fascist and Bork the nominee was not.

Bork himself watched the speech from a White House office where he had gone to make some telephone calls. "We were incredulous," he wrote in 1990. "Not one line of that tirade was true. It had simply never occurred to me that anybody could misrepresent my career and views as Kennedy did. The conventional wisdom in Washington then and for some time afterward was that Kennedy had made a serious tactical blunder. His statement was so outrageous that everyone said it helped rather than hurt me. I should have known better. This was a calculated personal assault by a shrewd politician, an assault more violent than any against a judicial nominee in our country's history. As it turned out, Kennedy set the themes and the tone for the entire campaign."

Bork had been chosen for his conservatism; his philosophy of judicial restraint would facilitate Reagan's social agenda, enhance the power of state governments, and weaken protections for minorities and women. So even if Bork found Kennedy's analysis distorted, he did not shrink from a fight that was basically ideological. Some of his allies did, because that was not the way the Senate was used to weighing court nominees. Kennedy himself had argued against considering judicial philosophy in 1965 during Thurgood Marshall's hearings. But Kennedy had gradually changed his mind and now felt that unless a nominee to the Supreme Court was committed to the "core values" of the Constitution, as Kennedy saw them, he should not vote to confirm. His opposition to Rehnquist the year before had been, for all the details like the Phoenix voting incidents, basically ideological, and he had collected thirty-three votes.

In 1986, Kennedy had made the argument that the Senate had a duty to consider a justice's philosophy. On July 24, 1987, Biden made it, telling the Senate it was clear from the Framers' debates and the *Federalist* papers that the Senate was meant to have the solemn task of preventing "the President from remaking the Court in his own image." He looked at the history of confirmation debates and at Andrew Jackson's efforts to put Roger Taney on the Court and Franklin D. Roosevelt's Court-packing plan and said, "We are once again confronted with a popular President's determined attempt to bend the Supreme Court to his political ends. No one should dispute his right to try. But no one should dispute the Senate's duty to respond."

Biden was in a complicated situation. He had launched a Presidential campaign in June, and one of his messages was that he was not a tool of special-interest groups, as Walter Mondale had appeared to be three years before. And, seeking to distinguish himself from Kennedy, Biden had told the *Philadelphia Inquirer* the previous November that while someone like Kennedy would vote against Bork if he was nominated, "I'd have to vote for him." His first comment on the actual nomination did not promise opposition to Bork. But within a week he told civil rights groups that he was the chairman, and he, not Kennedy, was in charge and would lead the fight. He asked them for time to make the announcement on his own. They agreed, but the story was promptly leaked to the *Washington Post* and *The New York Times*. That made it seem he had caved in to them.

Biden announced he would delay the hearings until September, which gave Kennedy and the opposition groups time to rouse their members. That is what Kennedy spent the summer doing. Podesta, a veteran of the 1980 campaign, had left People for the American Way on June 30; on July 2, Kennedy hired him to organize opposition. As the Senate left Washington for vacation on August 12, Kennedy sent a letter to 6,200 black political leaders, telling them, "Bork has been a lifelong opponent of civil rights. . . . Everything we have worked for in the past quarter century may be jeopardized if Bork is confirmed." He made hundreds of telephone calls to black political leaders and ministers, especially in the South, where Democrats depended on black support for reelection. Podesta recalled the message:

"This is really important, get your parishioners to write, get your members busy." Podesta said, "People weren't used to getting calls from Ted Kennedy. People took it so much more seriously." According to Podesta, Kennedy worked harder at organizing than he had in 1980; "If he'd made as many phone calls then, he might have been nominated."

Kennedy got up early, and one call woke Joseph Lowery, head of the Southern Christian Leadership Conference, at his group's convention in New Orleans, which had been focused on South Africa. At Kennedy's urging, the next session became an anti-Bork rally. Lowery told the SCLC, "His decisions would ignore the responsibility of the Constitution to guarantee the rights and to look out for the welfare of all the people." He promised the organization would take the case against Bork to Southern Senators who depended on black votes.

Some of the pressure Kennedy applied was more direct. He talked to Senators and sent them briefing books on Bork. When he worried about the vote of Senator Alan Dixon of Illinois, Kennedy called one of his allies, Edward Hanrahan, head of the hotel and restaurant workers in Chicago. Hanrahan then called Dixon, and so did Mayor Harold Washington. He called the moneymen for other Senators, too, and sometimes that pressure mattered. In general, the opponents were working harder than Bork's backers. Civil rights groups and unions were pushing their members. For example, the American Federation of State, County and Municipal Employees had Henry Griggs of its staff spend August and September making anti-Bork radio spots sounding like news items and calling them in—free of charge—to hundreds of radio stations which used them. Church services paused for minutes of letter-writing.

The White House was exhausted from a long battle over the Iran-Contra hearings. And because it chose to market Bork as a moderate in Powell's tradition, it was at cross purposes with his authentic conservative backers. Bork complained that the White House was unprepared for the public battle.

But its confidence and the opposition's fears were based on the hearings. Bork was expected to be a firm, intelligent witness, and there would be heavy television coverage that would affect public opinion. That summer, Senators on the committee, especially Biden, Kennedy, and Arlen Specter, a Pennsylvania Republican, studied Bork's writings and discussed them with law professors. Tribe played Bork for Kennedy and Biden to quiz in mock hearings.

Kennedy organized his summer learning into a careful speech delivered at Georgetown Law School on September 11, four days before the hearings began. A major focus was privacy, which was developing as a major weapon against Bork after being highlighted by a poll taken for AFSCME, the government workers' union, that summer. Bork's major argument—widely respected by scholars—was that the Griswold decision on Connecticut's law against contraceptives invented a right to privacy without sound constitutional basis. Kennedy finessed the constitutional argument by toting up the list of privacy decisions which Bork disdained, from custody cases to sending children to religious schools, from bans on teaching foreign languages to sterilization of repeat offenders, from zoning ordinances to abortion, and asserted, "Robert Bork's Constitution preserves no freedom at all for the individual against government interference with fundamentally personal human activities."

And he argued that on privacy, civil rights, economic power, and Presidential power, Bork was dangerous because of his contempt for precedent. As recently as January 1987, Bork had said that a justice should have "no problem whatsoever" in overruling Supreme Court decisions not based on the original intent of the Framers because such decisions had "no legitimacy." Kennedy translated: "If Robert Bork disagrees with a prior decision, that makes it a prime candidate to be overruled. A vast body of fundamental Supreme Court decisions would be placed in jeopardy by his search-and-destroy philosophy and his scorched-earth jurisprudence."

BORK'S PART OF the hearings, which began on Tuesday, September 15, ran an extraordinary thirty-two hours over five days, with Senators questioning him in unusually long thirty-minute blocks. Frequently the exchanges reached the level, as Linda Greenhouse wrote for *The New York Times,* of a "profound constitutional debate."

Bork's purpose throughout seemed to be to make himself appear more flexible, more thoughtful, and less smug than his writings suggested. In his opening statement, for example, he sought to show that he was no threat to precedent after all: "Overruling should be done sparingly and cautiously." But he rarely seemed to understand that these hearings were political rather than academic events, and that his technical, bloodless way of discussing cases that affected real people hurt his cause.

Kennedy gave him a prompt lesson in how to debate constitutional law in a television age. First he called him an "activist of the right" and complained that over twenty-five years, "Robert Bork has shown that he is hostile to the rule of law and the role of the courts in protecting individual liberty." Then Kennedy's turn to question Bork came at about 5 P.M., when television producers were deciding what to use on the evening news. The Senator hammered the judge. He interrupted him, wrapped questions inside speeches, and condemned him in sound-bite phrases like "I believe your clock on civil rights stopped in 1964" or "Your views would take us back to the days when women were second-class citizens." When Bork said the poll tax had not been very high, Kennedy replied by telling him, "You and I may not have to worry about where each dollar goes but a lot of Americans do."

Kennedy also asked hard questions. Seizing on Bork's condemnation of the birth control decision and his view that the right to privacy did not exist in the Constitution, Kennedy asked, "Doesn't that lead you to the view that you would uphold a statute requiring, say, compulsory abortion, if a legislature enacted it by a majority?" Bork's feeble answer, especially for someone who sought to serve on a court where judges constantly pepper lawyers with hypotheticals (as Justice John Paul Stevens had done recently on this very matter), was "I have never found it terribly useful, in testing constitutional theories, to use examples that we know the American people will never enact."

Biden passed Kennedy congratulatory notes after each set of exchanges saying, "Kennedy, six nothing," or "eight to nothing," but his own approach was very different. He never berated or interrupted Bork and kept asking him to take as much time as he needed to explain his views. But Biden's opening statement hit a central theme of Bork's jurisprudence when he reached for the majestic language of the Declaration of Independence, not the drier Constitution, and said, "I believe that all Americans are born with certain inalienable rights. As a child of God, I believe my rights are not derived from any government. My rights are not derived from any majority. My rights are because I exist." When Biden pressed him on the Connecticut birth control case and asked if marital privacy could not be justified, if not as the Supreme Court had done, then on some other grounds, Bork answered weakly, "I have never indulged in that exercise." It was a telling moment; Robert Bork's jurisprudence seemed an intellectual process without room for caring how courts affected people.

On that first day, Biden and Kennedy displayed the different but complementary approaches to defeating Bork which each had taken since July.

Kennedy was rallying the outside opposition; Biden was focused inside the committee and the hearing room.

A key issue on the second day of hearings was the shifting of Bork's views. When he said he favored a broader view of the First Amendment's protection of free speech than he had urged previously, and endorsed a decision he had once called "frivolous," he offended two of the four undecided members of the fourteen-man committee, Specter and Dennis DeConcini, an Arizona Democrat. Patrick J. Leahy, a Vermont Democrat, told reporters his answers amounted to "confirmation conversion." Then Specter damaged Bork again, by getting the judge to say there was no sound basis under his theory of the Constitution for the 1954 decision ordering the schools in the District of Columbia desegregated. "I have not thought of a rationale for it," Bork conceded to Specter. Then he begged Specter to believe that even though he doubted its reasoning, Bork would never "dream of overruling" the case.

On Thursday, Leahy and Specter confronted Bork directly with the charge of "confirmation conversion." Bork insisted he still held many controversial positions. Inside the room, that was plainly the most important issue, because it could change votes, and it dominated the newspaper coverage. But all three television news shows, the source of what most Americans learned, focused on Kennedy's accusing Bork of instinctively backing the executive against Congress and believing a President could do whatever he wanted. Kennedy led Bork through a series of laws and cases in which he had sided with the executive, including his view that members of Congress lacked the standing to sue a President (an issue that mattered especially to Robert Byrd, who was an uncommitted member of the committee), and then said:

"Judge Bork, the American people rely on the Congress to protect them from abuses by the Executive Branch. But, Judge Bork, whenever the Congress has tried to curb abuses, you always seem to side with the President. You broke the law in Watergate when you obeyed President Nixon and fired Archibald Cox. . . . You oppose limits on the national security power of the President, even when the issue is wiretapping and eavesdropping on American citizens. . . . You believe Congress can never use the courts to challenge the President when he abuses his power. You wrote that the War Powers Act was probably unconstitutional. . . . The Constitution calls for checks and balances. You seem to feel that when it comes to the relation between Congress and the President, instead of checks and balances the President has a blank check, and the Congress exerts no balance at all. You say you believe in original intent, Mr. Bork, but the Founding Fathers did not intend an all-powerful President."

Bork's answer quarreled with Kennedy's conclusions on specific cases. He never addressed the broader issue, his concept of Presidential power.

On Friday, Kennedy again dominated the networks. But this time he was playing on an issue that was central to the proceedings—doubts about Bork's sincerity. From the first day Bork had insisted he would not want to reverse settled cases, even if he thought them wrongly decided. Kennedy cited some of Bork's previous writings on the issue and then had aides play an audiotape of a question-and-answer session at Canisius College in Buffalo less than two years earlier, and the television cameras zoomed in to watch Bork listen to Bork. He was asked: "Now, the relationship between the judge, the text and precedent, what do you do about precedent?"

He replied, "I don't think that in the field of Constitutional law, precedent is all that important. . . . The Court has never thought that constitutional precedent was all that important—the reason being that if you construe a statute incorrectly, the Congress can pass a law and correct you. If you construe the Constitution incorrectly, Congress is helpless. Everybody is helpless. You're the final word. And if you become convinced that a prior court has misread the Constitution, I think it's your duty to go back and correct it. . . . I don't think precedent is all that important. I think the importance is what the framers were driving at, and to go back to that."

Kennedy pounced: "Judge Bork, in light of what we have just heard, how can anyone have confidence that you will respect the decisions of the Supreme Court with which you disagree?"

Bork first tried to justify his previous writings and then minimized the Canisius answer: "Senator, you and I both know that it is possible in a give and take question and answer period not to give a full and measured response. I have repeatedly said there are some things that are too settled to be overturned."

Bork damaged himself even further that day on the troubling issue of his concern for women. He had decided an Occupational Safety and Health Administration case in 1984 by ruling that the American Cyanamid Company did not violate the statute when it required women to undergo sterilization to work in jobs that exposed them to lead levels likely to harm a developing fetus. Bork held that OSHA was wrong to consider the choice "a hazardous condition of the workplace." Laurence Tribe, a severe critic of Bork's, later called the decision "defensible," and said "attempts to use it to show him to be a prosterilization ogre were terrible."

But the AFSCME poll showed the decision was political dynamite, and the anti-Bork groups used it fiercely. So did Howard Metzenbaum, who used it to show why women "are afraid of you." Bork explained the law, stressed that the decision was unanimous and had not been appealed. But then he ventured into emotions, an area he usually avoided, and said, "They offered a choice to the women. Some of them, I guess, did not want to have children," and "I suppose the five women who chose to stay on the job with higher pay and chose sterilization—I suppose that they were glad to have the choice."

The anti-Bork groups moved swiftly. Before the hearing was finished, the lawyer for Betty Riggs, one of the plaintiffs in the case, had got her to send a telegram to Metzenbaum, saying, "I cannot believe that Judge Bork thinks we were glad to have the choice of getting sterilized or getting fired. Only a judge who knows nothing about women who need to work could say that. I was only 26 years old, but I had to work, so I had no choice. . . . This was the most awful thing that happened to me. I still believe it's against the law, whatever Bork says."

The hearings finished on Saturday, a most unusual workday for a committee. There was a lengthy and thoughtful exchange between Specter and Bork on the usefulness of the "original intent" standard and whether it restricted the Constitution's protection to the rights the Framers had thought of, followed by efforts by Orrin Hatch to help Bork rehabilitate himself on the sterilization case, and by a telling Bork answer to Alan Simpson of Wyoming. Simpson asked the last question: "Why do you want to be an Associate Justice of the United States Supreme Court?" Bork answered: "I think it would be an intellectual feast, just to be there and to read the briefs and discuss things with counsel and discuss things with my colleagues."

On that exit line, Bork's turn at the witness table left a portrait of someone who viewed the Supreme Court as a way of gratifying his mind, not as a body whose decisions affect people.

There were seven more days of hearings. A group of opponents would be followed by a group of supporters. Law professors, former attorneys general and judges, cabinet officers and mayors, leaders of various lawyers' groups; each side assembled its most persuasive speakers. Two black leaders, William T. Coleman and Barbara Jordan, testified first, as if to remind Southern Democrats of whose votes had elected them. Two of the Watergate prosecutors testified against Bork and contradicted his claim that he had assured them immediately after Cox's firing that they should go on seeking tapes. Biden ran the hearings elegantly, even though in the middle of them, on Wednesday, September 23, he dropped out of the Presidential race. Disclosures that he had lifted a speech by Neil Kinnock, the British Labour Party leader, without attribution, plagiarized in law school, and made excessive claims about his academic record had sent his poll numbers plummeting.

But nothing was helping Bork. Public opinion polls showed that America had watched him in the hearings and had not been impressed. As the hearings continued, President Reagan complained of "ideologically inspired" attacks on his nominee, but he did not lobby hard personally, and the White House never mounted an organized offensive for Bork. When some right-wing groups bought ads, their personal attacks on Senators backfired.

The hearings ended on September 30, after Kennedy and Biden persuaded the outside groups opposing the nomination that if they testified, they might become the issue; some were very unhappy at the missed opportunity for attention and fund-raising, but they went along. The nomination effectively died the next day, when Specter announced his opposition, and so did three Southern Democrats who did not serve on the committee. On October 6, the committee voted 9–5 against the nomination; Bork had lost all four originally uncommitted Senators. Still, with rejection certain, Bork chose to force a full Senate vote, and some Republican Senators thought getting Southern Democrats on record against him might make them politically vulnerable. The debate began October 21, with fifty-four Senators announced in opposition.

Bork's supporters chose mostly to attack the attack on the judge as unfair. As the debate drew to a close, Kennedy disagreed, saying, "The hearings on this nomination were thorough—and fair. The American people have been involved—and they should have been because it is their Constitution and their Constitutional rights which are at stake." And he looked ahead, warning President Reagan not to act on his instinct to send up, as he had promised, another nominee equal to Bork. "If we receive a nominee who thinks like Judge Bork, who acts like Judge Bork, who opposes civil rights and civil liberties like Judge Bork," Kennedy said, "he will be rejected like Judge Bork." On October 23, the Senate voted 58–42 to reject the nomination. Six Republicans voted no while two Democrats voted yes.

After a brief, almost farcical near-nomination of Judge Douglas H. Ginsburg, a friend of Bork's on the court of appeals in Washington, which collapsed after it was revealed he had smoked marijuana in front of students at Harvard, Reagan finally settled on Anthony Kennedy, an appellate judge from California. He was not another Bork and was approved easily, sailing through a committee hearing in December after saying that the Constitution included rights not specifically enumerated and had a capacity for growth.

Those answers enabled Bork's foes, who insisted the fight had been about philosophy, to vote for Judge Kennedy and claim consistency. And after being confirmed in February 1988, Kennedy justified their hopes on the bench, voting to uphold *Roe v. Wade* and not to make flag-burning a crime or permit states to impose Congressional term limits. In those and other 5–4 decisions, he did not join the Court's conservative bloc, in which Bork surely would have been.

But the political legacy of the Bork nomination is one of all-out war on controversial nominees or candidates. The right was deeply frustrated by defeat. It could not acknowledge that the fight really was, at its heart, about philosophy and the country disagreed with Bork. It would not accept that the occasional false caricatures of the judge—such as attacks on him as a racist—were not fundamental. So the right counterattacked with the tactics it thought central in Bork's case, starting with campaign advertisements attacking Dukakis for a rape and murder committed in 1987 by Willie Horton, a murderer given a furlough by Massachusetts prison officials. By now, both the right and the left have martyred their political enemies. No end is in sight.

CHAPTER 29

Picking Up His Brothers' Mantle

Robert Bork was not the only conservative to discover in 1987 that Ted Kennedy played hardball. Rupert Murdoch, the Australian press lord who owned the *Boston Herald,* did not lose a job he coveted, but Kennedy cost him a Boston television station and the *New York Post.*

After Murdoch bought the *Herald* in 1982, it became intensely anti-Kennedy on its editorial page. It accused him of "self-righteousness," "Bolshevism," and "character assassination" of Rehnquist and "kow-towing to the labor bosses," and said he threw "charges of racism around like confetti." It quoted a fitness expert who put him in the "Worst Shapes Hall of Fame" and then reported on the front page "Ted Trims the Fat—all 31 lbs of it." One columnist, Norma Jean Nathan, studied his taste in blondes: "Call it constituent service or voter education, almost every weekend this summer there is the Senator schlepping into Hyannisport with a fresh blonde in tow." Another, Howie Carr, often called him "Fat Boy." Kennedy did not like it.

Ted rarely followed the Massachusetts political motto often associated with his older brothers and their pals—"Don't get mad, get even." This time he did, costing Murdoch a newspaper and a television station.

Since 1975 the Federal Communications Commission had prohibited cross-ownership of newspapers and television stations in the same markets. The rule was generally popular in Congress, where many politicians feared that monopoly ownership could be used to hurt or silence them. In 1985,

Murdoch bought three television stations in markets where he owned newspapers. He assured Congress he would comply with the rule, and he sold the *Chicago Sun-Times* within four months.

But he made no effort to sell anything in Boston or New York. The FCC gave him lengthy waivers of the cross-ownership rule. Mark Fowler, Reagan's chairman of the FCC, boasted at a retirement party that his greatest accomplishment as chairman was getting Murdoch the waivers. To Kennedy and some others in Congress, that seemed to prove that the Commission was ready to scrap the rule, as it had recently done with another that required broadcasters to show "fairness" by broadcasting opposing points of view. So in the spring of 1987, Kennedy got together with Senator Fritz Hollings of South Carolina, a friend who headed both the Commerce Committee, with oversight over the FCC, and the Appropriations subcommittee that provided money for it. Hollings introduced a bill in May to turn the ownership rules into law and held a hearing in July. Then in November the commission invited public comment on whether to eliminate the rule itself.

Hollings, with Kennedy's encouragement, then moved decisively. As head of his Appropriations subcommittee, he was part of the December House-Senate conference that rolled all the remaining spending bills into one large measure. Hollings inserted an amendment that prohibited the FCC from even considering a change in the rule and barred any new waivers to anyone who already had one. Murdoch, who had already become an American citizen to get his television licenses, was the only person with a waiver. Hollings cleared the amendment with the senior Republican on his subcommittee, Warren Rudman of New Hampshire, and with Representative Neal Smith of Iowa, who headed the House conferees. The provision was never debated in the conference, nor in the cursory floor discussion of the conference bill itself. It was "discovered" on December 30, and the *Herald* and the *New York Post* raised hell. A January 6 *Herald* editorial said of Kennedy, "When he goes after the Herald—and, make no mistake, his dead-of-night maneuver is a bullet aimed directly at the Herald—he does it sneakily."

The same day, Ed Koch, the mayor of New York, brought up Chappaquiddick when he complained at the National Press Club that Kennedy inserted the amendment "in the dead of night . . . and then, by the way, not to immediately own up to it—we've seen that before—and blame it on Hollings, that's an outrage." Koch worried about the future of the weakest of the New York or Boston Murdoch properties, the *New York Post,* which had endorsed him, and wrote that the amendment "has defamed our legislative process and undermined our freedom of the press."

Kennedy defended the move, saying the cross-ownership rule was a "cornerstone of the First Amendment and free expression . . . essential to our modern media-dominated society." He said Murdoch thought he was "entitled to be the only publisher in America who can buy a television station and obtain an exemption to keep his newspaper under the antitrust laws," and scorned the FCC as a place where "right wing ideology is determining policy and deregulation is running amok." He insisted his method was the only possible one. "Should I have told the *Herald* what I was doing?" he asked, and answered, "No, the first thing the *Herald* would have done was raise exactly this sort of self-serving hue and cry—and then get someone to block the legislation with a filibuster."

David Nyhan, a *Globe* columnist, cheerfully took Kennedy's side. "The people of Massachusetts know what Kennedy did. They also know that the *Herald* has repeatedly and consistently insulted and mocked Kennedy. The *Herald* puts Kennedy through the sheep dip every chance it can, and Kennedy got even. It's called settling a score." President Reagan took Murdoch's side, saying in his State of the Union message: "We have needlessly regulated our telecommunications industry. . . . The recent codification of the 'cross-ownership' rule" unfairly penalizes owners and "violates their First Amendment rights."

Senator Steve Symms, a very conservative Republican from Idaho, sought to repeal the Hollings-Kennedy measure on the Senate floor later in January, saying of Murdoch, "There is every reason to believe that he would have been able to obtain an extension of these waivers were it not for the furtive midnight measure designed to silence his critical voice."

Hollings had no patience with Symms. He said he had put the amendment in because the FCC was a "runaway animal," and had done it at ten-thirty in the morning a week before the bill was voted on. Senator Alfonse D'Amato, a New York Republican, said Murdoch was being punished because he was an Australian. Lowell Weicker replied, "My doubts have nothing to do with his citizenship. I just think he probably is the No. 1 dirt bag owner of any publications or electronic media in this Nation." Kennedy noted that D'Amato had been a member of the conference that adopted the amendment, implying that he should have paid more attention. The Senate killed the Symms amendment 60–30, on January 27.

But Murdoch went to court. Kennedy was surprised on March 29 when the U.S. Court of Appeals for the District of Columbia Circuit ruled that the part of the Hollings amendment which barred extensions of waivers of the cross-ownership rule was unconstitutional. Judges Stephen Williams and Laurence Silberman, two Reagan appointees, found that the Hollings amendment was directed solely at Murdoch, and thus was comparable to a bill of attainder, specifically barred by the Constitution. Judge Spottswood W. Robinson dissented, finding the amendment a proper step to "forestall evisceration of the cross-ownership rule. . . . Congress may deal with immediate threats as they arise."

Kennedy was surprised again when Murdoch, who had sold the *Post* in February, announced in April he would sell WFXT, Boston's Channel 25, saying that the uncertainty over the issue of waivers and the cross-ownership rule was not worth a prolonged fight. Owning the *Herald,* after all, was the violation of the cross-ownership rules that got to Kennedy.

SYMMS'S CROSS-OWNERSHIP PROPOSAL came as an amendment to Kennedy's Civil Rights Restoration Act, the first piece of legislation the Senate considered in the second session of the 100th Congress. After three years of Hatch's chairmanship of the Labor Committee, and one year consumed with Robert Bork, the Senate finally took up legislation to undo the Supreme Court's 1984 Grove City decision.

Kennedy managed the bill and opened debate on January 26 by arguing that the decision "placed at risk" much of the progress in civil rights made since the sixties, threatening not only gains by women, whose rights were at issue in the Grove City case itself, but also gains by blacks, the handicapped, and the aged. He insisted that a broad interpretation of the term "program or activity" had been Congress's intent in four major civil rights laws and was required to "protect the basic rights of millions of Americans to be free from federally subsidized discrimination."

The only real fight on the floor was over abortion. Since 1985 the nation's Catholic bishops had opposed any effort to overturn the Grove City decision that would reinstate the federal regulations that had existed since 1975. Those rules required recipients of federal aid to deal with abortion on the same terms as pregnancy. That meant that if they provided insurance that covered pregnancy, it had to cover abortion, too.

Senator John C. Danforth, a Missouri Republican, argued that it would be an "absolute outrage" to require institutions to pay for abortions when the Congress itself was not providing for them under Medicaid. He offered an amendment that allowed recipients of federal aid to refuse to pay for or perform abortions, though performing abortions was a dubious issue because the law already had a religious escape hatch. Danforth argued, "I really do not think we want to cram abortion down the throats of hospitals and colleges and universities that have the deepest religious and moral abhorrence at this practice." Bob Packwood, the Oregon Republican who had been an early advocate of abortion rights, said of supporters of the amendment, "Deep down in their hearts they are convinced that the world would be better off if women would not be in the marketplace, if they would simply stay at home with the children." The vote on Danforth's amendment was expected to be close, and Presidential candidates canceled campaigning to be on hand for the vote, but it carried comfortably, 56–39, on January 28.

Hatch, encouraged by the administration, offered a proposal that same day to limit the new law to educational institutions. Kennedy objected that such a proposal would allow discrimination against blacks or the handicapped by other bodies that received federal aid. Bob Dole, the Republican leader, who was back from New Hampshire for the abortion vote, agreed. Dole, whose right arm was crippled in combat during World War II, said the Grove City decision "has had a substantial detrimental effect on civil rights enforcement," particularly regarding the handicapped. "It is time to put this issue to rest," he said. Even though Hatch and Strom Thurmond argued that without Hatch's amendment, Reagan would veto the bill, it was defeated 75–19. Only Republicans voted for it.

The Senate passed the bill itself that same day, 75–14, with only Republicans opposed. Kennedy said the bill "closes a major loophole in our civil rights laws." But he called the Danforth amendment, which he had not spoken against on the floor, "a disturbing dilution of the protections which women currently enjoy." He said he hoped the House would discard it.

The House passed the Senate bill intact on March 2, by a vote of 315–98. Even though both houses of Congress had shown veto-proof majorities, Reagan vetoed the bill on March 16. He said it "would vastly and unjustifiably expand the power of the Federal government over the decisions and affairs of private organizations, such as churches and synagogues, farms, businesses and State and local governments. In the process, it would place at risk such cherished values as religious liberty."

Republicans like Senator Rudy Boschwitz of Minnesota, chairman of the

Republican Senate Campaign Committee, and Frank J. Fahrenkopf, Jr., chairman of the Republican National Committee, had warned that a veto would hurt the party in the 1988 elections. Once the veto was announced, Vice President George Bush, by then the front-runner for the Presidential nomination, gave it only the most formal support, emphasizing loyalty to Reagan, not the merits of the bill.

Those same political realities made it unlikely that many Republicans would change sides and vote to sustain the veto. The religious right worked hard on the issue. Jerry Falwell's Moral Majority, for example, sent a letter to ministers saying the bill could force churches to hire an "active homosexual drug addict with AIDS to be a teacher or youth pastor." The claim was false, but the heat it put on Southern Democrats was intense, although mainline Catholic, Protestant, and Jewish organizations supported the effort to override the veto.

As the debate closed on March 22, Kennedy said, "Never in the history of civil rights have so many phone calls done so much to distort so many facts. . . . The issue is discrimination, pure and simple. Opponents of this measure have left no stone unturned in their unseemly attempt to carve new loopholes in the law and provide greater leeway for bias and discrimination. The arguments of the opponents are awash in hypocrisy. They pay lip service to civil rights, but they refuse to practice what they preach. When the chips are down, they never met a civil rights bill they didn't dislike. . . . It has been 121 years since a President of the United States has vetoed a civil rights bill. Congress overrode Andrew Johnson's veto in 1867, and Congress should override Ronald Reagan's veto today."

It did. The vote in the Senate was 73–24. The vote in the House later that day was 292–133.

KENNEDY AND HATCH were together, for once, on a bill that mattered to labor, a measure designed to restrict the use of polygraph or lie detector tests. The House had first passed such legislation in 1985, but the Senate had not acted. The Kennedy–Hatch bill banned the use of polygraphs except for companies doing security work or producing drugs, or when an employer was investigating a specific crime which he had reported to the police. It also prohibited general snooping—outlawing questions about religion and sex.

Kennedy cited studies by the Congressional Office of Technology Assessment that indicated the tests were right less than 85 percent of the time. With two million polygraph tests given each year, he told the Senate on March 3, "you are talking about 260,000 honest and truthful Americans who are labeled liars and deceptive." He said the Office of Technology Assessment had studied who passed and who failed, and the answer was "If you are an altar boy, you probably will fail it. You would have a sense of conscience and potential guilt. But who passes it? The psychopaths, the deceptive ones. . . . Guilty psychopaths may escape detection because they are not concerned enough about misdeeds to create an interpretation of physiological responses." The Senate passed the bill, 68–24, on March 3, and after a conference report was adopted, including the Senate's exemptions, Reagan signed it on June 27.

Although the Labor Committee had no jurisdiction over the issue, Kennedy took a deep interest in hunger, which Robert had brought to the nation's attention. For months, Sarah von der Lippe, a Labor Committee staffer, worked with Ed Barron of the Senate Agriculture Committee to develop a bill that would undo many of the Reagan administration's deepest cuts in food stamps and school feeding programs. A consultant to the committee, J. Larry Brown of Harvard, brought a secret weapon to get Senate staff attention, a copy of a not yet released ABC television movie, *God Bless the Child*. The movie, shown to Congressional aides before ABC aired it in March, regularly brought tears as it conveyed the difficulty poor people and their children had in getting enough to eat.

Against this background, Kennedy introduced the "Emergency Hunger Relief Act" on March 2. "For the vast majority of recipients," he said, "food stamps run out well before the end of the month. Millions of other needy citizens would be eligible for food stamps, but they do not know they are eligible or how to apply." Patrick J. Leahy of Vermont, the new chairman of the Agriculture Committee, whose own interest in hunger was shown on his first day in charge in 1987 when he restored the word "Nutrition" to the committee's name, stepped back and cosponsored Kennedy's bill. The ABC movie, which beat out the Country Music Awards in the ratings, helped put the public behind the issue. When Leahy brought his own bill to the floor on July 26, he praised Kennedy: "He has tirelessly worked to raise public awareness of hunger in America." Most of the ideas in Kennedy's bill, he said, were incorporated in his own. It passed, 90–7, and President Reagan signed the conference report, providing about $3 billion over five years, on September 19, 1988.

TED WAS UP for reelection in 1988, and he asked his two oldest children, Kara and Teddy, to manage his campaign. Teddy, in particular, understood it was "a chance for me to really try to figure out whether this was going to be something that I was interested in getting involved with myself one day." It was not.

Their younger brother, Patrick Kennedy, by then a twenty-year-old Providence College sophomore, also chose politics that year, and for himself. He became the first of Ted's children to run for office. First, on March 8 he was elected as a Rhode Island delegate to the Democratic National Convention, pledged to Dukakis.

His career was soon threatened when he started to have trouble walking. He was admitted on April 18 to Massachusetts General Hospital. He had a tumor on his spine near the base of his skull, and it took surgeons nearly five hours to get around the spine to remove it. It proved to be benign.

The next month he hired a pollster to look into a race for a $300-per-year seat in the Rhode Island House of Representatives. The target was Frank Skeffington, a Providence undertaker and veteran Democrat who had only squeaked through the last time he faced primary opposition. The incumbent, whose name matched the one Edwin O'Connor gave to his fictional version of James Michael Curley in *The Last Hurrah,* was startled, telling reporters he did not "believe the Kennedys are going to try to knock off a true Democrat." Like Eddie McCormack twenty-six years before, he said, "I think you've got to wait your turn."

That argument meant no more to Patrick than it had to his father in 1962. There was an opening. Both Ted and Patrick say the father made no effort to convince his younger son to run, as he had with Teddy and Tip O'Neill's House seat two years before. Patrick recalled the 1988 candidacy as his own idea, though influenced by admiration for his father.

Patrick announced on June 29, twelve days before his twenty-first birthday. He campaigned energetically, challenged dubious voters off the rolls, and spent about $30,000. His parents and other relatives campaigned with him. Joan said going door-to-door reminded her, happily, of 1962. Ted would stop in Providence on Sunday afternoons on his way back to Washington from Cape Cod and go to small house parties with perhaps a dozen neighborhood residents, instead of the 50 to 150 he recalled from the 1980 campaign. He liked the questions better, too. He recalled one man who told him of the close-knit community where his son was running and its concern for its children: "And your son says that he will take an interest in this. And I'd like you to tell us what in your life, what in his life, would indicate to us that he will take that kind of interest?" Ted answered that he had got to know his son very well, especially when he was hospitalized for asthma attacks: "When he says things, what he will do. I was able to, at least, I think, give them some assurance that he would."

The race was not all smooth. Ted recalled a degree of bitterness Eddie McCormack's backers in 1962 had never shown. At the end of Catholic services, worshipers shake hands and say "Peace" to those in reach. But in Providence in 1988, Ted said, "you'd see people who were supporting his opponent would turn around and take their hand back, wouldn't shake hands with him in church."

Lots of Kennedys turned up on Election Day to help. Ted, standing near Skeffington, posed for Polaroid pictures with voters. On September 14, Primary Day, Patrick rolled up 1,324 votes to 1,009 for Skeffington. Ted said that night, "None of the victories I have ever had in my political life has meant so much as this one tonight." There was no Republican candidate, so Patrick was elected a state representative.

THE AIDS LEGISLATION, stymied for nearly a year because of threats of opposition from Jesse Helms, finally came to the Senate floor on April 27, 1988. It authorized spending of about $1 billion on counseling, treatment, education, and research.

This was the first major display of the Kennedy–Hatch team on difficult legislation. They had agreed to oppose all amendments, with Kennedy working to shut down the left and Hatch the right. But Helms proposed one that was too much for Hatch, and for most Senators in an election year. It barred any use of federal education funds that would "promote or encourage" homosexual activity. The problem was that gay men were the leading risk group, and that the message to deliver to them was that promiscuity, especially without using a condom, was deadly. It was adopted, 71–18, on April 28, after furious exchanges between Kennedy, who accused Helms of trying to kill the education program for AIDS, and Helms, who said, "Talk about misrepresentation. You have developed it into a fine art, sir!"

With fearful Senators now able to say that they had voted against homosexuality, Kennedy then came back with another amendment to say that education programs should stress the value of a "single, monogamous relationship" and that nothing in the act "shall restrict the ability of the education program to provide accurate information on reducing the risk of becoming infected with the etiologic agent for AIDS." Helms understood what Kennedy was doing, and was furious at being undercut. He said Kennedy was encouraging "shacking up. . . . In other words, just do it one at a time when you cavort around and cheat on your wife or husband." He said, "There is not one single case of AIDS reported in this country that cannot be traced in origin to sodomy. . . . We should not allow the homosexual crowd to use the AIDS issue to promote and legitimize their lifestyle in American society."

Then Hatch, up for reelection himself and urged by his staff to stay out of the fight, took on Helms. First, he said, "I have known homosexuals that I liked and I have known homosexuals that I disliked as human beings. Some of them are just as fine human beings as anybody on this floor, or in the U.S. Senate." To limit the spread of AIDS, it was important to encourage homosexuals to enter monogamous relationships. "We have a public health problem here and it affects primarily homosexuals. The fact that I disagree with their sexual orientation—and I do—does not stop me from having some compassion for the fact that AIDS is devastating that community. . . . Who are we to judge anyway? Let him who is without sin cast the first stone. . . . We have to tell homosexuals more than simply to become heterosexuals." He said he had counseled homosexuals that way, and his experience was that it made no "difference at all."

Hatch's eloquent support of Kennedy's amendment certified it as acceptable, and when Helms's attempt to kill it was defeated, 61–29, fifteen other Republicans joined Hatch in voting with Kennedy. The bill itself passed 87–4, with only Helms and three other conservative Republicans opposing it. After the vote, Kennedy told reporters, "Finally we have declared war on the virus and not on the victims in our battle against AIDS. Until we find the vaccine, we have, with this legislation, obtained the only vaccine which exists today, which is education."

The fight was hardly over. The House, where November's election concerned every member, took until September 23 to pass its version of the bill, which provided $1.2 billion for confidential AIDS testing and counseling. With Congress planning to adjourn three weeks later for the election, that gave Senate opponents like Helms the upper hand. Proponents abandoned the confidential testing in a House–Senate conference and then wrapped the bill into a catchall health measure reauthorizing aid for training nurses, raising spending levels for the National Institutes of Health, and requiring Senate confirmation of the commissioner of food and drugs. On AIDS, it included provisions for expanded home and community care of victims, for easier access to experimental drugs, and for a new national commission to develop AIDS policy. Before the bill was adopted without opposing votes on October 13, the nominal debate focused on the AIDS sections. Kennedy said the epidemic had brought "not only pain, suffering and death, but also confusion, fear and hatred." He promised to press again for confidential testing. He said that with this bill "the Senate has gone on record that a new and concerted national effort is required to mobilize the resources of this Nation to overcome the menace that AIDS poses to the American people and to all future generations." After urging by Hatch, Reagan signed the bill on November 4.

Kennedy campaigned earnestly for Dukakis all year, although in January in Iowa he sometimes stole the show while he tried to make the governor seem more human and less detail-dominated. E. J. Dionne wrote in *The New York Times*, "The signs said 'Dukakis,' but the nostalgia for the name Kennedy will not die." Dukakis was grateful for Kennedy's help, saying he did "anything I asked. Fund-raisers, appearances. Right from the beginning." Dukakis cruised to the nomination, although Jesse Jackson kept harrying him from the left. At the convention in Atlanta, Kennedy and Carter staged a well-photographed unity event at the Carter Library as an example to Jackson and Dukakis.

On July 19, Ted's nephew John F. Kennedy, Jr., introduced him to the convention, saying, "I owe a special debt to the man his nephews and nieces call Teddy, not just because of what he means to me personally, but because of the causes he has carried on. He has shown that an unwavering commitment to the poor, to the elderly, to those without hope, regardless of fashion or convention, is the greatest reward of public service."

Ted then joined in the convention oratory of ridicule against Bush, saying he hid from responsibility for the Reagan administration: "The Vice President says he wasn't there—or can't recall—or never heard—as the Administration secretly plotted to sell arms to Iran. So when that monumental mistake was being made, I think it is fair to ask where was George? . . . The Vice President, who now speaks quite fervently of civil rights, apparently wasn't around or didn't quite hear when the Administration was planning to weaken voting rights, give tax breaks to segregated schools, and veto the Civil Rights Restoration Act of 1988. So, when all those assaults were being mounted, I think it is fair to ask—where was George?" That line nettled Republicans. A couple of weeks later, Representative Harold Rogers of Kentucky came up with an answer at a political picnic at a place called Fancy Farm: "I'll tell Teddy Kennedy where George is. He's home sober with his wife."

AFTER THE SENATE fight over the Grove City bill, Kennedy and the civil rights groups wanted the House to go first on strengthening the law against discrimination in housing. Representative Hamilton Fish, a New York Republican, kept talks going between the civil rights groups and the National Association of Realtors (real estate agents), the prime opponent of the measure, and hammered out an agreement. The Realtors and the Reagan administration had objected to the bill's enforcement mechanism of settling disputes and imposing fines by an administrative law judge in the Department of Housing and Urban Development. Republicans from Vice President Bush down were concerned about going into the election with an anti-civil rights label. The solution reached in June, proposed by Penda Hair of the NAACP Legal Defense Fund, would allow either side (in practice usually the defendant) to take the dispute to a federal court proceeding instead.

When that deal was announced on June 21, the leading roles were played by Fish, Representative Don Edwards, a California Democrat, Ralph Neas from the Leadership Conference on Civil Rights, and Nestor R. Weigand, Jr., president of the Realtors. Kennedy was among the dozen lawmakers on hand to praise the compromise as a "giant step toward enacting legislation that will make the promise of fair housing a reality," while preserving the speedy, cheaper administrative law judge method in most cases.

Race was not the only issue involved. The only serious fight on the House floor was over how to prohibit discrimination against families with children. A survey by the Department of Housing and Urban Development found that half of all rental housing had some restrictions against children, with a quarter banning them altogether. After defeating an attempt to delete that section, the House passed the bill 376–23, on June 29. Bush promptly endorsed it.

Along with his chairmanship on Labor, Kennedy was also serving as acting chairman of the Judiciary Committee. Biden had been hospitalized for a brain aneurysm in February, and neither he nor Kennedy thought the next most senior Democrat without a chairmanship, Howard M. Metzenbaum of Ohio, had the temperament to run a narrowly divided committee.

But on this bill Kennedy kept the Judiciary Committee out of it, unsure of a majority because two committee Democrats were nervous and Biden was absent. Instead he sent Blattner to negotiate final details at the office of Secretary Samuel R. Pierce at HUD. Bush weighed in to help. The Senate was about to adjourn so Republicans could go to New Orleans and nominate Bush. The negotiations were spurred to conclusion on the weekend of July 30–31 after Kennedy persuaded Robert Byrd to announce on Friday the twenty-ninth that the bill would be on the Senate floor on Monday, August 1.

Before the Senate took up the House bill, Orrin Hatch had enlisted as a cosponsor with Kennedy and Arlen Specter, the Pennsylvania Republican. For all their alliances on other measures, Hatch and Kennedy rarely agreed on civil rights bills. But Kennedy has asked Muhammad Ali, the former heavyweight boxing champion, to intervene. Hatch, a boxer in his youth, admired Ali, who often supported Republicans. Ali told Hatch this was an opportunity to show that he was for all the people, not just conservatives, and Hatch acknowledged, "It was Muhammad Ali that really turned that bill around" for him.

In the Senate, the only attacks came over requirements that new multifamily units be built with such features as bathroom doorways wide enough for wheelchairs and light switches low enough to be reached from them. Those requirements were overwhelmingly approved. Before the final vote, Kennedy said, "What a beautiful moment it is. The twenty-year logjam on fair housing is finally breaking, and the promise of fair housing is about to become a reality." The consolation for waiting twenty years for adequate enforcement, he said, was that by 1988 Congress understood the burden of discrimination against the handicapped and against families with children. "In the course of the two decades of this debate, we have learned that separate is never equal in America—not in schools, not in jobs, and not in housing. We have tolerated separate but not equal housing for too long." The Senate passed the bill, 94–3, on August 2, and the House accepted the Senate version on August 8. Reagan signed it into law on September 13.

Just as the hunger bill was helped along by television, so was a Kennedy measure to require companies to give sixty days' notice before closing a plant that would cost at least fifty workers their jobs. Frequent news reports about the effects of sudden plant closings on workers who had just made a major expenditure or on middle-sized towns dependent on one employer made it popular with the public, but business and President Reagan fought it hard.

On May 2, Reagan told the U.S. Chamber of Commerce that the issue should be left to labor-management negotiations and a federal law on the subject would keep small companies from growing to the one-hundred-employee threshold of the law. "It's a shackle on smaller companies that want to take the leap and become large," he said.

Democrats attached the plan to a trade bill they thought the President would be reluctant to veto; they were wrong, and he vetoed it. On June 8, Kennedy said Reagan and business "fabricated horror stories." He said, "Workers deserve time to pull their lives together and plan for the future when their jobs are lost. . . . It isn't fair for management to know when plants will close but for workers to be kept in the dark. It isn't fair that management gets a golden parachute and workers get a closed plant gate." But only sixty-one Senators voted to override, and so the veto was sustained. On July 6, though, the Senate passed the measure again as a free-standing bill, 72–23. Election-year heat caused ten Republicans to change their votes in four weeks. After the House passed it by more than a veto-proof majority a week later, Reagan gave in, allowing the bill to become law without his signature.

Kennedy lost one major labor fight in 1988, an effort to raise the minimum wage, then $3.35 per hour, for the first time in seven years. On this issue, Democrats were divided, tired of casting pro-labor votes that could hurt them at home, especially if the bills were going to be vetoed anyway. It took Kennedy more than a year, until June 29, 1988, even to get a bill out of his committee. It called for an increase to $4.55 per hour over three years. In the fall campaign, Bush said he favored some kind of increase, and that led Democrats to bring the bill up on September 15, with Kennedy arguing that a raise was necessary to get millions of full-time workers above the federal poverty line. But Senate Republicans stood firm, arguing that an increase would actually cost lower-paid workers jobs. The real core of the argument was an effort by business and the Republicans to carve out a below-minimum training wage for workers in their first ninety days—and labor's intense resistance. Kennedy would not agree to a Hatch proposal for a sub-minimum, limited to three months, at 80 percent of the regular minimum wage. Republicans filibustered. Kennedy, as he expected, could not get the sixty votes required to end debate on September 22 and again on September 23. So Byrd pulled the bill. The Democrats thought they had an election issue, but it had little impact.

Kennedy and Simpson, following up on the 1986 law that dealt with illegal aliens, tried to reform legal immigration in 1988. Their goal, though they did not put it that way, was to remedy the unintended consequences of Kennedy's 1965 law. Its repeal of national quotas and emphasis on family unification had increased the percentage of Asians among immigrants from 8 to 45 percent in 1986, and reduced the share from Europe from 50 to 10 percent. Their plan was to increase the number of immigrants admitted because of education, occupation, and ability to speak English, but reduce sharply the numbers who qualified because their sisters or brothers lived here—a provision of almost no help to people from countries like Ireland. The Senate passed the bill March 15, 88–4, but the House was not ready for another immigration bill and buried it. Instead, Representative Brian J. Donnelly of Massachusetts pushed through a stopgap bill allowing fifteen thousand visas each for two years to citizens of countries thwarted by the 1965 act. As the Senate passed the bill without dissent on October 21, 1988, Kennedy said, "This measure is an important step in the right direction for Ireland and other countries suffering unfairly under our current immigration laws."

KENNEDY'S OWN REELECTION race was the easiest of his career. Teddy said his biggest problem was that his father "wasn't there a lot," because once Congress adjourned, late, he was stumping for Dukakis. Kennedy talked up his governor, but pointedly refused to join the Democratic clamor attacking Dan Quayle, Bush's running mate. He annoyed the Dukakis campaign by telling editors in October that Quayle was not dumb, and often left cheering messages on Quayle's home phone.

In the Senate race, Teddy filled in as a surrogate speaker, and Kara, who had been a television producer, worked on media. The Republicans nominated an amiable young conservative, Joe Malone, who attacked Kennedy for opposing the death penalty but not abortion, and even criticized him for effective pork barrel politicking for computer education funds for Massachusetts. Kennedy replied, "Any time that I can get funding to help education, whether it's in Massachusetts, New England or the country, I'm going to fight for it. And you can put any label on it you want."

Kennedy won going away, with 65 percent of the vote, but Dukakis was beaten badly, although his 46.5 percent nationally was better than Carter had done in 1980 or Mondale in 1984.

THIS FIRST CONGRESS with Democrats back in control of the Senate had been a spectacular success for Kennedy on domestic policy, where power led him to concentrate. In foreign affairs, he welcomed the immigrants from the USSR whom he had helped leave, and he organized enthusiastic Senate praise for the Chilean people and Ambassador Barnes after an October plebiscite chose democracy over military rule. Barnes appreciated their letter, especially its promise to support "normalization" of relations with the Chilean military once democracy was restored.

But at home, the win on Bork had been followed by two important civil rights victories. The Grove City bill restarted the effort to provide equality for women—especially in sports—in schools and colleges. The housing bill worked, adding enforcement tools to moral suasion. The AIDS measure broke through a political taboo and, with the lie detector bill, launched an intermittent but highly influential partnership with Hatch on tough issues. The hunger legislation and plant closing bills were deliveries on basic liberal commitments. The minimum wage fight was lost, but Kennedy promised to try again.

But as soon as Dukakis lost, Democrats began one of their ritual lamentations about being too liberal for the voters. For example, Al From, executive director of the moderate to conservative Democratic Leadership Council, said the latest defeat showed that "you need fundamental changes in the message." Kennedy answered back. Speaking at the John F. Kennedy School of Government at Harvard on November 14, he began by saying, "I'm here as one of those people called 'liberals.' " He said Dukakis had lost because he had hidden his liberal colors until the campaign's closing days. Kennedy said the campaign had been "eminently winnable."

To emphasize his argument, he cited the school's namesake: "I also know how my brother would have ridiculed the Republican attempts in the recent election to make 'liberal' a dirty word. They tried that trick in the 1960 election, too—and my brother rejected it out of hand. He embraced the 'liberal' label without hesitation. He didn't run away from it, he welcomed it. He campaigned proudly under its banner."

CHAPTER 30

Rights for the Disabled

The other thirty-two Senators elected in 1988 were sworn in when the Senate convened at noon on January 3, 1989. But Ted was in Providence, Rhode Island, instead, watching Patrick take the oath as a state representative with his hand on the Kennedy family Bible. "This is more important to me than any victory of my own," he said proudly, before flying to Washington to be sworn in himself late that afternoon.

The next week he was in a fight at a bar in New York. Early one morning he wandered into an establishment called American Trash and got into an argument with Dennis McKenna, a stranger. McKenna said something Kennedy considered an insult to his brothers' memory, and Kennedy threw a drink at him. When McKenna said that if Kennedy was not a Senator, he would "punch your lights out," Kennedy invited him to step outside, where they scuffled. Kennedy's friends got him into a cab about 4:30 A.M., and one of them told the *Herald* he was "feeling no pain."

The incident was fresh a month later when Kennedy voted to reject the nomination of former Senator John R. Tower to be secretary of defense. While mentioning that he attacked Tower's ties to lobbyists, reporters often pointed out that, as Johnny Apple wrote in *The New York Times,* Kennedy "has been accused, like Mr. Tower, of both heavy drinking and womanizing."

Seeking to aim his party, Kennedy laid out a huge agenda. Six weeks after Bush took office, Kennedy spoke at Yale, the President's college. On March 6, he argued that the Republicans had run out of ideas and the administration had "more power than purpose." Calling for action on areas like health insurance, wages, education, and community service, Kennedy said that "the creative initiative will pass over to the Democratic Party if we choose to take it."

His two biggest efforts of 1989 were carried over from 1988. First came the minimum wage increase and then a newer issue, civil rights legislation for the disabled. A 1973 law barred discrimination against the handicapped in programs with federal aid, but it lacked the broad protections against discrimination won by blacks and women. A young disability rights movement had managed to put that cause on the edge of the political field of play by winning endorsements from the two Presidential candidates, Dukakis and Bush.

Minimum-wage legislation was something Kennedy always associated with his brother Jack, who had pushed it unsuccessfully in the Senate in 1960, and then, as President, had secured an increase from $1 to $1.25 that was enacted in 1961. He threw himself into learning the statistics, the history, the arguments on both sides. But on this issue, Kennedy was dragging a reluctant labor movement behind him. Most union workers now made enough so that they would not be even indirectly helped by the increase, and unions feared a subminimum training wage would be adopted and would be used to keep wages down by employers like fast-food restaurants that did no real training but had high turnover. Moreover, quite a few Democratic Senators were unenthusiastic about supporting a bill that Bush might veto and that would stir unhappiness in small business.

Even so, Senator George J. Mitchell of Maine, elected as the new majority leader after Byrd decided to take the chairmanship of Appropriations instead, cosponsored the bill with Kennedy and gave it a priority number, S.4, meaning it was the fourth bill introduced when the Senate started work in late January. It would have raised the minimum wage, over three years, from $3.35 per hour to $4.65. That was about 39 percent, about as much as the cost of living had increased since 1981.

At a March 3 hearing, the first witness was Elizabeth Dole, Bush's secretary of labor and the wife of Bob Dole, the Senate minority leader, whom Bush had beaten for the 1988 nomination. She said the administration would go no higher than a $4.25 wage after three years, with a six-month training wage at $3.35. Kennedy objected that her own department said training for most minimum-wage jobs did not exceed thirty days. She said that even after the training, the lower wage was needed to preserve jobs. When the Labor Committee approved his bill the next week, Kennedy made it clear he would accept some kind of training wage, but only if it really included training and was not used to throw older workers off the job. On March 23, the House passed a $4.55-per-hour bill, with a two-month subminimum training period.

When the Senate took up the bill on April 6, Kennedy challenged the major argument against it, the contention that an increase in the minimum wage would cause employers to fire workers. He took the arguments the U.S. Chamber of Commerce had made whenever the wage had been raised since 1955, a litany of complaints that Congress was ensuring that low-wage workers would not get a raise but a pink slip. In fact, he told the Senate, overall employment went up after minimum-wage increases. "As I was reviewing the record of all the things the chamber has said, we could see that these are the same old tired arguments. . . . Each time the chamber has been wrong."

The next day he took another shot at the "crocodile tears" shed for minimum-wage workers, citing company after company where low-wage workers had stayed at close to $3.35 per hour since 1981, but chief executive officers had managed to double their compensation between 1981 and 1980: American Stores, Brunswick Bowling Lanes, and Food Lion. Then there was Giant Food, with a 472 percent increase; A&P, 388 percent; Greyhound, 261 percent; Kmart, 513 percent. He said it was silly to pay attention to arguments "about how devastating this is going to be in particular industries—which are primarily the minimum wage industries—when we see the kind of belt tightening, or belt loosening, by the CEOs."

Minimum-wage legislation has always been complicated, involving exceptions from one clause or another for particular industries or even businesses which have caught a Senator's attention. One way of minimizing opposition is by making concessions to particular Senators. So when James D. McClure, an Idaho Republican, sought to broaden the reach of the exclusion from time and a half for overtime enjoyed by rural irrigation cooperatives, Kennedy went along. He did not get McClure's vote in the end, but he minimized his opposition.

Then Kennedy wanted an exception of his own that mattered to Massachusetts's computer industry. The Internal Revenue Service had ruled that freelance computer experts paid by the hour were not independent contractors but really employees, even if they made $50 or $100 per hour. The

Labor Department had decided that meant they should be paid time and a half when they worked more than forty hours in a week. Senator David Durenberger, a Minnesota Republican, offered an amendment to deny the overtime rate to anyone who earned more than 6.5 times the minimum wage, or $29.57 per hour if the minimum wage got to $4.55. Kennedy and John Kerry jumped in to support him, and the amendment was adopted.

On April 11, with Kennedy's support, the Senate trimmed back his bill a bit, to the same $4.55 per hour and two-month training plan in the House bill. After rejecting an effort by Bob Dole to adopt President Bush's plan, the Senate voted on April 12 to adopt the modified Kennedy bill, 62–37.

But there was not a two-thirds majority in either House, and Bush knew it. On June 13, he vetoed the conference report, insisting on no more than $4.25 per hour over three years and no less than a six-month "training wage." It was his first veto, and it was upheld easily the next day when the House fell thirty-seven votes short of the total needed to override. The tally was 247–178.

Kennedy promptly reintroduced the vetoed bill, but during the summer he and House Democrats began looking for a deal. So did some Republicans, including Representative Bill Goodling of Pennsylvania, the senior House Republican on the House Education and Labor Committee, and Senator Pete V. Domenici of New Mexico, the senior Republican on the Budget Committee. One of the more obvious ideas was to increase the Earned Income Tax Credit, which would help poor families on low wages but not middle-class teenagers with summer jobs. But that would mean, under budgeting rules, that the federal government would have to come up with extra money, and the tax-writing committees showed little interest. When Congress returned from its August vacation, Kennedy started shopping a proposal for $4.25 after two years with a sixty-day training wage.

The House Education and Labor Committee adopted such a plan on September 19, despite threats of another veto. But nothing much happened for several more weeks until Lane Kirkland, president of the AFL-CIO, met with John H. Sununu, Bush's chief of staff, in late October. Kirkland urged a compromise on the training wage, and Republicans joined in, pointing out that the bill now met Bush's $4.25 demand and another veto would be hard to sustain. As Bob Dole put it, the Bush administration's push for a cut in the capital gains tax made it awkward for Republicans to be "holding up a 30 to 40 cents an hour pay increase."

On October 31, the administration agreed to $4.25 an hour over two years. Democrats and labor accepted a six-month training period at 85 percent of the minimum wage ($3.61 when the basic minimum was at $4.25). But they succeeded in limiting it to workers sixteen to nineteen years old and made the provision complicated and temporary—expiring in 1993. In the Senate, Orrin Hatch complained, "This training wage is so complex and so difficult to implement that I do not believe there is a small business in the country that is going to try it."

The House passed the bill on November 1, 382–37. The Senate followed a week later, on the day after Election Day, 89–8. It provided that the minimum wage would go to $3.80 per hour on April 1, 1990, and to $4.25 a year later. "The minimum wage," Kennedy said, "was one of the first—and is still one of the best—anti-poverty programs we have."

THE FIRST VERSION of the Americans with Disabilities Act had been introduced in April 1986, by Kennedy's Republican friend Lowell Weicker of Connecticut, who had a son with Down's syndrome. In 1988 he introduced it again with Tom Harkin of Iowa, a Democrat whose brother was deaf. Harkin had succeeded Weicker as chairman of the subcommittee on the handicapped in 1987. Their sweeping bill, produced by a Reagan-appointed commission on disability, allowed only the threat of bankruptcy as an excuse for not rebuilding facilities to make them accessible to the handicapped. Kennedy said years later that "it was, even in my terms, a big spending bill. They were going to retrofit every subway car in every part of the country and all the rest of it."

Although one hearing was held, there was no serious effort to move the bill in 1988. Instead it was a device around which the disabled organized in meetings in every state. Pat Wright, a Republican who headed the Disability Rights Education Fund, developed a campaign in which the disabled listed the problems they faced personally. Supporters of the bill deliberately sought an endorsement from Bush before asking Dukakis. Bush, who had worked with them to make sure the deregulation efforts in the Reagan administration did not weaken protections for the disabled, agreed at the urging of Wright, Boyden Gray of his staff, Representative Tony Coelho of California (the House majority whip), Kennedy, and others. On August 10, 1988, a week before he was nominated for President, Bush endorsed the bill, saying it was time to give the disabled the same protections as women and minorities.

Weicker was defeated for reelection in 1988, and Kennedy replaced him as Harkin's chief cosponsor. They quickly agreed on a strategy that abandoned Weicker's measure, which critics called scornfully the "flat earth bill," saying it required the world to be flat and accessible. Kennedy feared that even if they started there and compromised later, that label would stick. Harkin, who had started to negotiate with Hatch in December 1989, felt the same way. So did Coelho, their major House ally, an epileptic himself, who was the third-ranking Democrat in the House. They felt that it was important to compromise to get the Bush administration on board so the bill could move quickly in the Senate. They needed that momentum because in the House, aside from Coelho, most leaders were reluctant to push another vast civil rights bill. And this one was the first to impose substantial costs on private business, whose lobbying carried more weight in the House than in the Senate.

Working with the administration was bumpy. Aside from Bush himself and Boyden Gray, who as counsel to the President was busy with clearances and ethics guidelines, the administration at first was skeptical to hostile toward legislation that required employers to hire and accommodate handicapped workers, directed public and private institutions to ameliorate architectural barriers and ease telephone communications, and required transportation systems to adjust to the needs of the disabled. Sununu, the White House chief of staff and former governor of New Hampshire, and Samuel K. Skinner, the secretary of transportation, were the main obstacles. And, like any new administration, the Bush Presidency took longer to get organized than it had expected.

In March and April, drafts of the bill went back and forth between Harkin, Hatch, and the White House. On April 6, Harkin and Kennedy wrote to Bush, agreeing to wait two more weeks. The Senate backers expected the White House to join a May 1 press conference to support the bill. The White House did not come, but the bill was introduced on May 9, and hearings began that day.

The first hearing focused on costs. Disabled people told their stories. Justin Dart, Jr., who had been commissioner of the Rehabilitation Services Administration under Reagan, presented Harkin with boxes of the lists compiled by the disabled of what they faced. He testified that as "a fiscal conservative, an active Republican," and "a former CEO of both large and small enterprises employing persons with severe disabilities, I know that the modest expenditures required by ADA, almost all of which fall in the private sector, will constitute the type of investments in productivity that have already proven to be profitable to business and to taxpayers." Harkin illustrated Dart's point by showing inexpensive devices, like a $29.95 indicator light that enabled a deaf medical technician to perform tests, or a $49.95 telephone headset that allowed an insurance agent with cerebral palsy to write while talking.

On May 10, the hearing emphasized enforcement. Neil Hartigan, the attorney general of Illinois, said provisions for damages were essential to get the attention of business. "It won't work without damages," he said. Kennedy asked him if his experience with the Illinois disability law showed "that businesses are going to spend their lives in court," as some critics charged. Hartigan said it did not, because most businesses wanted to obey the law, but penalties could get the attention of the minority by "making it more expensive to break the law than it is to keep the law; in other words, by saying to the people who don't care, who have no conscience whatsoever, no concern about people living a full life—they are just in it for the buck—by saying to them, 'Look, you break the law, and we're going to throw the book at you.' "

Then Dole testified on May 16 in general support of the bill but with fears that some of its provisions might invite excessive litigation. But his chief message was that the administration wanted to support the legislation and needed a bit more time, perhaps another sixty days. He said he did not believe people were "dragging their feet," but instead "I think it is just a question of sort of getting their act together." He said he had told them it was time to focus on the bill. Timing was also the issue when the May 16 hearing began. Hatch said that while he did not think the committee should wait indefinitely for the administration, it could be ready in six weeks. Harkin and Kennedy agreed to wait.

Otherwise, that hearing focused on transportation, with several bus company and public transit spokesmen arguing that wheelchair lifts were too expensive. Harkin challenged their figures, and Kennedy said: "This legislation covers the areas of communication, public accommodation, and employment. If we don't have an effective program in the area of transportation, how meaningful is it that we pass the rest of the bill?" Tim Cook, of the National Disability Action Center, replied that "access to transportation is the key to opening up education, employment, recreation—and he [Kennedy] was correct when he said that the other provisions of the Act are meaningless unless we put together an accessible public transportation system in this country."

On June 22, the administration was ready, and sent Attorney General Richard L. Thornburgh to testify. Kennedy designated it a full committee meeting and presided, as he had at the first hearing. His welcome to Thornburgh was a firm explanation of the bill, calling it no more than "simple justice" for 43 million Americans and reminding everyone again that "President Bush shares my commitment to integrating disabled Americans into the mainstream of our society." Kennedy emphasized how much of the bill was drawn from or paralleled other civil rights measures. He contended that the bill's language, which required accommodation unless that imposed "undue hardship," was "a flexible concept which takes into account the size of the business and assures the ADA will not adversely affect small business." He closed by emphasizing the importance of enforcement mechanisms.

Thornburgh spoke of the President's personal commitment as well as the administration's to "a bipartisan effort to enact comprehensive legislation attacking discrimination in employment, public services, transportation, public accommodations and telecommunications." He listed several areas where the administration had problems with the bill—including cost, the scope of public accommodations to be covered, the scope of the requirements to buy buses with lifts, the precision of the terms—"reasonable accommodation" had to be made unless it would cause "undue hardship"—and the desirability of enforcement provisions such as punitive damages. He urged the committee not to rush ahead but to begin negotiating "immediately" to work out a bill that could be passed "expeditiously."

Negotiations began on June 27. Administration officials preferred to deal with Kennedy, perhaps because Harkin was up for reelection in 1990 and they did not want him to get credit for the bill. At first they were arguing over ground rules—about whether the disability community would be part of the talks and whether both sides would commit to sticking to any deal they agreed on. The haughty Sununu telephoned Kennedy at the Cape over the Fourth of July weekend and told him everything could be settled easily if he only removed Carolyn Osolinik, his chief civil rights counsel, from the team. Kennedy called her with congratulations on doing a good job. On July 6, Kennedy and Sununu agreed to oppose any changes once they made a deal, and to negotiate without the disability advocates present.

There were ten negotiating sessions in July at various levels and places. But one was especially dramatic, and significant. On July 28, Senators Harkin, Kennedy, Hatch, David Durenberger (a Minnesota Republican), and Dole met in Dole's office in the Capitol with Sununu, Thornburgh, Skinner,

Roger Porter of the White House staff, Osolinik, and Bobby Silverstein, Harkin's chief aide. The point was to see if all the remaining differences could be settled, or at least narrowed. Sununu complained that the bill would bankrupt small businesses by making them rebuild facilities to accommodate wheelchairs. Silverstein recalls telling him, "That's not what the bill says. The bill has a very low standard for businesses. It's called 'readily achievable' and you only have to make changes if it's easily accomplishable without much difficulty or expense, which is an extremely low standard."

After a few such exchanges, Sununu became enraged, shouting, "I've had it with you. What do you know about this? You don't know anything about business, you don't know anything about this and that, you staff people." Harkin was stunned; while he was trying to decide what to do, Kennedy stood up, his veins bulging, his face red with anger. He slammed his hand down on the table, leaned toward Sununu, and told him that yelling at staff was unacceptable. Kennedy said, "Don't you ever yell at my staff. You want to yell about something, you yell at me. You want to yell at me? You want to yell at me?"

The room stilled. Dole, who had little use for Sununu after his help to Bush in the New Hampshire primary the year before, rolled his eyes. Thornburgh made it clear to Sununu he had gone too far. And they actually started to negotiate, with Kennedy optimistically seeing agreement around the corner every time either side gave an inch.

The aftermath of the explosion also reflects one key part of the strategy of the bill's backers. They kept the incident secret until the bill was enacted the next year. All along, they had determined to emphasize only positive developments while reminding everyone of Bush's commitment to the bill. That would make it harder, they expected, for business to work against it.

Over the next few days, they reached a compromise that eliminated damages in lawsuits over the act, unless the attorney general filed the suit. In exchange for that concession by Kennedy, the administration accepted the bill's broad definition of places of public accommodation that were required to try to adapt to the disabled. The administration wanted to cover just restaurants, hotels, gas stations, and places of entertainment, as the 1964 Civil Rights Act had. Sponsors held out for all establishments that did business with the public, from drugstores to barbershops to banks to lawyers' offices to supermarkets. The bill would require the installation of ramps and elevators in most new buildings or during major renovations. Two years after enactment, employers of twenty-five or more workers would be barred from job discrimination against the handicapped whose disability did not prevent them from doing the job. Eventually the ceiling would drop to fifteen. After five years, new buses had to be wheelchair-accessible. The disabled who could not be discriminated against included not only the blind or the wheelchair-bound or the retarded, but also people perceived as flawed, like someone with a face scarred by old burns. Carriers of the AIDS virus were included among the handicapped, gaining them the protection against discrimination that had been dropped from the 1988 AIDS bill.

The deal was sealed only a few hours before the Labor Committee took up the bill on August 2. But word was out by 7:30 A.M., two hours before the meeting began, and people in wheelchairs began lining up to get in. Before the committee's 16–0 vote, White House aides distributed a statement saying, "The President endorses this legislation as the vehicle to fulfill the challenge he offered in his Feb. 9 address to the nation: 'Disabled Americans must become full partners in America's opportunity society.' "

Congress left for its summer vacation two days later, but Silverstein got the committee report done and filed on August 30, and Mitchell scheduled the bill for action starting Thursday, September 7, the Senate's second day back at work. Harkin managed the bill on the floor, but Kennedy was always nearby, often answering questions about the bill. In his opening speech, Kennedy said the bill could become "one of the great civil rights laws of our generation."

He said, "Disabled citizens deserve the opportunity to work for a living, ride a bus, have access to public and commercial buildings, and do all the other things that the rest of us take for granted. Mindless physical barriers and outdated social attitudes have made them second class citizens for too long. This legislation is a bill of rights for the disabled, and America will be a better and fairer nation because of it."

At the urging of Senator Bob Kasten, a Wisconsin Republican, the Senate debate was made comprehensible to the deaf in the television audience by a signing interpreter. Mitchell announced that unprecedented move, and he quoted Harkin's observation that "the only thing a deaf person cannot do is hear," and said the bill "would make it public policy that we will not, in our Nation, measure human beings by what they cannot do. We should, instead, value them for what they can do."

In the debate, there was a lot of talk about how the law would burden small business. Senator Dale Bumpers of Arkansas, chairman of the Small Business Committee, fretted over the opportunity for a lawsuit against a "mom and pop grocery store." Rudy Boschwitz of Minnesota, the committee's senior Republican, feared that small businessmen would be sued if they went on making the subjective hiring judgments they were used to making. But there were no serious attempts to change or defeat the bill. Too much momentum had built up. "No politician can vote against this bill and survive," conceded Nancy Fulco, a lobbyist for the U.S. Chamber of Commerce.

A central reason was Hatch's support. The backing of the ranking Republican on the Labor Committee certified the bill as reasonable. All year long, Hatch had been torn between his desire to help the disabled, whom his deep Mormon faith directed him to love and help, and his concern for small business. Reluctantly, he embraced the compromise bill on August 2, although he said it still "inadequately protects small business." On the floor, Hatch sought unsuccessfully to add a tax credit to help "small, small businesses" if they had to do something like spend $500 for a wheelchair ramp. He lost, angry at Harkin's lack of support, but said that he was going to "fight" to pass the bill anyhow.

The only prolonged floor fight was over who was covered as disabled. The bill defined a disability as "a physical or mental impairment that substantially limits one or more of the major life activities of an individual . . . or being regarded as having such an impairment." But Jesse Helms of North Carolina and Bill Armstrong, a Colorado Republican, worried about whether an employer's own "moral standards" might be overridden by the law. Helms quarreled with Harkin and Kennedy over treating HIV-positive people as handicapped, insisting they were almost certainly drug users or homosexuals or both. He asked whether transvestites were disabled, and put through an amendment that specifically excluded them.

Armstrong, meanwhile, had gone to the *Diagnostic and Statistical Manual of Mental Disorders,* published by the American Psychiatric Association, and he asked about homosexuality and bisexuality and "exhibitionism, pedophilia, voyeurism and similar." Not covered, said Harkin. But Armstrong and Helms were not satisfied. They wanted Armstrong's entire three-page list of sexual aberrations ("I could not believe some of the things that were in those three pages," Kennedy said years later) included in the bill. But eventually they agreed with Kennedy to limit their demand to one page of sexual items, plus pyromania, kleptomania, and compulsive gambling. The effect was to allow employers to discriminate against pedophiles, pyromaniacs, and the rest.

The Americans with Disabilities Act was debated for one long day. Around 11 P.M. it passed, 76–8, with only conservative Republicans voting no. Harkin and Kennedy came out into the Senate Reception Room, where lobbyists regularly gathered and sought audiences with Senators. Supporters jammed the room, and many of those who gave the Senators an ovation were disabled.

BEING CHAIRMAN OF the Labor Committee enabled Kennedy to influence the national agenda, but it sometimes created painfully insistent local expectations. For example, the leader of Boston's hotel workers, Dominic Bozzotto, didn't see why Kennedy couldn't just fix federal labor law to help his local out.

Bozzotto, president of Hotel Workers Union Local 26, had negotiated in December 1988 with thirteen Boston hotels to create a housing trust fund for his members, to help with down payments and security deposits in a city with a tight housing market. But the National Labor Relations Act, after experience with corrupt unions in the fifties, allowed very few kinds of trust funds. Housing was not one of them. So that contract specified that the law had to be changed for the new fund to take effect.

The AFL-CIO in Washington opposed amending the law, fearing that if it was changed at all, labor would end up losing more than it gained. That did not impress Bozzotto, who arranged protest demonstrations when Kennedy came home to Boston. For much of the summer and fall of 1989, Labor Committee aides were trying to make a quiet deal with Hatch's staff. Hatch's people had no objection to the housing trust idea as such, but they wanted to trade for a pro-management change in labor law. As the session neared its close, Nick Littlefield, a Boston lawyer who had succeeded Rollins as Labor Committee staff director that year, insisted they find another item for a trade.

The staffs could not do it. But late on the night of November 20, Kennedy pleaded with Hatch to help him out. In the hall at the back of the Senate he asked, "Please, Orrin, won't you please let me have this housing bill. Let's attach it to something that's going." Hatch answered, "I can't do it, Ted." Kennedy said, "Isn't there something in this world that you would like, that I could maybe help you with?" Hatch thought for a minute, and then said, "Yes, the downwinders."

Ever since he arrived in the Senate in 1983, Hatch had been trying to get the federal government to compensate cancer victims in southern Utah who lived downwind of the Nevada atomic test sites. The federal government, as Kennedy himself had exposed in a hearing in 1979, had callously failed to warn the inhabitants of St. George, Utah, and nearby places to avoid radiation in the fifties, but the courts had held it was not legally responsible for their illnesses. "It would be so great if we could do something about that," Hatch said then. "We both saw it as a fair exchange," Hatch reflected ten years later, "because this was the bill that helped him in his state and that was the one that really helped me in my state."

Actually, Kennedy got the best of the deal just then, because Hatch not only backed his bill but woke Helms up at 2 A.M. on November 22 to get him to drop his opposition. Then, by unanimous consent, the bill was passed and sent to the House, and Bozzotto said he was sorry he had called Kennedy's staff timid.

But Kennedy's backing was not enough to get Hatch's bill through the Senate in 1989. It was adopted the next summer, pleasing not only Hatch but also Wayne Owens, the former Kennedy aide who was back in the House representing Utah.

The housing trust measure, eagerly supported by the Massachusetts delegation, had little trouble in the House in 1990. Senator Kennedy had talked to Elizabeth Dole about it, and President Bush signed it in April. At a rally at the Park Plaza Hotel to celebrate the new law, Kennedy joked about Bozzotto's pressure: "When the bill got into the House," he said, "we told members either you pass or get a call or visit from Dominic."

NOTHING HAPPENED IN Washington that year that was remotely comparable to what happened in Eastern Europe, where the Iron Curtain fell apart. First in Poland and Czechoslovakia and then in East Germany, Communist regimes tottered and let their people escape to the West. Like most Americans, Kennedy watched on television.

Nowhere was the moment more dramatic than in Berlin on November 9, when guards stopped protecting the wall. Americans watched as Tom Brokaw of NBC reported from the scene: "What you're watching live on television is a historic moment, a moment that will live forever. You're seeing the destruction of the Berlin wall, the dividing line between East and West Germany. On the other side of the wall, young East Germans have rushed through the Brandenburg Gate. . . . They have been pulled up on this wall by other young West Germans who have come from this side. . . ." Things were changing as Brokaw ad-libbed his broadcast. He closed by saying, "And now once again at the Brandenburg Gate, with hammer and chisel they're trying to destroy the wall." Then he showed President Kennedy saying in 1963, "All free men, wherever they may live, are citizens of Berlin. And therefore, as a free man, I take pride in the words '*Ich bin ein Berliner.*' "

Ted Kennedy flew to Berlin to see for himself when Congress adjourned. Willy Brandt, who had been at President Kennedy's side in his triumphal visit in 1963, greeted him. On arrival on November 28, Kennedy said, "All of us have watched in awe in recent weeks as a new spirit of liberty has swept across Eastern Europe. Nowhere do these changes have greater symbolism for America than in the City of Berlin. In a sense, the Cold War began here, and now it is ending here. . . . I only wish that President Kennedy could have come here himself, to see this new day that is beginning."

Accompanied by Brandt, he visited the remains of the wall, where Volkspolizei still patrolled, and saw them taking pictures of its ruins, and of him. They drove through Checkpoint Charlie and prowled the streets of East Berlin.

At the Schoeneberg City Hall, where President Kennedy had spoken, Ted laid two white lilies, one for Jack and one for Robert, with whom Ted had first visited Berlin in 1962. He recalled Robert's speech, telling Berliners, "You are our brothers, and we stand by you." Then he recalled that President Kennedy, on June 26, 1963, saw the wall and then told a huge crowd, " 'Freedom has many difficulties and democracy is not perfect. But we have never had to put up a wall to keep our people in. There are some who say that Communism is the wave of the future. Let them come to Berlin. There are even a few who say that it is true that Communism is an evil system, but it permits us to make economic progress.' And President Kennedy responded again, '*Lass Sie nach Berlin kommen.*' Let them come to Berlin.' "

He and his staff had feared that it would sound corny, but in the end decided to close his speech with the same celebrated phrase Brokaw had shown: "*Ich bin ein Berliner.*"

Then Kennedy lunched with Brandt, current mayor Walter Momper, and several leaders of the suddenly formed East German political parties. They spoke about how suddenly change was coming, and the onrush of calls for German unity—not yet the policy in Bonn. They said things had gone too far for any military intervention to stop them, but they still had to work with the ruling Communist Party as it tried to reform itself, even though such cooperation angered their followers.

Kennedy told them there was not a U.S. politician who did not stand in awe of the East German opposition and its achievements. He said U.S. leaders identified with them. And he asked several questions about East Germany's future security policy. He wanted to know how East Germans felt about the continued presence of Soviet troops. He asked if East Germany expected the United States to protect it. He wondered if the country could afford to continue spending 9 percent of its budget on defense. Ibrahim Boehme of the East German Social Democrats said the actual, though secret, figure was 12 percent. Boehme also urged Kennedy to tell the Senate not to wait too long to start economic aid to Eastern Europe and the Soviet Union, too, which was nearing economic collapse. If Gorbachev failed, he said, that would lead to "chaos" in the Soviet Union and then to "apocalyptic" developments in Eastern Europe.

Afterward, Kennedy eagerly shook hands with a small crowd outside that was chanting "Ted-dy, Ted-dy," and then left for Geneva after seven hours in Berlin.

THE BERLIN JOURNEY was largely a personal pilgrimage; "Emotionally, I just wish my brother could have seen it," he said at the end of the day. But a few days later, Kennedy set out on a trip designed to produce publicity and focus attention on national health insurance and other health issues. In the Labor Committee, he had managed to win party-line approval that July of legislation to require employers to insure workers and their families and for a gradual expansion of Medicaid, beginning with children, to cover the rest of the uninsured public. But the committee action was just a gesture, for the bill was not stirring much support.

So on December 11, Kennedy began on a whirlwind four-day nationwide set of hearings by private jet. He started with a quick visit to Boston City Hospital, and then visited Bronx-Lebanon Hospital in the South Bronx. He opened that session by saying, "Health care should be a basic right for all, not an expensive privilege for the few." He called the health care system the "fastest growing failing business in America, because while its cost was soaring, 37 million Americans had no insurance and 60 million more had inadequate insurance."

The New York hearing focused on what the growing number of uninsured and the AIDS epidemic were doing to the hospital system. Dr. Stephen Lynn, director of emergency medicine at St. Luke's–Roosevelt and chairman of the Task Force on Overcrowding of the American College of Emergency Physicians, said patients were waiting for treatment in overcrowded emergency rooms for as long as a week before being admitted. A shortage of nurses meant regular beds went unused. He said, "When we add to that the increased number of poor and/or uninsured patients, persons with AIDS, and patients with drug abuse, the system that is overcrowded and overwhelmed begins to crumble. AIDS may be the straw that breaks the camel's back."

The next day in Los Angeles, Kennedy visited the Harbor–UCLA Medical Center. There he was told, according to the committee report, of "critically ill patients in need of intensive care who remained in the emergency room for as many as 3 or 4 days because of the lack of beds. During the committee's visit to Harbor–UCLA, a patient was being maintained on a respirator in the emergency room because the medical intensive care unit was filled to capacity; this patient had been in the emergency room for more than 24 hours."

The Los Angeles hearing itself began with accounts of people who had lost their health insurance. Jim Krause, a printer, told of how his insurer had gone bankrupt and another company had denied him coverage because of gastritis. A stroke and a heart attack exhausted his savings, and he was now uninsurable. Donna Van Tassel told of exhausting her savings to care for her aged mother because Medicare was inadequate.

Then Robert Gates, director of health services for Los Angeles County, tied the two problems together, and pointed out that they hurt everyone, not just the poor. With a growing population of uninsured patients, he said, ten of the county's twenty-three trauma centers had been closed because "hospitals find that it's simply not economical for them to continue, and that continuing on with their trauma service may threaten the entire existence of the hospital, and they can't allow that to happen. So they discontinue trauma services, and they are not available to anybody." There were similar problems for emergency rooms in downtown areas, he said, because "there are too many patients who simply are unable to pay."

In St. Louis on Wednesday, December 13, Kennedy heard again from uninsured patients. Brigett McDaniel, a student, reported being told by a collection agency that she should have thought of the bills before getting treatment for a tumor on the lung. Neal Baretta told of telling his uninsurable epileptic son that he could not play sports because they could not afford an injury. Kennedy asked, "What does Jason say to you when you tell him he can't go out and play with his friends?" Baretta replied, "It's hard to describe; you have to be there and see his face when he comes home. He came in the house last week and said, 'I want to play basketball.' And I said, 'You know, Jason, you can't.' "

Then Kennedy heard from the community health centers he had sponsored and protected. Betty Jean Kerr, executive director of the People's Health Center, said they treated not only the uninsured, but also people with only hospitalization insurance or those whose Medicare was inadequate. She said they had many seriously ill patients because "they avoid seeking health care from traditional institutions until an emergency arises." She said, "The current situation is absolutely untenable. We're weathering the storm as best we can, but, like most of our patients, we're on fixed—extremely tight budgets ourselves."

The final hearing was in Sparta, Georgia, where Hancock County had voted to raise property taxes four years earlier to reopen a hospital—a contrast to the declining availability of rural health care in many places. Here attention also returned to AIDS. Dr. Ted Holloway, director of the Southeast Health Unit of the Georgia Department of Human Resources, privately told committee staff of AIDS patients in small towns who waited "until it was dark so they could get into the car and people wouldn't see their sores." They would drive to Atlanta for treatment and come back at night when no one would see them. Holloway then testified, "All of our local hospitals have already admitted several patients. They cannot get them transferred out because there are no centers that are set up in Georgia to really serve the State for AIDS care, so I think it is going to be the straw that really breaks the back of the kind of faltering health system."

KENNEDY'S FIRST MAJOR legislative effort of 1990 was not in health but civil rights. In 1989 the Supreme Court had issued six major decisions that made it harder for workers claiming job discrimination to sue and collect damages. On February 7, along with thirty-three other Senators and 123 House members, Kennedy introduced legislation to undo those decisions and to allow victims of sexual harassment on the job to collect damages. The Bush administration welcomed parts of the bill. But Donald Ayer, a deputy attorney general, contended that employers would protect themselves from charges of discrimination under the act by adopting quotas: "so many blacks, so many Hispanics and be done with it." The Labor Committee approved the bill, 11–5, on April 5, despite a veto threat from Thornburgh.

Kennedy was looking for more support, and again he found a prominent Republican to work with. On May 17 he and Senator John C. Danforth of Missouri agreed to a provision they said would eliminate the risk that employers would use quotas. Under the original bill, if some requirement like strength or education turned out to keep minority or female employment down, employers could only have defended themselves by showing that the requirement was "essential to effective job performance." They softened the standard to demand the requirement only "bear a substantial and demonstrable relationship to effective job performance."

Negotiations with the White House dragged on, with Sununu sometimes encouraging and sometimes hostile. Boyden Gray, who said he had suffered discrimination as a white Anglo-Saxon Protestant on the *Harvard Crimson,* repeatedly blocked agreement, saying various formulations would lead to quotas. Mitchell impatiently brought the bill to the floor on July 10, and quickly forced a cloture vote to end debate. That vote succeeded, on July 17, but only by 62–34, and Dole accused Democrats of playing politics with civil rights and treating Republicans "like a bunch of bums."

Sununu accused Kennedy of bad faith, saying he'd made a deal and backed off. "I'm not unwilling to use the word 'renege,' " Sununu said. But William T. Coleman, Jr., the Republican former secretary of transportation whom the White House brought into the talks, blamed the administration for forcing a stalemate. Danforth, unhappy with Mitchell even though he was one of eight Republicans voting for cloture, said, "I, for the life of me, cannot understand how we have furthered the cause of getting this bill enacted into law."

The next day, July 18, the Senate passed the bill, 65–34. The total was two votes short of what would be needed to override the threatened veto. Kennedy scoffed, "Quotas, shmotas. Quotas are not the issue. Job discrimination is the issue." He said there was still a good chance that the bill, as it moved through the House and a Senate–House conference, could be modified to win Bush's support or the votes to override.

WHILE KENNEDY'S HEALTH care tour produced no groundswell for national health insurance, it led to action on AIDS and its impact on the health care system. The visits to Boston and New York showed him how AIDS overwhelmed big city hospitals. Two of Kennedy's staffers, Michael Iskowitz and Terry Beirn, came up with a rationale, stimulated by the 1989 San Francisco earthquake, of comparing the AIDS epidemic to the natural disasters on which Congress was always ready to spend money. They sought out Hatch, who agreed but was also especially concerned about the destabilizing effect AIDS was beginning to have on rural health care—a subject covered at the Sparta hearing.

Iskowitz and Beirn talked at first about a $25 or $30 million demonstration project, but Littlefield disagreed and said they should "go for broke" and ask for what they thought was needed. So they fashioned a bill authorizing $300 million for aid to the thirteen hardest-hit cities (those with two thousand or more AIDS cases) and another $300 million for the fifty states to develop AIDS care programs, especially early diagnosis and home care, another Hatch concern.

Getting the bill passed in a time of fiscal stringency and widespread fear of AIDS was a major enterprise, despite the breakthrough effort in 1988 and Congressional willingness to spend money for AZT, a drug that relieved AIDS symptoms. Two major public relations strategies helped.

First, Elizabeth Taylor was enlisted as an advocate. The actress was a cochairman of AMFAR, the American Foundation for AIDS Research, which had first put Iskowitz and Beirn in touch with Kennedy. Her appearance made the March 6 Senate news conference a "thirteen-camera day," as Littlefield described the television presence. She said, "We need money—we need lots of money and we need it now. Our cities and states are being choked by the enormous cost of caring for AIDS patients. They can no longer cope. They need disaster relief now." Kennedy played off her, saying, "This is not about resources. We can find the resources. What we need is the will." And that afternoon on the Senate floor he returned to the Iskowitz-Beirn message: "America responded within days to the California earthquake. We have pledged tens of billions of dollars to rescue the savings and loan industry. AIDS is a comparable disaster and we need to respond accordingly."

The second strategy was more instinctive. Ryan White, a Kokomo, Indiana, teenager who got AIDS from a blood transfusion in 1984 and was shunned and expelled from school, had become a national symbol of the randomness of the disease. President Reagan welcomed him. And he and his mother, Jeanne, had campaigned for AIDS understanding and legislation. But on March 29, 1990, he was having difficulty swallowing and entered Riley Hospital for Children in Indianapolis, knowing his situation was grave.

On April 4, the Labor Committee met to consider the Comprehensive AIDS Resources Emergency bill. It adopted the measure Kennedy and Hatch had introduced by a 16–0 vote. As he gaveled the result, Kennedy said, "This one's for you, Ryan." Kennedy then called Jeanne White at the hospital and had a second press release on the committee action issued, adding this statement to the originally prepared bill summary: "We hope this urgent measure can help relieve the suffering of Ryan and so many others with AIDS. Through Ryan's tremendous courage, America came to understand the pain of this disease. We take this action today in his honor."

Ryan died at eighteen on April 8, and a few days later the Labor Committee report on the bill said, "By dedicating this legislation to Ryan, the committee affirms its commitment to provide care, compassion and understanding to people with AIDS everywhere. Ryan would have expected no less." Helms, on the other hand, was arguing in North Carolina that this poor boy's memory was being distorted to further a homosexual agenda. So his mother, at Iskowitz's and Beirn's urging, wrote to every Senator, appealing for their votes by saying, "I believe Ryan would be proud to know that his struggle may help pass a bill that will aid thousands of others in their fight against this dreaded disease."

Mitchell had been unwilling to bring the bill up unless its backers could break a filibuster. But the Ryan White connection pushed it to sixty-four cosponsors, enough to invoke cloture. When the bill was first debated on May 14, Hatch took the lead, knowing the attack that would come soon from an incensed Helms. He began by citing White, "a courageous, uplifting, always optimistic, sweet, decent kind young man," and then anticipated the attack that would soon come from an incensed Helms, saying, "It is a mistake to condemn this bill or find fault with it based on anybody's lifestyle, because this is not a bill that should be characterized in any way other than as a public health bill. AIDS is a public health epidemic and we must treat those who are suffering from it."

Helms interrupted to complain, "I have never heard once in this Chamber anybody say to the homosexuals and drug users, who make up over ninety percent of the AIDS population, 'Stop what you are doing.' Does the

Senator realize that if they would stop what they are doing, there would not be one additional new case of AIDS in America." Hatch disagreed, saying for that to happen, there would also have to be a complete halt to sexual activity by everyone who had a potential infection and an end to blood transfusions. He said no Senator would refuse to jump in the water to save a drowning man or woman, and "without first asking how he or she happened to fall into the water." That was the approach the Senate should take to AIDS. He asked rhetorically, "Should we just let the disease run rampant because we do not agree with the morals of these people. For one, I do not. I do not condone homosexual activity, but that does not have a thing to do with this bill. AIDS is a public health problem."

Helms was hardly persuaded. In a speech which he repeated the next day when more Senators were there to hear it, he complained about "those who have so callously used the suffering of Ryan White to promote their political agenda." He said the "Hollywood and media crowd" were using White's death "to frighten the American public into believing that AIDS is waiting to happen to everyone, even if they do not engage in illegal and/or immoral activity." Helms, who said he did not care if he was called "homophobic," observed that as a Baptist, "I was taught a long time ago to hate the sin and try to love the sinner. But some of these people make the latter part of that tough." He said, "Ryan White would never have contracted AIDS had it not been for the perverted conduct of people who are demanding respectability."

Kennedy avoided taking Helms's bait, leaving it to the conservative Hatch to match moral prescriptions with him. But Kennedy outmaneuvered Helms with palliative amendments on issues like partner notification, needle exchanges, and banning high-risk blood donors. In each case, he offered enough cover for anxious Senators so he won overwhelmingly. When challenged on why this one disease should be singled out, he replied that others were already eased through Medicare and Medicaid, and that AIDS was a "classic public health epidemic," and one not spread evenly across the country, like heart disease or cancer.

As debate closed on May 16, he said, "By providing emergency relief to the cities hardest hit by AIDS, and substantial assistance to all States for more effective and more cost-efficient health care services for infected individuals, we are acting in the best interests of the Nation—by fighting AIDS and not people with AIDS. The AIDS epidemic is now a disaster as devastating as any natural catastrophe that has ever afflicted the Nation."

The bill passed, 95–4, on May 16. The House followed in June, and the conference report was adopted August 4 when Kennedy returned to the original Iskowitz-Beirn theme: "In terms of pain, suffering and cost, AIDS is a disaster as severe as any earthquake, hurricane, drought. . . . Health care institutions—and the vulnerable Americans who depend on these institutions—are in crisis. AIDS by itself is certainly not the only cause of these problems. But it is adding to the stress that is leading toward a total breakdown of our health care system."

KENNEDY RETURNED TO Chile on March 11, invited along by Vice President Quayle to the inauguration of Patricio Aylwin as the elected successor to General Pinochet. It was an unusual trip for him, not intended to push something along or to educate himself, but just to savor a change he felt he had been a part of. He was near the front when the Chileans arrived in the new Congressional Palace in Valparaiso and the hall erupted in shouts. Pinochet's supporters chanted, "Pi-no-chet, Pi-no-chet!" Aylwin's replied, *"As-se-si-no, as-se-si-no!"* (meaning "assassin") or *"De-mo-cra-cía, de-mo-cra-cía!"* Pinochet put down his sash of office. Aylwin was sworn in and donned a new one.

When he returned to Santiago that day, Aylwin promised to establish the truth about the repression of Pinochet's sixteen-plus years. Kennedy asked Aylwin why the inaugural ceremonies would take three days. (Monday the twelfth was devoted to a gala performance of Beethoven's Ninth Symphony.) The Christian Democrat answered, "In your country you can have the transition of power in an hour. But not in the minds of our people. It's not possible. They have to take time to be able to see this kind of a change."

The finale of the inauguration caught the spirit of that transition. It was held in the National Stadium, where thousands had been confined and hundreds murdered in the first days of Pinochet's coup. Poets and musicians who had been among the persecuted performed. Then mothers who had lost their children and husbands who had lost their wives strode in, holding pictures of the victims and dancing without partners. Aylwin spoke as the setting sun shone on the Andes, recalling, "During the hateful and blind days when force prevailed over reason, this stadium was for many countrymen a place of imprisonment and torture. We would like to say to all Chileans and to the world that never again will human dignity be violated; never again, fratricidal hate; never again, violence among brothers." Thousands of candles were lit in memory of the lost. Then fireworks, and finally the crowd came down from the stands and danced on the field.

THERE WAS NOTHING ceremonial about Kennedy's next trip. He went to Moscow on March 25, 1990, at a crisis point in the last years of the Soviet Union. Just as the Eastern European countries took Gorbachev's internal liberalization as a signal that they could escape Moscow's domination, Lithuania took Moscow's tolerance of those changes as an occasion to reassert the independence it had lost when the USSR annexed the Baltic states in 1940. Bush and Gorbachev had discussed the Baltic states at their summit meeting in Malta on December 3. Bush warned Gorbachev that Soviet violence against them would create a "firestorm" in the United States, but that if violence could be avoided, the United States government would be restrained in dealing with the independence issue because "we don't want to create big problems for you."

On March 11, 1990, Lithuania voted to declare itself independent. In Moscow the new Council of People's Deputies met for the first time the next day and elected Gorbachev as President, though 40 percent did not vote for him even though he was unopposed. He issued an ultimatum to Lithuania's new President, Vytautas Landsbergis, demanding withdrawal of the declaration by March 19. He did not withdraw it, and Soviet troops occupied the headquarters of the Communist Party (which backed independence) on March 23. Foreign diplomats and journalists were ordered out of Lithuania that day, in a move that suggested an imminent crackdown. On the twenty-fourth a convoy of Soviet tanks drove past the parliament building in Vilnius.

Kennedy saw Bush before leaving Washington, and when Jack Matlock, now the American ambassador, met him at the airport in Moscow, he sought a quick briefing before talking to reporters. Matlock said tensions were reaching a dangerous level. Kennedy then told reporters that a peaceful solution was essential but violence would endanger U.S.-Soviet relations.

At the Kremlin on March 26, Kennedy conferred with Gorbachev for ninety minutes. Gorbachev greeted what *Pravda* the next day called "his old friend" with praise of John and Robert Kennedy and said he would like to hire a biographer of President Kennedy to write his own. Kennedy opened the discussion of Lithuania, saying its situation imperiled progress in U.S.-Soviet relations. Gorbachev responded that he had warned Reagan, Bush, and other Western leaders that as he tried to change the Soviet Union, they would have to show "special attention, special patience, special responsibility," but the warning was being ignored. He said Bush imposed a "double standard" by claiming the right to invade Panama (to capture President Manuel Noriega in January, a move Kennedy had criticized). Gorbachev said he had not tried to upset U.S.–Soviet relations over that issue, while Bush and Kennedy were using Lithuania against him.

Gorbachev told Kennedy that there were strong forces in the Soviet Union that opposed his economic and political reforms, and they could use Lithuania against him. He said he was being criticized for continuing to supply Lithuania with oil. "I am keeping Lithuania alive, and no one else," he said. The Council of People's Deputies, after all, had just voted against Lithuanian independence. "Should I follow their mandate, or the Bush-Kennedy policy?" he asked.

Kennedy asked whether he would rule out using force in Lithuania.

Gorbachev did not answer directly. He said Americans did not understand what had happened there. He said Americans seemed to think that a group of people simply got together, raised their hands, and voted, and suddenly the world was a different place. In truth, he said, what had happened was nothing less than a "palace coup d'état." While the Soviet constitution did permit secession, it had to be carefully worked out to accommodate Soviet defense interests, the rights of Russians and Poles who lived there, and other questions. He said Landsbergis and his associates were not serious people. He said that they should be prepared to pay in hard currency for oil and other things they received from the Soviet Union.

Kennedy said that matters Gorbachev cared about in relations with the United States would be damaged if he used force. He asked if Gorbachev rejected force.

Again Gorbachev did not answer directly, but said he hoped the issue could be settled politically. But he warned that if things got out of hand, "extremists will undermine me." He said he was trying to develop a confederation in which the republics had new power. But in Lithuania, he said, "we know who these reactionaries are, who they worked with during World War Two." He said, "This is a tough time for me, and I cannot be undermined" as he sought to establish a new democratic Presidency that would prevent a return to the past, to a revival of a Russian czar.

Kennedy asked him about force five times, and finally Gorbachev said, "I pledge it will only be used if there is violence that threatens the lives of others." Otherwise, he said force was precluded and the Soviets would try every peaceful means to settle the dispute. He asked Kennedy to tell his colleagues in Congress that these were watershed times and not to rush things on the Soviet Union.

Kennedy then told a press conference, "President Gorbachev indicated to me that the position of the government was that there would be no use of force unless lives were threatened." Gorbachev generally held to that promise, but that very night Soviet troops in Lithuania broke into a psychiatric hospital to capture deserters from the Red Army.

The next day Kennedy had a very unusual meeting. He went to Lubyanka prison, the Dzerzhinsky Square headquarters of the KGB, to meet its boss, Vladimir Kryuchkov. It was Kryuchkov's idea, a first for an American Senator. Kennedy came away from the two-hour session impressed by Kryuchkov's sense of humor and his loyalty to Gorbachev and to economic reform. Kennedy wrote later, "Kryuchkov seemed open to promoting cooperation on numerous areas—drugs, crime, terrorism and even gun control. He seemed genuinely concerned over the rise in anti-Semitism and agreed to study a proposal I offered to send him on private chartered flights between Moscow and Israel. He struck me as a pragmatic realist who is willing to try and get things done."

But Kryuchkov was also blunt in explaining Soviet tactics in Lithuania. KGB troops were guarding power plants. Army forces were on the Polish border. Weapons in private hands were about to be seized. The Kremlin's point was to show that even if the Lithuanians called themselves independent, they exercised none of the basic powers of sovereignty, such as controlling borders or means of production. But Kryuchkov also told Kennedy that the Soviet leaders were very concerned about American opinion.

On Soviet television, in a speech to the Soviet Peace Committee, a government-sponsored group that denounced Western countries as warmakers, and in private meetings with Soviet politicians and generals, Kennedy stressed Gorbachev's commitment to a peaceful solution in Lithuania and the threat that violence would pose to U.S.–Soviet relations. "They have to take that into consideration," he told NBC News on March 28, "because the response in the United States would be extremely harsh."

After a meeting with Soviet refuseniks on March 28, Kennedy flew back to Washington, and he briefed Bush the next day. While the President thought Gorbachev's statement to Kennedy on force was a step backward, he heeded Kennedy's conclusion that Gorbachev felt under severe internal pressure to "cut Lithuania off, to treat it with an economic boycott, to effectively put the economic screws on Lithuania, which they can, and make it come to heel," as Kennedy put it to David Hoffmann of the *Washington Post*. Bush told Fitzwater to halt his criticisms of Soviet policy, and he sent Gorbachev a private letter, emphasizing that he was not trying to score propaganda points over Lithuania. Bush told Gorbachev he now had to defuse the situation. "I want to be sure the Soviets understand our position, and understand that we're not trying to make things difficult for Lithuania or the Soviet Union or anyone else," Bush told reporters.

Before those trips, another of Kennedy's heartfelt foreign policy goals had been achieved. On February 11, after more than twenty-seven years in prison, Nelson Mandela was freed. Kennedy talked to him on the phone the next day and invited him to Boston, where church bells pealed in welcome when Mandela arrived on June 23. At a luncheon at the Kennedy library, the Senator said, "We will not give up, we will not give in, until apartheid has been wiped off the face of the earth." Then he introduced his guest: "Nelson Mandela is the statesman of our time. He represents what courage and commitment is all about. . . . Nelson Mandela is today the true father of a new South Africa." After intense applause the South African said, "Right now, I consider myself an honorary Irishman from Soweto."

Mandela said his life had been touched by all three Kennedy brothers, but especially Ted. He recalled Kennedy's 1985 trip to South Africa and to Pollsmoor Prison: "It was indeed very frustrating to know that he was at the gates of the prison, but was unable to come in to give the message of hope, of strength, which as a black prisoner in a white South Africa, gave us a lot of strength and hope, and the feeling that we had millions behind us both in our struggle against apartheid but in our special situation in prison."

THE DISABILITIES MEASURE had to wend its way through four separate House committees, and without Coelho to guide it because he had resigned from the House in June 1989 as a junk bond scandal broke around him. His closest friend, Representative Steny H. Hoyer of Maryland, took over, but had none of his leadership status, so Coelho occasionally called in old debts to help move the bill along. But on May 22, the House passed the bill, 403–20. It was fundamentally like the Senate bill, with modest changes, including delays in enforcement of the antidiscrimination provisions against small businesses and requirements to make key rail and subway stations accessible. Another change, sponsored by Representative J. Dennis Hastert of Illinois, allowed commuter rail lines and subways to make only a single car per train accessible.

But the House had put one major obstacle in the way of final enactment, an amendment sponsored by Representative Jim Chapman, a Texas Democrat. Despite belated administration opposition, the House voted to allow employers to transfer workers with AIDS or HIV out of food-handling positions. The restaurant industry, while acknowledging that those diseases could not be transmitted through handling food, said the amendment was necessary because of public fear. The argument infuriated the disability groups, but Jesse Helms persuaded the Senate on June 6 to take the rare step of instructing its conferees to take the House language. A move to table it was beaten 53–40, after Bob Dole, ordinarily a supporter of federal help on AIDS, voted with Helms. The next day, Carolyn Osolinik, Kennedy's lead staffer, insisted at a staff meeting that codifying irrational fear and prejudice would be the beginning of the end and that Congress might go on to exempt one occupation after another, from pilots to child-care workers.

With majorities of conferees from both houses opposed to the Chapman amendment, it took only one two-hour meeting on June 25 to agree on a bill and drop that element.

But Helms kept trying. On July 11, with the conference report before the Senate, he moved to direct the conferees to go back and insist on the Chapman amendment. Kennedy, Harkin, their staffs, and the disability groups convinced Hatch that the Helms–Chapman provision could kill the entire bill. With the administration helping, Hatch proposed requiring the secretary of health and human services to compile a list of communicable diseases and indicate which could be spread by food handling. People with those diseases could be transferred. The proposal offered political cover to the Senators who had voted with Helms earlier, and he complained, fairly, "It will gut the Chapman amendment."

Hatch led the effort, as he had on the Ryan White bill. His conservative voice made it easier for Republicans and Southern Democrats to change position than Kennedy's or Harkin's would have. Kennedy did pitch in on the floor, attacking the argument that people's fears of contracting AIDS from food handling should be catered to. He said, "I come from a state where men feared witches and hanged women. That was in the 1690s. This is the 1990s, and I hope the Senate is mature enough and civilized enough to reject the appeals of witchcraft we have just heard and do the right thing." Helms's motion was defeated, 61–39, with Dole's help. Then Hatch's plan carried, 99–1, with only Helms voting no.

The conference quickly met again and adopted Hatch's plan. The House passed the final bill, 377–28, the next day.

When the Senate voted on July 13, the rhetoric started with proud statements on civil rights. Dole, for example, said, "Just as we have seen the walls go down in Eastern Europe, we are now witnessing some of our own walls crumbling, the wall of prejudice, isolation, discrimination and segregation."

But the chief sponsors told personal stories. Harkin began by using sign language to talk to his deaf brother. Then he explained to the Senate, "I just wanted to say to my brother Frank that today was my proudest day in sixteen years in Congress, that today Congress opens the doors to all Americans, that today we say no to fear, that we say no to ignorance, and that we say no to prejudice."

Kennedy followed: "Many of us have been touched by others with disabilities. My sister, Rosemary, is retarded; my son lost a leg to cancer. And others who support the legislation believe in it for similar special reasons. I cannot be unmindful of the extraordinary contributions of those who have been lucky enough to have members of their families or children who are facing the same challenges and know what this legislation means."

Hatch spoke of his brother-in-law, Raymond Hansen, who earned two engineering degrees despite contracting two kinds of polio, and "worked right up to the day he died, going into an iron lung each and every night just to be able to survive." Hatch called him "the greatest inspiration of a dogged determinist to do what was right and to make his life worthwhile of anybody I know in my life."

The bill passed, 91–6. Bush signed it before a large crowd on the South Lawn of the White House on July 26. The lawmakers who managed the bill were invited, though Weicker was asked and then told at the gate, on Sununu's instructions, not to come in. But Bush was not photographed with them, for fear of aiding Harkin's reelection race. The only lawmaker Bush mentioned was Bob Dole. The disabled were up front and on the podium. Kennedy sat in the bleachers with Teddy and Pat Wright and a few of her staff. Despite the obvious political slight, he said later, "It was a lovely, beautiful day."

THAT SUMMER Rose Kennedy was one hundred years old, and the family invited 370 people to a birthday party she observed from her wheelchair upstairs. The party was held on July 15, a week before her birthday, to coincide with an awards presentation by the Joseph P. Kennedy, Jr., Foundation for work against mental retardation. For the occasion, Ted offered quotes of hers which some family members doubted she had ever uttered. He told reporters she had recently asked him if the tennis racket he was carrying wasn't hers, because "I've been looking all around the house for mine," and had proclaimed herself "like old wine—they don't bring me out very often, but I'm well preserved."

But if he was kidding the outside world, or even himself, he was also caring. One nephew, Robert F. Kennedy, Jr., said that when he thinks of his uncle, he is "singing with my grandmother—'Rosie O'Grady,' her favorite song, and she would smile when he sang it." Others noticed the love, too. Bishop Tutu recalls staying at the Cape in May 1990 and going for a walk where he discovered "Ted, sitting with his mother, reading to her, and then they would be singing Irish ditties together."

Five weeks later, on Sunday, August 19, Steve Smith died of cancer. In Steve's last weeks, Ted had often been with him, trying to cheer him up, for Steve had become his last big brother—someone from whom a "well done" mattered, someone he could confide in. Ted was used to delivering eulogies, but two days later, Steve's was one of the hardest to get through. At the beginning, he faltered several times when he tried to say "When Jack left us, and then Bobby, too, there was a consolation in knowing we still had Steve." He sobbed, "I don't think I can get through this." Robert F. Kennedy, Jr., came up and stood behind him, and Ted pulled himself together and read most of his tribute. As he walked out of St. Thomas More Church in New York, he was working his jaw back and forth to keep from crying.

The next month there was another family event, simpler and happier. Joan joined Ted and four hundred others at Our Lady of Victory Church in Centerville when their daughter Kara married Michael Allen, a Washington architect, on September 8. Kara was thirty, a Tufts graduate who had been working for *Evening Magazine* on Boston's WBZ-TV. Michael, thirty-two, had one special entree with his father-in-law: before graduating from the Rhode Island School of Design he had sailed on the winning United States ocean racing team that won the Sardinia Cup regatta in the Mediterranean in 1980. Ted had the wedding cake decorated with a miniature of Jack's boat, the *Victura,* rigged in spun sugar. At the reception, in a tent on the broad lawn in front of the family house at Hyannis Port, Ted recalled Kara's birth during the 1960 campaign and said he had always regarded her as the "New Frontier."

At celebrations like this Ted was the accomplished host, taking Kara and her maid of honor for a sail, holding Joan's arm as they left the church, or dancing with Ethel. But to many in Washington he was perhaps better known for "boozing and womanizing," as a *Washington Post* profile put it that April, than for his ever-longer list of legislative victories. He did not drink too much to function on the Senate floor, but after dinner colleagues could smell the liquor on his breath. When a policy dinner at home in McLean was over, friends or staff stayed around to drink, leaving when a girlfriend turned up. Many mornings he would be up early anyhow to play tennis, but his staff tried hard not to schedule him for anything before ten.

CHAPTER 31

Battles in the Senate—and the Gulf

Congress stayed unusually late for an election year in 1990, adjourning eight days before Election Day. Much of its last month was spent wrapping up Kennedy's agenda. But his first fight came over a bill that was not before any of his committees, but the Senate Appropriations Committee.

When the Ryan White bill was signed by Bush, it authorized a total of $875 million in annual spending—$275 million for grants to hard-hit cities, whose numbers had grown to sixteen as the year went on, $275 million in grants for states to match for AIDS services, $305 million for state and private early-intervention programs, and $20 million for demonstration projects on treatment of children with AIDS and a study of the rural epidemic. But Harkin's Appropriations subcommittee voted no money at all for the new bill when it acted on September 12, though it provided that $490 million could be spent beginning in October 1991.

Harkin said the subcommittee just did not have the money to allocate in a tight budget year, for its existing programs were all popular and hard pressed. "I wish I had the money for it," he said. "It's a tragedy." The subcommittee's major increases had gone to three education programs—Head Start, Pell grants to college students, and public school aid for poor areas. The AIDS community and the cities that had expected help were angry. Kennedy argued furiously with Harkin, but got nowhere. Then he encouraged other Appropriations Committee members to shave money from other programs whose increases exceeded the rate of inflation. That produced a backlash from supporters of the other programs. The move was defeated on October 10. Kennedy issued a sharply critical statement, saying, "The real tragedy is for the thousands of people with AIDS in desperate need of treatment and care, who do not have time to wait." Harkin did find $49 million in new money for the bill, and ultimately a House-Senate conference allocated $221 million.

That was a setback, but Kennedy won a major victory on another spending bill and on half a dozen other important pieces of legislation. The spending success came on legislation to fund the National Endowment for the Arts, which was under severe conservative attack for two artists it helped exhibit in 1989. One was Andres Serrano, known for *Piss Christ,* a photograph of a crucifix in a jar of urine, and the other was Robert Mapplethorpe, who made bizarre homoerotic photographs. In 1989, Helms had succeeded in conditioning NEA spending on pledges by recipients not to use them for obscene art. Some groups cried censorship and refused the money. In 1990, Helms was running for reelection and vowing to fight obscenity.

The NEA, even if its leaders were sometimes obtuse about how its grants could outrage the public, was a major force in the arts and especially in spreading their impact. It appealed not just to Northeasterners like Kennedy and Claiborne Pell of Rhode Island but to Hatch of Utah and Nancy Landon Kassebaum of Kansas. They feared that the sort of restraints Helms wanted to impose would shatter the agency, but they needed to find a way to let Senators vote against them without seeming to advocate dirty pictures. So they came up with a proviso requiring artists to refund any NEA money they used on works later determined to be obscene by a criminal court. (After a prosecutor in conservative Cincinnati had failed to convict museum officials for showing Mapplethorpe's work, there seemed little risk of convictions.)

The Labor Committee adopted this approach, 15-1, on September 12, when it voted to reauthorize the agency for five more years. Hatch, once again serving as the conservative lightning rod for a fight with Helms, proposed the plan as an amendment to the spending bill on October 24. Helms cited the Cincinnati verdict as evidence that no federal money would be recouped, and said Hatch's plan only "gives political cover." He wanted a flat prohibition: "None of the funds appropriated under this act may be used by the National Endowment for the Arts to promote, distribute, disseminate, or produce materials that depict or describe, in a patently offensive way, sexual or excretory activities or organs."

Helms was right about the political cover; his proposal was beaten, 70–29, and Hatch's was adopted, 73–24, after Kennedy said, "Government officials are not the appropriate adjudicators of the arts. Art professionals have compiled an excellent track record in developing the peer panel system that is currently in place, we have no business substituting Congressional judgments for peer review. We must resist the calls to censorship—to return to our nation's regrettable periods of Comstock, McCarthyism and anti-intellectualism." Kennedy, Pell, Hatch, and Kassebaum and Representatives Pat Williams of Montana and Sidney Yates of Illinois managed to keep their obscenity prohibition in the bill that emerged from a House-Senate conference.

Another October success came on legislation to promote volunteerism and community service. Bush had raised the idea in his acceptance speech in New Orleans in 1988, hailing as "a thousand points of light" efforts by individuals, not government, to improve society. Kennedy was especially taken by a Boston program, "City Year," which brought poor and middle-class high school graduates together for a year of community service, starting each day with calisthenics in front of City Hall.

But Bush and the Democrats found it hard to work together, because Democrats wanted to include a demonstration program which gave volunteers some reward for service, and the administration argued that was not "true volunteerism." The Senate passed its bill in February 1990, but the House did not act until September, adding a provision the Bush administration found especially offensive, forgiveness of federal student loans in return for work. Conferees killed that idea, and they included a Kennedy–Hatch provision that formally established Bush's Points of Light Initiative Foundation.

In June, Kennedy addressed the City Year graduation, telling the volunteers they were another example of youth leading the nation, as it had on civil rights, Vietnam, and the environment. And when the Senate passed the three-year, $287 million bill on October 16, he said that while many of the nation's worst problems, from homelessness to pollution, could be diminished by local efforts, the bill's "most important result may well be its effect on young citizens in the earliest grades."

Kennedy described a Springfield, Massachusetts, program of connecting pupils with their community, with kindergartners folding napkins for feeding programs for the homeless, fourth-graders calling nursing home residents as phone pals, and seventh-graders doing pantomimes for the elderly. "No age is too young to begin meeting the needs of the community," he said. "By teaching our schoolchildren to help others, we will also be establishing the values that will keep America strong for generations to come." The Senate passed the bill 75–21. Bush signed it on November 16, though he offered "reservations about the wisdom of employing 'paid volunteers' to the extent contemplated."

On October 27 the Senate passed a bill giving the Food and Drug Administration much greater authority over medical devices, from tongue depressors to pacemakers. The manufacturers had originally opposed the legislation because of the expense and paperwork, and the FDA had feared the additional workload. But heart valves and pacemakers had malfunctioned inside patients and killed them, and the legislation gave the FDA power over devices comparable to what it already had over drugs and medicines, requiring not only premarketing testing but also follow-ups and reports of problems later. Kennedy had worked closely with Chris Dodd, whose state had many manufacturers, and with Representative Henry A. Waxman of California, chairman of the House Subcommittee on Health and Environment, and Bush signed the bill on November 28.

Congress passed legislation, pushed by Waxman and Senator Howard M. Metzenbaum of Ohio, requiring food manufacturers to display nutrition information and calories on packaged food. It provided grants to help states improve their emergency medical services. It authorized a large expansion of the National Health Service Corps, doctors who served in rural and slum areas in exchange for scholarships or loan cancellations. It expanded the Head Start program for preschoolers. In all, the Labor Committee produced fifty-four bills that became law in the 101st Congress, more than in any Congress since the 89th, which enacted Johnson's Great Society program.

One other end-of-session success was the immigration legislation Kennedy had been pushing since 1987 to increase visas issued to skilled immigrants and people from Europe who had been supplanted by Asian and Latin Americans since his 1965 bill. The House had ignored the proposal in previous Congresses and in 1989 when the Senate passed its bill. It was only after Kennedy intervened with the new Judiciary Committee chairman, Jack Brooks of Texas, that he agreed to act on the legislation in time to give it a chance. It was passed only on October 3. With elections imminent, the delay gave an advantage to Senator Alan Simpson of Wyoming, the Republican on Kennedy's three-man Judiciary Subcommittee on Immigration. Simpson wanted a firm annual limit on immigrants and a priority given to those who spoke English. In a House-Senate conference joined by administration officials, Simpson got his limit, 700,000 for the first three years and then 675,000, but lost his English-language preference.

Kennedy told the Senate before it passed the bill on October 26, "Our goal has been to reform the current immigration system—which has not changed for twenty-five years—so that it will more faithfully serve the national interest, and be more flexible and open to immigrants from nations which are now short-changed by current law. The provisions of this bill will accomplish those objectives, while also maintaining the priority we have traditionally given to those with family connections to the United States—and without departing from any of the basic goals of fairness established in the 1965 reforms."

The Senate vote was 89–8, but the House almost killed the bill over another demand Simpson pushed through in conference, an experiment with a forgery-proof driver's license, with identification such as a fingerprint, for employment purposes. Hispanic members called it the first step toward a national identification card and blocked the bill from coming up, but Simpson agreed to drop it, and the bill was passed October 27 and signed November 29.

Kennedy's biggest setback of the pre-election session came on civil rights. The House passed legislation to undo the 1989 Supreme Court decisions just before the August recess. After several leading blacks in his administration urged Bush to resume negotiations, he sent Sununu back to talk to Coleman and Kennedy in September. A House-Senate conference actually produced two reports, one on September 15 and another, after Hatch offered some modifications dealing with the administration's fears of quotas, on October 11. Coleman said the Hatch plan answered "every legitimate concern" of the administration. But the administration rejected it, and Hatch withdrew his support. Some in the White House seemed to want a deal; others seemed to believe the quota issue was good politics.

The Senate passed the second conference report on October 16, and the House followed the next day, but in both chambers the support remained below two-thirds. Still, Kennedy kept proposing changes, which could be accomplished through a concurrent resolution giving the enrolling clerk orders to rewrite the bill before it reached the White House.

Those efforts failed, and Bush vetoed the measure on October 22, saying "the bill actually employs a maze of highly legalistic language to introduce the destructive force of quotas into our Nation's employment system."

Civil rights groups and some of their House allies had scoffed all along at the possibility of a veto, feeling Bush would not take the political risk. But in fact there was little risk. The issue was legalistic and complex, and the most disputed definitions were almost incomprehensible to most Americans. There was no strong evidence that the pre-1989 interpretations of the law had led employers to choose quotas to avoid lawsuits, but politically it was still a potent issue, especially among white males who thought they were suffering discrimination. Helms used the issue powerfully in winning reelection.

Kennedy accused the administration of trickery, saying on October 24, "At several stages along the way, many of us genuinely believed that the White House was negotiating in good faith. But on every occasion when we felt a compromise was within our reach, the White House advisers always pulled back, and raised additional objections or submitted patently unreasonable new proposals." He said, "With this shameful veto, President Bush has placed himself on the wrong side of history and the wrong side of civil rights." By then he knew the issue was lost. Thirty-four Senators voted to sustain the veto, just enough. Sixty-six voted to override, including eleven Republicans (among them Rudy Boschwitz of Minnesota, who switched to the civil rights side after seeing his vote would change the result).

Even though several Republicans were uncomfortable sticking with the administration, there were two special reasons why the Bush administration held their votes. That fall it spent weeks locked in budget negotiations with Congressional leaders. Democrats were pressing for additional taxes, Republicans for spending cuts. While Representative Newt Gingrich, by then the House minority whip, deserted the President when he made a deal with some new taxes, Senate Republicans stood behind him.

Even more important, President Bush was building a coalition against Iraq and its President, Saddam Hussein, whose army had invaded the kingdom of Kuwait on August 2. Bush sent troops to bordering Saudi Arabia and vowed the invasion "will not stand," and won United Nations Security Council support for a ban on all trade with oil-rich Iraq. Some Republicans said the confrontation in the Gulf was another reason not to override the veto.

THERE WAS ALSO a Supreme Court nomination to be dealt with that fall. After Justice William J. Brennan retired in July, Bush nominated David H. Souter, a little-known federal appeals judge from New Hampshire. Liberal groups were doubtful, not so much because of his record either in four months on the First Circuit Court of Appeals or seven years before that on the New Hampshire supreme court. Fundamentally, they believed the Bush administration's assurances to conservatives that he would vote they way they expected. (Conservatives doubted that.)

Kennedy shared liberal groups' doubts, and he talked for hours with aides like Parker, Osolinik, and Blattner, and with Harvard Law School professors Laurence Tribe and Kathleen Sullivan. He questioned Souter on abortion—which Souter avoided discussing—and about many arguments he had made as state attorney general defending the ultraconservative governor Meldrim Thomson. He attacked some of his state decisions as legalistic or "heartless." But he was alone in voting against him on the Judiciary Committee on September 27, and he called Souter's sponsor, Senator Warren B. Rudman, to promise he would not wage an all-out fight. At that point, Rudman told Souter he would be confirmed.

On the floor on October 2, Kennedy said, "Judge Souter's reluctant comments, while ambiguous, suggest that in fact, he takes an excessively restrictive view of the right to privacy, and that he is likely to side with the Justices on the Court who are prepared to overrule Roe versus Wade, or leave it as a hollow shell. . . . I hope I am wrong. But I fear I am right. To a large extent, in spite of the hearings we have held, the Senate is still in the dark about this nomination, and all of us are voting in the dark. The lesson of the past decade of the Senate's experience in confirming justices to the Supreme Court is that we just vote our fears, not our hopes." Eight other Democrats joined Kennedy in voting against confirmation that day, but the vote was 90–9 in favor. And Kennedy's hopes, not his fears, proved right. Souter disappointed the conservatives on abortion, law and order, and issues involving federalism.

Kennedy had at least questioned Souter firmly and pressed him, usually without success, for his beliefs. Earlier in 1990 his examination had been perfunctory when the Judiciary Committee approved a nominee for the most important of the federal appellate courts. He took less than ten minutes on

February 6 to probe the views of Clarence Thomas, the head of the Equal Employment Opportunity Commission, when he came before the Judiciary Committee as a nominee for the Court of Appeals for the District of Columbia Circuit. Although Thomas had written about the importance of natural law as a legal principle superior to the Constitution, Kennedy focused just on job discrimination cases. Thomas testified, "I have always faithfully applied the law, I have always taken my oath extremely seriously. And even when I had significant personal differences, I have given precedent and priority to the law." Kennedy told him, "I do not think anyone could ask for better assurances."

In supporting Thomas, Kennedy was relying on the advice of William T. Coleman and Senator John C. Danforth, Thomas's patron since he had hired him out of Yale Law School to be a Missouri assistant attorney general. Kennedy was working closely with both of them on the civil rights bill, and that probably made him reluctant to look very hard. And, as Senator Howard M. Metzenbaum, the lone committee opponent of Thomas, said later, "A black person has a certain advantage with liberal Democrats." But Thomas, a black conservative who opposed affirmative action, was already being talked of as a successor to Thurgood Marshall if the civil rights giant retired from the Supreme Court. The severe standard Kennedy set for Supreme Court confirmations of voting one's fears, not one's hopes, implied at least a little scrutiny for prospective nominees for the courts of appeals. Years later Kennedy regretted his passive role and said the Thomas nomination just slipped by him. With the civil rights community quiet about him, Kennedy said, Thomas never got "on people's radar screens during that time. Hadn't gotten on ours." When Thomas was confirmed on March 6, only Metzenbaum and David Pryor of Arkansas voted no.

While supporting Bush's dispatch of troops to Saudi Arabia, Kennedy was among the most critical members of Congress over how they might be used. The Foreign Relations Committee prepared a resolution supporting what Bush had done, and supporting "continued action by the President in accordance with decisions of the United Nations Security Council and in accordance with United States constitutional and judicial processes, including the authorization and appropriation of funds by the Congress to deter Iraqi aggression and to protect American lives and vital interests in the region." Kennedy opposed it on October 1, saying that sentence amounted to a "blank check endorsement for future actions . . . a Tonkin Gulf resolution for the Persian Gulf."

He said, "If the President intends to go to war to achieve his objectives, then the Constitution requires him to . . . do so with the approval of Congress, in advance in specific terms, not by an after-the-fact invitation of an ambiguous resolution." Mitchell disagreed sharply: "This resolution is not an authorization for the use of forces, now or in the future." The vote was 95–3, with only Mark Hatfield of Oregon and Bob Kerrey of Nebraska siding with Kennedy.

On November 8, two days after Election Day, Bush ordered almost a doubling of U.S. troops in the Gulf, "to develop an adequate offensive military option." This time Kennedy was one Congressional voice among many who feared what he called "a headlong course toward war without giving sanctions a fair chance to work." He urged that Congress be called back into a special session—a case he made without success until the new Congress convened on January 3 with war clouds looming.

THE LAST TIME that a Congress had convened in such a clear atmosphere of impending war was fifty years to the day before, on January 3, 1941. On that occasion, Speaker Sam Rayburn said, "Great, and I may say, terrible things are happening in this world." He predicted, "In issues involving the safety and security and defense of America, I expect to see us act with practical unanimity, regardless of party." But times and trust had changed mightily since then, and Democrats were demanding that President Bush seek Congressional approval before making war. On the day Congress met, Bush told Democratic leaders he had no constitutional obligation to seek their votes. Kennedy was among those who strongly disagreed: "President George Bush is not King George Bush," he said. "He does not have the unilateral authority to take this nation into war. By refusing to seek Congressional authorization for offensive action, the President is acting unconstitutionally and irresponsibly. He may threaten Iraq with war in the Gulf, but he is also threatening America with our worst constitutional crisis since the Civil War."

Republican leaders persuaded Bush that he would be wise to ask Congressional approval—and that he would get it if he did. So on January 8 he asked Congress to adopt resolutions supporting "all necessary means" to oust Iraq from Kuwait if it did not withdraw by the United Nations deadline of January 15.

When Congress debated that issue, Kennedy urged patience: "War is not the only option left to us in the Persian Gulf. The President may have set January 15 as his deadline but the American people have not. Sanctions and diplomacy may still achieve our objectives, and Congress has the responsibility to insure that all peaceful options are exhausted before resort to war." He also reflected military concerns—not borne out by the war itself—of huge casualties, saying, "We're talking about the likelihood of at least three thousand casualties a week, with seven hundred dead, for as long as the war goes on."

But on Saturday, January 12, 1991, the Senate voted 52–47 to authorize Bush to go to war. Kennedy, along with forty-four other Democrats and two Republicans, voted no. Ten Democrats joined forty-two Republicans in support.

In a series of Labor Committee hearings, Kennedy had been arguing that the administration was devoting all its energy to Iraq and ignoring the United States, where unemployment had reached a three-year high of 6.1 percent despite recurrent administration predictions that the country was only suffering a brief recession. "If we can spend so much defending the throne of Kuwait," he asked at a January 7 hearing, "why can't we do something to defend the jobs of American workers? The sons and daughters of this country on the front lines in Saudi Arabia should not have to receive letters from home telling them that their parents are now standing in unemployment lines. . . . We may disagree with the President's foreign policy, but at least he has one. Right now we seem to have a half-administration, a half-presidency without any real domestic policy at all."

The war began on January 16 when bombs and missiles hit Baghdad that night. Five weeks of bombing and three days of ground warfare ended on February 27 with Kuwait freed, Iraqi forces routed, and Bush at awesome poll numbers (88 percent approval when the war ended, and still 67 in October). Saddam Hussein was still in power, but many leaders in the alliance believed he had been gravely weakened. (Eight years of trade embargoes later, he remains in control.)

But while Kennedy was mistaken about casualties (there were only eighty-nine Americans killed and thirty-eight missing) and, with years of hindsight, about the effectiveness sanctions might have had, his political sense of the "half-presidency" was borne out the next year, after many Democrats decided that running against Bush would be hopeless. The impression Bush gave of being uninterested in his own country while fascinated with foreign policy ultimately defeated him.

Still, in the early months of the session, there was little interest in legislating—except for additional benefits for the troops in the Gulf. Kennedy pushed successfully for a $50 million appropriation for counseling and child care services, and a guarantee of one month of health insurance for returning reservists without other coverage.

A few months later he seized on the record of 35,000 women in the American forces in the war, including helicopter pilots ferrying troops behind enemy lines, and led a successful effort to repeal the law banning women from flying combat missions. He worked with yet another Republican ally, Bill Roth of Delaware. Kennedy said on July 31, "The issue is not whether women should be shot at. They already are—five women died from enemy fire during the gulf war. The real issue is whether women can shoot back. The issue is not whether women should fly high-performance aircraft. They already do. Women serve as instructors for combat pilots. The real issue is whether we select our combat pilots based on ability or on gender."

DEMOCRATS WERE SO unnerved by Bush's poll standings, their failure to override any of his vetoes, small business lobbying, and their fear of being labeled a party of taxation that they took months to take a stand on even the simplest antirecession measures such as extending unemployment benefits beyond thirteen weeks.

On civil rights, House Democrats wanted to compromise less than they had the year before, and they introduced the previous year's first bill with the high-priority number H.R. 1. The administration's posture was equally rigid. Late one Friday evening in March, long after most lawmakers and their staffs had left, the Justice Department sent them a new proposal that made no references to the previous year's battles and no new concessions. One new provision, probably unconstitutional, would have barred jury trials in cases of sexual harassment. Kennedy denounced the whole proposal, saying on March 1 that it betrayed Gulf War veterans: "A stronger civil rights law is needed to give our returning forces the same freedom from discrimination at home that they had in the Gulf."

CHAPTER 32

The Turning Point

When Congress took its Easter recess in 1991, Kennedy went to Palm Beach. His sister Jean came with her children. Bill Barry, the former FBI agent who had been doing security work when Robert was shot and often handled security for Ted, brought his wife. It was the first time since Steve Smith's death that Ted had spent much time with his widowed sister. On March 29, Good Friday, they sat up late talking, mostly about Steve. Around midnight everyone turned in except Ted, who was restless and maudlin. He wanted company and asked Patrick and his cousin, William Kennedy Smith, who had already gone to bed, to join him for a couple of beers at Au Bar, a trendy but cramped club with loud music. They agreed.

Kennedy drank Chivas Regal and soda. When they left, they went back to the six-bedroom pseudo-Spanish mansion on North Ocean Avenue that Joe Kennedy had bought in 1933. Willy Smith, thirty, had returned with a twenty-nine-year-old woman who was a regular at the club, and he brought her out to the beach. They kissed and then had sex. He said it was consensual; she said it was rape, and that she screamed in protest. She called friends to take her home. She filed a complaint Saturday afternoon, but the police did not list the offense on the log reporters checked.

The police came by the house on Sunday afternoon, March 31, asking to speak to Smith and the Senator. Barry said they had left, although in fact while Smith left that afternoon and may have been gone by then, Kennedy was on the beach and did not leave until Monday. Kennedy started talking to lawyers on Sunday, and told Smith he did not want to hear the details but Smith should talk to a lawyer, too.

On Monday the Palm Beach police released some details of the woman's complaint, but without naming her or the alleged assailant. The local press charged cover-up, and the visiting tabloids started offering money for names. Elizabeth Ashley Murphy, an acquaintance of the complainant, had been at Au Bar that night, and the *Globe* paid her $1,000 for the name of Patricia O'Neil or Patricia Bowman (she used both). After the *Globe* published it, NBC News and *The New York Times* followed.

The first rumors had the Senator as the alleged rapist. He denied it, and Willy Smith issued a statement on Wednesday, April 3, saying news reports about what had happened at Palm Beach "are inaccurate and have unfairly embarrassed my uncle." Smith also denied being "involved in any offense."

On Friday the *Palm Beach Post* said Smith was the suspect, and the police confirmed it. But that took no pressure off Ted, because the same day the *New York Post*'s front-page headline proclaimed "Teddy's Sexy Romp," describing an account by another Au Bar patron, Michele Cassone. Patrick had danced with Cassone and brought her back to the house, and they had sat on the balcony drinking and talking with the Senator. After a walk on the beach, she and Patrick had returned to the house and were, as she testified, "making out" in his bedroom a bit after four. Then, Cassone told the *Boston Globe*, "the senator walked in and he had on his Oxford shirt, I don't know if he had jockeys on or what. I said to Patrick 'I'm out of here.' All I could see he had a long shirt. I got weirded out. He was there without pants. I saw his naked legs and said 'This was weird.' " The waitress, who described herself as a professional scuba diver and a bakery heiress, said the Senator did nothing suggestive. "He just had this really weird look on his face." She later testified that instead of leaving, she went out on the beach with Patrick and fooled around on a blanket, leaving when he tried to take her T-shirt off.

Jay Leno on the NBC *Tonight Show* lapped up the material. He asked, "How many other fifty-nine-year-old men still go to Florida for spring break?" Reporting on relief efforts for Kurds in Iraq that included shipping clothes, including "one million pairs of trousers," he asked, "Gee, on the way over couldn't we go a little bit out of our way and drop a couple pair where they are really needed, like at Ted Kennedy's house?" In the *Herald*,

Howie Carr compared Kennedy's claim that his family spent a "traditional" Easter weekend to the working-class Catholic tradition of praying at seven churches, and said, "Traditional? My conception of a traditional Easter weekend begins on Holy Thursday, and it involves visiting seven churches not seven ginmills. But maybe in Palm Beach . . . well, let's go down the list of the Sacred Ginmills, all of which serve sacramental wine by the glass, half-carafe or carafe. Au Bar . . . Taboo . . . Dempsey's . . . 264 . . . Chuck and Harold's . . . Bab's . . . Club Colette . . . yep, that's seven Palm Beach barrooms I can name off the top of my head, and all of them within easy staggering distance of St. Edward's."

Kennedy and his Senate staff tried to bury themselves in legislation. On the same day that the Palm Beach police said Smith was the suspect, Kennedy held a staff meeting to figure out what to do about a likely nationwide railroad strike. A Presidential emergency board had proposed a settlement that the unions actively disliked, most of all because it allowed carriers to reduce the size of train crews. Kennedy called Lane Kirkland, head of the AFL-CIO. He talked to experienced labor mediators like Theodore W. Kheel and William J. Ussery. They agreed with his view that it would be a mistake simply to impose the board's recommendations. That was the preference of the Bush administration, but the next Monday, Kennedy met Samuel Skinner, the secretary of transportation, who seemed open to compromise.

Kennedy's staff found a solution in a 1967 dispute, when a second board had been created, more or less as a place to appeal from the emergency board's ruling, but with final authority so the issue would not come back to Congress in a few more weeks. No one wanted to act too quickly and discourage further negotiations. But on Monday, April 16, the day before the strike was scheduled, Kennedy proposed that approach to Congressional leaders and John Dingell of Michigan, chairman of the House Commerce Committee.

The next day, nearly a quarter of a million workers struck against the ten biggest freight lines. Their walkouts affected commuters seriously only in California. But several major industries, from cars to coal, were threatened with serious shutdowns if the strike lasted. The House Commerce Committee met at 8 A.M., with various Senate staffs in one place, Littlefield looking over the shoulders of legislative draftsmen in the House, and Skinner and Dingell giving their approval in person. Kennedy won a last-minute skirmish with the White House, limiting Bush's choice of members of the new board. The bill passed 400–5 in the House shortly before 11 P.M. Kennedy and Hatch got it adopted by unanimous consent in the Senate a few minutes later, and Bush was awakened to sign it at 1:39 A.M. The rail workers went back on the job that morning.

George Mitchell, the majority leader, cited the strike legislation to reporters who asked if Kennedy's effectiveness had been dimmed by Palm Beach. He said Kennedy "was the person upon whom most senators relied for guidance." Other Senators challenged reporters. Alan Simpson said, "He's had his pain and many years of it. What's the purpose of heaping more on him when there is nothing to it." Arlen Specter asked, "What's the offense? Who's been charged with anything?"

But Kennedy's friends were worried, especially about his drinking. On Easter Sunday, after church and before hearing that the police wanted to talk to him, Kennedy stopped in at Chuck and Harold's. He had a bullshot, a head-clearing concoction of vodka, beef consommé, salt, pepper, and Worcestershire sauce. He followed that with another vodka drink, this time a screwdriver, and a Pierre, an obscure cocktail apparently made of equal measures of vodka, dry vermouth, orange juice, and peach brandy. He was there about forty minutes.

In April, Ted came over to Orrin Hatch to ask if they could send reporters to him if he wanted to persuade them of his effectiveness. Hatch said yes, but warned him he had to stop drinking. He recalled saying, "Ted, if you keep acting like this, I'm going to send the Mormon missionaries to you." Kennedy replied, "I'm just about ready for them." His Massachusetts friend Nancy Korman wrote to him, "During the last few weeks I have wished more than once that you were deeply involved with an Italian countess of a reasonable age. . . ." She also said, "As your friend, I do think it is time for you to seriously take care of yourself. I am not about to give you an Orrin Hatch speech, but I do think it is time to settle down to a life style that matches your professional achievements. That means high quality, thoughtful and surrounding yourself with the best people. . . . I can tell by your face and your speech pattern that this latest set of events has taken an enormous amount out of you, and personally, I do not want to lose you as either a friend or a Senator."

Friends said he was shaken by Palm Beach. In early June he told Jack Farrell of the *Boston Globe,* "You can't go through an experience like this and not make up your mind that you are going to be a little more attentive to your behavior." He insisted to Farrell that he did not have a drinking problem and would not seek help for one—but added, "That doesn't mean I am not going to be more attentive to it."

But that story passed nearly unnoticed, and the case itself had not gone away. Ted and Patrick were summoned for depositions, conducted at family offices in New York on May 1 and released to the press on May 14 over the objections of Willy Smith's lawyer. Smith had been formally charged with sexual battery, Florida's term for rape, on May 9.

Over two hours, Ted recalled a much shorter stay at the bar than others did, an hour and a quarter compared to three. He said his back had been bothering him, and that after chatting with Patrick and Cassone, he had a snack of cracked crabs and put on his nightshirt, which he said reached his knees. Then, he said, he had said good night to Patrick and Cassone and went to sleep about three and never heard any screams. Kennedy said he first heard that the police wanted to talk to Willy about a "serious offense" on Sunday afternoon, after Willy had left. He said his impression was that Willy was accused of "sexual harassment," not rape.

Patrick testified that afternoon. He said his father said good night in his nightshirt. Patrick testified he had asked Willy about his date, "How was she?" and whether his medical student cousin, who had once lectured him on using condoms, had used one himself. He quoted Willy as saying he had not, but instead "pulled out."

As these documents were released, police said they were investigating a possible obstruction of justice. On May 10 and 15, Kennedy denied that he had done any such thing. On May 16, Joseph L. Terlizzese, chief of police in Palm Beach, said the target was Bill Barry, whom he previously said had "deliberately misled" the detectives when they came looking for Smith and Kennedy on Easter Sunday.

Other Kennedys who had not even been at Palm Beach were having troubles, too. On May 14, when Patrick's and Ted's depositions were released, Joan Kennedy was arrested for drunk driving on the Southeast Expressway, the crowded main route between Boston and Cape Cod. Police said she had swerved from lane to lane, failed a field sobriety test, and had an open container of alcohol on the car seat. She came before Judge Albert L. Kramer, an intense foe of drunk driving, and he fined her $425, gave her a ninety-day jail sentence stayed on condition she stay sober for two years, and ordered her into an alcohol treatment center for two weeks. And their son Teddy, two weeks after earning a master's degree from Yale's School of

Forestry and Environmental Studies, entered an alcohol treatment facility in Hartford in June. "In my family and friends," he told the *Boston Globe* in 1993, "I've seen what can happen when people don't address the problem. I realized that nothing else could be right when that part was wrong."

IT WAS A tough legislative time, too. John H. Sununu, the White House chief of staff, had set the tone the previous November when he said, "There's not a single piece of legislation that needs to be passed. . . . In fact, if Congress wants to come together, adjourn and leave, it's all right with us. We don't need them."

In his first two years, Bush's stated willingness to seek a "kinder, gentler" America had invited Democrats as they pushed the agenda they had built up in the Reagan years. But most of the easy ones were done. And when Bush compromised with them in 1990 over new taxes, violating his acceptance-speech promise "Read my lips—no new taxes," the storm from the Republican right made him trim his moderate sails. With Bush's war-won popularity, Democrats were scared of crossing him.

They introduced measures like a health insurance bill that would require employers to "play or pay"—either provide health insurance or pay a tax and let the government buy it. On June 5, Kennedy said 34 million Americans lacked insurance, and even among those with coverage, "no American family can be secure that the health insurance they have today will protect them tomorrow." But Senator John Chafee of Rhode Island, leader of a Republican group worrying about gaps in health insurance, scoffed that the bill "does not have much of a chance" because neither Congressional Republicans nor the administration would accept its costs to employers.

The House passed the civil rights bill on June 5, with a majority well short of two-thirds. One bid for compromise, between civil rights leaders and the Business Roundtable, had failed in April when the administration and the *Wall Street Journal* attacked Robert E. Allen, chairman of AT&T, for getting involved. Allen cited a *Journal* column headlined "Big Business Shouldn't Sleep with the Enemy."

The strategy for getting a bill depended on leadership from John Danforth, and the only way for that to be effective was for Kennedy to keep his head down. So while Kennedy and his staff stayed in constant touch as Danforth tried to find a proposal the administration would buy, and sometimes rejected proposals, Kennedy's public comments were limited to brief statements of praise every time Danforth announced a new step. He left it to Danforth, who believed firmly that Bush himself wanted a bill, to accuse his aides of preferring an issue to a law. Kennedy, meanwhile, had a different, less satisfying task—keeping the other fifty-six Democratic Senators on board.

As unemployment rose, reaching 6.9 percent in June, Kennedy kept hammering at the recession issue. He introduced a bill to spend $2 billion on extended unemployment benefits for states with jobless rates over 9 percent and have Congress and Bush exempt it from the budget caps agreed to in 1990 by designating it as "emergency" spending. At a June 17 hearing in Dorchester, he heard economists warn that unemployment in Massachusetts, already at 9.6 percent, might go to 11 percent by December. Kennedy complained, "The White House has an economic policy for the Soviet Union, it has an economic policy for Kuwait, but there is no economic policy for the United States."

That evening, back in Washington, Kennedy went to a fortieth wedding anniversary dinner for two old family friends. Edmund and Doris Reggie had known the Kennedys since the 1956 convention, when they supported Jack for Vice President. Doris cast the only Louisiana vote for Ted at the 1980 convention. They called him "the Commander" for his sailing, and he sometimes sailed over to Nantucket in the summer and saw them at a summer place they had bought in 1982.

The dinner was at the Washington home of their daughter, Vicki, a partner in a law firm and a thirty-seven-year-old divorced mother of two children. In 1976, the summer she graduated with honors from Sophie Newcomb College in New Orleans, she had worked as an intern in Kennedy's mail room. She had seen him occasionally with her parents in recent years. He had been to her house for dinner before, bringing a date, and she set a place for one that night, too. But he turned up alone, and she teased him, "What's wrong, couldn't you get a date?" That horrified her mother, who thought Vicki had never learned how to talk to men. But he laughed, and went to help her pick vegetables for salad. She grilled steaks. When he left, he asked if he could call her to have dinner with him sometime. She said, "Sure," thinking it was no big deal.

Ted said the next year, "I had known Vicki before, but this was the first time I think I really saw her." The next day he sent flowers, and then called to invite her to dinner the following evening. They went out to dinner once or twice a week for the rest of that summer. Sometimes they ate at

Union Station so he could go back and forth to the Senate to vote, but she says that even though she had always followed politics closely, they hardly talked about legislation or Palm Beach or any of his professional life. She remembers kidding him, when his approval rating at home slipped to perhaps 47, that she had "never gone out with anybody whose approval rating wasn't at least 48."

Things got serious just after Labor Day. She recalled, "He called me and said, 'You've got these kids. I know you won't go out more than a couple of times a week. I really want to see more of you. So I am going to come to dinner at your house.' We started a really quite wonderful thing where he was at my house every single night for dinner. Some nights he would get there before I did, and he would be sitting there, playing with the kids," Caroline, who was five, and Curran, who was eight. He tried to amuse them with animal sounds, and read stories and joined in on puzzles or drawing. On Halloween he took them trick-or-treating in her neighborhood.

Vicki was sometimes embarrassed about how often her kids called for her after they had gone to bed and she and Ted were having dinner. "One night, I said, 'I really apologize. Look, I'm so sorry. It must not be a pleasant dinner for you because I have had to run upstairs five times.' He said, 'No, you don't understand, a child calling "Mother" is the most beautiful sound in the world.' "

IF HIS PERSONAL life was finally turning better, a decision by President Bush tipped his Senate career toward its worst moment.

On June 27, Thurgood Marshall announced his retirement. After Brennan stepped down, Marshall was now often a liberal minority of one. At eighty-two, he was overweight, short of breath, and suffering from glaucoma.

Bush, anxious to conciliate the Republican right, nominated Clarence Thomas. The appellate judge was the only black on Bush's short list of conservatives. His rise from childhood poverty (through the affirmative action opportunities he assailed when offered to other blacks) enraptured the President, who went from a prepared text calling him "the best man" for the job to an extemporaneous if unbelievable "the best qualified," a bizarre exaggeration of his minimal legal experience at little more than entry-level jobs before an undistinguished year on the appeals court. Announcing the nomination in Kennebunkport, Maine, Bush said something else nobody believed, that "the fact that he is black and a minority had absolutely nothing to do with this."

In tactical politics, which was the only level at which the nomination was weighed, the choice was brilliant. It appeased the right and at least initially divided the coalition of civil rights, labor, and women's groups that Kennedy had energized to fight Bork. Civil rights groups, in particular, were reluctant to oppose a black nominee. It took the NAACP, for example, four weeks to take a position against him, and even then it did not work hard, despite his criticism of civil rights decisions going back to *Brown v. Board of Education*. Meanwhile the White House was hard at work lining up votes and establishing a favorable image of Thomas. The silence of the civil rights groups and the enthusiasm of Thomas's patron, Senator Danforth, with whom Kennedy was still working on the civil rights bill, led Kennedy not to take a hard position early.

The right had no such hesitance. Floyd Brown, who had made the racist Willie Horton commercial that helped Bush win in 1988, made a new one, with the help of the National Republican Congressional Committee. This one attacked Kennedy over Palm Beach, Chappaquiddick, and his expulsion from Harvard. It said Joseph Biden had been "found guilty of plagiarism" in the 1988 campaign and Alan Cranston of California was "implicated" in the savings and loan scandal. Danforth and Thomas condemned the ad. So did Bush. White House aides said they had known Brown was making an ad, but not what it would say, and asked him to pull it. He refused, and their attacks guaranteed it would be shown again on news shows, Brown's main audience.

The momentum going into the hearings in September was all on the side of Thomas, Danforth, and the White House. Even so, Thomas did badly enough to raise new questions about him. The year before, Souter had hidden his beliefs and won confirmation anyhow. That was harder for Thomas, because in the late eighties, as he was currying favor on the right, he became a prolific writer and speaker for conservative publications. So his tactic before the committee was either to disown those articles completely or to say he had only been engaging in an intellectual exercise. Biden and Kennedy questioned him about a 1987 speech praising an article by Lewis Lehrman, "The Declaration of Independence and the Right to Life: One Leads Unmistakably from the Other." It maintained that natural law protects the right to life and therefore the Constitution must prohibit abortion. Thomas told Kennedy he disagreed with the article, but had praised it as "splendid" in order to get conservatives more interested in civil rights, which could also be defended by natural law.

His effort to avoid talking about abortion reached its peak on Tuesday, September 11, when Senator Patrick J. Leahy of Vermont returned to the 1987 speech. Leahy wanted Thomas to admit he had an anti-abortion view in that speech and in other writings. He insisted he had no opinion on *Roe v. Wade*. Then Thomas asserted, although he had been at Yale Law School when the case was decided, that it had never even been discussed around him when he was there. Playing on the humble-origins theme of his nomination, he told Leahy, "Because I was a married law student who also worked, I did not spend a lot of time around the law school doing what the other students enjoyed so much, and that is debating all the current cases and all the slip opinions. My schedule was such that I went to classes and generally went to work and went home." Leahy was unimpressed, saying he too had been married and working in law school, but had discussed current cases. Thomas insisted, "I cannot remember personally engaging" in discussions of *Roe*. It was an unbelievable assertion, and, as Kennedy put it years later, offered two grounds for rejection of the nomination: "Either he was lying or stupid."

As Thomas's testimony closed, Kennedy complained about contradictions between his past views and his testimony on everything from civil rights leaders to abortion: "The vanishing views of Judge Thomas have become a major issue in these hearings. If nominees can blithely disavow controversial positions taken in the past, nominees can say those positions are merely philosophical musings or policy views or advocacy. If we permit them to dismiss views full of sound and fury as signifying nothing, we are abdicating our constitutional role."

There were dozens more witnesses before the hearings ended on September 20. Only one, Erwin Griswold, former Harvard Law School dean and former Nixon solicitor general, focused clearly on Thomas's most obvious weakness, his lack of "any clear intellectual or professional distinction." Putting him on the Supreme Court, possibly for another forty years, said Griswold, was an "awesome risk."

That argument had no chance with the Senate, or even with the Judiciary Committee. Neither the civil rights groups nor Kennedy, their customary champion, had tried very hard.

But another issue was developing, through the persistence of a Kennedy aide, Ricki Seidman, an investigator for the Labor Committee who had been lent to Osolinik and Blattner. She had heard a tip about a charge which would sink Thomas if it was believed. She telephoned Anita F. Hill, a law professor at the University of Oklahoma, on Friday, September 6. Hill told her she would neither "confirm nor deny" that she had been sexually harassed by Thomas, but promised to call her back. On Sunday, Hill called Seidman and told her she had decided to talk about it, but was busy that day, and they agreed to talk the next morning.

Palm Beach hung over Kennedy. Prosecutors had announced they wanted to introduce testimony from three other women who accused Willy of raping them. The Senator started slipping in home state polls. He had to return to Palm Beach to testify before a grand jury investigating Bill Barry, the former FBI agent and occasional security aide, for obstruction of justice.

On Monday, September 9, Seidman came in to say that Hill was ready to talk to her. But Osolinik and Blattner told her that this was no time for Kennedy to be involved with a sexual harassment case—both for his own sake and because association with him would keep people from taking the charge seriously. She called Hill and persuaded her to talk to Jim Brudney, a Metzenbaum aide, whom Hill had known vaguely at law school. Metzenbaum's office, too, had heard the tip, and Gail Laster of their staff had called Hill the day before Seidman did.

On the tenth, the day the hearings started, Hill gave Brudney her story, but told him she feared becoming part of a circus. When Metzenbaum was told, he was not interested in having his staff pursue it and told them to pass it to Biden, the committee chairman. One of Biden's aides talked to Hill and believed her, but thought she would be unwilling to let the committee use her story to confront Thomas. The matter just hung there.

On the sixteenth, concerned that nothing was happening, Ranny Cooper, Kennedy's chief of staff, called Ellen Lovell, her counterpart in Leahy's office, and told her of the situation. Leahy's staff pushed Biden's staff to send investigators to Oklahoma, but were told the case was closed. On the nineteenth, after Leahy had protested directly to Biden, and Hill had told a Biden aide that her name could be used, Biden did move, asking the FBI to investigate. But Hill was still unwilling to talk to the FBI.

Finally, on Monday the twenty-third, Hill sent Biden's staff a four-page account of Thomas's efforts to date her while she worked for him at the Department of Education and the Equal Employment Opportunity Commission in the eighties, his discussions of pornographic movies, and his admission that it would ruin his career if she ever told anyone.

That led, finally, to action by Biden, who was trying desperately not to seem unfair to Thomas after telling him when the hearings began not to worry about the outcome. He told the FBI to check out the story and had his staff pass Hill's account on to the White House and to the staff of Strom Thurmond, the committee's senior Republican. Thurmond, a notorious groper himself, did not tell other Republicans. The FBI talked to Hill, then to Thomas, who categorically denied the entire story.

On Friday the twenty-seventh, with only the committee's Democrats and Thurmond aware of the FBI report (which some Senators got only that morning), the committee met to vote on the nomination. Any member could have forced a one-week postponement, or at least asked for a closed session to discuss the new charges. No one did. The vote was a 7–7 tie, but the nomination went to the floor anyhow. It could have been held in committee on a tie.

Republicans, sure they had the votes to win in hand, pressed for a vote on the floor that very weekend. Biden was willing, but Metzenbaum and Leahy objected because of Hill. By unanimous consent the vote was set for Tuesday, October 8, ten days ahead. For a week, somehow, the story stayed secret. Then on Saturday night, October 5, Tim Phelps of *Newsday* broke the story, followed the next morning by Nina Totenberg of National Public Radio, both making clear that the Senate had done nothing with the charges.

After a firestorm of protest, much of it directed at the Senate's sloth, the vote was postponed shortly before it was scheduled to occur on Tuesday so additional hearings could be held. But the postponement was just for one week, hardly time for a thorough investigation, and new hearings began Friday.

For three days the Judiciary Committee spun out of control on national television—reviving a public contempt of Congress that the thoughtful debate on the Gulf War had briefly stilled. The hearings were loaded against Hill as Biden caved in to Republican hardball. Thomas testified first and last on Friday. In the morning he said he "refused to supply the rope for my own lynching." Then Anita Hill testified for seven hours, describing in painful detail the humiliations she said Thomas had inflicted on her, and then undergoing another round at the hands of the committee. Arlen Specter and other Republicans, unrestrained by Biden and furious at attacks on a nominee they believed innocent, accused her variously of fantasy, resentment as a spurned woman, political conspiracy, and, after she had left, "perjury." The normally thoughtful Danforth, certain his protégé was being smeared, and the White House staff treated the issue as war. So did Thomas, who returned to the witness chair in prime time to call the proceedings "a high-tech lynching for uppity blacks" and offer a general denial of everything she said.

Overnight polls indicated more of the public believed Thomas than Hill.

Kennedy, who did some private negotiating over witnesses for Hill and her supporters, had almost nothing to say. With plainly inadequate time available, Democrats had Biden, Leahy, and Howell Heflin of Alabama do the bulk of the questioning. Kennedy occasionally joined other Democrats in complaining about Republican procedure and accused Specter of putting words in Hill's mouth. But he was all but invisible as Republicans insisted Hill's charges were disproved by the fact that she had willingly followed Thomas from the Department of Education to the EEOC.

On Sunday, October 13, the third day of the hearings, Kennedy finally spoke up. He said he hoped "we are not going to hear any more comments, unworthy, unsubstantiated comments, unjustified comments about Professor Hill and perjury. . . . I hope we are not going to hear any more comments about Professor Hill being a tool of the advocacy groups. . . . I hope we are not going to hear a lot more comments about fantasy stories picked out of books and law cases. . . . I hope we can clear this room of the dirt and innuendo, about over-the-transom informations, about faxes, about proclivities. . . . We heard a good deal about character assassination yesterday, and I hope we are going to be sensitive to the attempts of character assassination on Professor Hill. They are unworthy. They are unworthy."

He continued, "And, quite frankly, I hope we are not going to hear a lot more about racism as we consider this nominee. The fact is that these points of sexual harassment are made by an Afro-American against an Afro-American. The issue isn't discrimination and racism. It is about sexual harassment, and I hope we can keep our eye on that particular issue."

Kennedy's statement cheered Hill's followers briefly, but it had no impact on the rest of the day, which continued with Hill more on trial than Thomas. Witnesses accused her of having a crush on Thomas, and Danforth brought forth a psychiatrist who had never met Hill to suggest she was suffering from "delusional disorder." Seven women who had worked with Thomas testified he had never behaved remotely as Hill contended. At 2:02 A.M. on Monday, the committee adjourned, after deciding not to hear Angela Wright, another former EEOC employee who was ready to accuse Thomas of pressuring her for dates, just as Hill had claimed, and to say that Thomas had once asked her how big her breasts were. Wright had been subpoenaed, and had given a deposition by telephone, but the committee showed no interest in new charges.

Neither side could prove who was telling the truth in a classic "he said, she said" situation. But it was clear that while Thomas's foes had been eager to use Hill's testimony against a nominee they could not beat any other way, it was equally clear that, to friends, Hill had accused Thomas of sexual harassment many years earlier and that she had been a reluctant witness before the Senate, not a conspirator.

On October 15, the day of the vote, Kennedy's belated activity stirred two remarkable Republican responses. Kennedy went to the floor to denounce Hill's treatment before the committee. Specter, whose "perjury" accusation he had attacked in committee, hit back personally: "We do not need characterizations like 'shame' in this chamber from the Senator from Massachusetts." Hatch, of all people, went even further. In what everyone took to be a snide allusion to Chappaquiddick, he said, "Anybody who believes that—I know a bridge up in Massachusetts that I'll be happy to sell them."*

The Senate confirmed Thomas that night, 52–48. That was the closest margin ever for a successful Supreme Court nominee.

AFTER THOMAS, KENNEDY got the worst press of his career—some from his friends. Syndicated humor columnist Dave Barry said he had sat "through three full days of hearings with a bag over his head." Syndicated feminist columnist Anna Quindlen said his performance proved what she'd always believed—that personal behavior mattered to political fitness. She wrote, "He let us down because he had to; he was muzzled by the facts of his life." And the *Boston Globe*'s editorial page, a longtime friend, agreed. It said his "reputation as a womanizer made him an inappropriate and noncredible critic" of Thomas. It said Chappaquiddick, Palm Beach, "and other, periodic reports of reckless behavior by Kennedy have diminished his moral authority."

*Hatch, looking gravely embarrassed, told me a few minutes after making the comment that he had meant to say "bridge in Brooklyn." He had the *Congressional Record* changed accordingly.

These arguments followed two
Republic in June by James Carroll,
the first round of hearings by Mark
Both found Kennedy failing not on
their religion, too. As Shields put ... t
Catholic grown-ups, do not awaken
go out on the prowl." Carroll also ... to
causes Kennedy had carried had be ... n-
by their champion, and that while ... at
as the poor and AIDS patients, t
here are women." Nor was he ... ghts
showed skits in which every po ... end
as a drunk or he offered advice t
one option was suddenly appea ... state
on your boat, for some reason, ... their
away." ... years,

No matter how irrelevant
next week he got his civil rig ... recent
most Republican Senators had ... or the
Hill had been a victim, they ... ment of
as anyone else. This bill wou
cases. Second, Danforth's s ... s in the
harder for the White House ... hem and
some Republicans were als ... before, I
of David Duke, a white sup ... ike a better
a bill would help them. ... ndeavors are
Thomas, Kennedy and D
that he was the one that w ... n blessed with
kept the civil rights bill s ... aordinary broth-

The endgame took ... generation's sense
White House was still th
to the White House We ... ength of years and
ner of Virginia and Ted ... termined to give all
him he could not count ... od for almost a third
back to talking. Boyd
Thursday and was giv ... kie of the *Boston Globe*
for a long series of ta ... culpa," and his colleague
staff members shuttli ... comings and confronting

ing Point

the bills agreed to in 1991 and v
etween the agreed-on bill and a
support in September were e
ats by the White House. But af
twenty-fourth, Sununu pronounc
a symbol to women and minorities
s it was for the substantive changes
y provision, not a part of the 1989
islation, was to allow damages in
on religion, sex, or handicapped
be ordered only in cases of racial
alleging sexual harassment.
appy with that provision because
an $50,000 for small companies
ited a ceiling of no more than
harassment, not other kinds of
argued it was unfair that their
h limits at all. Kennedy, who
ninistration and Danforth had
at home after midnight to tell
get," as Ralph Neas, the civil
endorse the deal, they went
rst met with the other fifty-
if Danforth's efforts to line
er. He explained what the
visions were technical and
ct," or employment prac-
in hiring more men than
n as a height requirement
ard clearly reversed one
employers had to show
l practice, like a height
t it was up to workers
was a central sticking
than a year that if it

them was "a tantalizing tidbit, but there was little hint of what he thinks those faults are or of how he is working to confront them." Alessandra Stanley of *The New York Times* found it "tersely worded and unemotionally expressed," offered "without specifying what he had done wrong in his private life." Christopher Daly of the *Washington Post* wrote, "Kennedy stopped well short of offering a specific apology. Nor did he pledge specific steps he would take to restore his reputation."

But for someone who finds it as difficult as Kennedy does to discuss himself, it was a significant statement. And for supporters who wanted reassurance, it sufficed.

BACK IN WASHINGTON the next week, Kennedy said Danforth was a "profile in courage" as the civil rights bill was adopted by the Senate, 93–5, on October 30. This bill, he said, "is all the more satisfying because it involves a welcome restoration of the bipartisan coalition in Congress and between Congress and the Administration that has been responsible for so much of the historic progress we have made in the past half century." Bush signed it on November 21 after a last-minute dispute over a provision Boyden Gray wanted to include in the signing statement, ordering the federal government itself to end all affirmative action programs. Danforth threatened to boycott the signing ceremony unless it was dropped, and it was. Kennedy was the only Democrat to attend the Rose Garden ceremony. He praised Bush's hope that a "new civility" would be achieved on civil rights. But then Kennedy looked at Bush's final signing statement carefully and complained that it contradicted the words of the statute itself by offering employers more latitude on "business necessity."

Legislation to extend unemployment benefits was blocked twice by Bush—who wanted neither more deficit spending nor new taxes to pay for the cost—but a third try was passed and signed on November 15 with the unemployment rate at 6.8 percent, but it was largely a Finance Committee effort. National health insurance got at least a public relations boost on November 5 when Harris Wofford stressed that issue as he won an upset Democratic victory in a Senate special election in Pennsylvania.

TED WAS A part of the Willy Smith trial in Palm Beach long before he arrived to testify. When Roy E. Black, the high-powered Miami defense attorney Jean had hired for Willy, questioned prospective jurors on October 31, he found that they all cited Ted as the member of the family they least admired.

But in the courtroom at least, Ted was not on trial. Willy Smith was. And when the trial began on Monday, December 2, Judge Mary Lupo ruled in his favor on a critical point—refusing to allow testimony from three other women who said Smith had attacked them between 1983 and 1988. Prosecutors had argued their testimony would show that Smith used "a plan, a scheme in attacking these victims" and that the same MO, or modus operandi, had been used against Bowman. Black insisted there were no "strikingly similar" patterns, and Judge Lupo sided with him without saying why.

On December 4, Bowman testified. She said she had met Smith at Au Bar, found him pleasant, and agreed to give him a ride home when the bar closed at three. She said she agreed to a walk on the beach, but when Smith took his clothes off for a swim, she was frightened and tried to run away. Then he tackled and raped her, she said, while she screamed "No!" Afterward, she claimed, he told her that if she accused him of rape, no one would ever believe her.

Moira Lasch, the prosecutor, called Ted as a witness on Friday, December 6. She read woodenly from written questions, as Kennedy sometimes did in Senate committee hearings. Ted testified he had heard no screams and had seen nothing of Smith or Bowman after leaving Au Bar, and very little of either of them there. His account of the time spent at the bar was much shorter than others had said, "about an hour and a quarter, I imagine." His testimony was marginal to the trial.

But he was plainly pained as he talked about the evening. Lasch asked him why he had gone to Au Bar, and he answered:

"Well, there was no real special reason why Au Bar was chosen. . . . My sister Jean had invited the members of my family and the Barrys to be here over the Easter vacation, and we came here and this is really the first time that all of us had been together since the death of my brother-in-law, Steve Smith. And we were visiting in the patio after dinner, and the conversation was very emotional, a very difficult one, brought back a lot of very special memories to me, particularly for the loss of Steve, who really was the brother to me and to the other members of the family. And I found at the end of that conversation that I was not able to think about sleeping. It was a very draining conversation, a whole range of memories really came as an overwhelming wave in terms of emotion—"

Lasch interrupted, "Well, was it after that conversation– "

But Kennedy kept going: "—and at that time I said I can't possibly sleep. I noticed both Patrick and my nephew Will go by the glass windows, which are just adjacent to the patio. And I opened the door and I called for Will and Patrick and they answered and I asked them whether they wanted to go out. I needed to talk to Patrick or to William and they said yes they would—"

Lasch tried again: "Okay, was it after that conversation that you decided to go to Au Bar."

Kennedy said, "So we left that place and we went out. We went to Au Bar's. I wish I had gone for a long walk on the beach instead. But we did go to Au Bar."

When Roy Black cross-examined, Ted was even more emotional. Black got a barely audible "Yes" when he asked if Barry was "the man who knocked the gun out of Sirhan Sirhan's hand" in 1968. Black asked what the conversation with Jean and Barry was about.

"Well, I think I described it earlier," Kennedy said. For sixteen seconds he looked down, moving the muscles of his face, and finally said, softly, "I think I described it earlier."

Then, prodded by Black, he spoke haltingly of Steve again: "Very special to me. He was an extra brother, really. We lost a brother in the war. When Jean married Steve, we had another brother. And when Steve was gone, something left all of us. When we buried him."

SMITH TESTIFIED THE next week that Bowman had picked him up, that the sex on the beach was consensual, and that she had turned angry afterward.

He was acquitted on December 11. The jury deliberated for only seventy-seven minutes. Eight days later the grand jury dropped the obstruction case against Bill Barry.

CHAPTER 33

A President and an Ally

The recession dragged on into the election year of 1992, and Kennedy was at the center of Democratic Congressional efforts to shape the issues—a focus on education, health care, and unemployment and some family-friendly issues like requiring employers to give workers unpaid leave for family emergencies.

The election scene had shifted in the eleven months since the Gulf War, and Bush no longer seemed like a sure thing for reelection. His overall approval rating had fallen by half, to 44 percent. Only one American in five approved of how he was handling the economy, and in a January *New York Times*/CBS News Poll, he trailed a generic Democratic candidate for the first time. And so while Democrats had few illusions about passing legislation over a Bush veto—they had not done it yet—they felt issues like those would help their candidate, whoever he was, in November.

While Bush gave ground quickly on some issues, like extended unemployment compensation, the election politicking was obvious on the Senate's first major bill, an $850 million authorization for five years of grants to states to spur innovation in schools. On January 21, Kennedy told the Senate the bill encouraged a variety of local and state approaches by providing "funds to State agencies to encourage educational reform at the neighborhood school level." That day Bush tried to upstage the Senate by announcing that his budget would call for a $600 million increase in education spending. Kennedy scoffed, "All Republican Presidents want to increase the federal investment in education in election years." Orrin Hatch complained that Kennedy was playing politics. Hatch said, "Education should not become a political issue."

Then Hatch proposed a $30 million amendment to test private school vouchers in six states. Lamar Alexander, secretary of education, made the Republican case that "choice would unleash marketplace forces that would help make all schools better." Teachers' unions led the lobbying against it, calling it a threat to public schools. Kennedy agreed, asking, "Do we have a sufficient amount of public taxpayers' money to start utilizing it in private schools?" and sneered at the bill as "the Preparatory School Relief Act," contending the vouchers could be used at Andover, Exeter, or St. Mark's. That Hatch plan was defeated, 57–36, and the bill passed 92–6, on January 28, and was sent to the House.

When it came to reauthorizing the Higher Education Act, the critical fight was within the Democratic Party. For years the maximum level for Pell grants for low-income college students had exceeded the money appropriated to pay for them. In 1992 the maximum authorized was $3,100 but the real ceiling was $2,400. So both the Senate Labor Committee and the House Education and Labor Committee sought to turn the program into an entitlement, so that any student eligible for the maximum grant would get it. Bush would have vetoed such a bill, but it would have made an issue for the fall. But many Democrats, facing a $327 billion deficit caused in part by the growth of entitlements, did not want a new one, and the entitlement provisions, approved by both committees, were dropped on the floor in both houses.

With an election coming, both parties eagerly supported provisions that would help the middle class by raising income ceilings for Pell grants and creating a new, unsubsidized college loan program available to all students regardless of family income.

The issue that provoked the only real battle, one more between the administration and Congress than between the parties on the Hill, was over a proposal to have the government lend students money directly, instead of guaranteeing loans made by banks and loan agencies. Kennedy had first proposed that approach in 1978, with Republican Henry Bellmon of Oklahoma, but current budget accounting rules made it prohibitively expensive. The Congressional Budget Office had changed its approach, and Senators Paul Simon of Illinois and David Durenberger of Minnesota were pushing the idea. At a February 25 hearing, Kennedy argued that there was no question the loans could be made at lower cost to the government and students without banks as middlemen. The only problem, he said, was the Department of Education, which "remains a thinly staffed agency with questionable ability to undertake major new administrative responsibilities, let alone perform its current responsibilities adequately."

In conference, an experimental program, testing the idea at five hundred colleges, universities, and trade schools, was agreed to on June 16. A handful of the schools were to try another innovative feature, "income-contingent" repayment, which would enable graduates to choose low-paying jobs like teaching and simply pay back less of their loans. Lamar Alexander, secretary of education, urged Bush to veto the bill over the direct loan provision. But Congressional Republicans persuaded him that would be dumb politics because of all the other provisions in the bill. He signed it on July 23.

In the end, Congress voted the sham increase of raising the theoretical maximum Pell grant to $3,700—in a year when the budget squeeze forced the real maximum to fall to $2,300—and Kennedy, Claiborne Pell, Durenberger, Thad Cochran, and Chris Dodd all boasted of it when the Senate passed the final version of the bill on June 30.

On January 22, Kennedy pushed a national health insurance bill out of the Labor Committee on a 10–7 party-line vote. It was the "play or pay" measure that had been introduced the year before, altered just enough so it would be under his jurisdiction and not the Finance Committee's. After the vote, Kennedy said it "will guarantee health insurance coverage for every American, whether they are rich or poor, whether they are sick or healthy, wherever they work, and even if they lose their job."

The Bush administration countered with its own proposal, a system of tax deductions to help the middle class pay for health insurance and vouchers for the poor, allowing them to spend up to $3,750 for families of three or more. At a March 4 hearing, Kennedy scoffed at Louis Sullivan, secretary of health and human services, when he argued that Bush's tax package would bring insurance to many additional Americans. Stuart Altman, by then the acting president of Brandeis University, testified that few if any serious students thought the Bush plan would either curb costs or lead to universal insurance coverage.

Kennedy tried to sell the play-or-pay plan anywhere he could, even to the Health Insurance Association of America. On April 28 he told the group, "Your private sector role has unquestionably made the crisis worse. . . . The message sent out by too many of your member companies is unmistakably clear– don't insure anyone unless you think they won't get sick." He said "play or pay" had room for private insurance, unlike the Canadian single-payer system, but that inaction or the Bush plan would lead to government-controlled health care, because the public was fed up. But despite Kennedy's vow that the issue would come to the floor for a vote—he said he had George Mitchell's promise—it never did. Democrats were not united enough to risk a vote. When Mitchell saw that, he decided to invest the year in trying to build a consensus by getting small groups of Senators to work on the issues that troubled them, from rural health to small business to cost containment.

There may not have been another issue all year that was as heavily politicized as the question of whether the National Institutes of Health could provide federal dollars for research involving transplantations of tissue from aborted fetuses. The research seemed promising for treatments for Alzheimer's disease, Parkinson's disease, and diabetes, but Reagan had banned the practice five years before, yielding to arguments from the right-to-life movement that such a practice would encourage abortions. At first the fight was within the Republican Party. Bush was no longer just suspect to the Republican right but facing a primary challenge from Pat Buchanan, a television ideologue, stuck with the policy.

But Kennedy saw strong Republican support for a change and pushed legislation repealing the ban through the Labor Committee on a 13–4 vote. Three of the committee's seven Republicans joined all ten Democrats. One of them was Strom Thurmond of South Carolina, a staunch foe of abortion, who told the Labor Committee that his daughter Julie suffered from diabetes. On the floor, Kennedy cited not only Thurmond, but the weight of scientific opinion and the fact that fetal tissue had been used since the fifties with no impact on abortion rates. The Senate passed a repeal of the ban as part of an overall National Institutes of Health bill, on April 2, by an overwhelming 87–10 vote.

Bush responded on May 19 with an executive order creating "fetal tissue banks" to collect tissue from ectopic pregnancies and miscarriages, arguing that would provide enough material for research. When the House passed its bill anyhow on May 31, the vote was less than the two-thirds needed to override a veto.

Democrats who controlled a House–Senate conference made no move to compromise, and Representative Henry A. Waxman of California, the bill's chief sponsor in the House, said the strategy was to force Bush into a veto that would be unpopular to people concerned about the diseases that might be cured.

So when the conference report came up in the Senate, the political gloves were off. Kennedy said victims of diabetes, Parkinson's, and Alzheimer's "should not be pawns in the Bush Administration's craven capitulation to the demands of anti-abortion politics in this election year." He said, "There is no evidence whatsoever for the Administration's specious claim that more women will decide to have abortions if they know that fetal tissue will be available for transplantation research," noting that fetal tissue research continued, but just not involving transplantation. He closed sarcastically, saying, "Tissues and organs from human cadavers used for purposes of tissue or organ transplantation saves lives, and no one claims that it encourages murder."

Emotions remained high. Bob Smith, a New Hampshire Republican, cited his family in a way opposite to Thurmond's. He said his father, who suffered from Alzheimer's, "would not want an unborn child to lose its life for him." But while the Senate voted 85–12 to pass the conference report on June 4, and the House followed, Bush vetoed it and the House vote to override was 271–156, fourteen votes short of two-thirds.

KENNEDY ALSO GOT nowhere with various economic stimulus proposals, but he did succeed in preserving an urban summer jobs plan which Bush sought to kill. Three days of riots in Los Angeles followed the April 30 acquittal, despite videotaped evidence, of four white police officers charged with beating a black motorist, Rodney King. The immediate White House reaction, by Press Secretary Marlin Fitzwater, blamed the Great Society programs of the Johnson era for "many of the root problems that have resulted in inner-city difficulties."

The House passed an $822 million program of loans and disaster grants for Los Angeles and for Chicago, which had suffered a severe flood. Kennedy, who had struggled unsuccessfully to work out a national health insurance bill with Orrin Hatch, had him to dinner in McLean and persuaded him to join in support of $1.45 billion for summer jobs, summer Head Start, summer school programs, and a Justice Department program to help cities focus on high-crime neighborhoods. Kennedy assailed a "decade of neglect" that left the inner cities "festering." Hatch said, "I can't sit idly by and watch these riots and these kinds of vulnerabilities and not do anything about it."

A Bush veto threat and House Democrats' uncertainty led to a final bill of only $1.1 billion, but it did include $500 million for summer youth jobs, enough to hire 360,000 teenagers, and Kennedy backed it as a "small step" in the right direction. "None of us can ever afford to relax or think we have done enough," he said.

ON FRIDAY, MARCH 13, Caroline Raclin, Vicki's six-year-old daughter, went to school with a big secret she had promised to keep. She was a pupil at the Maret School, where a lot of *Washington Post* people sent their children. Around midday, one of those *Post* parents called Lois Romano, who wrote an impish but factual gossip column for the *Post*. Romano telephoned Paul Donovan, Kennedy's press secretary. She had called before, asking if Ted was getting married to Vicki; they were often seen together on Sunday mornings at Blessed Sacrament Church in Washington. He told her there was no truth to it, but she asked him to check with Kennedy, and he replied he could not ask Kennedy every time she got a tip. But he agreed to check and called Ranny Cooper, the chief of staff, who reached the Senator and confirmed it.

But Cooper and Donovan were at once delighted and panicked, pleased for Kennedy but worried about the family and friends, most of all his sisters, who did not expect to learn this kind of news in their morning paper. So about 9:30 P.M., Donovan called Romano back and told her, "The Senator has no comment on his private life." Romano laughed, knowing that meant it was true. But she did not have a column in the Saturday paper anyhow, and, lacking confirmation, Romano let it go.

Ted had proposed on January 13 at a performance of *La Bohème* at the Metropolitan Opera in New York. He and Vicki decided to keep it secret for a time, because they would not marry until her children's school year ended. But in March, Ted told his grown-up children. Vicki thought she should tell hers, too, and while she thought nine-year-old Curran would keep the secret, she doubted Caroline could. She couldn't.

Rather than leave the story for Romano, they decided to announce it. On Saturday, Cooper called in almost the whole staff to make phone calls or to find out where people would be so Ted could call them. Ted took Curran and Caroline ice-skating on the mall that morning. Then he went to his office, with Vicki and her children, to make calls and approve a statement that was sent to the newspapers.

A couple of weeks later, Ted spent a very different Easter from 1991's. He and Vicki visited his sister Pat, who had a home in the Virgin Islands. They went snorkeling at the Buck Island National Park. Ted pointed her toward a small piece of coral. She dove down to take a close look and found the diamond-and-sapphire engagement ring he had placed on the coral a few minutes earlier. She was delighted and he was relieved, because a big grouper had swum that way a minute before and Ted had worried he might swallow the ring.

They were married in front of the living-room fireplace in the house at McLean on July 3. While Vicki had won an annulment of her first marriage from the Catholic Church, Ted had not, and so a federal judge whose appointment Ted had sponsored, A. David Mazzone, conducted the ceremony. It was a small wedding with only about thirty family members present. Ted gave her a painting he had done of daffodils. They left that night for a week's honeymoon trekking in Vermont, and were first seen as a married couple at the Democratic National Convention in New York.

Before the wedding, Vicki joined Ted in hosting an emotional occasion honoring Mikhail S. Gorbachev, the former President of the former Soviet Union. He had presided over the end of the Cold War, and then over the dissolution of his country, which ceased to exist on December 31, 1991. The gala was held on May 15 at the John F. Kennedy Library, where for a decade Ted had conducted the nearest equivalent of state dinners available to an American who was not President. Nelson Mandela had been there two years before, and so had a variety of world figures, including the leaders of Ireland, Italy, Portugal, and just about any country with substantial representation in the Massachusetts electorate. Between Clarence Thomas's confirmation and his speech at Harvard the previous fall, for example, Kennedy had been host to President Mary Robinson of Ireland, and dozens of Irish-Americans could look out the windows behind her to Dorchester Bay, where their ancestors had first come to America.

But there were no votes available at this occasion, just emotion. Kennedy toasted his guest after the smoked scallops, spinach crepe spirals, and lobster salad and a cello interlude by Yo-Yo Ma. He said the fall of the Berlin Wall was "the defining moment for our time, and perhaps for our century," and it had happened because of Gorbachev. "Few individuals have the power to change the course of history. Even fewer actually accomplish that change. And fewest of all accomplish it for the better. Mikhail Gorbachev is one who did. He is the embodiment in our time of the famous phrase that President

Kennedy and Robert Kennedy both lived by, 'Some men see things as they are and ask Why? I dream of things that never were, and ask Why Not?' "

Gorbachev had toured the library before lunch and showed keen interest in exhibits and documents relating to President Kennedy. He responded to Ted and the audience of three hundred—Kennedys and Boston business, political, and academic leaders—that he had been inspired by President Kennedy's American University speech in 1963, an "intellectual breakthrough" that pointed the way to the end of the Cold War. He cited John F. Kennedy's words: "Peace need not be impractical and war need not be inevitable. Contacts among nations need not be transformed into an exchange of threats."

TED AND VICKI came to the 1992 Democratic Convention after the first race since 1964 in which Ted's candidacy had never been even a theoretical factor. Paul Donovan, then his press secretary, was never asked in 1992 if Ted might enter the race, even though most reporters thought the Democratic field looked weak. Kennedy's role in 1992 had been minuscule. In March he endorsed Paul Tsongas, the former Massachusetts Senator whose views of the economy were distinctly to Kennedy's right, as Tsongas tried to halt the progress of Governor Bill Clinton of Arkansas. At Tsongas's request, Kennedy called some union officials for him. But the labor leaders, a group who had backed Tom Harkin of Iowa, were wary of Tsongas's opposition to their bill to ban permanent hiring of strikebreakers. They stayed neutral.

Although Kennedy had occasionally dealt with Clinton since he won back the governorship of Arkansas in 1982 (and was again the first to call, with congratulations, that time), he was wary of any Democrat who sold himself as a centrist "New Democrat." Clinton had been an early leader of the Democratic Leadership Council, a group founded after the 1984 election that hoped to shift the party from the left to the center.

But except for Harkin, who dropped out in early March, the other Democrats were centrists, too. They pointedly identified themselves with the middle class, not the poor (while Kennedy tried to do both), and insisted they worried about the deficit. They sought to capitalize on a sense that Washington was utterly out of touch, with Bush unconcerned about domestic problems and Congress incapable of doing more than conducting absurd hearings like the Thomas-Hill show or bouncing checks at the House bank. In short, they told voters they were different from the familiar faces in Washington. Among Democrats, Kennedy's face and politics were more familiar than any other.

But on April 23, Kennedy endorsed Clinton, saying he was going to win and Democrats should rally behind him, and he joined in welcoming Clinton to Capitol Hill the next week when he came seeking support. Kennedy offered that view privately, too. In April, with Clinton running third in polls that showed him behind Bush and independent Ross Perot—an even more dramatic sign of the sense that politics was failing America—he encouraged Nancy Soderberg, his foreign policy aide, to join the Clinton campaign in Little Rock. He told her Clinton was going to win the Presidency.

Kennedy had a bit part at the convention as the Democrats returned to Madison Square Garden, the scene of his defeat and triumph in 1980. On July 15, he spoke after a videotaped tribute to Robert Kennedy. Ted said, "Perhaps more than any other leader in memory, my brother Bobby reached across the deepest divides of American life—black activist and blue collar, suburb and city, the young students on campus who protested the war and the young soldiers drafted to fight it." Then he praised the party's candidate for 1992, using a similar phrase to identify Clinton with Robert Kennedy: "There is one thing that matters most. He has sought to heal, to oppose hate, to reach across the divides and make us whole again."

LATER THAT NIGHT in his own acceptance address, Clinton locked in Kennedy's support when he promised, however vaguely, to "take on the big insurance companies to lower costs and provide health care to all Americans." When Bush accused Clinton of favoring socialized medicine "with the efficiency of the motor vehicles department and the compassion of the KGB," Kennedy sprang to Clinton's defense. On August 4, he joined a dozen other Senators at a news conference and attacked Bush's "tired old rhetoric of the right from the days of their battle against Medicare. . . . Ask any senior citizen whether they think Medicare is socialized medicine. We won that battle for health care for the elderly a generation ago, and the time has come to win it now for every other American."

There were other campaign issues on which Clinton's message matched Kennedy's. Clinton called for a large national service program. Clinton advocated family leave so workers could take time off in emergencies without risking their jobs. Clinton supported income-contingent loan repayments. Clinton wanted to lift the ban on federally funded research on fetal tissue transplants. By the fall, Kennedy's staff was doing all sorts of quick turn-around research for the Clinton campaign.

On only one of these items, family leave, did Congress help Clinton. It passed a bill that heightened the contrast between the candidates when Bush vetoed it on September 22, as he had rejected a similar measure in 1990. The 1992 veto message said that while leaves for family emergencies were a good idea, the federal government should not force them on companies, especially small businesses. Dodd managed the override attempt on the floor on September 24, and said Bush had missed a chance "to replace the rhetoric of family values that we have heard so much of in the last number of months with real policy that values families."

Kennedy backed Dodd, insisting the bill exempted small businesses and key workers. He said, "When a medical crisis strikes, when a new child is born or a family member is seriously ill, workers need leave." He closed with the argument "We know that Bill Clinton would sign it if he was now in the White House, and we know he will sign it if our effort fails today. So family leave is coming soon—the sooner the better." The Senate voted to override, 68–31, but the House fell twenty-seven votes short on October 1.

The school innovation bill died when Senate Republicans, anxious to save Bush from casting a veto that could be used against him, filibustered it to death on October 2.

As the session ended, Kennedy tried again with another NIH bill. It would continue the ban on transplantation research for a year to test whether the administration's tissue bank would produce enough specimens. On Friday, October 2, the Senate voted 85–12 to take up the bill, and over the weekend Hatch countered with a twenty-seven-month test period. Efforts to make a deal foundered when word came that the administration wanted no compromise and diehard foes of abortion threatened, credibly, to use every parliamentary device available to block passage in the Senate and the House, too. With Congress anxious to adjourn to campaign, Mitchell and Kennedy withdrew the bill, promising to bring it up when the Senate met in January. Kennedy said, "The greatest reason of all for hope is that Governor Clinton has assured me and the American people that as President Clinton he will immediately lift the ban," giving hope to sufferers from a wide range of diseases. He said, "The White House is being held hostage by the most extreme anti-abortion zealots in its party." Hatch replied that Kennedy and Mitchell had tried to "make a political football" of the issue by delaying bringing it up.

No broad health care bill ever came to the floor in either house, and while the Senate voted modest insurance reforms as part of two tax bills, the House rejected them. But Kennedy got a number of small, low-cost measures through Congress. The most imaginative was an effort to help the fifteen hundred community and migrant health centers. No additional money was available in the budget to help them spend more, so Kennedy, Hatch, and Representatives Ron Wyden of Oregon and Nancy L. Johnson of Connecticut figured out a way to reduce their costs.

Malpractice insurance charges for the clinics had risen between 30 and 40 percent in the last year—whether or not they had any claims against them. So the legislation directed the Justice Department to defend doctors and nurses at the clinics as though they were members of the U.S. Public Health Service. The clinics would pay the department's costs in defending them. If any claims succeeded, the clinics would pay them, but under the Federal Tort Claims Act, no punitive damages were permitted. Kennedy told the Senate the Congressional Budget Office believed that would save the clinics $45 million out of the $60 million they were now paying for malpractice insurance (about one-tenth of their federal subsidy) and enable them "to attract qualified obstetricians with full malpractice coverage and provide quality prenatal care to more low income women."

The savings, said Daniel R. Hawkins of the National Association of Community Health Centers, "will be sufficient to provide services to a half million additional people."

But an even larger impact on public health probably came from a measure designed to speed up the approval of new drugs by the Food and Drug Administration. For several years that agency had come under increasing attack for the time that it took to consider and approve new prescription drug applications. The pressure increased dramatically as research grew, especially in biotechnology. Attacks came from AIDS activists, the pharmaceutical industry, and Vice President Quayle's office. One administration solution was contracting with outside laboratories to investigate new drugs. Kennedy and Waxman had no use for that, fearing a lowering of standards. The obvious answer, increasing the agency's appropriations so it could hire more scientists, was practically foreclosed, not only because of the overall budget squeeze but because the FDA, for odd reasons of Congressional jurisdiction, came under the Appropriations Subcommittee on Agriculture, which was much more concerned with farmers than with pharmaceuticals.

The idea of charging the industry "user fees" for drug investigations had been floated occasionally, but the drug industry was understandably skeptical, doubting the money would actually be used to speed up investigations. It managed to label such proposals, as Mark Skaletsky of the Industrial Biotechnology Association acknowledged at a September 22 hearing before Kennedy, as a "tax on innovation."

During the summer of 1992, however, a consensus developed after complicated negotiations among the industry, the FDA, and staffers for Kennedy, Waxman, and the appropriators. The deal provided an escalating set of fees companies would pay for applications over the next five years. The fees would stop if Congress used them to make up for regular federal money. The money would be used to hire new investigators, and the FDA would commit itself to faster review. The bill had only a five-year life. It would have to succeed to be reenacted.

At Kennedy's hearing, everyone agreed that more ordinary federal money would not be forthcoming. Kennedy said, "The mismatch between accelerating pharmaceutical research and declining FDA resources is severely delaying the health benefits of medical advances, and some remedy must be found." But he sought the industry's explanation of why it was willing to foot the bill. Gerald J. Mossinghoff, president of the Pharmaceutical Manufacturers Association, explained: "It takes an enormous amount of resources to bring a drug to a point where it can actually go to the FDA and where we request approval. Studies done a couple of years ago indicated about a quarter of a billion dollars invested. For every year that that drug does not get approved and doesn't produce revenue, that's an enormous expenditure of opportunity costs that the shareholders of the company have to pay, that's making no income with an enormous investment."

In what had become known as "the gridlock Congress," this bill passed with remarkable speed and no opposition, on voice votes in the House on October 6 and the Senate on October 7, the next-to-last day of the session. Explaining the measure, Kennedy cited FDA commissioner David Kessler's promise that the $300 million raised through application fees, which would increase from $100,000 to $233,000 over five years, would enable him to hire six hundred new scientists and reduce the current twenty-month review time. Kennedy said, "FDA commissioner Kessler has developed a set of specific performance goals that will enable the agency to cut approval times dramatically—almost in half. The goal is to cut review times down to six months for applications for most breakthrough drugs, and twelve months for most standard applications. The Commissioner has also committed the agency to a timetable of eliminating current backlogs on applications. These actions will have a significant, long-term impact on health care in this country. Faster drug approval will save patients' lives, and benefit the industry, too."

In 1998, the FDA had kept that promise. The average time for priority approvals was 6.4 months; the average for all was twelve months.

KENNEDY WAS NOT deeply involved in that fall's elections, but he was thinking ahead to his own in 1994. The frequency of press releases pointing out what this bill or that bill did for Massachusetts started increasing. And over Labor Day weekend, he and Vicki did a marathon set of interviews at Hyannis Port.

When their engagement was announced, some articles had a skeptical tone, stressing his "still-frisky social habits," as the *Washington Post* put it, or "his tabloid-driven image as an aging rogue with an eye for blondes and a taste for scotch," in the *Boston Globe*'s words. *The New York Times* said later, "There have been suggestions by critics that this was a marriage of convenience for Senator Kennedy, that after the Palm Beach incident and his flawed performance at the Clarence Thomas confirmation hearings, he needs to project a new image, especially as he prepares for re-election in 1994."

Kennedy's staff set out to use the couple to fight that suspicion and prove that a happy marriage showed Kennedy was keeping his Harvard promise to do "better." They set up an elaborate schedule of back-to-back interviews so that competing organizations never had an edge on one another. The *Globe* and the *Herald* went one after another, as did the Boston television stations. So did NBC's *Today* show, ABC's *Good Morning America,* and CBS's *This Morning*. The house at Hyannis Port has two living rooms overlooking the ocean. While one was being used for a television interview, the next crew was setting up in the other.

The product of the interviews was a collection of warm, sometimes gushing portraits, a public relations home run. The *Today* show opened with Vicki telling how she got her engagement ring and closed with Ted answering whether he "got married for political reasons." He conceded that Vicki helped him politically. "But, you know, Vicki knows how much I love her. I have a sense about how much she loves me. My children know what it means for us to be together. Her children certainly have a sense of that. My sisters know. Her family knows. Her mother and father know. My friends know. And I think the people that see us together are coming to know that. And that's good enough for me." Vicki agreed: "I think everybody who's seen us together, people who care about us, know that this is a big-time real thing."

The newspapers glowed, too. The *Springfield Republican* noted with approval that Vicki brought Perrier to reporters and quoted her as saying, "I realized when he didn't come over for dinner there was a real void." The *Quincy Patriot-Ledger* said, "Reggie and Kennedy paint an elegant portrait of domestic tranquility and the ideal blended family of the '90s. Their feelings for each other, they say, are symbolized by the words of the Wordsworth poem that inspired Kennedy to paint the daffodil wedding gift: 'And then my heart with pleasure fills, And dances with the daffodils.' "

The *Globe* called her humor "wry" and her looks "smashing." Of Ted, it said, "He appears, after a decade of bachelorhood, to have settled happily into married life, bantering with Vicki, praising her politics and her personality, as well as her cooking, and integrating her smoothly into the conversation." It said she called him "both romantic and a caring stepfather. . . . She reaches for his hand and he surrenders it willingly."

THE COUNTRY PAID more attention to the 1992 election than usual. Voter turnout was up substantially for the first time since 1960. Ross Perot and his Reform Party had a lot to do with that, and not just because a three-way race was different and therefore interesting. He captured the generalized distrust/contempt for Washington, but could be a safe protest vote; he was not going to win. Clinton's campaign was tactically brilliant in its advertising focus on winnable states, and his message of concern for the middle class brought home many Reagan Democrats. Bush's campaign failed because it could not make Bush into something he was not—deeply interested in the American economy—and it could not make someone of his underlying decency approve going after Clinton on explosive subjects like his womanizing, or even on a clear issue of public policy like gays in the military.

CLINTON WON, AND that had a huge impact on Kennedy. For the first time in his thirty years in the Senate, Kennedy would be in a position of power during the administration of a Democrat committed to his own greatest goal, national health insurance, and in a relationship free of rivalry. By the time Kennedy had become chairman of Judiciary in 1979, health care was an issue that divided him from a Democratic President. Now it unified.

Kennedy pitched in to help, lending staff to work on projects and meeting regularly with leaders of Clinton's transition team, Vernon E. Jordan, Jr., the ultimate Washington insider, and Robert Reich, the Harvard economics professor who had been a Rhodes Scholar with Clinton and a key economic adviser during the campaign. Others in the transition team, like Tom Downey, who dealt with Health and Human Services Department issues, leaned on Kennedy's staff for help. Because of the range of his jurisdiction, "it was one-stop shopping," Downey said. Kennedy was pleased that the still vague national health insurance initiative was put in the hands of two old allies, Judy Feder, a former health aide to John D. Rockefeller IV, and Stuart Altman, and thought that would lead to support for the Senate Democrats' play-or-pay approach. Several of Clinton's favorite campaign plans would come before his committee, including enabling students to repay college loans with national service, family leave, and many education and training programs.

There was one post in the Clinton administration that Kennedy was more concerned about than any other—the embassy in Dublin. He wanted his sister Jean to be ambassador to Ireland. The ambassador's daughter as ambassador was strong family symbolism, and she asked Ted to help.

But if traditional patronage was the dominant motive, there was another as well. In the 1992 New York primary, Clinton had met with NORAID, a group that collected money for IRA prisoners, and had responded to the group's demand by promising to send a peace envoy to Ireland to bring all sides together and to get the United Nations involved. Clinton had a variety of Irish supporters, some pro-NORAID and others inclined, like Kennedy, to the peaceful path of Northern Ireland's John Hume and his Social Democratic and Labour Party. They felt that the American role had to be helpful, not dominant, and could not ignore Great Britain or the Protestant majority in Northern Ireland.

Jean was at least as devoted to Hume as Ted was. She called Hume a "visionary." Unlike Ted, who had never been to Northern Ireland, Jean had stayed with Hume and his wife, Pat, in Derry in 1974. She recalled "walking through the streets with him, and seeing all the bombed-out buildings and everything. It was very, very quiet, and there was nobody on the streets, except British soldiers and tanks. Their house—a window was being fixed because they had had some ammunition shots through their house."

It was a highly prized embassy among Irish-Americans, and there were other volunteers—Brian Donnelly, a former Democratic Congressman from Massachusetts, and Elizabeth Shannon, the widow of former ambassador William V. Shannon. Each of them had claims on Kennedy's support, but not against Jean. He started pushing it in November, and personally urged the President to appoint her. Before any decision was made—and it was not announced until March—Hillary Rodham Clinton called around Capitol Hill to find out how much it mattered to Kennedy. The answer was it mattered very much, and Jean got the job.

Long before that, Clinton had made another gesture that mattered greatly to Kennedy, quietly visiting the graves of his brothers at Arlington National Cemetery on the day before his inauguration. About twenty Kennedys, including Ted, were with him.

When Clinton actually became President, his first official act was to issue an executive order lifting the ban on fetal tissue transplant research. But that got far less attention than his intention to end the ban on gays in the military. Kennedy was one of Clinton's strongest supporters on that issue, too, though he thought the President had blundered by moving so quickly on an issue that was vastly more divisive than Clinton anticipated.

Clinton backed down on January 29, agreeing that no changes would be made in military rules for six months while the Defense Department, the Joint Chiefs of Staff, and Congress examined the issue—except that recruits would no longer be asked if they were homosexual.

But the issue was still fresh the next week, stoked by radio talk show hosts who urged their audiences to call Congress and protest. When Mitchell brought up the family leave bill on Tuesday, February 2, Republicans threatened to filibuster so they could force a vote against any change in the military policy about gays, including the questioning of recruits. The threat was potent, because Congress wanted to adjourn on the sixth for its second break of the new session.

There were really three debates going on. One concerned the family leave bill, the second was about gays in the military, and the third—inside the Republican caucus—was about how long and hard to push the second issue.

The family leave bill itself required employers of fifty or more workers to offer them up to twelve weeks of leave—without pay, but with health insurance continued if the worker already had it. While several Republicans, notably Christopher "Kit" Bond of Missouri, were among its cosponsors, others argued that it would cost jobs. Bob Dole predicted that tens of thousands would be lost, including up to seventy-one at that many Kansas companies which now employed exactly fifty and would fire one to get under the ceiling. No one really disputed the idea that it would be nice if employers had leave policies. But Republicans like the other Kansan, Nancy Kassebaum, followed Bush by saying the government should not tell them they must.

Chris Dodd managed the bill, citing a Small Business Administration study that said 300,000 workers had quit or been fired since 1990 because they could not take leave when they or family members became ill. Noting a Republican campaign claim that said the GOP promoted family values, Dodd called the bill "sound family policy that enables working people to deal with family emergencies without losing their job in the process."

Kennedy who stayed in the chamber at his back-row seat through most of the debate, said, "The first piece of legislation that will pass the U.S. Senate, with the strong support of President Clinton, who talked about putting people first, is an example of caring about the conditions of working men and women who are parents, wives, husbands, members of families—not just numbers—members of families, putting people first."

Republicans thought they could capitalize on the strong public unhappiness with Clinton's policy on gays. But at Kennedy's urging, Democrats attacked them for delaying a popular bill, accusing the Republicans of a commitment to gridlock. By Thursday, February 4, Republicans settled for a separate vote which they knew they would lose on restoring the old policy about gays in the military, and the leave bill passed that night, 71–27. The House followed later that night, 247–152, and Clinton signed it the next morning.

The debate had a new, strongly personal flavor. Two of the four new Democratic women elected in 1992, Dianne Feinstein of California and Patty Murray of Washington, told how they had been forced to quit jobs when they became pregnant, an experience no Senator had ever shared with colleagues before. Murray also told how, as a state senator, she was the only one of seven siblings able to juggle her time to care for her aging, ill parents. The bill meant a lot to her, and Kennedy did, too. Though they had never met, he came to her swearing-in party and said a few kind words of welcome, as he often did in trying to connect with new members. Murray's mother, who had led the family in months of mourning when President Kennedy died, was thrilled that he took the time for her daughter. When the bill passed, Murray went over to Kennedy to thank him for his efforts. He told her how when Teddy was undergoing chemotherapy, he would simply leave the Senate to be with him. "Some people can't do that," he said.

Dodd also thanked his friend Kennedy, and publicly, after passage of the bill which had been his cause since 1983. He said Ted had provided real help and taught him an important lesson about how Congress works. He said, "I remember about five years ago when I lost the first time on this matter and was feeling down about it. He pulled me aside and said let me tell you, first of all, a good idea takes time and, second, it's not uncommon for a good piece of legislation to take five or six years. He depressed me at the time, when he said five or six years, but he turned out to be extremely knowledgeable, almost to the year."

But in the same first few weeks of his administration, Clinton made a decision on national health insurance that turned out to have severely damaged his chances of getting anything passed. He scrapped the plans his transition team had worked on and turned the issue over to his wife and another Rhodes classmate, Ira C. Magaziner, a politically tone-deaf policy wonk. Magaziner had succeeded as a consultant offering new ideas to business but failed spectacularly when trying to reorganize Rhode Island's economy, and had found his ideas largely ignored when he tackled other projects involving government.

Kennedy and Mrs. Clinton were easy allies, and Kennedy got on better with Magaziner than did most leaders in Congress. Moreover, Ted and his staff had access to the immense, cumbersome task force structure Magaziner assembled. So he went along, swallowing doubts about the process, and with high praise for Mrs. Clinton. Describing her first visit to the Hill, he said in February 1993, "She knew exactly what the policy questions were. . . . She's got a good grasp" of the central issues of savings and managed care, and of how smaller issues, like Indian or veterans' health programs, could not be ignored.

In a cheerful interview with *The New York Times,* the newlywed reformer predicted passage of comprehensive health legislation, saying, "In politics, I think, as in life, a lot of it is also being at the right place at the right time, whether it's in love or politics, I mean that's certainly happened to me." The President made it the right political time because "you have a person that sort of understands the Presidency, and the role of the Presidency, what the

Presidency could mean in terms of domestic leadership as well as international leadership, and his own role in terms of history and having a feel for these issues. And you have a country that is in need, a need in the country, a President who understands it and believes it, and I think it's a very, you know, opportunity for progress."

Kennedy was hardly the only Congressional leader involved with the Clintons on the question. Senator Daniel Patrick Moynihan, the New York Democrat, had ascended to the chairmanship of the Finance Committee when Clinton made Lloyd Bentsen secretary of the treasury. He was never deeply interested in health care and thought welfare reform should come first. Health care, he felt, needed to be worked out with Senate Republicans to have a chance, and he assumed they would decide to cooperate. (Kennedy's approach more often is to try to find what it takes to unite Democrats first, and then compromise enough to bring enough Republicans into the fold.) Senator Mitchell, the majority leader, tried to keep the two committees from competing, which meant slowing Kennedy down, while handling other tough issues like Clinton's apparently contradictory tax-cutting short-term stimulus package and his long-term tax-increasing deficit reduction measure.

On the House side, there was skepticism about play-or-pay, as much because Senate Democrats had been unable even to bring it to the floor as for any reason. House leaders pressed the President to send up his own plan. Representative Dan Rostenkowski of Illinois, chairman of Ways and Means, was prepared to do his formidable best for the President. But he was the strongest voice in saying Clinton had to send up a real bill, not just a short statement of principles. In Magaziner's hands, that would prove to be a disastrous approach. Representative John D. Dingell of Michigan, chairman of the Commerce Committee, was at least as committed to national health insurance as Kennedy, having taken his father's seat and his father's issue in 1958, but he had a much less loyal committee behind him. Speaker Tom Foley, less enthusiastic about the whole idea than Mitchell, was content to let Rostenkowski and Dingell see what they could accomplish without any shove from him. He quickly rejected the idea of creating one special committee to deal with the issue—an approach that had worked when Speaker Tip O'Neill did it on energy in 1977.

Kennedy was more single-minded about the priority of health care and the need for speed than either his colleagues or the White House. He went to a Democratic Senators' retreat at Jamestown in late February to spread this message, as summarized in a February 20 memo from his health aide, David Nexon: "The bill needs to get up here fast and we need to move it quickly. If we don't keep the pressure up, this bill will get bogged down. The special interest groups are on the run now—even the Chamber of Commerce has said they will accept an employer mandate—but if this bill slows down, the whole dynamic will change for the worse. I personally think that the President should ask us to stay through August and pass his bill."

One even quicker route would be to make the health plan part of the budget reconciliation bill, which was the centerpiece of Clinton's economic policy. Reconciliation bills are filibuster-proof, so it takes only fifty-one votes to pass them. But Robert Byrd, minority leader for six years in the eighties, had chafed at the way Republicans dumped issues that had little to do with the budget or the deficit into reconciliation, and when the Democrats took the Senate back in 1987, he imposed the "Byrd rule," which sharply limited what could be added to reconciliation bills. Kennedy went to the White House in early March and pressed that option to the President, following Mitchell and Representative Richard A. Gephardt of Missouri, the House majority leader.

Clinton called Byrd, but Byrd rebuffed him, saying it would not be right to bend the rules, that the Senate was created to deliberate, and that a massive issue like health care reform deserved much more debate than the maximum of twenty hours permitted for the whole reconciliation bill. While a major part of a health plan—though one did not yet exist—could legitimately overcome the Byrd rule hurdle because it cut federal spending, no one would risk Byrd's opposition. Reconciliation was going to be hard to pass anyhow. Byrd also turned down Mitchell, Rockefeller from his own state, and Kennedy. He told Kennedy it was hardest to turn him down because Kennedy had done him the honor of offering the Vice Presidential nomination in 1980, implying that he was worthy of the Presidency itself.

By May, when Magaziner's hundred-day deadline was expiring, he was under fire from within. Proposals, like a national sales tax to pay the plan's costs, were leaked in ways to make the plan look bad, apparently from the Departments of Health and Human Services and the Treasury—which felt their views were being ignored. When Republicans ambushed Clinton's stimulus package by a filibuster, the administration aides who were most involved with the deficit-reducing reconciliation bill insisted that no distractions from health care be allowed to upset Democrats whose votes they needed, like

Californians fearful that a tax on wine might be part of the financing, or North Carolinians worried about tobacco. With final budget votes expected in late July or early August, health care was shelved until fall.

Though Kennedy griped to his staff about the delay, he had plenty of other legislation to handle. With the fetal tissue issue resolved by Clinton's executive order, passing the reauthorization bill for the National Institutes of Health was relatively simple. The Senate adopted it, 93–4, on February 18. The conference report did not even require a roll call vote when the Senate passed it on May 28. The emphasis was no longer on abortion, but on research on women's health issues. Kennedy explained the final bill as one that would end "the shocking lack of women in clinical trials. It will dramatically increase the resources for research on diseases of greatest concern to women—an additional $325 million will be available for breast cancer, an additional $75 million for ovarian, cervical and reproductive cancer, and an additional $40 million for osteoporosis research."

Kennedy used the reconciliation bill itself to push direct student loans, legitimate under the Byrd rule because they would reduce federal spending on subsidies to banks. Although the $500 million pilot program voted in 1992 had not yet been tried, President Clinton asked for a complete change-over to direct loans by 1997. Under reconciliation rules, Kennedy could probably have jammed it through, but he was looking ahead to health care, and in the Labor Committee he settled for 50 percent direct loans by 1997. The House version went to 100 percent.

In conference, he orchestrated a procedure in which the White House pushed Representative William D. Ford of Michigan, the chief House conferee, to back off from his support of the original administration position—to keep the support of Senator James M. Jeffords of Vermont on health care and on other issues that mattered to Ford, like national service. In the end, Jeffords and Ford were both satisfied with a plan that allowed 60 percent of the loans to come from the government in the fifth year, but also permitted any college or university to use them if it wished. That deal satisfied Republicans who thought direct loans would not work and Democrats who thought they would.

Kennedy suffered one major disappointment from Clinton that spring, with the failure of his effort to promote Stephen Breyer to the Supreme Court. A nomination became available to a Democratic President for the first time since 1968 with the retirement of Justice Byron R. White, with whom

Kennedy had dined on T-bone steaks, Coors beer, and political insights when he was chasing Colorado delegates for a week in 1960.

For a time in late May and early June, Breyer appeared to be the leading choice. Not only did he have Kennedy's strong backing, but Senate Republicans liked him, too, suggesting an easy confirmation in a period when Republicans were making nothing else easy. And he fit Clinton's cautious prescription for judges—a moderate to mildly liberal judicial philosophy. Kennedy had his staff prepare a press release hailing Breyer's nomination. It said, "His broad background and experience and his unusual skills as a consensus builder will make him an influential and effective member of the Supreme Court from the day he takes his seat on the bench."

Instead, on June 14, Clinton picked Ruth Bader Ginsburg, a federal appeals court judge in Washington whose cause had been promoted quietly but effectively up to the last minute by Moynihan. Breyer lost out because of revelations that he had failed to pay Social Security taxes for his house cleaner. That issue had doomed Clinton's first selection for attorney general, Zoë Baird, and other Clinton choices. But Breyer had also failed to impress Clinton in a personal interview. Ordinarily witty, Breyer was just out of the hospital after suffering a punctured lung and broken ribs in a bicycle accident. That barred him from flying to Washington. He was tired from a long train ride when he saw Clinton, and felt faint and had to lie down briefly while at the White House.

When Ginsburg was chosen, Kennedy enthusiastically praised her work in developing women's rights. At the White House, where Kennedy had rapidly become known as one of the President's major supporters, that reaction drew relief, because fear of offending Kennedy had complicated the choice. Clinton's chief of staff, Thomas "Mack" McLarty, told the *Globe,* "All of us were impressed." That Saturday, June 19, Clinton appeared at a $500,000 fund-raiser for Kennedy at Boston's Park Plaza Hotel. He cited Kennedy's central role on family leave and health care, saying that "every effort to bring the American people together across that which divides us, every effort has the imprint of Ted Kennedy."

On June 16, Kennedy had brought the national service bill out of the Labor Committee on a 14–3 tally, getting four Republican votes. Clinton was strongly for it even though it was smaller than he had asked, providing a $5,000 educational grant for each year a volunteer served, and authorizing only enough money for 25,000 volunteers in the first year and perhaps

150,000 after a few years, not the "millions" Clinton talked of. Fifteen percent of the program's funds were set aside for programs in elementary and secondary schools.

The committee majority was not safe for the floor, though. Since his April victory over the stimulus package, Dole had been leading the Republicans into one fight after another to deny Clinton victories over anything he might claim as a signature issue. And this was a signature issue. Clinton identified national service legislation as his third-highest priority, after the economy and health care, on July 20 in an interview on CNN's *Larry King Live* show.

Republicans also had serious quarrels with the bill itself, over its size and its cost, especially at a time when traditional student aid programs were being cut. Kennedy reduced the spending levels, to $300 million the first year and $500 million and $700 million the next two, but this did not satisfy Republicans. Still, Dole was under some heat as the Democrats charged obstructionism, and Kennedy read a letter from Dole's wife, Elizabeth, president of the American Red Cross, saying she "looked forward to the passage of the bill."

On July 28, after reporters for *The New York Times* and the *Washington Post* described Republican delaying actions—refusing to set a time for a vote and telling a reporter that a filibuster was under way and he had the votes to block cloture—as a filibuster, Dole glared up at them in the Senate press gallery. Speaking to them, not his colleagues, he said, "There hasn't been any filibuster. We've been negotiating." Kennedy replied that he thought "it is fair to make the observation that the process of delay is in effect."

Dole won the first cloture vote, on July 29, when forty-one Republicans voted not to halt debate. But later that day at least two of them bailed out, and on July 30 Dole gave up. Before the Senate voted on August 3, Kennedy told the Senate that thirty of the thirty-two amendments he had accepted were Republican ideas, and contended:

"At a time when partisanship is running high on the reconciliation bill, we have demonstrated that we can work together in a bipartisan manner on a bipartisan issue—providing more effective ways for more citizens to serve their communities. . . .

"The lesson of service to others learned in youth will last a lifetime, and produce a better, fairer and stronger America in the years ahead. In his inaugural address thirty-two years ago, President Kennedy emphasized this quality, and touched a deeply responsive chord when he called upon Americans of all ages to ask what they could do for their country. The best of the old frontier became the defining quality of the New Frontier. Citizens responded by the millions, and the spirit of America soared again, as it had so often in the past.

"We need to rekindle that attitude again for our own day and generation. . . . We do not have to compel citizens to serve their country. All we have to do is ask—and provide the opportunity."

Kennedy's argument of bipartisanship was strained. The bill passed, 58-41, with all but four of the no votes from Republicans. But Kennedy blinked back tears when one Republican supporter, Senator Jeffords, told a victory press conference, "I realize this must be a particularly wonderful moment for the chairman of my committee, Senator Kennedy, for he is finally seeing what his brother wanted for this to come to fruition."

A House-Senate conference quickly took the Senate spending levels, and the House passed the measure promptly. But bitterness over many issues, especially the final reconciliation bill, which passed 51-50 on Vice President Gore's tie-breaking vote, kept the Senate from acting until after the August recess. On September 8, it passed the final bill, giving Clinton a major victory.

For Kennedy, the highlight of the August recess was a sail with Jacqueline Kennedy Onassis, the President, and the First Lady off Martha's Vineyard. On August 24 they spent the day on the *Relemar,* a seventy-foot yacht owned by Maurice Tempelsman, Mrs. Onassis's companion of many years. As they were waiting for the Clintons, Jackie said, "Teddy, you go down and greet the President." He demurred; it was Maurice's boat and he was already there. Jackie insisted: "Teddy, you do it. Maurice isn't running for reelection." Ted went down and got in the pictures.

With his twenty-five-year dream of national health insurance in sight, Kennedy was almost boyishly eager to be with the Clintons. When Mrs. Clinton told him they were going to the Vineyard, he had telephoned Jackie to get her to invite them for a sail. He worried about not being invited to a dinner party they attended. Thank-you notes to the Clintons had to be just right. One went through nine drafts.

It was a cheerful day on the water, President Clinton later told Elsa Walsh of *The New Yorker.* "He talked a lot about his family's connection to the sea and how they all love the ships and the boats and all the things they have done." Clinton said it was "the most relaxed setting that I'd ever seen him in."

He had been carefully prepared by Nick Littlefield and David Nexon to talk about health care. He wanted to be sure the bill, when it came, was drafted to give the Labor Committee a big enough piece so that he would be a player at the end. This was a rare opportunity to be with both the Clintons and raise the issue. But in the event, Kennedy decided that this was not the time to talk about Senate jurisdiction, though he did tell reporters he brought up the question of a new Defense Department facility at Southbridge, a depressed town southwest of Worcester.

Clinton had given the nation's governors a taste of his plan in Tulsa eleven days before, and had been given a taste of how tough the opposition was going to be when John Motley, the chief Washington lobbyist for the National Federation of Independent Business, preceded him by savaging his ideas. Motley said requiring businesses to insure employees was like "tying small businesses to a sled and pushing them down a mountain, hoping they'll remain upright." The NFIB's grassroots lobbying later proved devastating, especially in the House.

On September 22, despite a TelePrompTer failure that showed him an old economic speech by mistake for a few minutes, Clinton gave a brilliant performance, seeking to establish his place in history by confronting and vanquishing a problem that had stymied past Presidents back to Franklin Roosevelt. "Tonight we come to write a new chapter in the American story," he said. "This health care system is badly broken and it is time to fix it. . . . At long last, after decades of false starts, we must make this our most urgent priority: giving every American health security, health care that can never be taken away, health care that is always there." Kennedy eagerly shook Clinton's hand when he finished, and his comment that night aimed at history, too. He said 1994 could bring "the most far-reaching improvement in social policy in this country since Social Security in the days of FDR."

While Moynihan sounded off from time to time, labeling the administration's cost estimates a "fantasy" the Sunday before they were made public, Kennedy tried to help the administration sell its program. He held dozens of hearings, examining the plan and pushing for support. After Mrs. Clinton appeared before Rostenkowski and Dingell, she testified before the Labor Committee on September 30. Ted welcomed her by producing the support of the first Republican to back the plan, Jeffords of Vermont. Mrs. Clinton thanked Jeffords, taking his support as evidence that "this is an issue beyond partisan politics." As it turned out, Jeffords was also the last Republican supporter.

The President's plan required employers to buy insurance for their workers (paying 80 percent of the premiums) through state-run alliances which would negotiate with insurance companies. The government would insure the unemployed, and the self-employed would buy their own, again through the alliances. The savings from the purchasing power of the alliances, from price controls on insurance, from managed care generally, and from cuts in future Medicare spending were supposed to pay for insuring the uninsured and for new drug benefits for Medicare patients. Insurance companies would be prohibited from denying coverage to sick people or excluding existing medical problems from coverage.

But its details and counterweights made the plan so complicated that it was hard to explain and easy to attack. In Clinton's September 22 speech, he took longest to explain the "simplicity" he said was in the plan. The administration had a slogan for a button, "Health care that's always there," but no two-minute explanation of how it would work. Republicans sold T-shirts with a complicated graphic that purported to map the system. The Health Insurance Association of America ran devastating television ads featuring Harry and Louise, a couple puzzled and worried over the plan; the ads were magnified by attacks on them from Mrs. Clinton. And in early October, President Clinton was diverted from a planned, if far-fetched, five weeks of selling by the deaths of American soldiers in Somalia, the embattled North American Free Trade Agreement, and problems in Bosnia and Haiti.

And while hearings could be held to discuss the plan, the Clinton bill itself, some 1,324 pages long, did not arrive until November 20, the last day of the Congressional session. When it arrived at the Senate, it ran into a jurisdictional buzz saw. Mitchell had set up a procedure in which committees claiming jurisdiction submitted written arguments to the Senate parliamentarian. The parliamentarian decided that much of the bill, including the requirement that employers insure their workers and its price controls on insurance, was within Labor Committee jurisdiction, though other elements like Medicare belonged to the Finance Committee.

But Moynihan was prodded into complaining by Bob Packwood of Oregon, his committee's senior Republican. Packwood had backed an employer mandate when Nixon proposed it in the seventies. But he was now under a serious investigation for having sex with staff members and was critically dependent on Dole's support, so he opposed the employer mandate and wanted it kept out of the hands of a committee friendly to it, like Labor. There was a confrontation among the three of them on the floor, with the angriest words between Kennedy and Packwood. In the end the bill was referred to no committee at all. That did not prevent committees from considering the issue and drawing up bills, near or far from Clinton's, but it was another measure of how hard it would be to make health care work in the Senate.

CHAPTER 34

Still Campaigning

Meanwhile an election battle was developing in Massachusetts. For the first time, Kennedy looked vulnerable to Republicans. Mitt Romney, a venture capitalist with his own money to spend, made it clear in November that he was going to run. A son of George Romney, the former governor of Michigan and secretary of housing and urban development, Mitt was a new face, rather than one of the tired party wheelhorses Republicans usually ran against Kennedy. In November 1993, Romney sent a letter to Republicans across the state telling them, "Like you, I am fed up with the wasteful spending and taxing in Washington, and I believe the time has come to send an urgent retirement notice to one of the worst offenders, Ted Kennedy."

Kennedy's big campaign event that fall took on the lingering women's issues about him, political and personal. The five women Democratic Senators brought the political message to Boston for a $100-a-plate lunch with about twelve hundred Democrats, mainly women. The personal message came from Vicki's obvious affection when she thanked them for supporting "Ted Kennedy, my husband, the love of my life, and the person who has consistently stood tall on all of the issues that all of us care so deeply about." The message of family was underlined, at least for careful observers, by the presence of Caroline Raclin, Vicki's seven-year-old daughter. Having children to come home to, instead of an empty house, was more than welcome. For

Caroline's First Communion, a few months later, he gave her the rosary that the Pope had given him fifty-five years before.

Patty Murray cited a litany of issues he led on: "health care, security, women's health care research, national service, freedom of choice, expanded student loans, comprehensive immunization for our children, AIDS research, the list goes on."

Carol Mosely-Braun of Illinois told of his personal kindness when she was suffering from the flu during budget votes, saying, "He said, 'Why don't you come up to my hideaway and relax?' " Vicki made her tea and took care of her. In a whispered aside, unheard by the audience but picked up by the wireless microphone he was wearing so the event could be used for campaign commercials, Kennedy said, "I'm sure glad you said Vicki was in that hideaway."

Barbara Boxer, who had flown in overnight from California, said, "There is a difference between a good vote and a great leader. Your Senator is a great leader. He's powerful not because of his name, but because of his conviction."

Dianne Feinstein told the guests to be proud of their Senator, and said he taught them "what it takes to be a serious legislator, to be really in the vanguard, to be unafraid, to stand up, to say it like it is, and to work your heart out. For all of those things, we salute you, we thank you, we support you."

Barbara Mikulski, the only nonfreshman among them, recalled Ted's helping to get her on the Appropriations Committee, and his insistence that she join Labor, too. She called him "one of the great Galahads of the United States Senate," and said, "We are pro-choice, we are pro-change, and we are pro–Teddy Kennedy. . . . We wanted to be there for Ted because Ted has always been there for us."

Kennedy closed, thanking Vicki: "She has changed my life, and brought extraordinary joy and happiness to it." He said he was proud to serve with the five Senators, citing their achievements as this session of Congress drew to a close.

After the lunch, they all flew back to Washington to fight for one last piece of legislation of special concern to women, legislation to make it a federal crime to bomb, burn, or blockade abortion clinics or to kill, injure, or threaten doctors and nurses who perform abortions.

The bill was partly a reaction to a Supreme Court decision outlawing injunctions against the mass protests that had become a widespread anti-abortion tactic. But it gathered momentum when Dr. David Gunn, a doctor who performed abortions in Pensacola, Florida, was shot and killed on March 10. An anti-abortion protester was charged with the murder.

Kennedy brought the bill out of the Labor Committee on June 23, but the full Senate did not act until November 16. He led the debate, saying, "This legislation will protect women, doctors and other health care providers from the tactics of violence and intimidation that are often used by anti-abortion activists. In the past fifteen years, more than one thousand acts of violence against abortion providers have been documented in the United States. Over one hundred clinics have been bombed or burned. Hundreds more have been vandalized."

The result, he said, was that "as clinics are burned down and doctors are intimidated, it becomes harder and harder for women to obtain a safe and legal abortion."

Opponents argued that while violence and murder were not to be tolerated, the bill violated the First Amendment rights of anti-abortion protesters. Kennedy insisted that if all they were doing was shouting or singing or praying, and not physically blocking a clinic entrance, the bill would not affect them. Unpersuaded, Strom Thurmond said: "I do not believe that the criminal and civil penalties contained in this legislation will have an incidental effect on pro-life expression. I believe that it will virtually eliminate such expression." He said the bill "raises the right of abortion above the Constitution," especially the right to assemble.

In a daylong debate, all five Democratic women Senators spoke for the bill. Boxer, who helped Kennedy manage the measure, said escalating violence had transformed doctors' offices from "safety zones to war zones." The two Republican women were silent, but all seven voted for it when the Senate passed it 69–30.

The vote came only after Kennedy asked Dole if he would vote for the bill if the penalties were reduced. Dole said he would, and Kennedy amended it to lower the maximum fines in the bill from $250,000 to $25,000 and cut the potential prison time for a first offense from one year to six months. Bodily injury could result in a ten-year sentence, a killing in life imprisonment. The penalties, lower than those voted by the House, prevented quick enactment before Congress adjourned. Dole said he had no doubt that Kennedy would preserve the Senate position in conference with the House. "Kennedy keeps his word," he said.

TED AND VICKI arrived in Dublin on December 29, 1994, for what they expected to be a simple family visit with his sister. But soon after their arrival, Ambassador Jean Kennedy Smith asked him a startling foreign policy question: What did he think about giving a United States visa to Gerry Adams, head of Sinn Fein, the political arm of the Irish Republican Army? The Senator dismissed the idea, saying, "He's a terrorist."

Adams had been denied a visa two months earlier. But John Hume had been talking to Adams. Hume had told Ambassador Smith that there was reason to hope that Adams would be a voice for politics, not bombing, in IRA councils. When she asked Albert Reynolds, the Irish prime minister, what he thought, he replied, "Why not? He's our best hope."

Reynolds and John Major, the British prime minister, had issued a joint declaration on December 15, pledging self-determination for Northern Ireland and promising negotiations with any groups that renounced violence. The British, too, had been talking to Adams, and there were hopes, strongest in Dublin, that the IRA might declare a cease-fire.

The day he arrived, Jean took Ted to lunch at the home of Tim Pat Coogan, a Dublin writer with unrivaled expertise on the IRA. Coogan told Kennedy a visa would enable Adams to see that there was American support for peace and for Ireland, though not for the IRA, while another denial would help the faction calling Adams a sellout.

The next day, having still made little headway, Jean took Ted for a courtesy call at Reynolds's office and then dinner at his apartment. "Tell Teddy what you told me about the visa," she urged. Reynolds said, "He's doing our best, and he's our best hope." He explained that the strategy behind it was to show the IRA that politics worked, that progress could be made without violence, and to enable Adams to lead the Republican movement toward peace.

Kennedy left Ireland on January 2, favorably disposed but still uncommitted. Eight days later he saw Hume at the funeral of Thomas P. "Tip" O'Neill and asked him if Adams should be given a visa. Hume said he should, and Kennedy went to work lining up support. When Adams, who had been invited to a February 1 conference in New York, turned up at Smith's embassy to apply for a visa on January 14, Kennedy was ready. He sent a letter to Clinton, signed by himself, thirteen other Senators, and twenty-eight representatives, telling the President, "It is important for the United States to facilitate the emerging dialogue as an alternative to violence."

The State Department, the U.S. Embassy in London, and officers in Jean's own embassy were furious that she had recommended a visa. So was the British Embassy in Washington, although Major knew that Reynolds thought it was a good idea. The embassy argued that Adams should commit himself to nonviolence before a visa was considered. While Kennedy lobbied the President from one side, Tom Foley, the Speaker of the House, told Clinton it was a bad idea.

The decision was not left to the State Department. Both Hume and Reynolds talked to Nancy Soderberg, the former Kennedy aide who was now the third-ranking official at the National Security Council in the White House. She and her boss, Tony Lake, supported the visa. Ultimately Clinton called Reynolds and sought his advice. Reynolds said later that he "recommended to the President that he give him a short visa." Clinton overruled the State Department and ordered Adams to be given a forty-eight-hour visa.

State was angry and the British were furious. The decision was widely portrayed as pandering to Kennedy and Moynihan so they would help on health care. Not letting down Kennedy was probably a consideration, but relations with Moynihan were too chilly to expect him to help, and his support for a visa was not all that strong anyhow. Clinton gave a different explanation to Conor O'Clery of the *Irish Times:* "It was one of those points where there had to be some tangible evidence that there could be a reward for the renunciation of violence and beginning to walk toward peace." And if Clinton had any doubts, he told Kennedy later, they vanished in his anger at the harsh attacks that followed in the British press.

Kennedy had been prepared to start marking up a health bill just after Thanksgiving. But Mitchell told him not to get in a race with Moynihan's Finance Committee, whose support was critical, too. Moynihan had no interest in a race. The Senate's only scholar was holding intellectually intriguing hearings, such as examining whether financing a bill in part by taxing ammunition would not serve multiple public health purposes, but doing nothing to move Clinton's plan along. He was waiting for Dole, or at least Packwood, to join in and work on a compromise. But Dole controlled Packwood, and so it was up to Dole.

The minority leader, by fits and starts, was moving away from support for universal health insurance. In the fall of 1993 he had cosponsored a plan offered by Senator John Chafee of Rhode Island that would require individuals, not employers, to arrange to be insured. Some days, especially back home in Kansas, he would worry aloud that something needed to be done because too many people lacked insurance. But on January 25, when Clinton used his State of the Union address to hold up a pen he said he would use to veto any bill that did not insure all Americans, Dole responded by saying the country had "no health care crisis," and that "the President's idea is to put a mountain of bureaucrats between you and your doctor."

So Kennedy soldiered on. He held hearings on the shape of the medical workforce of the future and the needs of academic health centers, what benefits package should be provided, and what the role of the states should be. He skirmished with opponents, blasting the U.S. Chamber of Commerce for abandoning in February its October support for requiring employers to provide insurance. Believing that Robert Reischauer, head of the Congressional Budget Office, was about to analyze the Clinton plan in a devastating way, describing the insurance purchasing alliances as government entities, Kennedy railed at him for half an hour one Sunday, asking how a minor staff official dared to thwart the President and indeed the will of the American people.

Throughout the first months of 1994, he kept meeting with Senators on his committee, mostly the Democrats. He and Littlefield and Nexon looked for ways to change whatever bothered Senators about the Clinton plan so that when the Labor Committee's turn finally came, he could count on them. He also wanted to produce a bill that would get more than the tepid support various constituency groups had been giving it. In the six months since the Clinton plan had been offered, its opponents had grown bolder while its supporters tended to assume that it would pass and felt free to campaign for their own particular interests. The most dramatic example was the American Association of Retired Persons pushing far harder to make sure any plan included long-term care than to ensure that there was a plan that Congress passed.

On May 9, Kennedy introduced his bill, cleared with Mrs. Clinton and designed to tackle some of the most serious problems in the original proposal. They hoped, vainly, that this set of changes would be taken as a signal of the administration's readiness to compromise.

Where the Clinton plan required all employers to insure their workers, Kennedy's exempted the smallest firms, with five or fewer employees, and allowed them to pay a 2 percent payroll tax instead. Where Clinton's made participation in insurance-purchasing alliances mandatory, Kennedy's only required states to set them up, allowing businesses and individuals to choose whether to buy through them or through traditional agents. Kennedy also proposed higher taxes on tobacco and large corporations and higher out-of-pocket payments by individuals than Clinton had urged. The Senator then proposed using the money for more generous benefits than the President suggested for women and children, for mental health and drug treatment, and for research and teaching hospitals, and for eliminating the deficit increase Reischauer had seen in the Clinton proposal.

Several elements of the Kennedy plan were taken from proposals by Senators who opposed the Clinton approach, with the hope of finding new backers or at least dimming their opposition. Senator Bill Roth of Delaware proposed allowing individuals to join the Federal Employees Health Benefits Plan (which Senators belonged to) and choose among its options. Kennedy took that idea. Drawing from Chafee's bill with its requirements that individuals get insured, Kennedy offered subsidies to low-income individuals directly. The mental health benefits dealt with a weakness seen in the Clinton plan by Senator Pete Domenici of New Mexico.

Health care was also a factor in another Supreme Court decision that was coming to a head. Justice Harry A. Blackmun, author of the Court's 1973 opinion in *Roe v. Wade,* announced on April 6 that he was going to retire. Mitchell immediately became the leading contender; he had announced on March 4 that he would not run for reelection to the Senate. Clinton had often said he wanted to put someone with experience in real-world politics on the Court, and Mitchell also seemed like a sure bet for easy confirmation, no small matter. But on April 12, Mitchell announced that he would not take the post, explaining that it would compromise his ability to get health insurance legislation passed. There was also a widespread belief that he wanted to become commissioner of baseball.

Kennedy quickly advanced Steve Breyer's name again. Clinton stewed over three possible nominees: Breyer, Secretary of the Interior Bruce Babbitt, and Judge Richard S. Arnold of the Eighth Circuit Court of Appeals in Little Rock. But Judge Arnold had cancer, and by the spring of 1994 Babbitt had made plenty of enemies in the West, so his confirmation was not a sure thing and getting a successor approved would be very hard. Clinton talked with cancer experts about Judge Arnold's leukemia. And on Saturday, May 7, Orrin Hatch, who had become the ranking Republican on Judiciary, told Clinton he would oppose Babbitt, fearing that his policies at Interior indicated he would try to make liberal law on the Court. Hatch told Kennedy of his message to Clinton, urging him to press the case for Breyer.

On Tuesday, May 10, Kennedy found an opportunity. He contrived to meet Clinton in a corridor at the Hyatt Regency Hotel, where they both had gone to speak about health insurance to the American Federation of Teachers. One reporter who had come with Kennedy from a health insurance hearing at the Senate stood by and took notes as Clinton praised Kennedy's new health insurance plan, saying, "You've got a great bill there." Clinton urged Kennedy not to "go higher" than five employees for the small business exclusion, and Kennedy thanked him for spending a few minutes with Patrick, who was running for the House, a few days before in Providence. Then Kennedy put his arm on Clinton's shoulders, turned him so the reporter could not hear, and told Clinton what a great selection Breyer would be.

Other allies of Breyer's played key roles, too. Lloyd Cutler, an old Washington hand serving a brief tour as White House counsel, urged Clinton to pick him. Richard Stearns, installed by Kennedy that winter as a federal district judge in Boston, knew the appeals court judge well. More important, he had been at Oxford with Clinton and had kept up with him ever since. Another federal judge, Joe Tauro, told Stearns about a recent witty lecture Breyer had given to a lawyers' group. Stearns located a videotape and sent it to Clinton to watch, with a covering note explaining that the healthy Breyer was someone quite different from the sick man he had met the year before. Clinton watched and agreed. And finally, the tax issue had disappeared; the Internal Revenue Service had written Breyer to say that the arrangements he had made with his house cleaner had not made him subject to Social Security tax.

Arnold's health eliminated him, but Clinton still thought Babbitt would make an excellent justice. On Friday, May 13, Stearns told the President that Babbitt would write fine dissents on this Court, but Breyer was used to leading other judges and would make law. Kennedy, who thought one reason he had failed to sell Breyer in 1993 was that Moynihan spoke to Clinton last, kept calling.

Clinton finally chose Breyer and announced it late that afternoon. When he announced the choice, he used Stearns's argument about Breyer, saying, "He has proven that he can build an effective consensus and get people of diverse views to work together for justice's sake." When a reporter asked why he thought someone with little political experience could reshape the Court, Clinton replied, "He obviously has a lot of political skills because of his reputation as a consensus builder on a court where most of the appointees were made by Republican Presidents. And look at the people supporting his nomination. He's got Senator Kennedy and Senator Hatch together. I wish I had that kind of political skill."

Breyer was approved, 18–0, in the Judiciary Committee on July 19, and 87–9 in the Senate ten days later, after Kennedy had said proudly: "Joseph Story, Oliver Wendell Holmes, Louis Brandeis, Felix Frankfurter—for nearly two centuries, Massachusetts has sent brilliant justices to the Supreme Court who have combined outstanding legal scholarship with a commitment to making the law work to enhance the lives of ordinary Americans." Breyer, he said, would "join that illustrious list of the finest justices ever to serve on our highest court."

BY THEN, HEALTH care had suffered a mortal wound in the House, though not everyone knew it immediately. On May 31, Dan Rostenkowski was indicted on corruption charges, including embezzling more than $500,000 from his official accounts "to benefit himself, his family and his friends."

Under House rules, he had to step down as chairman of the Ways and Means Committee, and health care lost its most skilled House lawmaker, a dogged Chicago pol with that city's deep understanding of party loyalty. On May 18 he had told his thirty-nine-member committee, "I will do whatever I need to to get at least twenty votes in this committee." He expected to get a bill to the floor, and pass it, including tax increases Clinton had been unwilling to ask. "This is a major program," he had said at the Harvard School of Public Health on April 22, "and if we're serious, we have to belly up to the bar."

In the Senate Labor Committee, Kennedy brought his bill up on Wednesday, May 18, hoping to finish by Memorial Day. Republicans were initially caustic. Hatch called it "nothing more than a pasteurized version of Clinton's blueprint for socialized medicine." Kassebaum said, "The Kennedy bill is rather like a casserole made from the leftovers of the previous evening's meal. The ingredients are mixed up differently, but they are the same ingredients."

But the next day the committee voted 17–0 to adopt a proposal that would provide nearly automatic cuts in benefits if costs proved higher than expected. Republicans had first sought to eliminate the benefits package entirely, but agreed to the compromise offered by Senator Jeff Bingaman of New Mexico, who had worked it out in advance with Kennedy.

On Wednesday and Thursday evenings, Kennedy and his wife flew to New York to visit Jacqueline Kennedy Onassis. Suffering from non-Hodgkin's lymphoma, she had been hospitalized for a month, and had chosen to go home to die. On Wednesday night, Kennedy told reporters standing vigil in a light rain outside her Fifth Avenue apartment building: "We all wanted to be here this evening. We were all distressed by the medical reports. We wanted to love Jackie. All the members of her family love her very deeply."

On Thursday, Ted and Vicki saw her again for more than an hour, and were in a plane returning to Washington when she died at 10:15 P.M., with her children at her side.

Obviously worn, Kennedy abandoned his effort to finish the health bill the next week. Instead, he delivered a eulogy at the funeral at St. Ignatius Loyola Church on Park Avenue on Monday, May 23. The service was closed to the press and to television cameras, but broadcast by radio.

He recalled her as both a national and a family figure:

"During those four endless days in 1963, she held us together as a family and as a country. In large part because of her, we could grieve and then go on. She lifted us up, and, in the doubt and darkness, she gave her fellow citizens back their pride as Americans. . . .

"Her two children turned out to be extraordinary, honest, unspoiled and with a character equal to hers. And she did it in the most trying of circumstances. They are her two miracles. . . . She once said that if you 'bungle raising your children, nothing else much matters in life.' She didn't bungle. . . .

"She made a rare and noble contribution to the American spirit. But for us, most of all, she was a magnificent wife, mother, grandmother, sister, aunt and friend. She graced our history. And for those of us who knew and loved her, she graced our lives."

She was buried next to President Kennedy at Arlington National Cemetery. The retired archbishop of New Orleans, Philip N. Hannan, conducted that service. He had read from Scripture and President Kennedy's speeches at his funeral in 1963.

AFTER CONGRESS CAME back from its Memorial Day vacation, Kennedy pushed his bill through committee, making occasional compromises with Republicans but not getting any of them except Jeffords to vote with him on final passage. When Kassebaum proposed amendments, he would sometimes ask if she would vote for the bill if he accepted her change. When she said no, he usually put the votes together to defeat her. Fundamentally, the bill his committee passed was close to the one he introduced. It raised the ceiling under which no insurance had to be purchased from five to ten or fewer employees.

While he waited to see what the other committees would do, Kennedy tried to stir public interest, which had flagged since Clinton announced his plan nine months before. He kept insisting on an employer mandate, the requirement that employers insure their workers, saying there was no other way to get everyone insured. One day he held a news conference with leaders of black organizations to argue that blacks were especially likely to go without insurance and needed an employer mandate badly. Each day representatives of groups supporting the plan met in the Labor Committee hearing room, and he would often speak to them.

On July 1, as Ways and Means and the Senate Finance Committee struggled to finish, he roused a crowd of 150, telling them, "If we have a battle for our nation's heart and soul, it is now, this is the time. And I don't believe that we have had a cause that has been addressed in this room that is as important as the cause that we will have on the floor of the United States Senate in some three weeks.

". . . Even after the last year and a half of assaults . . . of distortion, misrepresentation, opportunism, status quo profiteering, we are on the eve of that great national debate. And my friends, the outcome is really up to each and every one of you. . . . So we are here this morning to ask you to do what you have done so well in the past. We are asking you first of all to meet with your members of Congress or your Senators, during this Fourth of July break. We are here to ask you to talk to all of those in those Congressional districts and Senate, and say, 'Now is the time for the letters, now is the time for the phone calls.' "

Kennedy often told supporters how opinion in Congress could change quickly and cited past exploits like the plant-closing bill. He seemed sure that the public was behind universal health insurance, and that Republicans would not be able to stand up against public pressure. But in fact, when typical Americans thought of health care reform, they hoped for an end to the fear that they would lose the insurance they had or would discover it did not cover something serious. And by relying on the hope that managed care and managed competition would save billions, the Clinton plan and its variants, like Kennedy's and Mitchell's, all seemed to be promising that providing health insurance for another 37 million Americans would be free. That was a hard sell.

In retrospect, it is clear that by July 4, 1994, passing any health insurance bill was a long shot. None of the committees—Finance and Labor in the Senate and Ways and Means, Education and Labor, and Commerce in the House—had produced a bill that commanded solid Democratic support, let alone Republican backing beyond the solitary Jeffords. Commerce had not voted out a bill at all. So Mitchell and Gephardt were starting over, each trying to create a new bill of his own.

That did not make legislation impossible. Some of the most complicated lawmaking gets done in a few days behind closed doors. Mitchell had done it with the Clean Air Act in 1990. But then he had Republicans to work with. Dole had already lined up thirty-nine colleagues to vote with him against any Democratic plan, and had never been ready to work on a compromise. In the House, Newt Gingrich, seeing a Democratic failure on health care as the key to Republican gains in the House, had told his party not to offer constructive amendments that might make Democratic health bills better.

There remained the possibility that a group of Senators who called themselves the Mainstream Coalition might come up with a plan that could command the votes to succeed. The coalition was led by Chafee and John Breaux, a Louisiana Democrat. Other regulars included Democrats Bob Kerrey of Nebraska (who had been for a single-payer system in the 1992 primaries), Kent Conrad of North Dakota, David Boren of Oklahoma, and sometimes Bill Bradley of New Jersey. John Danforth of Missouri and David Durenberger of Minnesota were the constant Republicans. It was struggling behind closed doors to come up with a middle way to health care reform—not universal insurance, but something better than the country had.

July was a frustrating month for Kennedy. He was becoming convinced that Republicans would block anything so they could defeat Clinton in 1996. Still, he went on with the pep rallies and expressed confidence in public. He consulted and discussed compromises with anyone who would listen, but the Mainstream Coalition was not ready to talk to him. Chafee and the others had the unrealistic idea that it could expand from the center, getting additional Democratic support as it attracted more Republicans. An early alliance with Kennedy would spoil that.

On August 2, Mitchell came forth with his bill. It put off employer mandates until at least the next millennium, saying they could take effect in 2002 but only if 95 percent of the population was not insured by 2000. (The 1994 percentage was 85 percent.) It also dropped price controls on insurance. Mitchell said it was the strongest bill that could get through the Senate. Kennedy bit his tongue and praised the bill as "capable of achieving the goal of health security for all," saying it was "designed to achieve" universal coverage. Clinton, too, said "it provides for universal coverage." A blunter analysis came from Representative Jim McDermott of Washington, a single-payer advocate who did not really war against the Clinton plan. He said Mitchell "has run up the white flag to the insurance companies."

The bill came to the Senate floor on August 9. It was the first time the full Senate ever debated health insurance.

Mitchell began by saying his bill "will create universal coverage through a voluntary approach. . . . My approach is to build on an existing system of private insurance and expand it to those not now included: Americans who can't afford insurance, people with an illness that insurers won't cover, and people between jobs." He said it would expand the choices patients had by requiring employers to offer at least three options.

Dole responded by defending his bill, introduced that day, which included subsidies for the poor but was based largely on reforms of the insurance system, such as limiting insurers' ability to exclude existing medical problems from coverage. Those elements were common to the Clinton bill and its variants, and in January Dole had said it would be unwise just to do them and "save the tough stuff" for later. Dole said the basic issue was "will we trade in a health care system based on individual freedom for one based on government control."

Moynihan followed, praising Mitchell's bill but complaining that a provision designed to curb excessive supplies of specialists would hurt New York City and threaten medical progress. The longest speech came from Packwood, who said the Mitchell bill ignored marketplace realities. "Market forces work," he said. "Competition works. Price controls don't work."

Then it was Kennedy's turn. He said it was a historic moment for the Senate, and "a special moment for me. I introduced my first universal health care plan in 1970, and I have been working on the issue ever since." He said, "We must decide whether we will guarantee health insurance for every citizen, or whether we will continue to let millions of fellow citizens suffer every year from conditions they can't afford to treat, while millions more worry about losing their insurance." He ridiculed proponents of delay: "We need more time. We have to study it further. This is so complicated. Can't we just wait another year or maybe the year after that or the next century?"

"It is clear where the American people stand," he said. "The vast majority support universal coverage, despite months of irresponsible attacks by those who profit from the status quo. Most Americans support a system that shares costs between employers and employees as well."

He assailed Republican "naysayers" for making the "same old tired charges" about socialized medicine and bureaucracy against the Mitchell bill that they had made against Medicare in 1965.

"This legislation is the defining test for Congress today," he concluded. "This is the job the American people elected us to do, and I urge the Senate to get the job done."

Mitchell had told the Senate he would keep it on health care until it was passed or defeated, abandoning all of the August vacation if necessary. The debate droned on. Republicans held up thick copies of the 1,410-page Mitchell bill to show how complicated it was. Democrats responded, after a few days, by holding up their government health insurance cards to show they had something many Americans could not get.

As the first week ended, Mitchell said he would be happy to negotiate with Chafee, and Democrats made a point of arguing politely with members of the Mainstream Coalition, while answering attacks from other Republicans sharply. But when agreement was sought to allow a vote on a Chris Dodd amendment designed to speed coverage of children and pregnant women, Dole blocked it. Packwood said twenty-eight Republicans still had opening statements to make, and some would last three or four hours. Kennedy told reporters that was the equivalent of a filibuster. Packwood said it was the only way to educate the public.

In the second week, the Republicans allowed one vote, on the Dodd amendment. The Mainstream Coalition labored on. As its ranks expanded, its level of expertise declined, and it repeatedly missed the deadlines it announced for offering its plan. At a caucus luncheon on Thursday, August 18, Bob Kerrey of Nebraska explained the coalition's difficulties, but said Democrats should wait for it, because they were "on the cusp of a major political disaster." If they pursued the Mitchell bill, he said, they would lose the Senate. Then he got up to leave.

Kennedy, in a rage, shouted at the disappearing Kerrey: "Here it is. We are waiting for the Mainstream Group. We are taking hits. How long does it take to find out about cost containment? All of us who have been working in the committees and working on this issue for years know that when you get down to how to pay for it, it's difficult.

"And now we are watching the self-destruction of all of us because the

Mainstream Group can't make up their minds. How long do we have to wait? How long do you have to take while we are out there taking hits?"

Kerrey stormed back, and the shouting continued until they seemed to run out of breath.

The Mainstream Coalition (Dole called it "the midstream group") produced its plan the next day, Friday, August 19. The goal was insuring 92 or 93 percent of Americans by 2004, and in the process saving $100 billion. Just how sweeping subsidies for the poor would be was left uncertain. Still, Mitchell and Kennedy welcomed the effort.

The next week the Senate put aside health care to fight over a crime bill while Democrats kept trying to deal with the Mainstream Coalition. Democrats broke the crime filibuster and passed the bill on Thursday, August 25, and the Senate staggered off for a two-week recess. The next day Mitchell conceded defeat. There would be no comprehensive health care bill that year.

Still, they did not quit. The next Tuesday, Kennedy called Littlefield from the Cape and said, "Get off the train we have been riding on for two years and get ahead of the one that is leaving the station." Kennedy said he was going to be the last one to give up. So Littlefield went to work with Chafee's aides and Mitchell's. They were making an important, incremental bill out of the Mainstream Coalition proposal. It included, as he told Haynes Johnson, "real insurance reform, with real subsidies, real choice and offering federal employee benefits" and some help on long-term care.

On Sunday, September 18, *The New York Times* quoted Packwood as having told other Republicans in August, "We've killed health reform. Now we have to make sure our fingerprints aren't on it." The next day Mitchell asked Chafee how many Republicans he could get to vote for cloture on the latest bill. Chafee came back and told him there were only four—not enough, because some of the fifty-six Democrats would defect. On Tuesday, Republican leaders threatened Clinton with blocking an international trade measure they supported if health care was not dropped. Kennedy, Rockefeller, and Daschle pressed Mitchell to force a cloture vote, to underline those Republican fingerprints. Over the weekend, he decided not to, choosing instead to get the session over as soon as he could to help Democrats running for reelection. He announced that decision on Monday, September 26.

Kennedy went to the floor then to renew his constant allegiance to *his* issue. "I will never give up the fight for health reform," he said, "until the working men and women of this country know that years of effort and hard-won savings cannot be wiped out by a sudden illness. The drive for comprehensive health reform will begin again next year. We are closer than ever to our goal and I am confident that we will prevail."

ALONG WITH THE bad news on health care that summer, there was one stunning piece of good news from Ireland. The Adams visa had been worth the gamble. While everything went very slowly, the IRA was carefully considering a cease-fire, while going on with occasional violence that repeatedly made some Americans wonder if they had been gulled by Adams. But a group of prominent Irish American businessmen—Niall O'Dowd, Bill Flynn, Bruce Morrison, and Chuck Feeney—pressed on all summer, promising to help Sinn Fein in the United States if the IRA announced a cease-fire and halted its violence. O'Dowd in particular got encouraging messages, but it was still not clear if the cease-fire might be hedged about with conditions.

By mid-August, the IRA had decided on a complete cease-fire. But to make that work it wanted another visa—for Joe Cahill, a hard man, a convicted murderer and gunrunner. The IRA believed that Cahill, seventy-four, was the only one with the standing to convince American supporters that the cease-fire was the right decision. Prime Minister Reynolds tracked down Ambassador Smith in the South of France, and she rushed back to Dublin.

Smith would not stop calling. She yielded nothing to Ted in determination, and she called the President and persuaded her brother to call him, too. Clinton called Reynolds, pointing out Cahill's violent record. Reynolds told him, "There aren't any saints in the IRA." He told the President he was "absolutely certain" there would be a cease-fire, and was back on the phone to Washington within hours saying the phrase "complete cessation" would be used. Clinton ordered the visa issued, and Cahill flew to New York on August 30.

The next day the cease-fire was announced. Clinton called Reynolds to congratulate him. Five minutes later, Kennedy got through to say that "all our efforts were vindicated," Reynolds recalled. Kennedy issued a statement saying, "Today is a joyous and hopeful day for all of Ireland and for all the Irish people. . . . I hope today's historic announcement marks the end of the bloodshed and the beginning of a process that will lead, as soon as possible, to lasting peace, reconciliation and prosperity for all the people of Ireland." Then he congratulated Hume, Reynolds, Major, Adams, and "Ambassador Smith in Dublin who has worked effectively to reach out to all sides in this complex issue and encourage greater understanding."

The next spring, Reynolds, who had been on the floor at Madison Square Garden when Kennedy made his great convention speech in 1980, credited Kennedy with mobilizing support for the two visas. They were essential to the cease-fire, he said, by establishing Sinn Fein "as a legitimate political party." They were also important psychologically because they showed "the British didn't rule the world."

But there was no joy in Senator Kennedy's campaign for reelection. There had been signs all year that it would be tough, starting in March when Michael Kennedy, Ted's thirty-six-year-old nephew and this year's campaign manager, was soliciting signatures from nominating petitions. "There was a nastiness out on the street that was palpable. . . . 'He is a bum, throw him out,' 'It is time to go,' 'I have been with him every time, not this time.' "

Some of that was about the Senator himself. Though his personal poll ratings had improved since the depths of 1991 and 1992, Palm Beach had scarred him. But more came from the year's singular unhappiness with Washington and particularly with Democrats. Clinton had promised in 1992 that if he was elected, he and the Democratic Congress would get things done. By mid-1994, that was not happening. Kennedy, after thirty-one years in the Senate, was an obvious target if voters wanted change. In the summer it became clear that Mitt Romney had definite strengths as a candidate. He looked fresh and youthful. In a successful business career, he said he had created ten thousand jobs. He argued that prepared him to create jobs in Massachusetts, where the economy was not yet out of the recession. But most of all, Romney was ready to spend a lot of money.

Despite the warning signs, the Kennedy campaign was running as though this would be like his previous walkovers. The Senator himself had not gone through his customary precampaign weight loss regimen. He knew it made him grumpy, and he did not want to be grumpy around his new family. At sixty-two, when his brothers had not reached their fifties, he was fat and tired, not the youthful figure of Kennedy myth. The weight increased the pain in his back, and he lumbered; a Romney commercial caught him lowering himself gingerly onto a bench, looking old. The *Washington Post*'s David Broder called him a "whale," referring to his importance in the Senate, but the double meaning was there. Some of his workers named their effort the "save the whale campaign."

The campaign itself was trying to save money that summer. When Romney launched a television blitz in August, leading up to the Republican primary but hitting Kennedy on crime and welfare, Kennedy was not on the air fighting back. There were three arguments for that policy. Kennedy had never run ads attacking an opponent in a Senate race, and was reluctant to start. Second, pollster Tom Kiley said that Romney would inevitably surge, that it was the nature of American politics in the nineties for voters to be restive and consider new faces. Probably the most important reason was money. Michael Kennedy said after the campaign, in August, "We had a bit of a cushion, but clearly we did not want to go into debt, and I did not want to show my uncle a plan that had him going into debt."

The Kennedy campaign did have some weapons in reserve. Vicki was proving herself adept as a campaigner, and her picture in commercials helped with the understated message of Kennedy's personal redemption. She also brought a sort of experience to campaign deliberations that no one else could offer. In her law practice, she said, she had dealt with venture capitalists. Sometimes they produced jobs, as Romney argued he had. Sometimes they cut payrolls, and the campaign had to investigate Romney's business, she insisted.

Vicki prevailed, and in late spring the campaign hired The Investigative Group, Inc., a detective firm headed by former Senate Watergate counsel Terry Lenzner. It probed Romney's company, Bain Capital, and discovered the Ampad stationery factory in Marion, Indiana. Romney and his fellow investors had bought it that July, and the management then eliminated 41 of the 265 blue-collar jobs, slashed wages, boosted insurance premiums, and junked the union contract, which had two years to run. Shrum was ready to do commercials on Ampad and on the claim, implied in Romney's ads, that health insurance came with jobs at his companies.

But other family members, past and present, were creating new problems. Joan Kennedy hired Monroe Inker, Boston's toughest divorce lawyer, to reopen the financial settlement of their divorce, and he went to court on September 6. He insisted she was not being "vindictive" and trying to take advantage of his campaign circumstances, but it was plainly a smudge on the picture of a happy new marriage the campaign was using. After Inker complained repeatedly that Kennedy and his lawyer, Paul Kirk, had ignored his client, she agreed to put off the case until after the election on October 4.

A current family member caused an even greater problem. The *Boston Globe* and John Lakian, a long-shot primary opponent of Romney's, had both questioned him about the role of his Mormon faith in his political philosophy, especially concerning abortion, homosexuality, and the standing of blacks and women.

The Kennedy campaign officials thought they had agreed not to touch the subject. But on Sunday, September 18, Representative Joseph P. Kennedy II, the Senator's nephew and the campaign manager's brother, had a column in the *Herald* saying Mormons treated blacks and women as "second-class citizens." The Romney campaign demanded that its foes repudiate those comments, and the Congressman told the press, "I made a mistake" because he had not known that blacks had been eligible to be Mormon priests for fifteen years. But Michael Kennedy said he would make no effort to rein him in. "I've never been able to control my brother, but I would not deny him, either," he said. Then Kennedy called Romney to apologize and sent a letter of apology to several newspapers. The Romney campaign said he was just trying to fan the issue.

It sounded like a typical Boston campaign fuss, for Joe was widely regarded as a hothead. When Andy Hiller of WHDH-TV pursued Senator Kennedy on the subject on Monday, September 26, Kennedy avoided the issue several times. Finally, challenged on whether he was leaving Joe out alone, the Senator said it was proper to ask Romney where he had been on the issue of blacks in his church before its policy was changed to recognize blacks in 1978. On the twenty-seventh, Romney held a news conference to say that Ted was abandoning his brother's lesson of 1960. "The victory that John Kennedy won was not just for 40 million Americans who were born Catholics," he said, "it was for all Americans of all faiths. I'm sad to see that Ted Kennedy is trying to take away his brother's victory." The *Globe* agreed editorially on the twenty-eighth, saying Kennedy "hurts himself and discredits his family history." On the twenty-ninth, Kennedy said, "I believe that religion should not be an issue in this campaign. The way to make that so is to stop talking about it."

One last problem: Kennedy had been in Washington, working on health care and education bills, and dashing up to Boston to campaign. He said soon after that he felt he "didn't do either one well," always worrying if he was spending too much time on one or the other. He told Senators he was afraid he might lose.

It was October 5 before Congress reauthorized the first major federal aid bill, the Elementary and Secondary Education Act of 1965. That bill had been aimed at disadvantaged children, but the formula for distributing what had reached $6.6 billion in appropriations spread the money so thinly that the law had more legislative support than real impact. Kennedy and the Clinton administration tried to change the formula to send more money to schools in poor areas, but made only modest progress. No Senator wanted his state to get less money; no House member wanted his district cut. In the end, Kennedy had rounded up twenty Republicans to help break a Jesse Helms filibuster aimed at fostering prayer in school. The bill passed, 77–20.

That was the last bill for what Kennedy called "the Education Congress," which also enacted direct student loans, a measure designed to involve business and labor to work with schools in preparing students for work, and the "Goals 2000—Educate America Act." That measure, enacted in March, was probably the most far-reaching, because it sought to set national standards of educational achievement. When the bill passed, Kennedy predicted it would offer "a framework for high academic standards, locally developed and implemented with our support." In practice, because of Republican concerns about rules made in Washington, the act played a major role in the development of statewide tests in many parts of the country.

Kennedy used the education legislation heavily in campaign advertising. But it also focused on local accomplishments, especially his efforts to get a Defense Department accounting school in Southbridge. That community, on hard times since the American Optical Company closed a factory in the mid-eighties, had won a nationwide competition under the Bush administration to get one of five major new accounting facilities, with up to five thousand jobs. But Clinton's Defense Department scrapped the plan in early 1993 just before the winners were to be announced, to make sure new facilities went to places that had larger bases closed.

Kennedy complained that Southbridge had played by the rules and had been treated unjustly. He lobbied the President and the Pentagon, and in May 1994, Defense decided it needed a school to train its accountants and Southbridge was the place. This offered perhaps 750 jobs, but it was so welcome that Florence Chandler, the town manager, made a campaign spot: "I called Senator Kennedy and he went to the President and to the Pentagon and he made the case for us over and over. He never gave up. Today in Southbridge there is hope and it was Senator Kennedy who made the difference."

STILL, THE SITUATION seemed grim when the Kennedy high command met at the Senator's apartment on Marlborough Street in the Back Bay on the evening of Sunday, September 18. Kiley discussed a poll he had completed three days earlier, showing that Kennedy's lead had shrunk by 23 points. In August, Kennedy held a 59–33 percent lead. Now it was 50–47, within the margin of sampling error. Depending on how the numbers were analyzed, Kennedy might even be a point or two behind. Kiley had expected Romney to pull about even after winning the September 20 primary, but this was worse news than he had expected.

One obvious response was to step up television advertising, and Kennedy agreed to spend whatever was needed. Ultimately, that meant taking out a $2 million mortgage on his home in McLean. As word that he was in trouble spread from public polls released in the next few days, friends like Tom Harkin started raising money for him. Hillary Rodham Clinton came to raise a quarter of a million dollars on September 23. Six days later, Clinton himself raised three times that much in McLean.

But the immediate, public news was bad. As Romney's supporters cheered his Republican primary victory on Tuesday the twentieth, he gave the speech of his campaign. Its sheer energy conveyed much of his message of change. He called Kennedy out of date, unaware that "the world has changed" in the thirty-two years since he first ran. When Kennedy went to the Senate, Romney said, "the big problem we had in our schools was kids chewing gum. Now the problem is, kids shooting guns and carrying guns to school. The answers of the past just aren't working anymore. . . . Thirty-two-year-old social problems like our welfare system that Ted Kennedy helped create just isn't working for us."

"Ultimately, this is a campaign about change," he said. "It's about the change in direction for this country and this state. It's about the answers and the solutions, recognizing that the world has changed. It's also the difference, it's change in the way we do business in Washington. . . . You have to learn how to identify problems, figure what the solution might be, build consensus among a lot of people, and then take action. But in Washington, people don't do that a lot."

In contrast, Kennedy was flat when he claimed his nomination at a hastily arranged appearance at a badly lit bar near his headquarters. He had spoken that morning at a funeral for longtime immigration aide Jerry Tinker, who had died suddenly. He had been negotiating over education. His plane was late, and he broke a small bone in his foot on the cobblestones outside the bar. Inside, he said the campaign would be about "who is better able to fight for the working families of Massachusetts." But on television that night, he didn't look like much of a fighter.

Things began to change the next morning. Starting at 5:30 A.M., new Kennedy commercials went on the air. An early-summer wave had touted

Kennedy's accomplishments, with a particular focus on health care, education, and jobs he claimed to have created or saved. This round attacked Romney, accusing him of a "negative campaign" marked by ads that were "misleading" and "deceptive." And then it hit him with the accusation that his companies provided no health insurance for "most American workers," or, in another version, "many American workers." Referring to Romney's own gains, it concluded, "Eleven million dollars for himself, no health insurance for American workers, and a misleading campaign for the Senate."

Those ads were followed on September 29 by a battery of devastating spots that put to work Lenzner's research with the workers who had lost their jobs in Indiana after Romney's firm took over their company in Indiana.

One woman said, "I would like to say to Mitt Romney, if you think you'd make such a good Senator, come out here in Marion, Indiana, and see what your company has done to these people." A man said, "I'd like for him to show me where these ten thousand jobs that he's created are." Another man said, "If he's creating jobs, I wish he could create some here instead of taking them away." As his own polls began to show that the ads were hurting him, Romney condemned their "continuous distortions." He said, "I have lived my entire life to help build businesses and to help people."

The ads were bolstered by a striking homecoming of former Kennedy aides. On October 1, Ranny Cooper took time off from the New York public relations job for which she had left Kennedy's office in December 1992 and gracefully started managing the campaign, a shift accomplished without affronting Michael Kennedy, who shared his office with her and spoke for the campaign while she remained her unquotable self.

A few days later another former staff chief, David Burke, puzzled by Kennedy's use of the Mormon issue, called around and learned that Kennedy was in some trouble. Burke, now a wealthy Dreyfus Brothers executive, called Kennedy and offered to help. A few days later, Kennedy called back. Burke, who had headed ABC News and CBS News after leaving the Senate, thought Kennedy was going to ask him to help prepare to debate Romney. Instead the Senator asked him to ride with him around Massachusetts, helping keep him on schedule, on message, and relaxed. Burke started October 10. Cooper and Burke were the most dramatic additions, but dozens of former aides turned up for a weekend or a week, to advance an event or knock on doors.

Cooper and Burke walked into a fight within the campaign over whether to debate Romney. Kennedy's campaign had accepted two invitations from colleges at hours when few would be watching, and demanded that Romney agree to them. Romney wanted better times and a chance for candidates to challenge each other directly.

Shrum opposed giving any ground to Romney. While the press still treated the race as a toss-up, he felt the new ads and Kennedy's in-state efforts had put him in control and saw no reason to change the equation. Some of Kennedy's aides in Washington also opposed a debate, fearing that Kennedy would do badly. Burke argued that they were foolish to worry, that Kennedy would do fine. Cooper worried that if there was no deal, the only thing on which television would quote Kennedy would be why he wasn't debating—hardly the campaign's message of choice.

On Wednesday, October 12, Boston's newspapers forced his hand. The *Globe* and the *Herald,* which hardly ever work together, jointly invited the candidates to two debates on Boston, on October 25 and November 1. Kennedy agreed to the first, but insisted on a western Massachusetts debate at Holyoke on October 27, and Romney agreed to that.

Kennedy was enjoying himself on the campaign trail. In Worcester he brought oversize checks to a biotechnology firm and city officials to represent federal money he had helped get them. At a GTE plant in Taunton, he told how much being on the Armed Services Committee meant to him, and how his commitment to advanced military technology meant jobs at the plant. In Boston, he grumbled about poor advance work when a women's road race blocked a street, but then blushed when many of the runners swerved in his direction to cheer him on. They seemed to agree with *Globe* columnist Ellen Goodman, who wrote, "At the polls, I'm not looking for a husband, but a policy-maker."

He chatted easily with reporters at lunch about Ireland, and then gave an old-fashioned stump speech at the Dartmouth branch of the University of Massachusetts. The election, he told students, "isn't about me. It's about who is going to represent you, whether that representative is going to fight for jobs, whether that representative is going to fight for education, whether that representative is going to be able to go to the Senate of the United States and has the experience and has the know-how to make a difference for Massachusetts. That is the issue. Am I going to have your help?"

It was an election about him, but he was answering the questions voters had, especially whether he was washed up, and perhaps never better than in Faneuil Hall on Sunday, October 16. In a thirty-minute speech that lasted nearly an hour because it was interrupted fifty-seven times for applause by an overflow crowd of Democrats, he said, "I stand for the idea that public service can make a difference in the lives of people." Promising to fight for universal health care, he said, "Our opposition constitutes an honor roll of reaction—Reagan, Bush, Gingrich, and Romney. They have ignored, resisted, and delayed, but in the end they will not defeat the basic justice of guaranteed private health insurance that protects working families and the middle class from increasing costs and arbitrary denial, and sudden cancellation."

Kennedy promised more help for education, a higher minimum wage, and more jobs. "I will stand up in the Senate with President Clinton—not with Phil Gramm and Bob Dole. I will help to break filibusters against progress, not join filibusters. I will gather support on a bipartisan basis, just as I did on the education bill. But I will never surrender to the intense, intemperate, irresponsible obstruction of those who oppose anything and everything that President Clinton decides to propose."

Clinton, perhaps more popular in Massachusetts than anywhere else, helped too. He came to the John F. Kennedy Gymnasium at Framingham High School on October 20 to sign the Elementary and Secondary Education Act and then to cheer Kennedy on at a rally.

But Kennedy's next appearance at Faneuil Hall was plainly going to be critical. The first debate was Romney's last chance to halt Kennedy's gains. The reluctance of the Senator's campaign to debate helped stamp Kennedy as an underdog. One *Globe* story said debates "are widely seen as fraught with danger for the aging, sometimes tongue-tied Kennedy." The *Herald*'s Howie Carr called him "incoherent" and suggested his understanding of "a 'sound economic policy' means only buying every fourth round."

OUTSIDE THE HALL, Romney's suburban supporters carried signs, but union Democrats were there, too, occasionally slamming their sticks down on the Romney people's feet. Inside the hall, the candidates stood behind wide lecterns, insisted on by Kennedy's side to hide his girth.

In the first part of the debate, with reporters asking questions, Romney seemed smoother, though his description of a Senator's duties included "traveling from state to state and around the country to bring jobs to Massachusetts." He surprised Kennedy by telling him that the Merchandise Mart, which the Kennedy family owned, provided no health insurance for part-time workers, just like many of Romney's businesses.

Though Kennedy's supporters booed when a reporter asked how he dealt with personal failings, Kennedy was ready for it. He recalled his Harvard speech and said, "Every day of my life I try to be a better human being, better father, a better son, a better husband, and since my life has changed with Vicki, I believe the people of this state understand that the kind of purpose and direction and new affection and confidence on personal matters has been enormously reinvigorating. And hopefully I am a better senator."

He was also ready when Romney asked him about a story in the *Herald* charging that the Senator had profited from a no-bid contract with Washington's Mayor Marion Barry under which minority ownership rules were waived. Kennedy said he knew nothing about it because his finances were in a blind trust, then hit the issue out of the park, saying, "Mr. Romney, the Kennedys are not in public service to make money. We have paid too high a price in our commitment to the public service." Romney, incredibly, did not let it drop but complained that Kennedy brought up his family too often.

As the debate wore on, Romney seemed increasingly rattled and Kennedy increasingly confident. Kennedy hammered Romney for talking about health care without knowing what his plans would cost. Romney complained that he did not have the Congressional Budget Office to figure it out, as Kennedy did. When Kennedy challenged him on how he would balance the budget, he tried to be sarcastic about how it would be "wonderful to take it through piece by piece." Kennedy snapped, "That's what you do as a legislator."

After that debate, the race was over. Romney did a little better two days later in Holyoke, but it didn't matter. Kennedy had already reassured Massachusetts.

His campaign pace slowed a bit, but there were still interesting events. On October 28, the Jewish community threw a thank-you rally for him at Brandeis University. It focused on his success in getting refuseniks out of the Soviet Union. Many were there, including Natan Shcharansky. Boris Katz recalled, "Sixteen years ago I was in Moscow. It was a desperate time. Senator Kennedy intervened on our behalf. He helped get our freedom." It was one of the best days of the campaign for Kennedy personally. He thanked the refuseniks for coming: "Knowing them has touched my heart with fire. They are extraordinary men and women."

The next day, Ted and Vicki gave a long interview to Natalie Jacobson of Boston's WCVB-TV, and along with the usual political arguments, he offered some unusual insights. She asked him if he had been particularly lonely before Vicki. He answered, "Probably. I don't think I probably knew how lonely I was until I met Vicki. I had no idea of ever having a serious relationship like this in my life. And all of that went aside when I met her, and she became a part of my life."

She asked him what he was least proud of in his career, and he stressed Vietnam: "I wish I had been more aggressive and more active in ending the war sooner."

One of the few answers that came easily and quickly was when she asked, "What do you think are your personal strengths?" He replied, "Perseverance."

And Kennedy, a continual public supporter of Clinton, offered the mildest of criticism after praising the President on health care, on gun control, and on Ireland, and for his decisiveness and determination. He said, "I suppose the only suggestion I would have is governing. I'd like to see him govern. . . . Governing is what people would like to have him do. I think he is so interested in so many issues, and thinks he can make a contribution to so many different issues." She asked, "Do you think he needs to focus?" He said, "That is something I would like to see him do more of."

On Election Eve, Kennedy spoke at the Harry S. Truman Society's outdoor rally in West Roxbury, a successor to the G&G delicatessen event he had shared with his brothers in 1958. In theory, candidates spoke in the order they signed in, but Kennedy was invited to jump the line. He said, "I am looking forward to the outcome of that election tomorrow, and I want to tell every one of you, that the day after tomorrow, I am going to begin again to fight for a health care reform, to fight for jobs and programs. I am not going to sleep, I am going to continue that fight until there is the kind of economy, the kind of jobs, the kind of education, the kind of health care, the kind of respect for the individual that is so much a part of our Democratic tradition. Will you help me?"

They did. The next day Kennedy got 1,265,997 votes to 894,000 for Romney, 56.7 percent of all ballots cast and 58.6 percent of the major-party vote. Kennedy carried voters of all ages and all income and educational groups, and 63 percent of women as well as 53 percent of men.

CHAPTER 35

The Counterrevolution

Kennedy won reelection in 1994, but he and Senator Chuck Robb of Virginia were the only Democrats who won high-profile races. The Speaker of the House, Tom Foley, was defeated. The three-term governor of New York, Mario Cuomo, was defeated. Jack Brooks, the House Judiciary Committee chairman, was defeated. Jim Sasser, the Senate Budget Committee chairman who had seemed likely to succeed Mitchell as Democratic leader, was defeated. Republicans won every open Senate seat and entered the 104th Congress with a 53–47 majority. Kennedy lost his chairmanship to Nancy Kassebaum. But the most spectacular change was in the House. Republicans gained fifty-three seats and took control for the first time in forty years, and Newt Gingrich became Speaker. For Ted there was one bright spot in the House: Patrick was elected as a Representative from Rhode Island.

For Democrats generally, the result was a disaster. President Clinton went into a deep funk, seeing the election as a referendum on him and listening as his old Democratic Leadership Council colleagues said the problem was that his administration was too liberal.

The obvious interpretation was that the country had moved to the right, and House Republicans claimed they had a revolution. Certainly the voting electorate had moved to the right, as conservative Republicans turned out in higher numbers than usual while many Democrats stayed home. And Clinton was a piece of that, though probably not as much as Congress itself. Just as the 1992 election had not been a move to the left as much as a rejection of Washington's leadership, which that year meant President Bush, the 1994 election was a rejection of the only Washington target the voters could hit directly—a Democratic Congress. Still, if the Republicans kept their promises—balancing the budget and generally reducing the role of government—better than the Democrats had kept theirs, they could solidify their hold on Congress, capture the White House again (dismissing the Clinton Presidency as a Perot-driven accident), and indeed solidify a conservative shift.

Kennedy could not accept the idea that there had been any decisive ideological shift. He talked with professors and politicians, probing the meaning of the election, and came up with this interpretation:

1. Low turnout, with less than 40 percent of voting-age Americans voting, enabled Republicans to capture Congress with the votes of about a fifth of the public. "Some mandate!" as he put it in January to the National Press Club.

2. Democrats lost not because they failed to shift with the political winds, but because they tried to shift. His own victory, sticking to his guns, was the right course. (He never acknowledged that Massachusetts was a special case.) "If Democrats run for cover, if we become pale carbon copies of the opposition, we will lose—and deserve to lose," he said at the Press Club. He revived a line he had used in 1981: "The last thing this country needs is two Republican parties."

In that Press Club speech, he defended Clinton against critics like former Representative Dave McCurdy of Oklahoma, saying, "Blaming Bill Clinton by some in our party comes with ill grace from those who abandoned him on critical votes on the last Congress, then ran from him in the campaign—and then lost, often by wide margins. Now they come forward with a strategy discredited by their own defeats."

Before he made those arguments publicly on January 11, Kennedy had made the same arguments to an audience of one—President Clinton. Fearing that Clinton would heed the Democratic Leadership Council view and try to get ahead of the Republicans with budget cuts of his own, Kennedy went to the White House and met with the President in the Treaty Room on December 13. Clinton's description in 1999 of what Kennedy told him at that meeting in 1994 conveys both what Kennedy said and how personally Clinton took the election:

"When I got beat—when we lost the Congress, I mean—he was not at all despondent. He said, 'You know, we've just got to live with this. They convinced people that the health care system was too much government and no one has felt the benefits of your economic plan yet.' And he said, 'But we've still got a lot we can get done.' And he said, 'We can articulate the differences and require them to deal with you, and get us to stick with you.'

"He said, 'Just stay with it, stay with what you believe in for children and education.' And he said, 'Let's take some incremental health care.' He said, 'I even think we can raise the minimum wage.'

"Somebody telling me that, I thought they had a screw loose. But he was right. I mean, he was right.

"One thing we have in common is the way we keep score on ourselves is whether we're actually doing something. This is not a position for us as much as it is a mission. My despair was that I was afraid after the '94 election that I wouldn't be able to do what I was elected to do. And he said, 'Let's just take a deep breath, settle down, and go back to work.' "

Kennedy told the President that the issues that had worked for him in the election in Massachusetts could work for the President for the next two years. Kennedy told Clinton he had sold himself as a fighter for working families, and Clinton could, too. With a mind toward the conservative advice that was starting to come to Clinton from pollster Dick Morris, Kennedy said a more traditional agenda would get the Democratic base of unions, minority groups, women, gays, educators, and the health care community behind him, while it would be hard to succeed in 1996 without their enthusiastic support.

After he argued his case, Kennedy left behind a copy of the talking points he had worked out with Nick Littlefield and Carey Parker. One urgent paragraph said:

"*Your budget is a political document.* The budget is a political document, not a policy document. Accordingly, it should avoid making cuts in core programs like jobs and job training, education and health care which would neutralize Democrats' ability to attack expected Republican cuts in these programs. Consider every proposal in terms of themes of whose side we are on. You may have to veto reconciliation bill, if Republicans pack whole agenda into it, as they did in 1981. Need to have prepared a political case to justify veto, if necessary."

Some specific points included:

"Education. Resist all cuts in student aid. Fighting to keep higher education (both four year and two year community colleges) affordable is a great middle class issue. Make every dollar taken away from students a dollar taken by Republicans.

"Republicans will cut Medicare for balancing the budget or for tax cuts and 'no cuts in Medicare except for health care reform' will be a great 'wedge' issue if we can keep distinction clear.

"Democrats are for higher wages and new job opportunities. Republicans are for cuts to pay for tax breaks for the rich.

"The DLC and Bob Reich's ideas on reducing corporate welfare and tax expenditure are worth looking at—especially those that can be characterized as 'corporate welfare.' Even if the savings turn out to be relatively small, they are an effective antidote to Republican plans."

It is hard to measure the impact of any single conversation on someone who listens to as many people as Clinton. But this one seems to have been significant. Clinton once said it "helped me never to despair." George Stephanopoulos, who was fighting the same fight inside the White House, said at the time, "Clinton was going back and forth. It pushed Clinton away from Medicare cuts." Leon Panetta, then Clinton's chief of staff, said Kennedy "always left an impression on the President." The President's "guts were with Kennedy," and "Anybody who could buck up his spine for a good fight on the issues that count for Democrats deserves some credit in that very rough period."

Two days later, Clinton gave a speech in which he proposed a series of steps which he called a middle-class bill of rights, and talked of spending cuts in the Departments of Energy, Transportation, and Housing and Urban Development—not in Education or Health and Human Services. He said he would not support proposals "that weaken the progress we've made in the previous two years for working families, ideas that hurt poor people who are doing their dead-level best to raise their kids and work their way into the middle class." That attitude lasted for some time. In his State of the Union address on January 24, he said his budget "protects education, veterans, Social Security and Medicare."

BEFORE THAT FIRST marker on how Clinton would approach the Republican Congress was laid down, Rose Kennedy died. Four of her surviving children, but not Rosemary, and dozens of grandchildren gathered at the house in Hyannis Port, and were there with a priest when she died of complications of pneumonia on January 22 at 5:30 P.M. She was 104.

The next day, they told affectionate stories about her, with the children and older grandchildren dominating because they remembered her from the years before she suffered debilitating strokes in the early eighties. Some of those recollections, later written for a book so the youngest would have something to remember, focused as much on Ted as on his mother. Douglas Kennedy, Robert and Ethel's tenth child, recalled his uncle Ted and his grandmother at the Cape: "Every morning he would structure his day around her. He was constantly talking to her. Tell her the whole story of the day in a jolly, gregarious way, and he is describing to her what happened. That made me as proud of being a member of this family as anything."

On the twenty-fourth, a chill morning, thousands lined the roads and looked down from overpasses as her funeral cortege brought her seventy-five miles, from the Cape back to St. Stephen's Church in the North End, where she had been baptized in 1890 in the Presidency of Benjamin Harrison. Several hundred mourners stood outside in the cold, listening to the service through loudspeakers.

Like the obituaries published the day before, Bernard Cardinal Law stressed her connection to the nation's history. In his homily, he said, "Few lives have been so intertwined with the joys and sorrows of our nation's life as has hers." The setting of St. Stephen's was as historic as the occasion, epitomizing the annals of Boston. America's first great architect, Charles Bulfinch, designed the redbrick church in Italian Renaissance style for Congregationalists in 1802. A church member, Paul Revere, cast its bell. When the potato famine drove the Irish to Boston, the Congregationalists built a new church on Beacon Hill and sold the old one in 1862. The next year Rose's father, John F. Fitzgerald, was baptized there. By the time Rose was born, Italian immigrants dominated the North End. The church was decorated with paintings of saints. In 1965, Richard Cardinal Cushing had it restored to Bulfinch's austerity. The Stations of the Cross, carved in walnut, were the only adornments on the walls. The only color, as the mourners entered, came from abundant bouquets of red and pink roses.

The absent sister was recalled by Eunice Kennedy Shriver, who said Rose "would never let us forget our special sister, Rosemary." Cardinal Law read an Apostolic Blessing from Pope John Paul II. Kara was a pallbearer. Vicki and Teddy were among the relatives who read from the gospel and from Rose's writings. Teddy quoted her autobiography, *Times to Remember:*

"The most important element in human life is faith. If God were to take away all His blessings, health, physical fitness, wealth, intelligence and leave me with but one gift, I would ask for faith—for with faith in Him and His goodness, mercy and love for me, and belief in everlasting life, I believe I could suffer the loss of my other gifts and be happy."

Ted stressed her faith, too. "She was ambitious not only for our success," he said, "but for our souls."

He began his eulogy with a funny story, about a letter she had sent him correcting his grammar in an interview: " 'Dear Teddy,' she wrote in the note, 'I just saw a story in which you said: "If I *was* President . . ." You should have said, "If I *were* President . . ." which is correct because it is a condition contrary to fact.' "

He said, "She sustained us in the saddest times by her faith in God, which was the greatest gift she gave us—and by the strength of her character, which was a combination of the sweetest gentleness and the most tempered steel."

"Jack once called her the glue that held the family together," he said. "We learned a special bond of loyalty and affection—which all of us first came to know in the deep and abiding love that Mother shared with Dad for fifty-seven years. From both of them together, we inherited a spirit that kept all their children close to each other and to them. Whatever any of us has done—whatever contribution we have made—begins and ends with Rose and Joseph Kennedy. For all of us, Dad was the spark. Mother was the light of our lives. He was our greatest fan. She was our greatest teacher." His voice broke, as it did when he said he expected that, in heaven, she would "welcome the rest of us home someday." He said, "Mother's prayers will continue to be more than enough to see us through."

After the casket, covered by a white pall with a purple cross, was returned to the hearse, he stood outside shaking hands with hundreds of the mourners, and then went with his family to a private burial service at Holyhood Cemetery, where Rose was buried beside Joe.

Before the eulogy, Ted was able to take Communion from Cardinal Law—and to be seen taking it by a vast television audience—because the Catholic Church had granted him an annulment a couple of months before. He and his office never discussed it, but Joan said years later she had not opposed it, and that the ground Ted had cited was that his marriage vow to be faithful had not been honestly made.

THE NIGHT OF the funeral, President Clinton gave his State of the Union address. As he had told Kennedy he would the night before when he called to offer condolences, Clinton proposed an increase in the minimum wage. He said, "You can't make a living on $4.25 an hour." While he did not propose a specific increase, he observed that by the end of the week, members of Congress would have earned as much since January 1 as a minimum-wage worker would make by December 31.

Clinton had promised an increase in the minimum wage during his campaign, but labor agreed to forgo it to help pass health care reform. Since that had died, the wage increase was back on the table, but not sure of even Democratic support in Congress.

Kennedy had backed his friend Chris Dodd's last-minute candidacy to be minority leader in place of Sasser. Dodd's supporters said his experience and combativeness would be needed in the minority. But Tom Daschle of South Dakota had been running since March, and his supporters said he would bring change. Daschle won by a single vote on December 2. Undaunted, Kennedy went to work on Daschle—with whom he had worked well on health care on the Senate floor—to sell him and his team on the Kennedy view of what Democrats should stand for.

Some of them thought that the idea of proposing new legislation was unrealistic, and that the best they could do would be to defend against Republican initiatives. Some thought the minimum wage was absolutely the wrong political signal, identifying the party once again with the poor.

On January 31, 1995, Kennedy came to defend his wage idea to a joint meeting of House and Senate Democratic leaders. It was on the House side of the Capitol, and he was late. The issue was being discussed when he walked in. He was prepared for objections from Senators like John Breaux of Louisiana and Wendell Ford of Kentucky. But he got angry when he heard John Kerry, his junior colleague from Massachusetts, offer doubts. "If you're not for raising the minimum wage," he said, "you don't deserve to call yourself a Democrat."

That sort of appeal, and less emotional arguments about how polls showed the public really thought people who worked deserved decent wages, carried the argument. Daschle said he would support an increase if Kennedy got more conservative Democrats to support it. So it was up to Kennedy to get a package the leaders, the White House, and labor would all support. In a couple of weeks, he brought them together, not behind his original proposal of $1.50 over three years, with indexing, but 90 cents over two years, without indexing.

By mid-March, while House Republicans were steadily rolling over the minority and passing the programs in their "Contract with America," Senate Democrats were finding their feet. First, with Robert Byrd leading the fight, on March 2 they defeated a constitutional amendment to require a balanced budget.

Their next victory came on a labor issue. Republicans tried to overturn an executive order by Clinton barring companies that hired strikebreakers for permanent jobs from federal contracts.

Kennedy led a filibuster, making it clear the fight was about politics. On March 15, he lumped the striker measure with a variety of House bills, cutting the school lunch program and college aid and threatening wage levels on federal projects:

"What is it about working families that Republicans have it in for them? Why is it that our Republican leadership in the House of Representatives and here today on the floor of the U.S. Senate, virtually in lockstep, wants to deprive them of some legitimate rights? . . . What is it about their children that we want to cut back in terms of Medicaid? What in the world have they done, except be the backbone of our country? . . . Make no mistake about it, this is the first battle . . . and we are not going to let this stampede that may have gone over in the House of Representatives run roughshod in the U.S. Senate."

Every Republican but Jeffords voted for cloture, and so did five Democrats, but with only fifty-eight votes and no prospect of getting to sixty, Dole gave up and shelved the striker measure.

FOR WEEKS, SENATE Democrats had been worrying about how to deal with the Republicans' first effort to cut domestic spending, a bill to take back money that had been appropriated the previous fall. It would rescind some of those appropriations, and it was called a rescission bill. On March 16 the House passed a bill which rescinded $17 billion in appropriations for the current year, hitting heavily at education.

All through February, Kennedy urged Democrats to resist all cuts in education. In late March, Daschle brought the caucus behind a plan to restore $1.3 billion in cuts affecting children. It put money back into seventeen programs—for child care, Head Start, national service, the Women, Infants, and Children feeding program, and a variety of school aid programs. On March 31, Daschle said whatever had to be cut, it was essential that "we protect our children and the investment that we need to make in children." Kennedy followed, citing the specific programs where money would be restored. "When you have decent, good, effective education programs and you cut back on them," he said, "you are basically increasing social costs and decreasing revenues in the long term for this country. It makes no sense at all."

Daschle's amendment might have carried. Dole could not be sure how many moderate Republicans would vote with the Democrats, at least on some parts of the proposal. So Dole withdrew the rescission bill so he could negotiate with Daschle. Instead, he called up a tax bill on which the House had refused to include a $3.6 billion Senate measure that sought to prevent wealthy Americans from abandoning their citizenship in order to avoid taxes. Kennedy asked why Congress was "preserving $3.6 billion for a dozen very wealthy individuals who renounced their citizenship, and we have no time for these hard-working Americans. . . . They are playing by the rules and are honored to be citizens of this country. But we have no time to consider them. . . . That is shameful. . . . It is wrong, it stinks."

In the end, Dole agreed to support $800 million of Daschle's $1.3 billion in restored spending. He was anxious to get the bill passed, the Senate adjourned, and himself to Russell, Kansas, the next Monday to announce his third try for the Presidency.

BUT EVEN WITH some Democratic successes, the President's pollster, Dick Morris, kept telling him that Republicans were winning a major propaganda war. After the Senate killed the constitutional amendment to require a balanced budget, House Republicans decided to balance it on their own. Morris saw that goal as more important than any of its components, like Medicare or college aid, and warned Clinton that not making a credible promise to balance the budget could cost him reelection.

Morris was almost unique in the polling business. He had both Democratic and Republican clients, including Senator Trent Lott, the conservative Mississippian who had ousted Alan Simpson as deputy leader the previous December. Consequently, he enjoyed little respect among professionals of either party, who saw him as the stereotype pollster with no principle but winning, someone who thought politicians should use polls to decide what they believe in. But he had helped plot Clinton's way back into the Arkansas Governor's Mansion in 1982. The President trusted him and brought him into the White House mix in the fall of 1994.

The budget argument became intense in May. Morris pushed for a specific commitment by Clinton to balance the budget in a specific number of years. Aides like Stephanopoulos argued this would be seen as treachery by Congressional Democrats. On May 19, the President told a radio interviewer he was going to propose a timetable for a balanced budget. But he delayed while waiting for figures he could defend, which frustrated Morris. Daschle and Richard Gephardt, the House Democratic leader, met with the President and urged him not to do it. For once, Kennedy could not get an appointment to see the President. So on Thursday, June 8, he went to a routine White House ceremony honoring police officers that he would normally have ignored. Kennedy took a front-row seat and caught the President's eye, and Clinton invited him to talk in the Oval Office.

Clinton recalled that Kennedy argued that the President's proposing a balanced budget would blur the increasingly clear choices that were developing between the parties, especially since it could not be done without cuts in Medicare and education. Clinton argued that he was only doing on the budget what Kennedy himself had done in the seventies and eighties with anticrime measures, getting a political monkey off his party's back. Kennedy said, according to Clinton, "It will look like two Republican parties and people will vote for the real one." The talking points Kennedy had brought with him said Democrats were gaining support, but "if you change signals now, you'll leave us high and dry in Congress. You'll give the Republicans cover. You'll help them get off the defensive and go on the offensive again. Inevitably, they'll claim that you're finally on board."

A new budget would become only a starting point for further compromise and retreat, Kennedy argued. And, with an argument aimed against Morris's idea of "triangulation," Kennedy's talking points warned, "I know there are some who want you to run *against* liberal Democrats as well as conservative Republicans. But we're your base. This somersault would be especially resented, because just when we feel we have a solid majority of the country on our side, you yank it out from under us."

Clinton said that the budget issue enabled Republicans to block many domestic policies; once he had it behind him, he could be the progressive President that Kennedy was always urging him to be. Clinton said that once he had a proposal out there, he would be credible when he attacked Republicans over Medicare. Kennedy said he would hold him to doing just that.

Kennedy left after an hour, discouraged. He told Littlefield he did not think the President would make a balanced budget speech "this week."

When the President did speak the next Tuesday, he proposed balancing the budget in ten years. He contended he could cut $128 billion from Medicare, with only hospitals and doctors, not patients, losing anything.

Kennedy issued a brief, tepid statement saying Clinton's plan was "far better" than the Republicans'. But Kennedy said the Medicare and Medicaid cuts left him "concerned that any significant reductions in these programs without health reform will impose unacceptable burdens on senior citizens and the needy."

That comment was no rave like John Breaux's "The President has been reborn," but it was much gentler than the comments of House Democrats. Representative David Obey of Wisconsin, the senior Democrat on the Appropriations Committee, said, "I think most of us have learned that if you don't like the President's position on a particular issue, you simply need to wait a few weeks." (His Republican chairman, Bob Livingston of Louisiana, agreed.) Patricia Schroeder of Colorado said, "He's thinking about himself and Presidential politics." Rosa DeLauro of Connecticut called it the "stupidest" thing Clinton had done.

Clinton has said that he thinks history proved him right on the question of offering a new budget plan then, but he makes two favorable points about Kennedy's protest. One is that it is hard for Presidents to find people who will tell a President something they think he does not want to hear. "I appreciated the fact that he was so aggressive about it," Clinton said. The second is that Kennedy kept his complaints between the two of them and did not join other Democrats in piling on, which he said was unusual in Washington, where "everybody knows that one good way to be quoted in this town is to hit the President."

BESIDES RAISING THE minimum wage, which in 1995 was more a Democratic statement than a potential law, Kennedy advanced one other important initiative that year. This one was bipartisan and could be enacted.

Kennedy had sat next to Nancy Kassebaum through nearly fifty health care hearings and nearly three weeks of committee votes in 1994. When the election reversed their positions, she was now chairman and he was the ranking Democrat on Labor. He set out in December to develop legislation that would seize on their areas of agreement. They talked for months, but found their surest consensus was in changing some of the rules of health insurance. Their bill guaranteed Americans the right to buy insurance, though not the rates, but said any policy a company sold had to be available to everyone. It limited how long an insurer could deny coverage for a specific preexisting medical condition. The limit on exclusions for preexisting conditions made health insurance more portable. Existing practices kept people from leaving jobs with health insurance because they could be denied coverage even if their next employer had a similar plan. A major impact of the bill was on big companies that insured themselves. States were barred from regulating their insurance, which covered about 60 percent of all workers who got insurance from their jobs, but the federal government could.

The bill was introduced on July 13. Senator Kassebaum, who said she expected opposition from the insurance industry, said, "While insurance alone will not cure all of the ills in the nation's health care system, it will require private insurance companies to compete based on quality, price and service instead of by refusing to provide coverage to those who are in poor health and who need it most."

Kennedy, who in the past had argued against a piecemeal approach to health care, said, "This bill will make life better for millions of people." A couple of weeks later, when he was asked at a breakfast with health reporters why he was backing a partial measure, he grinned and replied in a stage whisper, "I'm eating a little crow about incremental health care."

Senator Kassebaum pushed the bill out of the Labor Committee quickly, on a 16–0 vote on August 2. But she was right about insurance industry opposition. Insurers complained that requiring them to insure everybody would force rates up. Acting for them, various Republican Senators blocked the bill with "holds" (a signal that a Senator would deny unanimous consent to bring a measure up and at least a theoretical threat to filibuster). No more of it was heard in 1995.

The first session of the 104th Congress ended chaotically, but with a solid defeat for Republicans, who had begun it with the highest hopes.

Convinced they had won a national mandate, rather than merely an opportunity, the Republicans overreached. Speaker Newt Gingrich talked of shutting down the government if Clinton did not bend to Republican wishes.

And his party called for $270 billion in cuts in the anticipated growth of Medicare spending, a sum embarrassingly close to the $245 billion it proposed in tax cuts. That enabled Democrats to portray them, as Kennedy did on October 25, as trying "to take from the needy to give to the greedy . . . , to propose these deep cuts in Medicare to pay for tax breaks for the wealthy." Gingrich gave Democrats more ammunition the next day when he said Republicans expected that under their legislation, traditional fee-for-service Medicare would "wither on the vine."

But none of their substantive policies cost them as badly as shutting down the government. With all the legislative restrictions that House Republicans put on regular spending measures, on issues from abortion to the environment, only two of the thirteen appropriations bills had been passed by the beginning of the new fiscal year on October 1. A stopgap spending bill was passed to keep the government running until November 13. But when that one ran out, House Republicans insisted on adding new elements, like higher Medicare premiums, to a new version that also required Clinton to commit himself to a balanced budget within seven years. Clinton immediately vetoed it, along with an extension of the national debt limit similarly decorated with unusual riders such as limits on how the Treasury could shift money between accounts—Secretary Robert E. Rubin's way of avoiding a default on government bonds.

Most government departments then shut down most functions for six days, laying off 800,000 workers. Many Republicans talked casually about how the public would never miss it. They were wrong. New Social Security and passport applications were not taken. Tourist attractions like the Washington Monument and national parks were shut. Some medical research halted. Overwhelmingly, polls showed the public blamed the Republicans. Then in a supremely egotistical moment at a late-November breakfast with reporters, Gingrich complained that he had been treated badly when President Clinton took him along to Yitzhak Rabin's funeral in Israel. He claimed that he and Dole had been snubbed by the President on the flight (though a photograph showed him with the President at a card game aloft) and forced to exit by the rear door. He said he knew it was "petty," but he retaliated by inserting provisions unacceptable to Clinton in the stopgap spending bill. The *New York Daily News* cartooned him on its front page as CRYBABY, in diapers, and Democrats carried copies onto the floor for the television cameras.

Although the government reopened after six days when a new, uncluttered stopgap spending measure was passed, it shut down again on December 16. Whatever electoral value Republicans had found in running against the federal bureaucracy, the 250,000 laid-off workers were somebody's neighbors, going without paychecks at Christmas, and good material for human-interest stories in newspapers and television.

House Republicans would later complain that it was Clinton who shut the government down by vetoing their bills. But at the time they were cheerfully playing a very high-stakes game of chicken, believing that a President known for changing his stance and backing down on some nominations would cave in. But, as Gingrich acknowledged in January, Clinton and his aides were "tougher than I thought they would be."

The White House started running television ads attacking the Republicans, and in January they backed down, after a three-week closure, and negotiated stopgap spending measures that eventually ran for the rest of the fiscal year.

WITH THE BUDGET fight over and Democrats more confident, there was a moment for Kennedy and Clinton to return to health care. The President called for enactment of the Kassebaum-Kennedy bill in his State of the Union address on January 23, 1996.

His emphasis led to scheduling the bill for action, but not quickly or directly. Encouraged by Kassebaum, Judith Havemann of the *Washington Post* started writing articles day after day about the "holds" which the Health Insurance Association of America had gotten Republican Senators to put on the bill.

Kassebaum and Kennedy went to the Senate floor on January 25 to call for action. She described the bill and said she was "absolutely committed to passing health insurance reform legislation this year—either as a freestanding bill or as an amendment to another vehicle."

Kennedy was blunter. He said, "The only opposition to this bill comes from those who profit from the abuses of the current system. . . . A few Senators have placed secret holds on the bill in an attempt to kill it. They know that if the legislation is brought to the floor of the Senate, it will pass overwhelmingly. . . . So I join Senator Kassebaum in urging Majority Leader Dole to bring this bill to the floor. It is time to break the log jam. The American people deserve action—and they deserve it now."

ABC's *Nightline* program caught Dole campaigning in New Hampshire on January 31 and asked him why he did not call up the bill. He said there were "lots of holds on it." He said he did not know why. But the next week Dole gave in and agreed to call the bill up by May 3.

BUT REPUBLICANS GAVE no ground on the minimum-wage legislation. Kassebaum held one hearing on December 15, 1994. She opened by saying, "I think it is terribly important for us at this juncture to keep the level that it is at $4.25 an hour, where low-skill workers can move into jobs and then advance." Secretary Reich ridiculed employers' arguments that raising it would cost jobs. He said, "If we had a minimum wage of three dollars or two dollars or fifty cents or twenty-five cents or a dime, we would still have some employers who would say, 'If you raise the minimum wage, we are going to have to lay people off.' " Kennedy, challenging a pro-business economist, interrupted and shouted at him to argue that his study of job losses after the most recent minimum-wage increase overlooked the 1991 recession as a cause.

Clinton returned to the minimum wage in his State of the Union address, saying it was near a forty-year low in purchasing power: "$4.25 an hour is no longer a minimum wage. But millions of Americans and their children are trying to live on it. I challenge you to raise the minimum wage."

In late March, Dole had the Republican nomination for President wrapped up. His strategy was to show that he could get things done in Congress where Clinton had failed.

That made him vulnerable, because Democrats could force votes on amendments to bills he wanted to pass, or simply stall. But while Democrats kept threatening to attach the minimum-wage increase to any bill they could find, they delayed. There was always a good reason why one bill or another just wasn't the right one. Some other important issue was involved, or some Democrat desperately wanted the underlying bill to pass and the minimum wage would mess things up. Finally, on March 26, they were ready.

Dole was in Kansas on the twenty-fifth, announcing his candidacy. That evening, Senator Lott, the Republican whip, promised the Democrats they could offer an amendment to a measure to create some new park areas. That bill, in turn, was an amendment to a bill to redevelop the Presidio Army base in San Francisco. But the next morning, Dole, who knew what was coming, withdrew Lott's promise. All morning Democrats talked about the minimum wage, but Dole kept the Senate in a session permitting nothing but talk. Republicans charged the Democrats were just paying off the AFL-CIO, which had endorsed Clinton for reelection on the twenty-fifth.

When the Senate came back from lunch that afternoon, Dole and his staff made a rare parliamentary slip, and Kennedy pounced on it. Dole offered two amendments in rapid succession, believing he had forestalled any Democratic proposals. But he forgot that the underlying Presidio bill was also open to amendment, and Kennedy and Kerry, who was running for reelection that year, rapidly offered two almost identical versions of the minimum-wage measures, while Dole stood by, angry that he had been caught napping. "That won't happen again," Dole said to his staff, loud enough to be heard in the gallery.

Dole withdrew the bill to avoid taking a position that was disliked either by voters if he voted no or by Republican business supporters if he voted yes. But Daschle had filed a cloture petition. And so Democrats got a test vote on March 28. It showed that a majority of the Senate wanted to increase the minimum wage. All forty-seven Democrats and eight Republicans voted with Kennedy, and four of the Republicans said openly they thought the minimum wage should be raised.

After Congress came back from its Easter vacation, Democrats persisted in trying to attach minimum-wage proposals to various bills. On April 17, Dole pulled an immigration bill because of the minimum wage. That same day, thirteen House Republicans announced they would vote for an increase to $5.25 per hour. Seven more Republicans defected on April 17, and Gingrich told a Republican caucus that they would not be able to avoid a vote.

In the Senate, Dole suggested trying a gasoline tax reduction instead of a minimum-wage increase, but Democrats filibustered and killed it on May 9. The Senate, Kennedy said, was stuck in the "Doledrums." With traditional spelling, he said, that word was defined by one dictionary as "a state of inactivity or stagnation," and by another as "a vessel almost becalmed, its sails flapping in every direction" or "a region of unbearable calm broken occasionally by violent squalls."

The next week, Dole made it clear he knew his strategy of trying to run for President from the Senate was not working. On May 15 he announced he would resign from the Senate the next month. Democrats had balked his hope of running as the man who could get things done on big issues that mattered to ordinary Americans. He came across instead as the leader who could not get the Senate out of gridlock, not battling Clinton on big issues but squabbling with Kennedy and Daschle on parliamentary questions and filibusters.

DOLE WAS ALSO running into difficulties with the health insurance bill. Here Dole was not throwing himself in front of an onrushing truck, as he did with the minimum wage. He wanted the bill to pass, but he wanted to steer it toward a Republican goal, Medical Savings Accounts. These would be tax-free savings accounts set up to pay medical expenses, available to people who bought health insurance policies with high annual deductibles, such as $3,000 for a family. Critics argued, as Kennedy did in the Senate on April 18, that the MSAs would only be used by wealthy people who could afford up to $3,000 in medical bills and by young people sure they would not get badly ill. The result, he said, would be that the "general insurance pools are going to include the sicker people, and the premiums are going to go up for everyone." That, in turn, would cause many employers and individual people to drop their health insurance altogether.

Kennedy and Kassebaum had agreed to oppose any amendments the two of them did not agree on. So while Kennedy made the tough speeches, she offered the motion to kill Dole's MSA proposal. The April 18 roll call was one of those rare Senate votes when the outcome is not clear in advance. Vice President Gore sat in the chair, ready to break a tie by casting a decisive vote against MSAs.

Dole pulled out all the stops. He persuaded one Republican whose vote he expected to lose, Ben Nighthorse Campbell of Colorado, to take an early plane home to miss the vote. Freshman Bill Frist of Tennessee waited until what seemed the last minute, when a majority had sided with Kennedy and Kassebaum, and voted with Dole. When everyone who was still in Washington had answered, the tally stood at 53–45 against Dole. But he kept the roll call going, as he lobbied Republicans, in full view of the galleries, to switch their votes. Bill Cohen of Maine was the only one who changed, and so the vote was 52–46. Final passage was delayed until the next week so everyone could be there to cast a rare 100–0 vote for the bill.

The House had already passed a bill with a heavy emphasis on MSAs, so the logical step would be to appoint Senate conferees and let the negotiators and the administration fight it out. But two days later, Dole proposed a Senate set of conferees, most of whom had backed his MSA provision. Kennedy objected. Without unanimous consent, conferees could be filibustered, one name at a time, so neither side sought a vote.

Even now, after Dole had announced he would resign, Kennedy was making his Presidential effort look bad. Dole told reporters testily, "I don't have to work it out with Ted Kennedy." But Kennedy came to the press gallery to say Dole had told him on the floor, during a quorum call while Senators tried to figure out what to do next, that he was "not that crazy" about MSAs, but needed them for political reasons.

There were no formally appointed conferees, but discussions between both houses suggested the House would be willing to settle for an experiment with MSAs, and that Clinton would accept one. But there was a chasm between different proposals for experiments. The administration suggested trying a few states. The House idea was to allow all workers in companies with more than fifty employees, an estimated eighty million workers, to try them. Dole wanted a bill to crown his career, but did not work hard on it. When Kennedy complained of inaction, Lott said, "Ted Kennedy does not rule the world."

On June 11, Dole's last day after thirty-five years in Congress, he spoke of issues he had worked on, colleagues he had admired. Known as the fiercest of partisans, this time Dole cited nineteen Democrats and eleven Republicans, including Kennedy for his work on the Americans with Disabilities Act. When the speech was over, Kennedy was among the throng who clustered around to say goodbye, shaking hands and talking in Dole's ear. Dole had long since conceded that if the Democrats' tactics hurt his effort to campaign from the Senate, it was no more than he would have done with their positions reversed. And no matter how hard Kennedy and Dole fought over health insurance and the minimum wage, Kennedy and Dole could do business. On May 2, Kennedy had proposed a bill creating a national historical park in New Bedford dedicated to whaling—and the Nicodemus Historical site in Kansas to commemorate where newly freed slaves lived in dugouts after the Civil War. Dole got the bill passed.

IRELAND WAS PROVING a problem in 1996, too. The British government continued to insist that the IRA begin disarming itself before any peace talks could begin. Then the IRA ended its cease-fire after seventeen months with a bomb in London that wounded one hundred on February 9. Kennedy called the development "ominous." Gerry Adams, welcomed to the White House St. Patrick's Day party the year before, was excluded in 1996. On April 17, Kennedy met with Protestant loyalist politicians and pointedly praised them for not responding to the IRA with violence of their own.

Jean Kennedy Smith was formally reprimanded by Secretary of State Warren Christopher on March 7 for retaliating against two embassy officials who opposed her support for the 1994 visa for Adams. They had disagreed with her recommendation over the formal State Department dissent channel, but investigators found she excluded them from embassy functions and arranged unfavorable personnel evaluations that blocked their hopes for promotion. In Dublin, Ambassador Smith issued a statement denying that she had ever retaliated against dissenters and said she had "strongly recommended for promotion" the other two of the four dissenters in her embassy.

Ted had nothing to say about Jean's reprimand. But when another relative, closer at hand, got into scraps, he was delighted. He was eager to hear from a reporter how Patrick had tangled with Gerald B. H. Solomon of New York, the chairman of the House Rules Committee, over assault weapons. On March 22 the House passed a bill to repeal the 1994 ban on manufacturing nineteen semiautomatic handguns and rifles and high-capacity ammunition clips. The repeal was not expected to be acted on in the Senate; the House voted on it because the National Rifle Association wanted a vote for campaign purposes.

Representative Kennedy objected, telling Solomon that "families like mine all across this country know all too well what the damage of weapons can do. You want to add more magazines to the assault weapons so they can spray and kill even more people. Shame on you! There are families out there. You'll never know, Mr. Chairman, what it's like because you don't have someone in your family who was killed. It's not the person killed, it's the whole family that's affected." Solomon, equally furious, said Patrick had questioned his integrity, and he was supporting the bill because of *his* family: "My wife lives alone five days a week in a rural area in upstate New York. She has a right to defend herself when I'm not there, son."

Ted tried to have lunch with Patrick every Tuesday. He told Chris Black of the *Globe* he admired the way Patrick stayed involved with the state government back in Rhode Island, thereby having more impact than he could have as just a minority-party freshman in Washington. For his part, Patrick enjoyed being more than just a front-row spectator of his father's political life, but being on a more equal footing as a member of the House.

Ted took Patrick along on Air Force One to Rabin's funeral in November 1995, and they sprinkled handfuls of dirt from his brothers' graves on Rabin's. On the way back, at a refueling stop at Shannon, they were drinking Irish coffee with Moynihan, Dodd, and Representative David Obey. Kennedy nodded toward Patrick and said to his contemporaries, "I really think that these young people ought to work their way up just the way I did."

Another family issue ended when Joan dropped her demand for an increased divorce settlement. In a statement released through her lawyer on June 25, 1996, she said she had been "assured protection in the event of the Senator's death," presumably in his will. She also said reports that she had received between $4 and $5 million from the 1982 divorce were "grossly exaggerated," and that she received only "about one-fifth of what was reported."

THE DEPARTURE OF Dole, a master lawmaker when it suited him, actually led to the disintegration of the Senate logjam and gave Kennedy a chance to get his priority bills passed. Lott, who succeeded him, had no personal face to lose if progress was made on the minimum wage and health insurance. Democrats could no longer savor embarrassing the Republican nominee-to-be.

From the start, Lott wanted to get things done to establish his credentials as a leader. In a striking conversation related by Dick Morris in *Behind the Oval Office,* soon after Lott was elected leader on June 12, Lott told Morris, "We have to get the minimum wage off the floor so it doesn't block everything."

On June 25 he agreed to a vote on July 9, with the wage increase coupled with a variety of tax breaks for small businesses similar to those in the House version of the bill, passed May 23 by a 281–144 vote. The Senate tax package totaled $14 billion.

After the defeat of a Republican amendment to exempt businesses with gross sales of $500,000 or less from paying their workers the new $5.15 hourly wage, the bill was passed easily, 72–24, on July 9.

But it did not move on immediately to enactment. Don Nickles of Oklahoma, a diehard opponent of the increase who had been elected to succeed Lott as majority whip, blocked the bill from going to conference. He said he would stall until Kennedy agreed to a conference on the health insurance bill.

That measure was being negotiated in June and July, but largely between the White House and Congressional Republicans. Kennedy worried that the people from the White House, especially Morris, were making it appear to Republicans that they desperately wanted a bill, any bill.

Lott was trying to get the problem solved, too, even though he talked tough publicly and said the delay was all Kennedy's fault. Republicans even ran ads claiming that Kennedy was blocking health insurance. In late July, Lott suggested that Kennedy see if he could cut a deal with Bill Archer of Texas, the chairman of the Ways and Means Committee, who was convinced that MSAs would solve many of the nation's health care problems.

Kennedy did not know Archer. But he had Carey Parker do some quick research and learned that Archer had a ninety-four-year-old mother. So he got a copy of a new edition of Rose's autobiography, *Times to Remember,* and inscribed it to Archer's mother. Archer said later he was touched, though he was not sure his intensely Republican mother would read anything about a family of Democrats. Their first meeting, on July 25, was businesslike but left them some distance apart. Still, Archer was impressed that Kennedy was the Senate's leader on the issue without even being a member of the Finance Committee, and thought he used staff well.

The next day they made a deal. They allowed an experiment limited to 750,000 Medical Savings Accounts in the next four years. "He was a tough bargainer," Archer said the next week. "I feel like I am a tough bargainer, too. We resolved it, I think, as gentlemen."

Kennedy then agreed to let conferees be appointed to settle other differences between the House and Senate bills. On July 31 those conferees announced a deal. That day the minimum-wage bill conferees agreed, too, providing a 50-cent raise to $4.75 that October 1, and another 40 cents to $5.15 on September 1, 1997.

There was one final snag on the health insurance bill over a provision Lott had inserted at the last minute. It was a two-year extension of a patent on Lodine, an anti arthritis drug. Extending the patent would bar competition from a generic version and keep the price up. The rules of both House and Senate bar conferees from inserting material beyond the scope of either bill before them, but the rules are often ignored.

Before the House Rules Committee, Representative Pete Stark of California said Lott's action could be a "serious" ethical breach. But the House passed the bill 421–2, on August 1.

The problem was more serious in the Senate. Senator Paul Wellstone of Minnesota, already furious that the conferees had dropped the Senate provision for equality of treatment between physical and mental ills, threatened to filibuster the conference report unless the Lodine provision was stricken. With vacation bags packed, the Republican convention only ten days off, and a recess scheduled to begin August 2, Wellstone could do it.

Lott was vague at first about where his Lodine idea had come from. Then he said he had put the measure in at the request of Senator Arlen Specter and Rick Santorum of Pennsylvania, where the drug was made. But *Congress Daily,* a spunky Capitol Hill newsletter, reported that Lott's legislative aide, Vicki Hart, was married to Steve Hart, the Washington lobbyist for American Home Products, which owned the company that made Lodine. Susan Irby, Lott's press secretary, said Vicki Hart had foreseen the "potential conflict of interest and recused herself" from staff discussions of the drug provision. "She had no dealings with it whatsoever," Irby said.

Lott was angry over Stark's attack on his integrity. He said he had no personal interest, but was only trying to be fair, since a Lodine competitor had received a recent extension. But the Harts' connection and the fact that other committees that had looked at the Lodine case saw no merit in it made his position shaky.

Kennedy was angry, too. He went to the Senate floor and complained that he had been rushed to sign the conference report, and had done so under the impression there were no special provisions. "There was never any mention of any special interest provision for Lodine," he said. He was also angry that Irby had told reporters the conferees knew it was being done out of fairness. "It certainly was not done for fairness," he said. "It was slipped into the bill without telling anyone because it is not fair and it is not deserved."

By now everyone wanted the insurance bill passed. It was among the talking points House Republicans had been given to boast about when they went home. But if it was going to be done that day, a way to do it with minimal embarrassment for Lott had to be found. A full confrontation on the floor would not shield Lott. Daschle, Kennedy, and Wellstone were in

Daschle's office in early afternoon, talking about what to do, when Lott walked in.

He suggested that the Democrats settle for a losing vote on a point of order, so they could say they tried. Kennedy refused and asked why Lott would not let them win a vote to direct the enrolling clerk to remove the Lodine provision from the bill. Lott said he did not think he could. He would not ask his members to take a dive, and many did not want to be on record on what looked like a pro- or anti-Lodine vote. Kennedy had a solution:

Let the presiding officer rule against Wellstone's complaint that the clause was outside the scope of the bill. Then let Wellstone appeal the ruling of the chair, but make sure nobody asked for a roll call, and have the chair declare Wellstone's side the winner. Once that was done, pass the concurrent resolution instructing the clerk to leave the Lodine provision out. And do it quickly, so the House could pass the resolution before leaving for vacation.

Lott agreed, and that is what the Senate did that afternoon.

So on August 2, 1996, the biggest lawmaking day of his life, Congress sent two major Kennedy bills to the President for signature. The minimum wage was a Democratic cause alone, but the insurance bill was traditional Kennedy-Republican collaboration. For the first time in thirty-four years, it was a bill with his name on it. He was the only one who put the committee chairman's name first and called it the Kassebaum-Kennedy bill.

THOSE TWO BILLS were part of a pattern of collaboration that developed that summer between the White House and Congressional Republicans. Both sides wanted the voters to think they had done enough to deserve reelection, putting their own interests ahead of their party allies. It was no help to Dole when President Clinton signed a welfare bill that gave the states much greater power to cut people off benefits. Nor were Congressional Democrats, especially in the House, pleased to see the President cost them the chance to run against a Congress they would have accused of doing nothing except shutting the government down.

But while Congressional Democrats played almost no role in the battle over welfare legislation, where Kennedy joined Moynihan in futile opposition based on fear that children would suffer, Kennedy was always central to the wage and health insurance bills except for a brief period when Morris seemed to think he was in charge. Daschle was grateful, saying, "He was really like this enthusiastic freshman looking for more work and demonstrating his capacity for taking on even more roles. But instead of a freshman I had somebody with thirty years' experience across the table."

Kennedy came down from Hyannis to Washington for the signings of the wage and insurance bills, but not the welfare measure, of which he had said, "To label this legislation 'reform' is no more accurate than to call the demolition of a building 'remodeling.' " But at the Democratic National Convention in Chicago—where the "in" event was a reception sponsored by his nephew John F. Kennedy's magazine *George*—he hailed the Clintons. He compared Gingrich's attacks on Hillary Rodham Clinton to the criticism of Eleanor Roosevelt, and said, "We love her for the enemies she has made."

He recalled the young Bill Clinton shaking President Kennedy's hand in 1963, and said, "And now he is the young President who has taken up the fallen standard—the belief that America can do better."

KENNEDY HAD ONLY a few other legislative initiatives in 1996, but one of them involved the broadest reach for a Republican ally of his career. He and Lauch Faircloth, a Jesse Helms protégé from North Carolina, sponsored a tough bill to punish burning of churches. Kennedy needed a Republican ally to get action; Faircloth needed credibility among black voters. There had been at least thirty black churches burned in the South in a year and a half, and their bill made it much easier for the FBI and the Alcohol, Tobacco and Firearms Bureau at Treasury to investigate.

When they introduced the bill on June 19, Kennedy said, "We again speak with a united voice in introducing bipartisan legislation to address this alarming recent epidemic of church burnings." Faircloth added, "If we in Congress can't agree that church burning is a despicable crime, then what in the world can we agree on?" Their ideological distance impressed staffers for other Senators; "If Kennedy and Faircloth agree on this, I don't even have to read it," they said, according to Tom Perez, who worked on the bill for Kennedy. The bill passed a week later, 98–0, and the House took it on a voice vote on June 27. Clinton signed it on July 3.

But Kennedy was on the losing side on two votes he regarded as civil rights matters shortly after Congress returned from its convention recess. In their major victory of the 104th Congress, social conservatives passed a bill defining marriage as only between one man and one woman. The act, titled the Defense of Marriage Act, told states that did not want to recognize same-sex marriages in other states that they did not have to, despite the full faith and credit clause of the Constitution, which usually requires one state to accept another's legal judgments.

Sponsors said the bill was necessary because Hawaii courts might soon validate gay or lesbian marriages. Once President Clinton announced in May he would sign the bill, it was assured of passage. The Senate adopted the House bill with an 85–14 vote on September 10. Kennedy called the bill unnecessary, saying states could already deny the validity of same-sex marriages and calling it "a mean-spirited form of Republican legislative gay-bashing cynically calculated to try to inflame the public eight weeks before the November 5 election."

Later that day he failed by a single vote to get the Senate to adopt a bill banning job discrimination against homosexuals. He said, "Discrimination against gay and lesbian people for characteristics they do not control or reflect their deep personal identity, that are irrelevant to their ability to do their job, and that provoke irrational animus among some of their co-workers is the classic case for federal intervention." The bill lost, 50–49. The lone absentee was Senator David Pryor of Arkansas, at the Little Rock bedside of an ill son. He said he probably would have voted for the bill. The narrow vote raised the hopes of gay and lesbian groups for the next Congress.

Finally, Kennedy delayed the Senate's adjournment for three days by fighting a measure to help Federal Express keep its truck drivers out of unions. This measure was another last-minute insertion in a conference report on a different subject, a reauthorization of the Federal Aviation Authority. That bill included lots of money for local airport improvements, a pre-election bounty many members were anxious to claim quickly. The provision defined Federal Express as an airline—although it had thirty times as many truck drivers as pilots—and therefore it had to be unionized nationally if at all. Its chief competitor, the United Parcel Service, had its air operations treated as an airline but its truck drivers could unionize locally.

Although it was barely mentioned on the floor, Federal Express owed its clout to campaign contributions and the corporate jets it made available to Senators. Still, that fact dominated newspaper accounts of Kennedy's fight to filibuster the conference report to get the Federal Express section removed.

The debate was often testy. Kennedy treated it as another "shameful Republican maneuver." His friend Fritz Hollings, the South Carolina Democrat, said it was his amendment, and "I have been a Democrat since 1948. I think you were just learning to drive at that time." Hollings also called

Kennedy's arguments "extraneous garbage" and said, "The distinguished Senator from Massachusetts has just spewed out such a bunch of nonsense that it is hard to know where to begin."

Unfazed, Kennedy insisted, "Key Republicans in Congress have conspired with Federal Express to amend the Railway Labor Act in order to deprive Federal Express workers of their right to form a union."

Hollings and John McCain of Arizona said that the law had been inadvertently altered to change the status of Federal Express, and all they were doing was correcting a mistake. Kennedy said, "It is important to look beyond the legal technicalities and talk about what is really at stake here. Hundreds of truck drivers in the State of Pennsylvania want to join the United Auto Workers and bargain with Federal Express over the terms and conditions of their employment. Federal Express is trying to deny those employees their right to organize. That is basically the issue."

It was a lively battle. Exploiting his colleagues' desire to go home to campaign, Kennedy thought he might win until Daschle told him he would vote with Federal Express. Daschle, sixteen other Democrats, and forty-nine Republicans invoked cloture on October 3. The filibuster ended. The bill passed and the Senate adjourned.

CHAPTER 36

Health Care, Bit by Bit

Clinton won reelection comfortably in 1996. House Democrats picked up nine seats, narrowing the Republican majority to nineteen. But the Democrats lost two Senate seats, giving Republicans a 55–45 majority. And several moderate Republicans, like Alan Simpson and William Cohen, had left.

January 20, 1997, Inauguration Day, was a cold gray day, but the Kennedy family was out in force with enough tickets for three generations. About half the family's representatives were under ten. In Ted's Senate office, there were forty gallons of clam chowder, diapers being changed, and a general air of festivity. Two days before, the patriarch had taken the children to a mall to meet Barney, the purple dinosaur of educational television.

Kennedy was working on the 105th Congress even before the 104th adjourned. Drawing on a Massachusetts program enacted in 1996 as the Commonwealth retreated from its own universal health care plan, on October 1 he and John Kerry introduced a bill intended to make sure all ten million uninsured children in America were covered. They said it could be done with a 75-cent-a-pack boost in cigarette taxes, with states contracting with insurance companies to offer coverage and using federal grants to subsidize poor children. Kennedy's staff prepared the bill. But Kerry introduced it because the publicity would help him in a tight race for reelection against

Governor William Weld. With help from Clinton, who got 61.5 percent of the vote in Massachusetts, Kerry won with 52.2 percent.

Kerry was a perfectly good cosponsor during an election campaign, but to get a bill passed Kennedy needed, once again, a Republican. At one level that seemed easy, for all the health care bills of 1993 and 1994 had emphasized coverage for children, who were both politically appealing and cheaper than adults to insure. In December and January, Kennedy met with Senators Jeffords of Vermont, Mike DeWine of Ohio, Hatch of Utah, Chafee of Rhode Island, Christopher "Kit" Bond of Missouri, Bill Frist of Tennessee, Specter of Pennsylvania, and Olympia Snowe and Susan Collins of Maine. All had shown some interest in children's health in the past, and all told him they supported his goals, but none agreed to cosponsor.

After the health insurance and minimum-wage battles of 1996, Lott had made it clear to Republicans that he, not Senator Kennedy, was going to be in charge of the Senate. Jeffords, in some ways the most obvious choice, because of both his own moderate inclinations and his likely ascent to the chairmanship of the Labor Committee opened by Kassebaum's retirement, was particularly vulnerable. But more conservative Republicans were threatening to vote against him for chairman, so he lay low.

Kennedy was pushing the idea publicly even before he had a cosponsor. He even appeared as himself on an episode of *Chicago Hope,* a television program about a hospital. He was shown running a Congressional hearing in which he got to offer a forty-five-second speech calling for broader health care coverage.

Hatch was another logical choice for an ally. Mormon-dominated Utah had no sympathy for tobacco. While Hatch had left the Labor Committee, he now had a seat on Finance, which would ultimately have to deal with any taxes. He had got used to criticism for working with Kennedy. In February, Hatch agreed to try to develop a bill with Kennedy, though he insisted that the cigarette tax increase had to be smaller than 75 cents, and that therefore all uninsured children could not be covered.

Kennedy always went to Hatch's office for meetings; it was a sign of respect he had shown other Senators for thirty-four years. In the last couple of years, this usually meant starting by listening to one of the patriotic or religious songs Hatch had started writing lyrics for, and which he played on an elaborate sound system in his office. Often Kennedy had Littlefield, who has a fine baritone, respond with a show tune like "The Girl That I Marry" or something from Gilbert and Sullivan. But the final details were hammered out in writing, to avoid misunderstandings, with the Senators sending letters back and forth to each other from their homes while staff pressed them not to give up.

Hatch wanted a five-year, $20 billion program, with half the money going for deficit reduction, so he could say he had moved Kennedy toward the center. Never deeply troubled by deficits, Kennedy wanted $25 billion to cover five million children at $1,000 a year for five years. They compromised on $30 billion in taxes over five years, with $20 billion for children and $10 billion for deficit reduction. That required raising the federal cigarette tax by 43 cents a pack, from 24 cents to 67 cents.

They proudly announced their plan to great applause at the Children's Defense Fund's legislative conference on March 13, 1997. Hatch told the group, "Some have referred to us as the odd couple of the United States Senate. I like to think of us as the dynamic duo."

Just how dynamic was clear only four weeks later, when Kennedy and Hatch introduced their legislation. On April 7, Hatch announced seven other Republicans as cosponsors. Three had been on Kennedy's list: Collins, Jeffords, and Snowe. But Hatch had found four more: Senators Robert F. Bennett of Utah, Ben Nighthorse Campbell of Colorado, Ted Stevens of Alaska, and Gordon H. Smith of Oregon.

Lott was plainly annoyed, saying, "This is not last year, and a Kennedy big-government program is not going to be enacted." And the next day three of the Republicans backed off—Campbell, Collins, and Stevens, saying they only favored the insurance, not the tax to pay for it. Hatch was particularly disappointed when Bennett, his state's junior Senator, followed on Friday, April 11. Bennett said he opposed the insurance. Hatch said, "I am disappointed, but I accept whatever my colleague wants to do. As for me, I am going to fight my guts out for these kids." Republican leaders were mobilizing against the bill. Don Nickles immediately distributed a yellow card with talking points against it, including the untrue assertions that it created a new entitlement and "mandates abortion funding for teens." But in the House, Nancy L. Johnson of Connecticut, a Republican with a long interest in health care whose ties to Gingrich had almost cost her reelection, introduced a companion bill to the Kennedy–Hatch measure.

Some conservative groups aimed directly at Hatch. Phyllis Schlafly's Eagle Forum sent speakers to Utah to attack him. A group called Citizens for a Sound Economy spent $35,000 for a week of radio ads in Utah attacking the bill. An actor playing Kennedy thanked Hatch for backing a "liberal" bill. Hatch was already annoyed at attacks from the right on various subjects, and particularly angry at a cover of *National Review* that ridiculed his religion by calling him a Latter Day Liberal, a clear sneer at the Mormon Church, whose formal name is the Church of Jesus Christ of Latter-Day Saints. Hatch answered the radio ad by calling the sponsors "knuckleheads" and saying, "They're going about it in the wrong way if they think they are going to influence me like this. It just makes me dig in and work even harder."

Meanwhile another approach was developing. Senators Chafee and Rockefeller, the West Virginia Democrat, proposed a $16 billion expansion of Medicaid, the federal health program for the poor, so it would cover more children. Parents of perhaps three million eligible children either were ignorant of it or spurned it as "welfare." The subsidies in the Kennedy-Hatch proposal were aimed at children from families with incomes above the Medicaid eligibility level. The Chafee-Rockefeller bill, whose authors were both on the Finance Committee, appealed to many of the outside groups because Medicaid benefits could be regulated from Washington. But many governors disliked the idea because states had to share the costs. Kennedy and Hatch agreed to support the Medicaid measure, calling it complementary to their own.

Neither children's health nor any other specific program preoccupied the White House that spring. Clinton was reaching for a place in the history books by trying to negotiate an agreement with Republicans to balance the federal budget for the first time since 1969. The pace of those talks intensified in late April.

Some Democrats, even though they did not know just what agreements were being reached, were furious. The House Democratic whip, Representative David E. Bonior of Michigan, said on May 1, "This deal does not speak to the heart of what Democrats are all about in this country, and that is to step up and go to bat for working families, the kids who need a health care package, and education." Representative Barney Frank claimed he sent a letter to Clinton, "addressed to the *Democratic* President of the United States, and it came back 'addressee unknown.' " In the Senate, Daschle complained, "To cut these kinds of deals, to me, is the worst kind of governing." But Senators generally were milder on the substance. Kennedy said, "I want to see who the winners and the losers are. That should be the measure."

On May 2, thanks to a last-minute windfall in economic estimates about how a strong economy would increase government revenues, Clinton and the Republicans agreed on a plan that was expected to balance the budget by 2002. The five-year deal included Republican tax cuts, more spending for education—and $16 billion for children's health initiatives. But there was also $16 billion in reduced Medicaid payments, of which perhaps half would affect children, so the net addition was modest.

Kennedy and Hatch decided to push ahead, by adding authority for their program to the budget resolution, the blueprint by which the Clinton-Republican agreement would be transformed into Congressional policy. For weeks the Children's Defense Fund had run ads in *Congress Daily* saying the issue was between kids and cigarette companies. The ads pictured Joe Camel, the amiable cartoon character whom R. J. Reynolds used to intrigue children with smoking, and "Joey," a sweet little tyke. Kennedy raised money for the ads in April with a personal appeal to Lew Wasserman, the retired head of MCA, the entertainment conglomerate. He gave $100,000.

But those battle lines invited Republican opposition. The tobacco industry had invested heavily in the Republican Party, giving it $8 million for the 1996 election, compared to $2 million for Democrats. The Republican leaders were not ingrates, and on May 20 their Senate Policy Committee produced a paper arguing against a tax increase, saying it would hurt states that depended on tobacco tax revenues. "Even if one believes that decreased demand for tobacco is positive from a societal view, it still has negative fiscal aspects for the states," it said. Lott joined in, telling the Senate there was already enough money for children's health and he hoped the Senate would not be "duped." When Nickles was asked if tobacco donations affected Senate Republican policy, he said there was no connection: "Oh, no, no, no. It's a concern about having an open-ended entitlement program."

The next day Kennedy and Hatch offered their proposal—as an amendment to the budget resolution—to direct the Finance Committee to raise taxes by $30 billion and have $20 billion of it spent on children's health.

Hatch said the issue was whether the Senate "would stand up for children and against Big Tobacco. Senators, who do you stand with? Joe Camel? Or Joey?" Kennedy said, "Our amendment will make the Hatch-Kennedy children's health insurance plan part of the budget. Our goal is to make health insurance accessible and affordable for every child. The plan is financed by an increase of forty-three cents a pack in the cigarette tax. That increase has the additional important benefit of reducing smoking by children."

Senator Pete Domenici, the New Mexico Republican who headed the

Budget Committee, and Senator Lott insisted the additional $20 billion was not needed. Lott called it "excessive." He also argued that "this is clearly a deal-buster" because Republicans had been assured of a net tax cut of $85 billion in their deal with Clinton. This amendment would cut that to $55 billion.

Then Lott startled Kennedy by telling him he had talked to Clinton and the President had agreed to "work to try to get Democrat votes . . . against making this change in the budget resolution." Kennedy said he was puzzled because Lott said "the President is supporting his position while the Vice President is on his way up here to vote for our position."

Lott was correct. After he saw that Hatch and Kennedy had the votes to win, Lott had called Clinton to tell him that this was a "deal buster" and he would pull the budget agreement from the floor unless Clinton leaned on Democrats to vote against the amendment. Clinton had Franklin D. Raines, his budget chief, and John Hilley, his chief lobbyist, start calling Democrats. And Gore, whose staff had told Kennedy's he would be there to cast a tie-breaking vote if needed, was not there. He drove up to the Capitol and drove back before the vote.

Late in the afternoon, when it was clear that enough votes had been changed so Lott would win, there was a final flurry of debate before the vote. Hatch said, "I think the President and the people in the White House have caved here, people who we had every reason to believe would be supportive of children's health." Kennedy said, "It is more important to protect children than to protect the tobacco industry." Domenici said, "The best thing we can do for kids and for children is to balance the budget." Lott said he had talked to the President again, and Clinton realized "this is a deal breaker."

Lott won, 55–45, with the support of eight Democrats. Five came from states that did not grow tobacco. Hatch claimed the President and his aides had turned at least five votes around, enough to make the difference, though other estimates suggested only four Democrats had shifted. Lott had changed some Republican minds, too.

Kennedy was angry: at the White House position, at the fact that he'd learned it from Lott on the floor, and because his calls to Clinton and Gore had not been returned. And while he was not as caustic as Hatch, he came to the press gallery to tell reporters he believed the administration was "mis-taken," and promised to offer the proposal "again and again until we prevail." When he was asked if he thought he could count on the President's support on those attempts, he turned to Hatch and said, "Why don't you answer that?" Hatch said, "Yes," and then Kennedy said, "I would hope he would."

The next day the White House issued statements saying that it was all a big misunderstanding and the President would support the bill some other time. But Kennedy felt that the budget resolution, on which decisions take only fifty-one votes, had been their best chance. He was still annoyed twelve days later, when he was interviewed for this book, saying Clinton's support would have made the difference and Clinton could have got the credit Lyndon Johnson got for Medicare.

The next battlefield was the Finance Committee. On June 18, Hatch offered the Hatch–Kennedy plan with its 43-cent cigarette tax increase, but Roth ruled the amendment out of order. The next night, at a closed session, Hatch proposed a 20-cent-a-pack tax increase, an amount that would raise $15 billion over five years. When he only sought to require that $8 billion be spent on children's health, with $7 billion available for other uses, he got his proposal adopted.

Hatch said later that Kennedy chewed him out on the phone for half an hour that night, complaining that Hatch had surrendered on both the size of the tax increase and the purpose for which its proceeds would be used, and had done so without consultation. Hatch argued that he got what he could pass, and that children's health now would get $24 billion, or $4 billion more than he and Kennedy had sought in the first place. "I seized the moment," he said, "and it worked." On the Senate floor the next week, Kennedy tried to raise the tax increase back to 43 cents, but was defeated 70–30. Hatch voted no, as he had promised Finance Committee members he would when the 20-cent increase was adopted.

Kennedy and the outside groups worried that the administration would not push hard for the whole $24 billion in conference. He worked with Hillary Rodham Clinton to maintain support for that level of spending, which almost equaled the $25 billion he had told Hatch in March that he had to have. The House had stopped at $16 billion. With strong administration support, the $24 billion stayed in, although at the last moment the tobacco industry succeeded in shaving a nickel off the per-pack increase, with a 10-cent hike in 2000 and 5 cents more in 2002. Conferees found the money for children's health elsewhere and kept a $24 billion, five-year block grant program.

When the Senate came to vote on that measure on July 31, Kennedy claimed a great victory. He said the $24 billion was enough to "give every child access to affordable health insurance." He credited the President, the First Lady, Daschle, Marian Wright Edelman, head of the Children's Defense Fund, and Hatch, "whose courageous leadership was indispensable."

He called the measure "the most far-reaching step that Congress has ever taken to help the nation's children, and the most far-reaching advance in health care since the enactment of Medicare and Medicaid a generation ago."

THERE WERE PERSONAL milestones passed in 1997, too. Ted sold his home in McLean and moved into the District of Columbia to be closer to his stepchildren's school. He got $6 million, about $1 million more than the asking price, in an auction in the hot Washington real estate market. One condition of the February 5 deal was that they move out quickly, and Vicki dropped out of her law practice to house-hunt. On March 1, at a seventeenth-birthday dinner for Chelsea Clinton at "21" in New York, Vicki was telling Hillary Rodham Clinton about her search. The First Lady, alluding to accusations that the Clinton campaign had sold invitations to the White House for campaign contributions, said, "You can stay in the Lincoln Bedroom." They ended up renting a large house in northwest Washington, and, in 1998, buying a home in the Kalorama embassy area, featured, like the house in McLean, in *Architectural Digest*.

Eddie McCormack died on February 27, and Ted was a prominent speaker at his funeral, recalling their 1962 contest. "I thought I was reasonably well prepared," he said, "because I had three or four advantages I'd worked hard to acquire—a brother who happened to be President, another who happened to be attorney general, a number of post office and other patronage appointments that happened to be available to dispense, and, of course, a dad who happened to have deep pockets." In their debate, Kennedy said, "some people say Eddie came on too strong. Others still say he was right on the mark. I agree with them both." McCormack, who said years later that Ted had become a better Senator than he would have been, had raised money for Ted's 1994 campaign, the Senator recalled. "Eddie could see it was my toughest race since 1962."

KENNEDY ALSO FOUGHT a long war of attrition in 1997 on legislation intended to speed up decisions by the Food and Drug Administration, or, in some cases, to take power from the agency. While the user fees enacted in 1992 had succeeded as promised in reducing review time for new applications, the time it took to get a drug through the development phase to that point had increased.

The Republican 104th Congress had made reduction of federal regulation one of its highest goals, but had been unable to pass FDA legislation in 1995 and 1996.

In 1997, Jeffords proposed legislation coupling the universal desire to renew the expiring user fee legislation with a series of proposals designed to speed up the approval process, and in some cases to farm out testing of medical devices to private laboratories. Kennedy was always suspicious of that idea, saying it would "turn over reviews of medical devices to private companies selected and paid by the very industry they are supposed to regulate." Some of the provisions were uncontroversial, such as extending expedited reviews to other life-threatening diseases besides cancer and AIDS. It expanded access to drugs still under investigation for patients with no alternatives.

Kennedy fought losing battles in the Labor Committee on the issues of outside testing, national rather than state labeling of nonprescription drugs, and cosmetics and food labeling done without FDA review. On many of those issues he did not hold the votes of Democrats whose states had major drug industries, including Dodd of Connecticut, Mikulski of Maryland, Wellstone of Minnesota, and Murray of Washington.

After the budget deal was finished and Congress returned from its August vacation, Jeffords brought the bill to the floor on September 5. But Kennedy delayed it, fighting for more changes even though he had already won some during the summer, including exclusion of medical devices with the most serious risk potential, such as pacemakers, for testing outside the FDA itself. He opposed shifting regulation and labeling of cosmetics to the FDA from the states, saying the federal agency had neither the expertise nor the resources to deal with cosmetics. A compromise was reached on September 11 under which state regulation continued, but could be preempted by the FDA. But Kennedy could not prevent passage of the element he most strongly opposed, directing the FDA to consider only the use manufacturers said they intended for a medical device and not other likely uses. That provision remained when the Senate passed the bill on September 24, 98–2, with the only noes coming from Kennedy and Jack Reed, a Rhode Island Democrat.

With the help of the Clinton administration, Kennedy won that fight in conference with the House. The final bill allowed the FDA to examine the safety of medical devices for expected uses, and not just those on the manufacturer's label. The conferees also narrowed further the types of devices that could be tested outside the FDA, excluding those the agency felt must be examined using clinical data.

When the Senate passed the final version of the bill in November, Kennedy hailed it for providing "responsible reforms that will protect the public health while expediting the process of bringing safe and effective products from the laboratories to the marketplace to the bedside of the patient." The original bill, he argued, depended on the false image of the FDA as a "regulatory dinosaur, . . . a myth perpetuated by those who want to undermine the FDA to advance an extreme antigovernment ideological agenda, or to obtain higher profits at the expense of patients' health."

IN JULY THE Irish Republican Army restored its cease-fire. In September that led to Sinn Fein's admission to the Belfast talks, over which George Mitchell presided with patience honed in the Senate. The political situation and the diminished security risks made it seem both useful and safe for Kennedy to decide to visit Northern Ireland. But by the time he arrived in January, sectarian killings had resumed, and the talks in Belfast seemed to be making little progress.

He went to John Hume's Derry, and spoke at the University of Ulster on January 9. "Today," he said, "we stand at a defining moment in the modern epic of this land. The talks that are about to resume offer both a challenge and an opportunity. In the coming crucial weeks, the parties will determine whether this is a genuine way forward, or just another failed station on the way of sorrows."

Kennedy said Irish Americans had "a long enduring desire to see peace and prosperity take root here. Our commitment embraces the welfare of all the people of Northern Ireland—and when we say 'all,' we mean all."

He reached out particularly to Irish Protestants, saying that while the contributions of Catholics who fled the potato famine were well known, "it is often forgotten that more than half of the forty-four million Americans of Irish descent today are Protestants" whose ancestors fled persecution as Presbyterians. Eleven Presidents, from Andrew Jackson to Bill Clinton, had that Scots-Irish ancestral connection.

"Unionists who often feel afraid of what the future may bring," he said, should recall that they shared a heritage with "the pioneers who helped build America, and now you can build a better future for this island."

He called on Northern Ireland to learn from the example that "the diversity of America is America's greatest strength, and the diversity here can be your greatest strength as well."

Above all, he said, the "two communities in Northern Ireland must reach out and do what must be done—and join hands across the centuries and chasms of killing and pain." He cited two couples in the audience as examples. Michael and Bride McGoldrick's son, Michael, was a Catholic taxi driver murdered in 1996 in Drumcree. John and Rita Restorick's son, Stephen, was the last British soldier killed in Northern Ireland, shot by a sniper at a checkpoint in South Armagh the same year. Both families condemned violence and forswore retaliation.

Ted read from a letter his own father wrote in 1958, after the deaths of Joe and Kathleen, to a friend whose son had died:

"When one of your children goes out of your life, you think of what he might have done with a few more years and you wonder what you are going to do with the rest of yours. Then, one day, because there is a world to be lived in, you find yourself a part of it again, trying to accomplish something - something that he did not have enough time to do."

He closed by saying, "Too many lives of too many sons and daughters of this land have been cut short. We must dedicate ourselves to accomplish for them what many 'did not have enough time to do'—a lasting peace for Northern Ireland."

He went on to Belfast, and offered the message more privately to Irish politicians, including David Trimble, the Unionist leader who was one of the Protestants he had sought out ever since the Adams visa was issued.

On Good Friday, April 10, 1998, because of the encouragement of Mitchell, Clinton, and Prime Ministers Tony Blair of Britain and Bertie Ahern of Ireland, the politicians summoned the courage to make concessions to each other which many of their followers distrusted, and to establish a framework for a new foundation of government in Northern Ireland.

KENNEDY'S AGENDA FOR the second session of the 105th Congress had begun developing in the first. The next health care initiative, a "patients' bill of rights," was actually introduced in February 1997. It was designed to loosen the controls various managed care plans had on emergency care, choice of physicians, and diagnostic tests. Lott strongly discouraged Jeffords from working with Kennedy, but warned the industry to "get off your wallets" if it expected Republicans to continue to block the measure.

In July 1997, he called for another rise in the minimum wage, to bring it from $5.25 per hour in September 1997, to $7.25 by 2002. He said that big an increase was still needed to return the minimum wage to the purchasing power of 1968.

In June the tobacco industry reached a tentative settlement of a raft of antismoking suits with the states for a proposed $368.5 billion. The deal was made with forty state attorneys general. The industry promised to curb its advertising, but the settlement was contingent on Congress providing immunity from future class-action suits or punitive damages. Kennedy pressed Clinton not to endorse the deal, and in November he called for a sharp increase in cigarette taxes—$1.50 per pack over three years—without immunity from lawsuits.

Kennedy pulled this agenda together in a speech at the National Press Club on December 11. He shifted the minimum-wage proposals to three 50-cent boosts, to reach $6.65 an hour by 2000. He called for a cut in the payroll tax, financed by an end to the cap on the tax, which reached no income above $65,000. He urged that the $20 billion that could be raised from a $1.50 increase in the tobacco tax not be scattered among federal programs, but spent half on medical research and half on childhood development and child care.

On health care, he urged both the patients' bill of rights and a straightforward but enormously difficult step toward universal coverage—requiring every employer to contribute to health insurance for his or her workers. He said that would take only a ten-page bill. Every business, he said, was already required to pay the minimum wage, obey child labor laws, provide safe working conditions and workmen's compensation, and contribute to Social Security and Medicare. "It is long past time for every business to contribute to the cost of basic health insurance for its workers," he said.

None of those proposals got very far in 1998. Republicans filibustered against tobacco settlement legislation after the industry concluded it was being asked for too much.

The Senate voted down a minimum-wage increase of $1 over two years on September 22. Clinton had endorsed a raise in his State of the Union address, but no serious outside pressure campaign had been put together, and the 55–44 vote was no surprise.

The patients' bill of rights had more resonance with voters, and had support from many House Republicans, led by Representatives Charles Norwood of Georgia, a dentist, and Greg Ganske of Iowa, a physician, who symbolized how much doctors had come to hate the managed care revolution. The House actually passed a bill in July, but it did not allow patients enrolled in federally controlled health plans to sue them—a measure Democrats insisted was essential.

On October 10, with Congress about to adjourn, Daschle forced a vote in the Senate on whether to bring up Kennedy's bill. It was defeated 50–47, and Kennedy contended, "Big money and powerful special interests kept us from debating this issue in the United States Senate."

Kennedy's other initiative of 1998 was in education. In the Press Club speech, he warned of a coming shortage of teachers, two million were needed in the next ten years to cope with retirements and the need to reduce class size. The federal government should provide half those teachers, he said. Congress should "bring one hundred thousand new, well-trained teachers into our public schools each year for the next ten years."

The number had been taken from a piece of the 1994 crime bill that promised federal help to put 100,000 additional police officers on the streets. Clinton adopted the teacher proposal in his State of the Union address, but that did not help much.

Just a few days before that speech, the *Washington Post* and the *Los Angeles Times* had broken the story that Clinton was being investigated for possible perjury or obstruction of justice in connection with an affair with a White House intern, Monica S. Lewinsky. While Clinton's State of the Union speech commanded unusual attention, the audience was drawn less by what he said than by wondering if he was about to resign.

Throughout the year, Kennedy and Patty Murray, a former teacher who was facing a tough reelection race in Washington, tried to attach the plan to budget and appropriations measures. On April 22, Murray told the Senate how she had been overwhelmed by being unable to help one troubled child when she had twenty-three other first-graders who needed attention. "We could have made a difference simply by having fewer students in the classroom," she said.

Kennedy ridiculed the Republican preference for vouchers that could be used in private schools and said, "We ought to give a helping hand to the local communities. We are not interested in superimposing some Federal solution, some 'new bureaucracy,' those old clichés. I have listened to the same clichés for thirty-odd years. You would think they would have some new ones, talking about 'the new bureaucracy,' 'one size fits all,' 'Washington doesn't know everything. . . .' " The issue, he said, was simply that "class sizes are too large. Students in small classes in the early grades make much more rapid progress than students in larger classes. . . . The benefits are greatest for low-achieving minority and low-income students."

The Murray proposal was defeated, 50–49, that day and it never was passed on the Senate floor.

But as the Congressional session neared an end, Republicans were anxious to adjourn and campaign. Polls showed their moves to impeach Clinton were backfiring with the public. But Democrats were happy to delay adjournment and castigate the Republicans as interested only in investigations, not legislation.

When the administration made it clear that money for teachers was one of the prices of Clinton's signature on a big spending bill that would permit adjournment, Republicans agreed to discuss it. Daschle sent Kennedy to an unusual four-man conference. The others were Representative Bill Goodling of Pennsylvania, chairman of the House Committee on Education and the Workforce, Representative Bill Clay of Missouri, that committee's ranking Democrat, and Senator Slade Gorton of Washington. Gorton was the GOP's Senate point man on block grants, and Lott picked him over the more obvious selection of Jeffords, chairman of the Labor Committee.

Murray was back home campaigning, but Kennedy brought his staff together with hers and Barbara Chao of the White House. They agreed they would oppose block grants because they would not necessarily reduce class size, and would reject merit pay and testing because teachers' organizations were strongly against them. When the conference began, the Republicans proposed $1.2 billion worth of large block grants to the states, to be used for almost any education purpose. Kennedy reacted amiably and said Democrats would look at the proposal and come back with their own idea. Instead, he came back with the Democrats' original plan.

But he explained how Republicans could argue that it met their purposes, too. The money could indeed be used for special education teachers, one of their prominent concerns, and 15 percent could be spent on staff development and teacher training. By October 14 the Republicans agreed, and even got their press release out claiming victory before the last details were set.

Kennedy and Clinton had got an appropriation of $1.2 billion for hiring teachers to reduce class size in the early grades. They also attained a formula by which most of the money would go to needy school districts. But one problem remained: the money was only for one year. The program could not continue beyond then unless Congress passed an authorization bill, and six months into the second session of the 106th Congress, nothing had been done about that bill.

CHAPTER 37

The Job You Asked the People For

The consuming issue in Washington all through the fall and winter of 1998–1999 was impeachment. On August 17, President Clinton testified before a grand jury about his affair with Monica S. Lewinsky, a White House intern. On September 11, Kenneth W. Starr, the independent counsel appointed four years before to look into Whitewater, a murky Arkansas land deal in which the Clintons actually lost money, urged the House to impeach President Clinton for perjury and obstruction of justice, for lying about the affair to the grand jury, and for trying to cover it up.

Kennedy, with his own experience of a censured private life, talked often to the President, cheering him up and urging him to keep focusing on his job, not his attackers. He made one very specific contribution, urging the President to add Greg Craig to his legal staff and urging Craig to accept the offer. Craig, who had known Clinton at Yale Law School, joined the weary lawyers on September 15. He provided fresh energy. Unlike Clinton's principal defense attorney, the tin-eared David Kendall (another Yale Law contemporary), Craig had an instinct for how things sounded politically, and the public presentation of Clinton's defense improved sharply.

All year long the general public had told pollsters that Washington should forget about the President's sex life. But Congressional Republicans paid more attention to their core voters, who despised Clinton, and they made his sins an election issue. That cost them six seats in the House. In the Senate, Republicans came out even, despite early expectations of gaining seats, perhaps even the five that would bring them to a filibuster-proof total of sixty.

One widespread interpretation of the election was that it would warn House Republicans off impeachment. But they drew a very different lesson: that they must hurry and act before their majority narrowed in 1999. Along party lines, the House voted on December 19 to impeach Clinton for high crimes and misdemeanors, calling on the Senate to try him and remove him from office.

Hardly anyone expected the Senate, with only fifty-five Republicans, to assemble the sixty-seven votes required to convict. But the Senate has a way of seeing itself as the center of anything it touches. Collectively it wanted to avoid the rancor and raw partisanship the House had displayed—a perception House Democrats enhanced to discredit a process in which their views were ignored.

But that was not the objective of the White House, where the lawyers had been surprised by the House vote to impeach. They feared that if the House managers found some damaging evidence and surprised them with it, Democrats might melt away and the President could be ousted. For four days in January the Senate wrangled over how to proceed. Democrats wanted no witnesses and a tightly paced schedule. The House managers wanted to call as many as they thought would help their case. The White House wanted an early vote that would divide on party lines, lock in Democrats behind Clinton, and make Senate proceedings look to the public like a continuation of the House's party-line lynching.

But the Senate avoided partisan meltdown, and Kennedy played a crucial role. Lott and Daschle scheduled a closed-door caucus of one hundred Senators for January 8. They met in the ceremonial Old Senate Chamber. Robert Byrd appealed to his colleagues' institutional pride, saying, "The White House has sullied itself. The House has fallen into the black pit of partisan self-indulgence. The Senate is teetering on the brink of the same black pit." Chris Dodd recalled that this was the room where Senator Charles Sumner of Massachusetts was nearly caned to death for his antislavery views in 1856, and said the same "toxic" feelings of that time had been spread by the Lewinsky case and threatened the Senate with the same venom that had poisoned the House and the Presidency.

There was a sense of failure in the room as the two-and-a-half-hour meeting neared its close. Senator Slade Gorton, a Washington Republican, announced he no longer favored the compromise he had proposed with Senator Joe Lieberman, a Connecticut Democrat.

A move toward conciliation came from an unlikely source, Senator Phil Gramm, a conservative Texas Republican. He began by quoting Daniel Webster's classic appeal to national commitment, his March 7, 1850, speech, "The Constitution and the Union." As Congress fought over extending slavery to territories in Kansas, Nebraska, and the West, Webster said, "I wish to speak today not as a Massachusetts man, nor as a northern man, but as an American, and a member of the Senate of the United States. . . . I speak for the preservation of the Union. Hear me for my cause." Gramm said the Senate had to serve the Union that day, too. He argued it would be unjust to foreclose the House managers, in advance, from any chance to call witnesses, and in any case, a single Senator could always force a vote on calling them. But on other issues, he said, the differences between the two parties were trivial.

Kennedy, a proud successor to the seat held by Webster and Sumner, was moved by the room and its history. He was struck by Gramm's argument and jumped up to say the Texan was right about the degree of agreement. Kennedy said there was already enough accord on questions of timing and the order of business to get to first base and second base, and why not defer the witness issue: "We could see later on how we're going to get to third."

They worked out a schedule that afternoon, with the help of Gorton and Lieberman, and the Senate avoided looking like the House. And even when later votes followed party lines on the path to Clinton's acquittal, the atmosphere was never as rancid as it had been in the other body.

That was Kennedy's institutional contribution, a moment across party lines in the tradition of his alliances with Howard Baker, Jacob Javits, Peter Dominick, Hugh Scott, Mark Hatfield, Bob Dole, Orrin Hatch, Alan Simpson, John Danforth, and Lauch Faircloth. His interest as a Senator diverged from the White House, where the lawyers now feared that the Senate's 100–0 vote might mean that their safety net of Democratic votes to block conviction was frayed.

KENNEDY HAD ONE other central role during the impeachment season. He talked to Clinton often when the President was gloomiest. In those talks, he reflected on his own life, when, like Clinton and for the same reasons, he had often been in an unwelcome limelight.

In an interview on January 29, 1999, before the trial was over, President Clinton described Kennedy's message to him:

"You couldn't have a better friend. I mean, he is loyal. People have been loyal to him, and understanding, and he's had to ask for forgiveness a time or two. And so he gives as good as he's gotten on that.

"His advice is always simple. It's just sort of get up and go to work, just keep going, and remember why you wanted the job in the first place.

"He's a very tough guy and he understands that if somebody accuses you of something that's true, maybe you're your own worst enemy, and you have to hope that when people add up the score, there will be more pluses than minuses. And if somebody accuses you of something that is not [true], then it will probably get sorted out sooner or later, and there is very little you can do about it except do the job you asked the people for."

ON FRIDAY, JULY 16, 1999, Kennedy had just finished another battle over the patients' bill of rights, losing the votes but, he thought, winning the argument before the public. Robin Toner of *The New York Times* came by his office to interview him for a profile of him as the liberal lion roaring again, reveling in Republican attacks on "Teddy Kennedy's" bill.

He was optimistic about the bill, amused by the personal attacks, and enjoying the Senate, telling her, "If you want to get things done in the Senate you can't afford to be personal." But he was equally excited about the weekend, when Robert's last child, Rory, would be married at the family compound, and the family would be together. But that reminded him that John Tunney had once told him of his own father's comment that large families get together only at formal events like weddings and funerals. So, after Michael's funeral in 1998, Kennedy had decided to do something about that. He told Toner about a reunion in the summer of 1998 at the Cape, when "we had 105 Kennedys," 53 of them age thirteen or younger. "We had two or three great days. All they were primarily interested in was fingerpainting and music. They love to dance. But we had Mass, my mother's prayers, and all kinds of things."

He flew up to the Cape that afternoon as the family gathered.

But there was no celebration. John F. Kennedy, Jr., his wife, Carolyn Bessette Kennedy, and her sister, Lauren Bessette, all died when the plane John flew up from New Jersey crashed into the ocean near Martha's Vineyard, where Lauren was to be dropped off.

All day Saturday and into Sunday the family nourished waning hopes that John, Carolyn, and Lauren would be found alive. The wedding tent was used instead for Mass, for prayers for the missing. On Sunday evening, the Coast Guard said there was no hope, and on Monday Ted flew across Long Island Sound to comfort John's sister, Caroline Kennedy Schlossberg, and her children, as television helicopters whirred overhead to record the "news." The three deaths had stimulated incessant television coverage, labeling John "America's prince" and the best hope of his generation.

The bodies were recovered from the ocean on Wednesday, July 21. Ted and his sons were on the USS *Grasp* when they were raised to the surface. The bodies were cremated and buried at sea by the Navy on the twenty-second. There would be no grave to gawk at.

Ted and John had been close over the years. John served with Ted on the board of a favorite family memorial to the President, the Institute of Politics at Harvard's John F. Kennedy School of Government. With Ted's encouragement, like a senior partner bringing along a protégé, John was scheduled to become chairman in the fall of 1999, succeeding Ted's college chum John Culver.

On Friday, July 23, a private Mass was held in New York, at St. Thomas More church on East 89th Street. No television cameras or reporters were present among the 350 mourners, but President Clinton, his wife, Hillary, and his daughter, Chelsea, were there. The President had followed the search closely and, at a news conference, defended the unusually extended recovery effort as appropriate, considering the family's contribution to the nation. The Kennedys, the Bessettes, and the Clintons got up from their pews to fill the aisle in an extended kiss of peace.

Ted gave the eulogy for John with humor and love. He began by telling a story that poked fun at himself: "Once, when they asked John what he would do if he went into politics and was elected President, he said, 'I guess the first thing is call up Uncle Teddy and gloat.' I loved that. It was so like his father. . . .

"We dared to think," Ted concluded, taking a phrase from the Irish poet William Butler Yeats, "that this John Kennedy would live to comb gray hair, with his beloved Carolyn by his side. But, like his father, he had every gift but length of years."

Ted's role that week in July, as so often over thirty-one years, was as the head of a large, exciting, and often troubled family. Some of those 105 Kennedys, like his niece Maria Shriver, see him as the proper patriarch, reminding the family of its heritage. She said, "It's important that my children know who he is, who his brothers were, Grandma and Grampa, the role of the Irish. He gives so much of the credit to Grandma and Grampa—he takes that role very seriously." Kerry Kennedy Cuomo thinks of him and good times: "He would get us to sing songs and tell stories. He would make us feel that this is a great and wonderful place to be and share his love of the sea—and of fun." Others, like Robert F. Kennedy, Jr., remark on how he seems constantly available to the family, within hours or minutes, a lesson to them all: "If anybody in my family calls, I always get back to them instantly."

And nephew Joe said, "When things are going badly he is somebody who you can turn to. . . . Our family has gone through too many crises, too many sad times, as other families have. But every time it happens, he is the guy that people look to."

THE ADVICE KENNEDY gave Clinton about the pluses and the minuses reflects how he has lived his own life at the center of American politics for a third of a century. Politically, he has towered over his time, and even if his failures outside the Senate have fascinated the supermarket tabloids, the talk show comics, and millions of their followers, his successes have affected almost every American, even the dwindling minority who recognize the name Kopechne.

Take the minuses. At the bridge at Chappaquiddick his conduct caused Mary Jo Kopechne's death. Whatever happened in the hours before and after the accident—the stuff of controversy for three decades—is trivial compared with that one blunt, sad fact.

Politicians and journalists used to argue that the personal lives of public men were irrelevant, that those details should be put aside in a separate box, taped and sealed, while only their civic life was judged. The accident at the bridge changed that, for the better and for the worse, but forever. Chappaquiddick, as the voters understood it, excluded him from the Presidency he might have won. Twenty-two years later his inadequacy in the Clarence Thomas confirmation fight was a direct result of Palm Beach, and the attention Palm Beach got was a direct result of his well-earned reputation for drinking and pursuing women half his age.

Take the pluses. He speaks often of civil rights as the great unfinished business of the nation, but since the 1982 Voting Rights Act extension, no one has done more to finish it. In some years he resisted retreat, in others he advanced the concept of equality to include the disabled and the dignity of women in the workplace.

His commitment to national health insurance has been less successful, although there is great potential in the children's health legislation of 1997. But the nation's health is vastly better for his efforts, starting at the neighborhood health center at Columbia Point in 1966 and continuing through cancer research, health maintenance organizations, quicker drug approvals, and portable health insurance.

Ted Kennedy helped make the draft fairer and then end it in the Johnson and Nixon years. The elderly people who receive Meals on Wheels owe him. So do the children who read to them through national service programs. Kennedy protected redistricting and advanced cleaner elections. He played critical roles in the eighteen-year-old vote and the abolition of the poll tax. If he failed to keep Clarence Thomas off the Supreme Court, he was central to G. Harrold Carswell's defeat. He blocked Robert Bork, but at a lasting cost to the civility of political dialogue.

His almost unnoticed role in foreign affairs runs from Vietnam to the Soviet Union, from Bangladesh to Chile, from Biafra to China, from South Africa to Chile to Ireland. He affected American relations with the world, occasionally through confrontation with the administration of the time, sometimes as a spokesman who conveyed American unity, and consistently as an advocate of the ideals of the Declaration of Independence. He would never grant that the ideas of Philadelphia were too advanced for Soweto, Moscow, Santiago, or Belfast.

With these issues and others, he stands out for perseverance. Twice in his career his party has lost control of the Senate. Other Democrats quit once they lost their chairmanships to Republicans. Kennedy seems to thrive as much on the complexities of getting things done in the minority as on the partisan delights of thwarting the majority. But it is not just persistence. He has an instinct for the rhythms of the Senate, a special knack for finding a critical Republican ally (even if he or she ends up with most of the credit), and an optimist's willingness to settle for half a loaf, or even a slice, today and work on getting the rest in the next Congress.

Still, many people think of him as a doctrinaire liberal, a spokesman—even an effective one—for a cause whose time has gone. That is much too simple. There was nothing "liberal" about denying bail to dangerous criminals or ending parole in the federal system. Airline deregulation contradicted the liberal orthodoxy that required as much control of big business as possible. For all the discomfort of cramped seats and awful food, and the loss of service to small cities with their subsidized flights cut off, most Americans fly today. That is because airline deregulation made fares much lower than they were in 1978.

That liberal image is fixed by more than his ideology. His following (and how many other Senators have a national following?) is passionate; it feeds on political red meat, and he has no hesitance in serving it. His status as the Kennedy who has aged recalls his brothers, who died young and are recalled as almost wholly liberal, when they, too, were more complex. The way America learns about the Senate has changed since 1962. Newspapers cover governmental process less than they used to. Television hardly covers it at all. There may be a picture when a Kennedy appears with a Faircloth, but no cameras are around to record the quick concession in a conversation with Bob Dole that clears a bill.

How should Kennedy be ranked in the history of the Senate? First, how do you rank Senators? About the only statistical measurement is length of service, which is almost always necessary for influence. He ranks eighth, but that is a measure that puts Strom Thurmond first and Carl Hayden of Arizona second, proving its limited value. Counting bills passed would be another statistic, but since Congress passes vastly more measures than it used to, it is one that hardly holds validity over time.

But even though the measure is subjective, it is worthwhile trying to compare accomplishments in terms of major legislation passed. Henry Clay, with the Missouri Compromise and the Compromise of 1850 to his credit, is in a class of his own, though the first was accomplished while he was Speaker of the House, not a Senator. In this century, Robert F. Wagner of New York, author of the National Labor Relations Act of 1935 and the Public Housing Act of 1937 and a forceful advocate of causes from unemployment insurance to antilynching legislation, is probably Kennedy's most credible rival.

There are Senators of great moments, who turned their party around and served the nation. Arthur H. Vandenberg internationalized the heart of the Republican Party. Everett M. Dirksen got Republicans who were distrustful of national government to acknowledge that the time had come for civil rights laws. J. William Fulbright's opposition to the war in Vietnam became a rallying point as the country changed its mind. Kennedy ranks with them for his battle, from the minority and against a popular President, to keep the nation from abandoning civil rights in the eighties.

Not all moments are equal. Henry Cabot Lodge's defeat of United States accession to the League of Nations and Richard B. Russell's many years of thwarting civil rights legislation required great skill and dedication. Lodge and Russell led causes with intense support. But they were wrong.

There were others who combined Kennedy's success in the Senate with national political leadership. Not John F. Kennedy, nor Richard M. Nixon; neither had an important Senate career. Lyndon Johnson and Harry Truman were not national leaders when they were chosen for the Vice Presidency. The better comparison is with two other Senators who did not become President, Hubert Humphrey and Robert A. Taft.

Humphrey led his party, first nationally and then in the Senate, "to walk forthrightly into the bright sunshine of human rights," as he said at the 1948 convention. In the Senate, he managed the 1964 Civil Rights Act on the floor. The Humphrey–Hawkins Act of his second Senate tenure was a model of his party's commitment to jobs, a cause to which Wagner had first called it. Until he served as Lyndon Johnson's Vice President, Humphrey owned the hearts of American liberals.

Taft served only fourteen years in the Senate, but he was an authentic contender for the Republican nomination three times. He spoke for its Midwestern roots, but he was not at its extreme conservative edge. His major legislative monument, the Taft–Hartley Act, was less severe than the House would have made it, and he favored federal aid for housing and education. He was known as "Mr. Republican."

Kennedy has affected his party's course as long, or longer, and at least as deeply. His legislative accomplishments are greater, and many have been won in the minority. He was less of an insider in running the Senate than was Taft, and probably Humphrey. But while the Senate is no longer run by anyone, Kennedy, as Trent Lott complains, often looks as if he is running it.

Ultimately comparisons and lists from different eras are intriguing but unsatisfying. Times and even institutions like the Senate change. Robert C. Byrd, who knows the Senate's history and how it has developed better than anyone, probably measured him best in 1997. As Kennedy completed thirty-five years in the Senate, Byrd said his career had been distinguished by a timeless talent for accommodation despite constant goals that cannot be halted by "even the strongest headwind." Imagining Kennedy in the Senate at other times, from the First Congress of 1789 through to the Depression, Byrd said, "Ted Kennedy would have been a leader, an outstanding Senator, at any period in the nation's history."

Kennedy has often been dismissed as an eloquent anachronism, the last liberal of a conservative age, overmatched by the hopes created when his brothers died young, perhaps someone to be pitied when his personal flaws could be read as running away from the excessive demands of political inheritance.

He deserves recognition not just as the leading Senator of his time, but as one of the greats in its history, wise in the workings of this singular institution, especially its demand to be more than partisan to accomplish much. Of all his legislative accomplishments, only the minimum-wage bills were passed without Republican help early on. A son of privilege, he has always identified with the poor and the oppressed. The deaths and tragedies around him would have led others to withdraw. He never quits, but sails against the wind.

Notes

Abbreviations used in the notes:

af	author's files
BEG	*Boston Evening Globe*
BG	*Boston Globe*
BH	*Boston Herald*
CQA	*Congressional Quarterly Almanac*
CQWR	*Congressional Quarterly Weekly Report*
CR	*Congressional Record.* Citations are to volume and session. The first session of the 99th Congress is cited *CR* 99:I. Page numbers are from the printed bound volumes through the 102nd Congress, which ended in 1992. Thereafter, because the Government Printing Office has not published annual indexes for sessions in 1993 or later, some citations are to the daily edition.
EMK	Edward M. Kennedy
GRFL	Gerald R. Ford Library
HHH	Hubert H. Humphrey Papers, Minnesota State Historical Society
int	interview (with author unless otherwise noted)
JCL	Jimmy Carter Library
JFK	John F. Kennedy
JFKL	John F. Kennedy Library
JFKL-EMK	John F. Kennedy Library–Edward M. Kennedy papers
LAT	*Los Angeles Times*
LBJ	Lyndon B. Johnson
LBJL	Lyndon Baines Johnson Library
LOC	Library of Congress
NA	National Archives
NPMP	Nixon Presidential Materials Project at the National Archives facility in College Park, Maryland
NYHT	*New York Herald Tribune*
NYT	*New York Times*
OH	oral history

PPP	*Public Papers of the Presidents,* published by the Government Printing Office, Washington, D.C.
RFK	Robert F. Kennedy
RRL	Ronald Reagan Library
WES	*Washington Evening Star*
WHCF	White House Central Files
WHSF	White House Special Files
WP	*Washington Post*

Congressional hearings, cited by committee, subject, and date, are published by the Government Printing Office, usually in the year in which they are held. If cited committee hearings were not published, that is noted.

CHAPTER 1 A DECISION FOR 1984

page

3 Message, scene, Larry Horowitz int 6/25/92.

3 Memorandums, Peter Goldman and Tony Fuller, *The Quest for the Presidency 1984* (New York: Bantam, 1987), pp. 59, 65.

4 "worrying about whether," *Philadelphia Inquirer* 6/28/82; Rules, Goldman and Fuller, *Quest for the Presidency 1984,* pp. 56f.; New Orleans, campaign structure, Mark Siegel int 3/12/92, Horowitz int 5/29/92, EMK int 9/10/92.

4 Staff, Jack Leslie int 4/9/92; Reagan, Horowitz int 5/29/72; nomination valuable, Bill Carrick int 5/27/92, EMK int 9/10/92.

4 Obligation, Stephen M. Smith, Jr., int 9/16/92; polls, Horowitz int 5/29/92, EMK Jr. int 1/31/96.

5 Cousins, Smith Jr., int 9/16/92; EMK Jr. speaks, *New York Daily News* 12/5/82; concerns, Kathleen Kennedy Townsend int 6/19/92, EMK Jr. ints 1/31/96, 2/13/96.

5 "two Republican parties," *NYT* 6/28/82; O'Neill, Thomas P. O'Neill, Jr., with William Novak, *Man of the House* (New York: Random House, 1987), p. 327 (O'Neill confused dates; Kennedy's records show the meeting was September 1982); Iowa, Bob Shrum int 3/10/92.

5 Family, EMK int 9/10/92.

6 Losing influence, Horowitz int 5/29/92; great Senator, John Tunney int 3/6/95.

6 Deletion, Shrum int 3/10/92, *New York Daily News* 12/5/82.

6 Priest, *New York Daily News* 12/2/82; gallant defeat, Kathleen Kennedy Townsend int 6/19/92.

6 Joe to EMK, Goldman and Fuller, *Quest for the Presidency 1984,* p. 66, Joseph P. Kennedy II int 1/31/95; "That's it," Horowitz int 5/29/92.

7 Patrick, Joan, EMK int 9/10/92; EMK and Patrick, Melissa Ludtke int 5/4/92.

7 Staff meeting, speech, Shrum int 3/10/92.

7 Drayne, author's recollection.

7 English arrival, Shrum int 3/10/92; call to Mondale, Goldman and Fuller, *Quest for the Presidency 1984,* p. 68.

7 Reporter, author's recollection.

7 EMK called, Pat Caddell int 1/1/98; Caddell, *Nightline* transcript 12/1/82; press conference, *BG* 12/2/82.

7 Tactics, *NYT* 12/2/82.

7 Ludtke to EMK 12/1/82, af.

7 "fulfilling," *BG* 2/14/81.

CHAPTER 2 BORN TO A DYNASTY

page

10 Joe Jr., EMK int WGBH 7/25/91; Jack, EMK in *BH* 10/22/79; comics, EMK, ed., *The Fruitful Bough: A Tribute to Joseph P. Kennedy* (privately printed, 1965), p. 201; Sermon on the Mount, EMK int 9/10/92; nursery school, Rose Fitzgerald Kennedy, *Times to Remember* (Garden City, N.Y.: Doubleday, 1974), p. 133.

10 Dodge ball, EMK int 9/10/92; Rosemary's impact, Rose Kennedy, *Family Weekly* in *WES* 8/3/75; influence on health care politics, EMK int 4/17/95.

10 Zebra, EMK int WGBH 7/25/91.

10 Fight, elevator, riding, EMK int WGBH 7/25/91.

11 Papal coronation, EMK int 10/9/95, *NYT* 3/24/39, Doris Kearns Goodwin, *The Fitzgeralds and the Kennedys* (New York: Simon & Schuster, 1987), pp. 579–81.

12 Joe Sr. letter, Joe Jr. as delegate, Goodwin, *Fitzgeralds and the Kennedys,* pp. 558–73, 586–98.

12 Teddy and Pegler, *Yonkers Herald Statesman* 11/5/40.

12 Father's letter, *Boston Record American* 10/14/62.

13 Public service, EMK int 9/10/92; "exciting talk," William Peters, "Teddy Kennedy," *Redbook* 6/62.

13 "We don't want any losers," Rose Kennedy, *Times to Remember,* p. 142; Teddy's sailing, pp. 268, 270.

13 Riviera, Laurence Leamer, *The Kennedy Women: The Saga of an American Family* (New York: Villard, 1994), pp. 284–85; parachute, Rose Kennedy, *Times to Remember,* p. 487; tossed overboard, EMK essay in JFK, ed., *As We Remember Joe* (privately printed, 1945).

14 Visiting churches, Luella Hennessey, as told to Margot Murphy, "Bringing Up the Kennedys," *Good Housekeeping* 8/61; St. Luke, *BG* 9/22/70; religion, Rose Kennedy, *Times to Remember,* p. 161.

14 Candy, EMK in *Fruitful Bough,* p. 205; universe, Eunice Kennedy Shriver int 12/13/93; plumber's son, EMK int 9/10/92.

14 Portsmouth Priory, EMK int 10/9/95; checking up, EMK essay in Patricia Kennedy Lawford, ed., *That Shining Hour* (privately printed, 1969), p. 300.

15 Benet, EMK, "The Spark Still Glows," *Parade* 11/13/83; Burke etc., EMK, ed., *Words Jack Loved* (privately printed, 1977)

15 Movies, EMK int 11/27/92.

15 Priests, EMK, *Fruitful Bough,* p. 206.

15 Pressure, JFK int with James M. Cannon, 4/5/63, Cannon files.

16 Boarding school, EMK int with Natalie Jacobson, 10/29/94.

16 Honey Fitz and Kennedy children, Jean Kennedy Smith int 5/4/95; visits with Honey Fitz, EMK int 9/10/92, EMK introduction to Nancy Sirkis, *Boston* (New York: Viking, 1965).

16 Fessenden, Daniel B. Burns int 2/17/95.

17 Cranwell, Burns int 2/17/95; mother on Catholic schools, Rose Kennedy, *Times to Remember,* p. 162.

17 Horsing around, Frank Cahouet int 3/23/95; EMK letters 12/12/46, 4/25/48, Doris Kearns Goodwin papers, File "Milton," JFKL; Boogies, Hugh Chandler int 3/14/95.

17 School activities, tennis, yearbook comment, *The Orange and Blue,* Milton Academy, 1950; run to huddle, EMK int 4/17/95; football, Herbert Stokinger ints 4/13/95, 4/22/95.

17 Bobby and Milton, EMK int 4/17/95.

18 Unassuming, John Culver int 2/1/93.

18 Spanish exam, EMK int 4/17/95.

18 Harvard standard, McGeorge Bundy int 11/11/93.

18 Careful, Grace Warnecke int 12/15/94; playboy, Arthur M. Schlesinger, Jr., *A Thousand Days: John F. Kennedy in the White House* (Boston: Houghton Mifflin, 1965), p. 79.

18 Korea, *BG, BH, NYT* 6/25/51; father's reaction, Rose Kennedy, *Times to Remember,* p. 141; intervention, Burton Hersh, *The Education of Edward Kennedy* (New York: Morrow, 1972), p. 82; draft rules, EMK int 4/17/95.

19 Holabird, Army behind him, EMK int 4/17/95; Camp Gordon, EMK int 4/17/95, Matty Troy int 3/15/94.

19 Political questions, Culver int 2/1/93.

19 Grades, James MacGregor Burns, *Edward Kennedy and the Camelot Legacy* (New York: Norton, 1976), pp. 46, 49.

20 Courses, EMK int 4/17/95.

20 Mass, Vicki Kennedy int 6/28/95; golf, rugby, Lester David, *Ted Kennedy, Triumphs and Tragedies* (New York: Grosset & Dunlap, 1971), pp. 56, 61; basketball, friends, EMK int 4/17/95.

20 RFK at Hoover Commission, Arthur M. Schlesinger, Jr., *Robert Kennedy and His Times* (Boston: Houghton Mifflin, 1978), p. 108; Davenport game, EMK int 4/17/95, EMK essay in Lawford, ed., *That Shining Hour,* p. 301.

21 JFK watched games, EMK int 10/9/96.

21 Yale games, father, EMK ints 9/10/92, 4/17/95.

21 Algeria trip, Fred Holborn int 1/27/94, Hersh, *Education of Edward Kennedy,* p. 99.

21 JFK urged, EMK int 10/9/96; times with Tunney, parties, studying, John V. Tunney int 3/6/95; studying, Warren Rogers, "Ted Kennedy Talks About the Past, and His Future," *Look* 3/4/69; speeding, Hersh, *Education of Edward Kennedy,* p. 104; RFK quote, David, *Ted Kennedy,* p. 81.

22 JFK asked Humphrey, EMK int 10/9/96.

22 Reuther, Burns, *Edward Kennedy and the Camelot Legacy,* p. 80; other speakers, JFK, EMK int 4/17/95; Humphrey letter 3/14/59, HHH 150.H.9.9 (B), Box 33.

22 Moot court, Tunney int 3/6/95.

22 Reed opinion, vote, outdid RFK, Tunney int 3/6/95; Haynsworth vote, EMK int 4/17/95.

23 Meeting, Joan Kennedy memo, Rose Kennedy, *Times to Remember,* pp. 430–31; beauty contests, commercials, Lester David, *Joan: The Reluctant Kennedy* (New York: Funk & Wagnalls, 1974), pp. 38, 42.

23 Stowe, Betty Hannah Hoffman, "What It's Like to Marry a Kennedy," *Ladies' Home Journal* 10/62; cozy week, Joan Kennedy int 1/18/96.

23 Courtship, Joan Kennedy int 1/18/96; learned to paint, Melody Miller int 2/8/95.

23 Proposal, Rose Kennedy, *Times to Remember,* pp. 431–32.

24 Politics never discussed, William Peters, "Teddy Kennedy," *Redbook* 6/62; political weekend, "I didn't have a clue," Joan Kennedy int 1/18/96.

24 "support my daughter," David, *Ted Kennedy,* p. 72; ring, Leamer, *Kennedy Women,* p. 473.

24 Jitters, Joan Kennedy int 1/18/96, Leamer, *Kennedy Women,* p. 474.

24 Faithful, Leamer, *Kennedy Women,* pp. 477f, Joan Kennedy int 10/12/98.

24 Wedding, *NYT* 11/29/58, David, *Ted Kennedy,* pp. 75–77, Leamer, *Kennedy Women,* pp. 477–78, Joan Kennedy int 1/18/96.

CHAPTER 3 THE APPRENTICE

page

25 Drivers, Theodore C. Sorensen int 6/17/96.

25 JFK talked him up, Thomas Winship OH 7/8/64, JFKL.

25 EMK campaigning, Jim King int 2/14/94.

25 Press setting vote target, Burns, *Edward Kennedy and the Camelot Legacy,* p. 64; G&G, Kenneth P. O'Donnell and David F. Powers with Joe McCarthy, *Johnny, We Hardly Knew Ye* (Boston: Little, Brown, 1972), p. 42, Jean Stein and George Plimpton, *American Journey: The Times of Robert Kennedy* (New York: Harcourt Brace Jovanovich, 1970), pp. 44f.

25 "Flowering," Harold H. Martin, "The Amazing Kennedys," *Saturday Evening Post* 9/7/57; Joe as source, Rose Kennedy, *Times to Remember,* p. 433.

26 Toasts, EMK, "The Spark Still Glows," *Parade* 11/13/83.

27 "chain store," Hubert H. Humphrey, *The Education of a Public Man: My Life in Politics* (Garden City, N.Y.: Doubleday, 1976), p. 152; Kara's birth, David, *Ted Kennedy,* p. 83; energy, Healy int 7/9/94, *BG* 3/27/60.

28 Ski jump, *NYHT* 5/16/61; sharpshooter, Theodore C. Sorensen, "Ted Kennedy's Memories of JFK," *McCall's* 11/73; *Caroline,* EMK int 10/9/95.

28 EMK in West Virginia, William H. Honan, *Ted Kennedy, Profile of a Survivor* (New York: Quadrangle, 1972), p. 9; poverty, EMK int 9/10/92; Joan, Joan Kennedy int 12/12/96.

28 Humphrey on RFK and FDR Jr., Humphrey, *Education of a Public Man,* p. 363, Humphrey, Memo 4/2/68 11:35 A.M., HHH 148.A.4.7 (B), Box 412, File VP Memos for the Record 1968.

28 Moved, Joan Kennedy int 12/12/96.

29 Impressed, Frank Mankiewicz int 7/5/95; dubious, Fred Dutton int 7/25/95.

29 Arizona, EMK int 10/9/95; Denver, *Rocky Mountain News* 10/2/59, *Time* 10/12/59.

29 Utah, Oscar McConkie, Sr., int 3/3/95, EMK int 10/9/95.

30 Roll call, Theodore H. White, *The Making of the President 1960* (New York: Atheneum, 1961), p. 169, Hersh, *Education,* p. 137.

30 Texas, Ralph Yarborough telegram to EMK 11/9/60, Yarborough papers, University of Texas, Austin, Box 2R507, File EMK Massachusetts; bronco, *Miles City Star* 5/14/92.

31 Arizona, Joan Kennedy int 12/12/96; New Mexico, EMK int 2/3/97; RFK declines, Ted decides, EMK int 10/7/93.

31 "Up for grabs," Thomas B. Morgan, "Teddy," *Esquire* 4/62; "cake," Milton S. Gwirtzman int 4/3/93; "opening," Eunice Shriver int 12/13/93.

31 Meeting with JFK, EMK ints 10/7/93, 10/9/96, 2/3/97.

32 Church organized, EMK publicity, Frank E. Moss int 3/3/95; JFK and Africa, *BG* 12/11/60; Abidjan dinner, Frank E. Moss diary, LOC Former Members Oral History Project, Box 8; EMK toast, EMK int 10/7/93, Moss int 3/3/95.

32 Healy story, *BG* 12/18/60; McCormack hope and rejection, Edward J. McCormack int 5/15/92.

32 JFK doubts, Lawrence F. O'Brien, *No Final Victories: A Life in Politics from John F. Kennedy to Watergate* (Garden City, N.Y.: Doubleday, 1974), p. 142; O'Donnell opposition, Arthur M. Schlesinger, Jr., to author 1/20/95, EMK int 10/7/93.

33 Father's intense support, Robert F. Kennedy, *Robert Kennedy in His Own Words: The Unpublished Recollections of the Kennedy Years,* ed. Edwin O. Guthman and Jeffrey Shulman (New York: Bantam, 1988), pp. 328–29; medals, Jim King int 2/14/94.

33 Africa speech notes, NA, RG 46, Judiciary Committee, Subcommittee on Refugees and Escapees, Box 152, folder United States Senate Committee on Foreign Relations Africa; Cancer Crusade, EMK int 10/7/93; Latin America, *BG* 9/17–21/61; Morrissey, *NYHT,* 5/16/61; community singing, praises coffee, *BG* 11/3/61; Faneuil Hall, *BG* 7/5/61.

33 Swearing in, Kevin H. White int 10/7/93.

34 Apartment hunting, Joan Kennedy int 1/12/96; dresses, *NYHT* 4/17/62; jewelry, "forever house," "Ted's away," Hoffman, "What It's Like to Marry a Kennedy," *Ladies' Home Journal* 10/62; JFK swimming, Joan Kennedy int 9/19/98.

34 Hennessy case, EMK int 10/7/93, Robert H. Stanziani int 11/11/93; victories, *BG* 11/21/61, Joe Laurano int 2/8/95; general experience, Newman Flanagan int 2/9/95.

34 EMK for AG, McCormack for governor, *BG* 11/3/61, *BH* 11/20/61; attitude on McCormacks, Joe Laurano int 2/8/95, John E. Nolan int 3/30/95.

35 Poll, *BH* 2/1/62, Stewart Alsop, "What Makes Teddy Run, and Why Did the President Let Him?" *Saturday Evening Post* 10/27/62; poll invented, Gerard Doherty int 3/24/95.

35 Itinerary, *NYT* 7/18/61; speech, *BH* 1/22/62; dinner quarrel, *Boston Record American*

10/26/61; "Some 200 million," *BG* 9/17/61; stayed in bed, David, *Joan,* p. 114; catch up, Joseph Laurano int 2/8/95.

35 Locke-Ober, Gerard Doherty int 11/18/94; EMK says he is running, *BG, BH* 12/6/61; tells Smith, Smith lends Gwirtzman, Milton S. Gwirtzman OH 1/19/66, JFKL, p. 26.

35 Oval Office visit, EMK int 10/9/96, President's appointment calendar, JFKL.

35 Feared cheating question, Benjamin C. Bradlee, *Conversations with Kennedy* (New York: Norton, 1975), p. 67; *Meet the Press* transcript 3/11/62; confusion pleased JFK, EMK int 1/6/93.

35 Announcement, McCormack reaction, *BEG* 3/14/62.

36 Aide helps, *Time* 3/23/62; tells columnist, Richard Starnes column 1962.

36 Rumors, conversations with Maguire, JFK editors, Healy int 7/9/94.

36 Cheating story, *BG* 3/30/62; *Boston Record* 3/30/62.

37 Writing of story, Healy int 7/9/94.

37 "On his own," *NYT* 3/15/62; reliving campaigns, EMK int 10/9/96.

37 JFK attitude, April 27 meeting, Gerard Doherty int 11/18/94, Dan Fenn int 1/6/95, Arthur M. Schlesinger, Jr., to author 1/10/95 and personal journal 4/27/61; assumed names, Maurice Donahue OH 3/27/76, JFKL, p. 18; patronage, Paul Driscoll, "Jack Was Asking About You," *New Republic* 6/4/62; Sorensen help, Theodore C. Sorensen papers, JFKL, Box 5; coaching, John Culver int 2/1/93; Hartigan, *Wall Street Journal* 6/5/62.

37 Academics, Murray B. Levin, *Kennedy Campaigning* (Boston: Beacon Press, 1966), pp. 146–47; Howe letter, Mark deWolfe Howe papers, Harvard Law School Library, MS Box 13, folder 5.

38 Reston, *NYT* 3/11/62, 9/20/62; editorial, *NYT* 9/19/62.

38 McCormack attitudes, McCormack int 5/15/92.

38 Daily routine, Doherty int 11/18/94; three-fourths of delegates, Levin, *Kennedy Campaigning,* p. 54; shout, Stewart Alsop, "What Makes Teddy Run"; 5 A.M., Gwirtzman int 1/30/93; roofer, *BG* 6/10/62; plane, parade, Andrew Vitali int 1/26/95.

38 Joan's role, Doherty int 12/13/96, Joan Kennedy int 12/12/96.

38 Bounce back, John Culver int 2/3/93.

39 Parochial aid, Charles Haar int 6/1/93; academic support, Culver int 2/1/93.

39 Doherty count and Boland speech, Levin, *Kennedy Campaigning,* pp. 65, 66; McCormack count, *BEG* 6/8/62.

39 Overture to McCormack, McCormack int 5/15/92, Thomas P. O'Neill, Jr., int 11/23/92.

39 Spending, Levin, *Kennedy Campaigning,* pp. 233–84; polls, Joe Napolitan int 2/16/95.

40 Debate challenge, Doherty int 11/18/94, Edward Martin int 12/17/92, *BG* 8/10/62, *Boston Record* 8/14/62.

40 Robert's warning, John Seigenthaler int 9/27/95; Berlin convoy, Levin, *Kennedy Campaigning,* p. 195; scored on nuclear weapons, *BG, WES* 8/28/62.

40 McCormack quotations, Levin, *Kennedy Campaigning,* p. 210; EMK startled, *NYT* 8/28/62; crowd, *WP* 8/28/62; McCormack pointed, John Culver int 2/1/93.

40 EMK shaken, *WES* 8/28/62; sock McCormack, Doherty int 11/18/94; no handshake, *BG* 8/28/62.

41 Reactions, Eddie Martin int 12/17/92.

41 JFK calls, telephone logs, JFKL, Gwirtzman int 1/30/93.

41 Poll, Joe Napolitan int 2/16/95, Napolitan, "Survey on reaction to Kennedy–McCormack debate," 8/28/62, af.

42 Talk shows, bakery, EMK int 4/17/95; "Ted, me boy," *NYT* 9/3/62.

42 "Pillow fight," *BG* 9/6/62; poll, Napolitan, "Survey on Voter Reaction" to second Kennedy-McCormack debate, af.

42 Avoid Cuba, Theodore C. Sorensen int 6/17/96; Eisenhower, *NYT* 10/16/62.

CHAPTER 4 THE PRESIDENT'S BROTHER

page

43 Presidential advice, Russell visit, John Culver ints 2/1/93, 4/16/93; Russell quotation, *NYHT* 8/4/63.

43 William S. White, *Citadel: The Story of the U.S. Senate* (New York: Harper, 1957), p. 2; Kennedy read it, Theo Lippman, *Senator Ted Kennedy: The Career Behind the Image* (New York: Norton, 1976), p. 2.

44 Senate atmosphere, Charles Ferris int 4/10/93.

44 Washington preconceptions of EMK, Robert Massie, "New Taft and a Young Kennedy Go to Washington," *Saturday Evening Post* 1/19/63; Win Turner, quoted in William V. Shannon, "Emergence of Senator Kennedy," *NYT Magazine* 8/22/65, *NYHT* 8/4/63.

45 Kara, *BG* 1/10/63.

45 Joan, Sue Seay, "A New Mrs. Kennedy in Washington," *Look* 2/26/63; annoyed, Marcia Chellis, *Living with the Kennedys: The Joan Kennedy Story* (New York: Simon & Schuster, 1985), p. 39.

45 Impressions, Russell Long int 8/9/94; Birch Bayh int 2/3/95.

46 Party, Thomas P. O'Neill, Jr., int 11/23/92.

46 Avoiding attention, Culver int 2/1/93; freshmen, *Boston Sunday Advertiser* 7/28/63; Press Club remarks, *Boston Traveler* 2/14/63.

46 Prayer breakfasts, EMK int 10/9/96.

46 Impression on Southerners, Eastland summons, Jerry Marsh int 1/14/93; subcommittee assignments, EMK int 10/7/93.

47 Eastland's advice, EMK int 10/7/93.

47 Local focus, "your problems," EMK int 10/7/93; "Some pipeline," Bradlee, *Conversations with Kennedy*, p. 129.

47 Fire hoses, dogs, *NYT* 5/4/63.

48 University, *NYT* 6/12/63.

48 Choosing sides, Richard Reeves, *President Kennedy: Profile of Power* (New York: Simon & Schuster, 1993), p. 522; "color blind," *PPP Kennedy 1963*, p. 238; Evers murder, *NYT* 6/13/63; trial, *NYT* 2/6/94; bill introduced, *NYT* 6/20/63.

48 EMK and march, EMK int 10/9/96; King text, William Safire, ed., *Lend Me Your Ears: Great Speeches in History* (New York: Norton, 1992), pp. 499–500.

49 Russian note-takers, *BEG* 9/16/63; Belgrade text, JFKL-EMK.

49 Coup, Reeves, *President Kennedy*, p. 604; Madame Nhu lunch, *BEG* 9/13/63; Joan, *BEG* 10/11/63.

50 Cables, Rusk to U.S. Embassy Belgrade 9/13/63, Eric Kocher to Rusk 9/15/63, JFKL; Kocher's meeting with EMK, Kocher, letter to author 1/12/94.

50 Advertising Club text, JFKL-EMK.

51 American University speech, *PPP Kennedy 1963*, pp. 463–64.

51 Test ban speech, *CR* 88:I, pp. 16549, 16553.

51 Honey Fitz imitation, Bradlee, *Conversations with Kennedy*, p. 168; battlefields, EMK int 10/9/96.

51 Swims, talks, EMK int 10/9/96.

52 "Don't trust," Bob Shrum int 4/8/94.

52 November 22 Senate debate, *CR* 88:I, pp. 22679–92.

52 Riedel, EMK reactions, William Manchester, *Death of a President* (New York: Harper & Row, 1967), p. 197.

52 Phones dead, car trip, Hooton, Milton Gwirtzman OH 1/19/66, JFKL, p. 42.

53 EMK reaches RFK, Manchester, *Death of a President*, p. 199.

53 EMK talks to mother, Manchester, *Death of a President*, p. 199; Joan, p. 355; trip to Hyannis,

pp. 256, 371; delay telling father, p. 372; tears TV wires out, p. 373; calls Martin, Martin int 12/17/92.

53 Return to White House, Manchester, *Death of a President,* p. 554.

53 Events of 11/24/63, Manchester, *Death of a President,* pp. 518–43; Mansfield eulogy, *CR* 88:I, p. 22695.

53 Visit to Rotunda, Gwirtzman OH, p. 42; length of line, Manchester, *Death of a President,* p. 563.

54 Hatless, Manchester, *Death of a President,* p. 576.

54 Worldwide pause, Manchester, *Death of a President,* pp. 584–85.

54 Cushing, Communion, Hannan, Manchester, *Death of a President,* pp. 586–88.

54 Cushing in English, Manchester, *Death of a President,* p. 589.

54 Eulogy decision, Manchester, *Death of a President,* p. 601.

55 Birthday party, O'Donnell and Powers with McCarthy, *Johnny, We Hardly Knew Ye,* p. 42.

CHAPTER 5 CRASH

page

56 Idlewild, *NYT* 12/25/63; Boston dinner, *BH* 3/16/64.

56 Macmillan, Pompidou, William vanden Heuvel int 4/6/95.

57 Ireland, Edward Behr, "A Day of Joy and Sadness," *Saturday Evening Post* 7/11–18/64.

57 Maiden speech, *CR* 88:II, pp. 7375–80.

57 Joan, voice broke, *BEG* 4/9/64; Douglas et al., *CR* 88:II, p. 7380.

57 LBJ, Doris Kearns, *Lyndon Johnson and the American Dream* (New York: Harper, 1976), p. 191; filibusters, *CQA 1964,* p. 368; Dirksen role, *NYT* 4/10/64.

58 Indianapolis, *Indianapolis Star* 4/26/64; "President Kennedy believed," *WES* 5/18/64; "stood for," *WP* 5/18/64; parade, Alan Lupo, *Liberty's Chosen Home: The Politics of Violence in Boston* (Boston: Little, Brown, 1977), p. 146.

58 RFK on Vice Presidency, Schlesinger, *Robert Kennedy and His Times,* p. 653; EMK on office, O'Donnell and Powers with McCarthy, *Johnny, We Hardly Knew Ye,* p. 392; May meeting, p. 395; RFK announces, *NYT* 8/23/64; first choice, Robert Mann, *The Walls of Jericho* (New York: Harcourt Brace, 1996), pp. 433–36; LBJ excludes cabinet, *NYT* 8/1/64.

59 Ovation, *NYT* 8/28/64.

59 Vote to end filibuster, *NYT* 6/11/64; passage, *NYT* 6/20/64.

59 Minimum visibility, Civil Aeronautics Board Report 2-0058, 9/10/63, Vol. 302, pp. 42–43, Exhibit 3730, NA; EMK account, *BG* 10/20/64.

60 Bayh account, *BG* 6/21/64.

60 Arrival at hospital, Martin int 12/17/92; Johnson interest, Name File Edward M. (Ted) Kennedy 1963–64, LBJL.

61 Medical situation, *BG* 12/1/64.

61 Operation considered, rejected, Rita Dallas with Jeanira Ratcliffe, *The Kennedy Case* (New York: Putnam, 1973), pp. 284f; William V. Shannon, "How Ted Kennedy Survived His Ordeal," *Good Housekeeping* 4/65.

61 Transferred, *NYT* 7/10/64; Senate business, Andrew Vitali int 1/26/95.

61 Rehabilitation, movies, Vitali int 1/25/95; animal stories, Christine Sadler, "Coming of Age of Joan Kennedy," *McCall's* 2/65.

62 Comics, EMK, *Fruitful Bough,* p. 201, candy, p. 205, taskmaster, p. 203.

62 LBJ visit, Martin int 6/2/93.

62 Campaigning, Joan Kennedy int 12/12/96, *BEG* 10/8/64, *BH* 10/12, 10/23, 10/28/64, *BG* 10/20/64; "fun," *BH* 10/26/64.

63 Fretted over RFK, John Kenneth Galbraith int 2/6/95; "casting a shadow," William V. Shannon, *The Heir Apparent: Robert Kennedy and the Struggle for Power* (New York: Macmillan, 1967), p. 75.

63 Sense of purpose, George McGovern int 4/14/94; pain, Vitali int 1/25/95; seminars, Beer int 7/2/93, Galbraith int 2/6/95, Lippman, *Senator Ted Kennedy,* p. 28.

63 Health care costs, Doherty int 11/18/94.

63 Visit to grave, Martin int 12/17/92, *BEG* 12/16/64; EMK on hospitalization, Shannon, "How Ted Kennedy Survived His Ordeal," *Good Housekeeping,* 4/65.

CHAPTER 6 THE LAWMAKER

page

65 Longevity, Gwirtzman int 4/3/93.

65 Seeking issue, RFK contrast, David Burke int 6/3/93.

65 Cattle prods, *NYT* 2/11/65; whips etc., *NYT* 3/8/65.

66 Jackson death, *NYT* 2/28/65; Reeb, *BG, NYT* 3/12/65.

66 "American problem," *PPP Johnson 1965,* Book 1 pp. 281ff.

66 Selma singing, *NYT* 3/16/64.

66 March, Liuzzo, *NYT* 3/27/65.

66 Rauh advice, Burke int 6/3/93; Kennedy speech, *CR* 89:I, p. 10081.

67 Rauh meetings, Marshall, EMK int 10/7/93; students, Charles Haar int 6/1/93; Freund, Howard, Jeanus Parks int 2/22/95.

67 Poll tax decision, *Breedlove v. Suttles* 302 U.S. 277.

67 Sunflower registration, U.S. Commission on Civil Rights, "Voting: 1961 Commission on Civil Rights Report," p. 275; Committee debate, Senate Judiciary Committee unpublished transcripts, NA.

68 Kennedy speech, *CR* 89:I, pp. 7882ff; Rauh authorship, Burke int 6/3/93; Freund letter 4/21/65, Jeanus Parks files.

68 LBJ reluctance, Katzenbach to LBJ 5/21/65, LE/HU 2-7, LBJL, Evans and Novak, *WP* 5/16/65; leadership bill, *NYT* 5/1/65, Katzenbach int 5/12/93; "purists," Lawrence F. O'Brien OH 7/24/86, LBJL, p. 61; angered, EMK int 10/7/93; Rauh et al., Rauh to Walter P. Reuther 5/14/65, Rauh papers, LOC, Box 26.

68 More support, Katzenbach int 5/12/93; mutual unhappiness, foreknowledge of defeat, Rauh to Reuther 5/13/65.

69 "The question," *CR* 89:I, p. 10081.

69 Mansfield, *CR* 89:I, p. 10079; Humphrey view, letter to EMK 4/30/65, Humphrey correspondence, LBJL, FG 440, Box 346; *WES* 5/12/65; *NYHT* 5/16/65; *WP* 5/12/65.

69 Supreme Court, *Harper v. Virginia Board of Elections* 363 U.S. 663, 679.

69 "He believed," Parks int 2/22/95.

70 "Defining aspect," EMK int 9/10/92.

70 JFK cause, Robert F. Kennedy, introduction to new edition of John F. Kennedy, *A Nation of Immigrants* (New York: Harper, 1964), p. ix; 1965 quotas, *BH* 8/29/65; "indefensible" JFK, *A Nation of Immigrants* (New York: B'nai B'rith, 1959), p. 34; Eastland acceptance, Mike Manatos to Lawrence F. O'Brien 8/16/65; WHCF, Name File Edward M. Kennedy 1967, LBJL; EMK assignment, Manatos to O'Brien, Dale de Haan int 12/22/93.

70 EMK tells Honey Fitz story, Dale de Haan int 2/22/95; his speech, *CR* 54:II, Appendix, pp. 45–49.

71 The evidence that the speech was never given lies in its appearance in the Appendix of the *Congressional Record,* where inserted material was published then, and in the fact that Fitzgerald did not vote on the day the speech was given nor on the previous or following days; Cleveland veto, John Henry Cutler, *"Honey Fitz," Three Steps to the White House* (Indianapolis: Bobbs-Merrill, 1962), p. 64.

71 EMK effort, Dale de Haan int 12/22/93; Irish, Burke int 6/3/93; lobbying, Irish diplomat int 5/8/95.

71 Committee action, *CQ* 9/10/67, p. 1837.

72 Holland-EMK exchange, *CR* 89:I, p. 24776–77.
72 Holland-RFK exchange, *CR* 89:I, p. 24778.
73 "Waiting," William vanden Heuvel and Milton Gwirtzman, *On His Own: Robert F. Kennedy, 1964–1968* (Garden City, N.Y.: Doubleday, 1970), p. 64.
73 Visited Joe Kennedy, Milton S. Gwirtzman int 1/30/93; "never let," *WP* 8/31/65.
73 "After my father," EMK int 10/7/93; surprised at weakness, George S. Abrams int 6/17/92, Burke OH 12/8/71, JFKL.
73 Morrissey experience, 10/13/65, unpublished Judiciary Committee executive session, Box 22 Sen 8E2/22/15/5 NA.
73 EMK pressed, Joseph A. Califano, Jr., 6/10/90 draft of *The Triumph and Tragedy of Lyndon Johnson* (New York: Simon & Schuster, 1991), Burke int 6/3/93; Katzenbach warning, *BG* 10/1/65; confirmation likely, Katzenbach int 5/12/93.
73 Morrissey, Katzenbach to LBJ 9/2/65 and Watson to LBJ 9/13/65, Diary Backup 9/24/65, LBJL; a favor, Katzenbach int 5/12/93.
74 Dirt, LBJ to Hoover, William Sullivan int by Arthur M. Schlesinger, Jr., JFKL; RFK letter to LBJ 9/2/64, WHCF FG530/ST21, LBJL; Laitin briefing, WH Press Office Files, Box 8, 9/26/65, LBJL.
74 "nauseous," *NYHT* 10/28/65; Wyzanski, *NYHT* editorial 10/4/65; "backyard fence fights," *BG* 10/1/65; Cushing, *NYHT* 10/11/65; McCormack, *BG* 9/28/65.
74 Dirksen lulls, Burke int 6/3/93; Kennedy leaves, *BG* 10/13/65; Jenner, Segal, *Baltimore Sun* 10/13/65.
75 Morrissey, Dodd, Smathers in hearing, unpublished Senate Judiciary Committee transcripts 10/13/65, NA; Kennedy coaching, John McAvoy int 11/23/92; Dirksen account, *BG* 10/14/65.
75 Legislature, *BG* 10/15/65; O'Donnell, Bob Healy int 7/9/94; register to vote, *BG* 10/18/65.
75 Tydings view, Joseph D. Tydings int 6/15/93; Tydings speech, *CR* 89:I, p. 27108; anger, McAvoy int 11/23/92, Tydings int 6/15/93.
75 RFK lobbied, Evans and Novak, *NYHT* 10/24/63; Capri, *Il Mattino* of Naples, Italy, 5/28/61, with LOC translation, Everett McKinley Dirksen Library, Gwirtzman int 1/30/93.
76 "hate," Neil MacNeil, *Dirksen: Portrait of a Public Man* (Cleveland: World, 1970), p. 267.
76 Unconvincing, *Baltimore Sun* 10/17/65; "no basis," Katzenbach to Eastland 10/18/65, John W. Macy Office Files, LBJL; Morrissey a liar, Katzenbach int 5/12/93.
76 No early lobbying, O'Brien OH XIII 9/30/86, LBJL; "Work hard," Califano draft.
76 RFK, former Senate aide int 1992.
76 Told LBJ, Califano draft; gallery full, Dirksen summons, *Baltimore Sun* 10/22/65; thirty-two present, *CR* 89:I, p. 27936.
76 "local law school," *CR* 89:I, p. 27936; sobs, *Baltimore Sun* 10/22/65.
77 RFK decision, cable Gwirtzman to EMK, NA, RG 46, Judiciary Committee, Subcommittee on Refugees and Escapees, Indo-China hearings, Box 152, folder Vietnam 1965; Morrissey letter, Macy Office Files, Morrissey, Francis X., LBJL.

CHAPTER 7 TO VIETNAM—AND BACK

page
78 Arrival statement, *BEG* 10/23/65.
78 Foreign issues, George S. Abrams int 11/27/92; refugees, Burke int 6/1/93.
79 U.S. in Vietnam, 1956–1963, Bruce Palmer, *The 25-Year War: America's Military Role in Vietnam* (Lexington: University Press of Kentucky, 1984), pp. 10f.
79 LBJ plans, Robert S. McNamara with Brian VanDeMark, *In Retrospect: The Tragedy and Lessons of Vietnam* (New York: Times Books, 1995), pp. 120–22; no second attack, Jim and Sybil Stockdale, *In Love and War,* rev. ed. (Annapolis: Naval Institute Press, 1990), pp. 3–25; McNamara quoting the North Vietnamese commander Vo Nguyen Giap: "He said def-

initely it did not occur," NBC *Today* show 11/24/95; EMK to Yarborough 8/6/64, Yarborough papers, University of Texas, Austin, Box 2R507.

80 Troop strength, McNamara, *In Retrospect,* p. 169; EMK letter, *BG* 5/6/65.

80 Opening statement, Senate Judiciary Subcommittee on Refugees and Escapees, hearings on Vietnam 7/13/65, p. 2; transportation, temporary, 7/14/65, pp. 34–35; voluntary agencies, 7/14/65, pp. 39–83; coordination, 7/20/65, pp. 95–96; planning, p. 105; Poats on black market, camps, 7/27/65, pp. 121–22.

81 Payments, Poats, Hearings, p. 125; one million refugees, Frank H. Weitzel memo, p. 343, pressing administration, Abrams int 11/27/92; new AID program, *NYT* 8/31/65.

81 Briefings, Dale de Haan notes, NA, RG 46, Judiciary Committee, Subcommittee on Refugees and Escapees, Box 151, Trip Files 1965–78 Indochina, folder 1965 study mission.

81 Tran Ngoc Ling, Marks, EMK "Notes from Diary Far East Trip," NA, RG 46, Box 151 Trip Files 1965–78, folder Southeast Asia Trip, pp. 3–5, 7–8.

82 Purple Hearts, *Boston Record* 10/26/65; tie clips, huts, *NYT* 10/25/65; blankets, EMK, "A Fresh Look at Vietnam," *Look* 2/8/66.

82 Military missions, *NYT* 10/28/65; dancers, former government official int 1993.

82 Westmoreland, EMK "Notes from Diary," pp. 14–15.

82 Saigon statement, *Saigon Daily News* 10/27/65; Lowell Tech, *BG* 11/16/65.

83 Peabody, *Boston Traveler* 11/17/65; Merrimack Valley, *BG* 11/22/65; Brockton, *BG* 12/16/65.

83 Doubts, EMK int 4/17/95; *Look* article, Abrams int 11/27/92; Tunney and Fall, EMK int 4/17/95, Tunney int 3/6/95.

83 Tran Van Do, EMK "Notes from Diary," p. 15; RFK statement, *NYT* 2/20/66.

84 Humphrey, *NYT* 2/21/66; "Ho," *Chicago Tribune* 2/21/66; *Meet the Press* 3/6/66.

84 Bombing pause, resumption, McNamara, *In Retrospect,* pp. 218–31; reactions, *BG* 2/7/66.

84 Huntington role, Lippman, *Senator Ted Kennedy,* p. 75; ABC appearance, *BG* 6/13/66.

85 Random systems, *BG* 6/30/66.

85 Withhold federal aid, *BG* 6/16/66; Boston School Committee, *BH* 6/30/66.

85 EMK to SCLC, 8/8/66, EMK-JFKL.

86 Columbia Point project, Seymour S. Bellin and H. Jack Geiger, "Actual Public Acceptance of the Neighborhood Health Center by the Urban Poor," in Robert M. Hollister, Bernard M. Kramer, and Seymour S. Bellin, eds., *Neighborhood Health Centers* (Lexington, Mass.: D. C. Heath, 1974), pp. 199–212; developing countries, Sar A. Levitan, "Healing the Poor in Their Back Yard," in Hollister et al., p. 53; staffing, H. Jack Geiger int 5/18/95; impact, Muriel Rue int 10/12/95.

87 Urging, Joseph T. English int 5/11/95; EMK visit, Geiger int 5/18/95; EMK impressions, Burke int 12/4/95.

87 Frustration with health efforts, English int 5/11/95.

87 Legislation, Timberlawn meeting, English int 5/11/95.

87 EMK text 12/11/66, EMK-JFKL.

88 Coverage in 1995, *America's Health Centers* (Washington, D.C.: National Association of Community Health Centers, 1995), p. 1; Kennedy and Weicker, James R. Kanak int 10/12/95.

88 Omaha, *NYT* 9/25/66; Douglas, Paul H. Douglas, *In the Fullness of Time* (New York: Harcourt Brace Jovanovich, 1972), pp. 262f, *Chicago Tribune* 10/29/66, *Chicago Daily News* 10/28/66, *Chicago Sun-Times* 10/29/66; Bruce Biossat column, North American Newspaper Alliance, 10/25/66.

88 1966 Massachusetts campaign, *BG* 6/10/66, Martin F. Nolan, "Kennedy at Home Has a Few Problems," *New Republic* 6/25/66, Nolan, "Teddy and Eddie Revisited," *Reporter* 9/8/66.

89 Levin book changes, Jack Mendelsohn to RFK 8/1/66, RFK Senate Papers, personal correspondence 1966, JFKL; EMK to RFK, same file.

89 Rio Grande, *NYT* 6/30/67; Palestine, *Los Angeles Times* 12/1/66.

89 Private talks, Meg Greenfield, "The Senior Senator Kennedy," *Reporter* 12/15/66, *BG* 9/23/67; POWs, *NYT* 5/10/66, 5/12/66; trials, *CR* 89:II, p. 15851.
89 "class bias," *BG* 10/7/66; Press Club speech, *BG* 1/13/67.
90 EMK reaction on lottery, Dun Gifford int 2/2/96.
90 Hershey, *NYT* 3/21/67; Marshall, *NYT* 3/23/67.
90 College students, *CR* 90:I, p. 12471; "black army," prohibitive costs, p. 12472.
90 "Chance," *CR* 90:I, p. 15429; "filibuster," p. 15434.
90 EMK plan, letters, *CR* 90:I, p. 15751; "nitpicking," p. 15756.
91 EMK expected defeat, Russell comments, Gifford int 2/2/96.

CHAPTER 8 ALLIANCES AND COALITIONS

page
92 Black opinion, *Wesberry v. Sanders* 378 U.S. 1.
93 Celler, *CR* 90:I, pp. 11071–73; Conyers's call, Flug's attitude, James Flug int 4/3/92.
93 Variations in districts, *CQWR* 5/15/66, p. 699; committee vote, *BG* 5/24/67.
93 RFK role, Howard H. Baker, Jr., int 9/11/96; Baker attitudes, Baker int 1/13/94; Baker speech, *CR* 90:I, p. 10416.
94 Sent Senators report, Hersh, *Education of Edward Kennedy,* p. 268; reporters, labor, *NYT* 6/12/67; EMK speech, *CR* 90:I, pp. 14778ff; Baker speech, p. 14784.
94 Lobbying colleagues, *NYT* 6/12/67.
94 Conyers criticism, *NYT* 6/26/67; House fears, *CR* 90:I, p. 17627, *CQWR* 6/30/67, pp. 1105f.
94 House vote, *CR* 90:I, p. 30251; EMK speech, p. 31698.
95 Baker speech, p. 31700.
95 Senate vote, *CR* 90:I, p. 31712; Broder analysis, *WP* 11/14/67.
95 Inadequacy, Luella Hennessy quoted in David, *Joan,* p. 114; Joan Kennedy int 10/1/98.
95 "responsibility," *National Enquirer* 6/18/67; parties, Bermuda, Geneva, *Women's Wear Daily* 5/1/67, 5/12/67, 5/17/67.
96 McLean home, *WP* 5/19/67, *Architectural Digest* September–October 1973.
96 "not from emotion," *BG* 9/24/67; White, and Goodwin help, Kevin White int 10/27/94; fund-raising, *BG* 11/9/67, *Real Paper* 6/12/74.
97 Meeting, John P. McEvoy int 11/23/92.
97 Americans for Democratic Action speech 3/4/67, EMK-JFKL.
97 100,000 casualties, *NYT* 5/8/67; "agitating," R. W. Komer to Manatos 3/30/67, LBJL; hospital promises, *NYT* 4/7/67, 1/19/68, Robert S. McNamara to EMK 8/17/67, NA, RG 46, Box 61 labeled 3E2/11/4/1.
97 EMK influence on doctors' trip, *NYT* 7/31/67; doctors' report, Report of the Vietnam Medical Appraisal Team, E. Barrett Prettyman, Jr., private papers.
98 "Hospitals appear," Senate Judiciary Subcommittee on Refugees and Escapees, hearings on Vietnam 9/21/67, p. 277; "manipulation," "ripping," p. 283.
98 "sick and injured," hearings 10/4/67, p. 4; "final cure," p. 25.
98 "How can you," hearings, p. 10; doubles, p. 9; "stick," p. 25.
99 Luce, hearings 10/10/67, pp. 62–80.
99 "cavalier," hearings 10/11/67, p. 129; "bomb a village," p. 136.
99 "progress," hearings 10/16/67, p. 153; U.S. takeover, p. 176.
99 Avoiding reporters, *Baltimore Sun* 1/2/68; Khe Sanh, John M. Levinson int 9/30/95.
100 Bomb, Levinson int 9/30/95.
100 Tin, Prettyman int 3/1/64, EMK int 10/9/95; payments, Robert R. Komer to LBJ 1/13/68, Presidential Appointments File for 1/24/68, LBJL.
100 Cemetery, Thomas S. Durant int 10/13/95.
100 Komer, Prettyman diary; unresponsive, Zorthian int 7/24/95.

100 Correspondents, R. W. Apple, Jr., int 5/17/95; Danang, *WP* 1/6/68; Phu Cuong, Vann arguments, John Sommer int 9/21/95.
101 Vann drive, Sommer int 11/18/93, Levinson int 9/30/95.
101 Biggest impression, EMK int 10/9/95; Quang Ngai, Westmoreland, H&I fire, Levinson int 9/30/95; H&I fire, N. Thompson Powers int 8/14/95; sham, EMK int 10/9/95.
101 Bomb threat, *NYT* 1/13/68; drinks, Guam, Levinson int 9/30/95.
101 "conclusion," EMK int 10/9/95.
101 Boston speech, *BG* 1/26/68.
102 Meeting, David W. Burke int 2/4/96, Memcon 1/25/68, William Leonhart to LBJ, WHCF, CO312, LBJL.

CHAPTER 9 THE LAST BROTHER

page
103 "Shape up," *Face the Nation* 1/28/68.
103 Tet military impact, Stanley Karnow, *Vietnam* (New York: Penguin, 1991), p. 547; credibility, Clark Clifford with Richard Holbrooke, *Counsel to the President: A Memoir* (New York: Random House, 1991), pp. 473ff.
104 McCarthy announcement, *NYT* 12/3/67.
104 Breakfast, Frank Mankiewicz int 7/5/95.
104 Katzenbach request, Katzenbach int 5/12/93.
104 Walks, Lawford, *That Shining Hour,* pp. 304f.
105 EMK on RFK's running, Jack Newfield, *Robert Kennedy: A Memoir* (New York: E. P. Dutton, 1969), p. 211, Schlesinger, *Robert Kennedy and His Times,* p. 833, Burke int 6/3/93, EMK int 4/13/96.
105 Fear of RFK assassination, EMK int 4/13/96.
105 Senate gym, George S. McGovern int 4/14/94.
105 RFK on cities, EMK int 4/12/96; Kerner report, *NYT* 3/2/68.
105 RFK kept asking, EMK int 4/12/96.
106 Dinner, Goodwin to RFK 2/13/68, RFK Senate Correspondence, personal file 1964–1968, JFKL.
106 Dutton meeting, Fred Dutton int 7/25/95.
106 Students, Curtis Gans int 4/20/96, Paul Wieck, "McCarthy: Alive and Well in New Hampshire," *New Republic* 3/2/68, *NYT* 3/4/68.
106 California announcement, Newfield, *Robert Kennedy: A Memoir,* p. 232.
107 "certain," Haynes Johnson int 4/14/96; Goodwin told McCarthy, Goodwin int 5/5/96.
107 "victor in defeat," *LAT* 3/13/68.
107 Corridor, RFK int with Cronkite, *NYT* 3/14/68; Manchester, *LAT* 3/14/68; McCarthy meeting, Hersh, *The Education of Edward Kennedy,* p. 296, Theodore C. Sorensen, *The Kennedy Legacy* (New York: Macmillan, 1969), p. 139.
107 Meeting, television, Schlesinger, *Robert Kennedy and His Times,* p. 850.
108 Congratulations, EMK int 4/12/96.
108 Commission, Sorensen, *Kennedy Legacy,* pp. 135–37.
108 Meeting, rejection, Clifford, *Counsel to the President,* pp. 502–5.
108 Planning the meeting, Curtis Gans int 1/30/96; told McCarthy, Blair Clark int 1/30/96; formal explanation, EMK int 4/12/96.
109 Meeting in Green Bay, Clark int 1/30/96, Goodwin int 5/5/96; "Abigail said," Schlesinger, *Robert Kennedy and His Times,* p. 856.
109 RFK announcement, *NYT* 3/17/68.
109 Troubled, David W. Burke int 4/12/96.
109 Gifford, K. Dun Gifford int 2/2/96; Flug, Flug int 4/14/96; RFK and unions, EMK int 4/12/96.

109 Doherty role, recruiting, certification, Doherty int 11/18/94.
110 Branigin, Marvin Watson to LBJ 3/27/68, LBJL.
110 Sandwiches, Doherty OH 2/3/72, JFKL, Doherty int 11/18/94.
110 LBJ withdrawal speech, *PPP Johnson 1968–69,* Book 1, pp. 369f.
110 Humphrey worries, memo 4/2/68 11:35 A.M., HHH 148.A.4.7 (B), Box 412, File VP Memos for the Record 1968.
110 O'Brien, O'Brien, *No Final Victories,* pp. 232–36; Meany, Humphrey, Memo for the Record 4/2/68.
111 RFK speech in Safire, *Lend Me Your Ears,* pp. 197–99; EMK speech, *Rocky Mountain News* 4/6/68.
111 Tub conversations, Hoffa, EMK int 4/13/96.
112 Unruh, Burton, John Jacobs, *A Rage for Justice: The Passion and Politics of Philip Burton* (Berkeley: University of California Press, 1995), p. 156; labor leaders, EMK int 4/13/96.
112 Party, *San Francisco Chronicle* 6/6/68.
113 RFK speech, *RFK: Collected Speeches,* ed. Edwin O. Guthman and C. Richard Allen (New York: Viking, 1993), pp. 401–2.
113 Burke reaction, Burke int 4/13/96; Burton orders, Jacobs, *Rage for Justice,* p. 157; place on helicopter, Seigenthaler int 9/27/95.
113 Surgery, bulletins, Jules Witcover, *85 Days: The Last Campaign of Robert Kennedy* (New York: Putnam, 1969), pp. 282–90.
114 EMK appearance, Mankiewicz int 7/5/95.

CHAPTER 10 A SUMMER OF ENTREATIES

page
117 Lowenstein, Honan, *Ted Kennedy,* pp. 129–30; in 3/25/96 interview, Honan said Lowenstein described the encounter to him.
117 Corbin, Gwirtzman int 4/3/93; Washington talk, *NYT* 6/7/68.
117 "someday," Tunney int 9/25/96.
118 With coffin, *NYT* 6/8/68; comforting, Gifford int 12/10/96; reporters, Witcover, *85 Days,* p. 295.
118 Hearse, John Rooney, quoted in Stein and Plimpton, *American Journey,* p. 11; car, rosary, missal, *Syracuse Herald-Journal* 6/7/68; Pan Am Building, *NYT* 6/8/68; driving, John Culver int 4/16/93; fears, *BG* 9/30/68.
118 Eulogy, *NYT* 6/9/68.
119 Train, Stein and Plimpton, *American Journey,* passim; deaths, *NYT* 6/9/68.
119 Sailing, Gifford int 12/10/96; Joan told children, David, *Joan,* p. 173, Joan Kennedy int 10/1/98.
119 Thanks, *BG* 6/16/68.
119 Harris Poll, *NYT* 6/25/68; Hunphrey questioned, *NYT* 6/21/68.
120 Note, Humphrey to EMK 6/27/68, HHH 150.FD.3.6 (F), Box 875.
120 Leaving politics, Archbishop Philip Hannan quoted in *New Orleans Times-Picayune* 6/10/68, John V. Tunney int 3/6/95, *BG* 10/11/68; absent from Senate, *NYT* 7/24/68; gun legislation, Johnson, *NYT* 6/7/68; gun letter, *BG* 6/24/68.
120 Family sailing, Gifford int 12/10/96; Spain, *BEG* 6/6/68; beard, *BG* 9/30/68; memorial, *BG* 7/23/68; Daley call, vanden Heuvel int 5/11/95, EMK int 10/9/96.
120 RFK–Daley, Mankiewicz int 7/5/95; Shapiro, Hughes, *NYT* 7/22/68; Curtis, Dempsey, King, McKeithen, *WES* 7/22/95; Docking, *Topeka Daily Capital* 7/22/95.
120 Connally, *LAT* 7/24/95; Daley, John Criswell to Jim Jones 7/27/68, LBJL.
121 EMK statement, *NYT* 7/27/68.

121 O'Brien called, Carl Solberg, *Hubert Humphrey: A Biography* (New York: Norton, 1984), p. 351, O'Brien OH XXIII 7/21/87, LBJL; Nixon, *NYT* 8/8/68.

121 O'Brien announcement, EMK call, *NYT* 8/21/68.

121 Carry on, James Wechsler memorandum 4/15/71, describing 4/13/71 conversation with EMK, Rauh papers, LOC, Box 37, Kennedy correspondence folder.

121 EMK text, *NYT* 8/22/68.

122 Compromise, Solberg, *Hubert Humphrey,* p. 352; might run, *NYT* 8/23/68; Michigan, *Detroit News* 8/23/68; Johnson, Califano, *Triumph and Tragedy of Lyndon Johnson,* p. 319; Daley called, Peter Maas, "Ted Kennedy—What Might Have Been," *New York* 10/7/68.

122 McLean meeting, *BG* 9/30/68, 10/11/68, vanden Heuvel int 5/11/95.

122 Smith-Daley, Maas, "Ted Kennedy—What Might Have Been."

122 EMK thinking, Tunney int 3/6/95, Maas, "Ted Kennedy—What Might Have Been."

123 Chaos, safety, Burke int 12/12/96.

123 Missouri bus, Richard G. Stearns, "The Presidential Nominating Process in the United States: The Constitution of the Democratic National Convention" (dissertation presented 6/71 for M.L. degree, Oxford University), p. 166.

123 Unruh arrival, Daley's thoughts, William Daley int 1/13/93.

123 Unruh statement, *Chicago Tribune* 8/25/68.

123 Breakfast, *Chicago Tribune* 8/26/68.

124 Hart, Levin, *Detroit News* 8/26/68, *Detroit Free Press* 8/27/68, *Chicago Tribune* 8/26/68; Smith-Daley conversation, DiSalle, Lowenstein, Lewis Chester, Geoffrey Hodgson, and Bruce Page, *An American Melodrama: The Presidential Campaign of 1968* (New York: Viking, 1969), pp. 570-71; Humphrey not told, Solberg, *Hubert,* p. 357.

124 Smith, Chester et al., *American Melodrama,* p. 570; Daley, William Daley int 1/13/93; DiSalle, EMK, Unruh, *Chicago Tribune* 8/27/68; estimate, Hyman int 6/20/96.

124 "No one," *NYT* 8/28/68; Smith convinced, Maas, "Ted Kennedy—What Might Have Been"; Smith in favor, Mankiewicz int 7/5/95; Smith opposed, Dutton int 7/25/95; Hyannis call, Seigenthaler int 9/27/95.

124 Votes solid, Walter F. Mondale int 3/27/97, Humphrey, O'Brien, Solberg, *Hubert,* p. 358; California, *Chicago Tribune* 8/28/68.

125 Russell Long, "long hill," vanden Heuvel int 5/11/95.

125 Meeting, Goodwin, "The Night McCarthy Turned to Kennedy," *Look* 10/15/68, Maas, "Ted Kennedy—What Might Have Been."

125 Smith's evening, Maas, "Ted Kennedy—What Might Have Been."

125 Smith to EMK, EMK and Hyman, Hyman int 6/20/96; too young, *Look* 3/4/69.

126 EMK to Burke, Peter Collier and David Horowitz, *The Kennedys: An American Drama* (New York: Summit Books, 1984), p. 366, Burke int 6/3/93.

CHAPTER 11 A LEADER WITH ROBERT'S CAUSES

page

127 O'Donnell, Hyman, Hyman int 6/20/96; Shriver squelched, Max Kampelman int 6/29/95, Harris Wofford, *Of Kennedys and Kings* (New York: Farrar, Straus & Giroux, 1980), p. 431, *Detroit News* 8/26/68; Shriver's standing, Mankiewicz int 12/16/96, vanden Heuvel int 5/11/95.

127 Bayh, *NYT* 10/4/68; Culver, *BEG* 10/23/68; Chicopee, *BG* 9/10/68, *Boston Record* 9/10/68.

127 Humphrey to EMK 9/13/68, HHH 150.F.3.6 (F), Box 875.

128 Boston appearance, *NYT* 9/20/68, *BH* 9/20/68, Hersh, *Education of Edward Kennedy,* p. 354; Edgar Berman, *Hubert: The Triumph and Tragedy of the Humphrey I Knew* (New York: Putnam, 1979), p. 199.

128 Salt Lake City speech, O'Brien, *No Final Victories,* pp. 259–61; Chester et al., *American Melodrama,* pp. 645–50.
128 Nixon not debating, *BG* 10/11/68.
129 EMK as President, Gifford int 8/12/96, EMK to Gifford 9/30/68, Gifford files.
129 Wallace, *BH* 9/20/68, *BG, Worcester Evening Telegraph* 10/11/68; promoted, John Herbers int 9/8/96; New Bedford text, af.
129 Audience, *BG* 10/25/68.
129 Marriage to Onassis, *BG* 10/21/68, Leamer, *Kennedy Women,* pp. 641–43.
130 Ethel, Leamer, *Kennedy Women,* p. 639; Ted's visit, *WP* 10/19/68; prenuptial agreement, Willi Frischauer, *Jackie* (London: Michael Joseph, 1976), pp. 209–10, EMK int 10/9/95.
130 Biafra, EMK, *CR* 90:II, pp. 27848f; Russell, p. 27849; Mansfield, p. 27850.
130 U Thant, *NYT* 9/26/68, EMK int 10/9/96.
130 EMK to Rusk 10/11/68, EMK to Clifford 12/17/68, EMK to Nixon 11/15/68, NA, RG 46, Box 9, folder #1 Biafra VIP Correspondence.
130 EMK 12/6/68 speech, NA, RG 46, Box 9, folder Press 1968.
131 Sun Valley trip and Hart, Lippman, *Senator Ted Kennedy,* pp. 112f; impulsive, EMK int 4/17/95; Ferris call, Burke int 12/12/96.
131 Long's inattention, Mike Mansfield int 1/27/93; drunk, author's observation; embarrassed, Charles D. Ferris int 4/10/92.
131 Friendly Senators, *BG* 1/4/69; no weighing, Burke int 6/3/93, Gifford int 2/2/96; threats, EMK int 4/17/95; delight, Burke, Lippman, *Senator Ted Kennedy,* p. 113.
131 Muskie, *BG* 1/4/69; Humphrey, *Chicago Daily News* 1/7/68; Long unaware, Burke int 6/3/93, *BG* 1/4/69.
132 Eastland, Stennis, Ferris int 4/10/92; cooperation, Russell B. Long int 8/9/94.
132 Announcement, *NYT* 12/31/68.
132 Ferris help, Burke int 6/3/93; Muskie, Mondale, and McIntyre, *WP* 1/4/69; outside help, Hersh, *Education of Edward Kennedy,* pp. 360f; Ferris as source, James S. Doyle int 9/22/96; announcement, *BG* 12/31/68, *NYT* 12/31/68.
132 Muskie nominates, minutes of Democratic conference 1/3/69, af; running against EMK, Long int 8/9/94; vote, *WP* 1/4/69; Long comments, *BG* 1/4/69.
133 EMK on whip's job, *BG* 1/4/69; Long on EMK, *U.S. News & World Report* 1/13/69.
133 "stepping stone," *NYT* 1/4/68; Hart to EMK 1/9/69, Philip A. Hart papers, Bentley Library, University of Michigan, Box 74; Salinger, *NYT* 1/13/69; desk, Lippman, *Senator Ted Kennedy,* p. 118.
133 Mansfield, *Kennebec Daily Journal* (Augusta, Maine) 1/5/69.
134 Rules, Wayne Owens int 8/10/94.
134 Whip office, role, Owens int 8/10/94.
134 Lead in relief, *CR* 91:I, p. 1472; praises Nixon, *NYT* 2/9/69; Ferguson appointment, *NYT* 2/23/69.
134 North Andover, Wiesner call, Gifford int 2/2/96; "folly," *NYT* 2/2/69.
135 Compromise, Richard H. Rovere, "Letter from Washington," *New Yorker* 3/22/69; entering wedge, Robert Rothstein, "Nixon's ABM: Very Thin Indeed," *New Republic* 3/29/69, Edmund S. Muskie, "Blocking the ABM," *New Republic* 6/7/69.
135 Percy, *Chicago Tribune* 4/5/69; Cooper, step back, Gifford int 12/10/96.
135 Cost, Jerome Wiesner and Abram Chayes, eds., *ABM: An Evaluation of the Decision to Deploy an Antiballistic Missile System* (New York: Harper & Row, 1969), p. xx; crossroads, p. xiv.
135 "sophistication," John W. Finney, "Winds of Change in the Senate," *New Republic* 4/5/69; "old system," Philip M. Boffey, "ABM: Critical Report by Scientists Brings Sharp Pentagon Rebuttal," *Science* 5/16/69.
135 Draft, NPMP, POF Box 30, annotated news summaries about 2/7/69; job discrimination,

NYT 4/2/69; demonstrations, Alexander Butterfield to John D. Ehrlichman 4/10/69, NPMP, WHCF HU Box 23 HU 3-1, Civil Disturbances—Riots 4/1/69—4/30/69.

136 Television coverage, NPMP, POF Box 30, news summaries; ABM, H. R. Haldeman diary 4/8/69, CD-ROM edition.

136 Muskie, *Newsweek* 3/17/69; Healy, *BG* 3/16/69; Reston, *NYT* 5/21/69.

136 Dreams, *BG* 3/23/69; White House, Mass, attention, David, *Joan*, pp. 127, 131, 135; Boston Pops, *BG* 5/12/69.

136 EMK to McGovern, McGovern int 4/14/94; irrational, vanden Heuvel int 5/11/95; Alaska flight, Burton Hersh, "The Thousand Days of Edward M. Kennedy," *Esquire* 2/72.

137 Gunfire, Bob Bates int 9/21/94; Loeb, *Memphis Commercial Appeal* 4/5/69.

137 Abernathy, EMK, *BG* 4/5/69, *Memphis Commercial Appeal* 4/5/69, *Nashville Tennessean* 4/5/69.

137 Robert's plan, *Anchorage Daily Times* 4/8/69; schedule, *Anchorage Daily News* 4/9/69; politics, *Chicago Tribune* 4/10/69; pullout, *NYT* 4/11/69.

138 Stevens, pressure, *Anchorage Daily Times* 4/10, 4/11, 4/12/69; legislation, Lippman, *Senator Ted Kennedy,* p. 94; visits, *Time* 4/18/69.

138 Mondale, Senate Labor Subcommittee on Indian Education, hearings on Alaska 4/11/69, p. 457; EMK, hearings, pp. 463f.

138 Arctic Village stop, *BH* 4/12/69.

138 Flight, John Lindsay memo 4/69, Hersh, *Education of Edward Kennedy,* pp. 379–80; "Eskimo power!" *Life* 8/8/69.

139 Admired Chavez, Lindsay memo 5/22/69; security, Hersh, *Education of Edward Kennedy,* p. 366; advice, decision, *Life* 8/8/69.

139 Sirhan letter, *BG* 5/22/69, *LAT* 5/22/69, *Newsweek* 6/2/69.

139 Calexico, *Imperial Valley Press* (El Centro, Calif.) 5/19/69, *LAT* 5/19/69.

140 Space speech, JFKL-EMK.

140 Nixon reaction, Nixon papers, NPMP, news summaries May 1969.

141 Furious, Gifford int 2/2/96; McGovern, *BG* 3/18/59.

141 Kill totals, *NYT* 5/21/69, later modified to 269 enemy and 55 Americans killed; Zais, *BG* 5/23/69; abandoned, Karnow, *Vietnam,* p. 316; pullout predicted, *NYT* 5/21/69.

141 EMK attack, *CR* 91:I, p. 13003.

141 Scott, *CR* 91:I, p. 13004; Dirksen, Mansfield, p. 14484.

142 Rough Rock, *CR* 91:I, pp. 16400ff.

142 "mission," calls, Kathleen Kennedy Townsend int 11/3/97.

142 Mohbat article, *BG* 6/8/69.

142 Mansfield "judgment," *WES* 7/18/69.

142 Tired, Thomas P. O'Neill, Jr., int 11/23/92.

CHAPTER 12 CHAPPAQUIDDICK

page

143 Lyons chided, Hersh, *Education of Edward Kennedy,* p. 390.

143 Clams, EMK, Inquest into the Death of Mary Jo Kopechne, Edgartown District Court, p. 17; rental, Joseph F. Gargan, Inquest, p. 208; driving, John B. Crimmins, Inquest, pp. 334f.

144 Howie Hall, Gargan, Inquest, p. 210; light breezes, Dun Gifford int 12/10/96; finishes ninth, *BEG* 7/24/69; hired boat, Paul F. Markham, Inquest, p. 285.

144 Checked in, EMK, Inquest, p. 17; 7 P.M., p. 18; soaked, p. 22; served drinks, p. 23.

144 Neighbors, Jack Olsen, *The Bridge at Chappaquiddick* (Boston: Little, Brown, 1970), p. 92; hors d'oeuvres, radio, Charles C. Tretter, Inquest, p. 146; steaks, Gargan, Inquest, p. 214;

two drinks, teasing EMK, *BEG* 7/24/69; EMK two drinks, EMK, Inquest, p. 45; Kopechne hardly drank, Newberg, Inquest, p. 623; three or more, John J. McHugh, Inquest, p. 275.

144 Departure, EMK, Inquest, pp. 26ff; Crimmins, Inquest, pp. 345f.

145 Turned right, EMK, Inquest, p. 29.

145 Police statement, *BG* 7/20/69; girl in car, diving, resting, EMK, Inquest, p. 52.

145 Concussion, James A. Boyle, Inquest, p. 8; excerpts from doctors' affidavits, David, *Ted Kennedy,* pp. 224–26; effects, James E. T. Lange and Katherine DeWitt, Jr., *Chappaquiddick: The Real Story* (New York: St. Martin's, 1992), p. 82.

145 No lights, EMK, Inquest, p. 56; summoned Gargan and Markham, EMK, pp. 56f, and Raymond LaRosa, p. 102.

146 Gargan, Markham dive, EMK, Inquest, pp. 59f, Gargan, pp. 227–32, Markham, pp. 299, 304–5; swam to Edgartown, EMK, pp. 59f, Gargan, p. 243, Markham, p. 309.

146 Time, Russell E. Peachey, Inquest, p. 449; morning, Ross W. Richards, Inquest, pp. 259–61.

146 Burke, Markham, Inquest, p. 324; Marshall, EMK, Inquest, p. 73; Wagner, *NYT* 2/12/80; Joan Kennedy int 10/1/98.

146 Fishermen, tragedy, Olsen, *Bridge at Chappaquiddick,* pp. 111f, 123; tides, license plate, Dominick J. Arena, Inquest, pp. 574f.

146 Diving, John Farrar, Inquest, pp. 535f; Arena, Olsen, *Bridge at Chappaquiddick,* p. 128.

146 Examination, Donald R. Mills, Inquest, pp. 432–35; prepared for burial, clothing, blood, Eugene Frieh, Inquest, pp. 515–18; blood alcohol, McHugh, Inquest, p. 272.

147 Arena and EMK, Arena, Inquest, pp. 579ff.

147 EMK statement, *BG* 7/20/69.

148 Not involve others, Markham, Inquest, p. 327.

148 "No comment," George W. Kennedy, Inquest, p. 481.

148 Marshall, Arena, Inquest, pp. 587f; evade reporters, Olsen, *Bridge at Chappaquiddick,* pp. 159f; mumbling, George W. Kennedy, Inquest, p. 487; tells father, Dallas with Ratcliffe, *Kennedy Case,* pp. 338f.

148 Women, Esther Newberg, Inquest, pp. 419–21.

148 Steele, Dinis, Killen, Olsen, *Bridge at Chappaquiddick,* pp. 167–69.

149 Dinis's office, Mills, Inquest, p. 476.

149 Release statement, Olsen, *Bridge at Chappaquiddick,* p. 175.

149 "Imperiled," *NYT* 7/21/69; "stars," Mondale int 3/27/97; Nixon, Haldeman diary 7/19/69.

149 "End of Teddy," Haldeman diary 7/20/69; "another front," William L. Safire, *Before the Fall: An Inside View of the Pre-Watergate White House* (Garden City, N.Y.: Doubleday, 1975), p. 149.

149 Advisers, Hersh, *Education of Edward Kennedy,* pp. 408–13.

150 Examination, Lange and DeWitt, *Chappaquiddick: The Real Story,* pp. 72f.

150 EMK talked, Newberg, Inquest, p. 423, Leo Damore, *Senatorial Privilege: The Chappaquiddick Cover-Up* (Washington: Regnery Gateway, 1988), p. 140.

150 EMK comments, Damore, *Senatorial Privilege,* p. 143, *NYT* 7/23/69.

150 EMK inattentive, Hersh, *Education of Edward Kennedy,* pp. 409, 411, Dallas with Ratcliffe, *Kennedy Case,* p. 340, Lange and DeWitt, *Chappaquiddick: The Real Story,* pp. 72, 120; Smith, Hersh, *Education of Edward Kennedy,* p. 413; Steele, Olsen, *Bridge at Chappaquiddick,* p. 201.

150 Clark, Steele, Arena, Olsen, *Bridge at Chappaquiddick,* pp. 210–13, 224–32; Dr. Brougham, David, *Ted Kennedy,* p. 225.

151 Plea, *NYT* 7/26/69.

151 Testimony, *NYT* 7/26/69.

151 "my statement," *NYT* 7/26/69.

151 Eunice, Gwirtzman int 4/3/93; simple, Gifford int 8/12/96.

151 Broadcast text, *NYT* 7/26/69.
152 Massachusetts reaction, *NYT* 7/26/69, 7/27/69.
153 Saw a car, Christopher F. Look, Inquest, pp. 495ff; tides, Lange and DeWitt, *Chappaquiddick: The Real Story,* pp. 73f.
153 Kopechne survival, Lange and DeWitt, *Chappaquiddick: The Real Story,* pp. 85–89.

CHAPTER 13 A STAFF TO LEAN ON

page
155 Mansfield to EMK, Mike Mansfield int 1/27/93; Democrats and Republicans, *NYT* 8/1/69; Andrews, Nixon to EMK 8/4/69, Nixon papers, NPMP; press, Haldeman diary 8/4/69.
155 "Ever-ready eye," *Newsweek* 7/28/69; "ring of conviction," *Time* 8/1/69; "roving eye," Cornelia Noland, "Washington's Biggest Male Chauvinist Pigs," *Washingtonian* 9/72.
156 1972 disclaimer, *NYT* 8/1/69; medical research, *BG* 9/19/68; Nixon reaction, Alexander Butterfield to Nixon 9/23/69, NPMP.
156 Floor, Wayne Owens int 8/10/94; clean water, EMK to Phil Hart 10/31/69, Philip A. Hart papers, Bentley Library, University of Michigan, Box 74; reporters, Gifford to EMK 1/21/70, Gifford files.
156 Rules, Scotch, Owens int 8/10/94; bitterness, Robert C. Byrd int 2/12/93.
156 Father's death, Dallas with Ratcliffe, *Kennedy Case,* pp. 349–52.
157 "reminder," Ann Gargan, *NYT* 11/21/69.
157 RFK essay, *NYT* 11/21/69.
157 Knelt, *BG* 11/21/69.
157 Dinis, Owens int 8/10/94.
158 Boyle's conclusions, Inquest, pp. 11f.
158 Kopechnes, Gifford int 8/12/96.
158 Richardson, *NYT* 1/16/70; censored Ferguson, Dale de Haan to EMK 1/70; cable on Biafra, John Scali, ABC News broadcast 1/21/70, Elizabeth Drew, *Atlantic* 6/70.
159 260 tons, Trueheart, General Accounting Office Report 8/25/70, all in records of the Judiciary Committee Subcommittee on Refugees and Escapees, NA, RG 46, Box 9.
159 Kissinger worries, Haldeman diary 1/10/70, 1/21/70; Nixon and Wilson, Haldeman diary 1/11/70; Nixon and EMK, *BG* 1/24/70.
159 Reuther, John Carlova, "Reuther's Strategy for a Health-Care Revolution," *Medical Economics* 7/21/69; Reuther and EMK, Max Fine int 6/9/95, EMK int 4/12/96.
159 Boston University speech, JFKL-EMK.
160 "living memorial," *NYT* 5/15/70.
160 Health care bill, *CR* 91:II, p. 14338.
160 Tax reform, Minutes, Senate Democratic Policy Committee luncheons 4/15/69, 5/13/69, 5/27/69, af; RFK, Edwin L. Dale, Jr., "The Realities of Tax Reform," *New Republic* 2/15/69.
160 Anti-Haynsworth motives, Flug int 4/3/92, John P. Frank, *Clement Haynsworth, the Senate and the Supreme Court* (Charlottesville: University of Virginia Press, 1991), pp. 133ff; doubts, EMK int 4/17/95.
161 Donovan, *LAT* 1/20/70; Roeder broadcast, Richard Harris, *Decision* (New York: Dutton, 1970), pp. 15f, 26, Ed Roeder int 1/25/97.
161 Meeting, Flug memo, Harris, *Decision,* pp. 36f.
162 EMK questioning, Senate Judiciary Committee, hearings on confirmation of G. Harrold Carswell 1/27/70, p. 32; Flug, Harris, *Decision,* p. 58.
162 Flug, Eastland, Burke ints 6/3/93, 12/12/96.
162 March 19 speech, *CR* 91:II, p. 8081; Morrissey, p. 8381.
163 Broader issue, *CR* 91:II, p. 10366.

163 Flug role, Harris, *Decision,* passim, Roeder int 1/25/97, Burke int 6/3/93.
164 "If Congress can," Archibald Cox, "The Supreme Court: 1965 Term," *Harvard Law Review,* vol. 80:91, p. 107.
164 Fears, Burt Wides int 1/16/97.
165 Lowenstein, Railsback, Senate aide int 1/3/97.
165 Magnuson role, Robert C. Byrd, *The Senate 1789-1989* (Washington, D.C.: Government Printing Office, 1994), vol. 4, p. 684; Kennedy as handicap, Senate aide int 1/30/97.
165 War, armor, *CR* 91:II, p. 6931.
165 Letter, *WP* 3/17/70.
166 Nixon opposition, Haldeman diary 3/17/70.
166 Staff role, pay, EMK int 10/9/95.

CHAPTER 14 VICTORY AND DEFEAT

page
167 "Discontent," EMK statement 5/8/70, af.
167 Baltimore speech, *BG* 5/7/70.
168 Flag, *Wakefield Item* 7/6/70.
168 "Love affair," *BG* 7/30/70.
168 Calls, *WP* 2/7/71; Teddy asks, Betty Hannah Hoffman, "Joan Kennedy's Story," *Ladies' Home Journal* 7/70.
168 Door, rifle, McGovern int 4/14/94; balloon, Hoffman, "Joan Kennedy's Story."
168 Marriage, Betty Hannah Hoffman, "Joan Kennedy Today," *Ladies' Home Journal* 8/70.
168 Boats, Gifford int 12/10/96.
169 Sandwich, Dutton int 7/25/95.
169 *Today* show, *NYT* 5/30/70; *Meet the Press, BG* 10/5/70.
169 Lawrence, beer, Hoffman, "Joan Kennedy Today"; South Boston, *BG* 3/28/70.
169 Reassurance, *NYT* 8/27/70.
169 Blouse, *BG* 9/23/70.
170 Pianist, David, *Joan . . .*, p. 205; "used," Joan Kennedy int 12/12/96; "alcoholic," Leamer, *Kennedy Women,* p. 651.
170 No voters asked, Paul Kirk int 1/19/97, *NYT* 10/31/70; campaigning, Vietnam, *WP* 9/20/70.
170 Belmont, *BH* 9/1/70; staff meeting, *BH* 8/27/70.
170 Packwood, *BEG* 4/23/70; abortion on demand, *BG* 10/17/70.
170 Headquarters, computers, Ann F. Lewis int 2/13/97.
171 White ulcer, Kevin White int 10/27/94, *BG* 10/19/70, 10/25/70.
171 Detective, Haldeman diary 12/5/70; *The People,* Nixon news summaries 11/70, NPMP, NA; plotting, Haldeman diary 12/5/70, 12/7/70, 12/12/70, 12/19/70.
171 Byrd running, *BG* 11/7/70; EMK argued, Martin Nolan int 5/10/93.
171 "The title," *BG* 11/15/70.
171 Europe, Thomas F. Eagleton int 7/7/95.
171 Byrd-EMK conversation, Byrd int 2/12/93.
171 Jackson on EMK, Max Kampelman int 6/19/95.
172 Jackson meeting, Owens int 8/10/94.
172 POWs and Fulbright, EMK int 4/17/95; no new names, *NYT* 12/23/70.
172 "humiliation," *WES* 1/22/71; "astonishing," *BH* 1/22/71; Nixon interpretation, Haldeman diary, 1/21/71.
172 Twenty-eight Senators, Gridiron, Warren Rogers, "Kennedy's Comeback: Will He or Won't He?" *Look* 8/17/71.
173 Best things, Byrd int 2/12/93.
173 Lasker role, Benno Schmidt int 12/14/93.

173 Kennedy gets subcommittee, Richard Rettig, *Cancer Crusade: The Story of the National Cancer Act of 1971* (Princeton: Princeton University Press, 1977), p. 120; "afford," p. 109.
174 Enough, John D. Ehrlichman to Nixon 12/16/70, NPMP, WHCF, Subject Files HE Box 8, folder [EX] HE 1-3 Cancer; $100 million, *PPP Nixon 1971,* p. 53.
174 $1.2 billion, new agency, *CR* 92:I, p. 341.
174 Basic research, Schmidt int 12/14/93.
174 "American people," Senate Labor Subcommittee on Health, hearings on Conquest of Cancer Act 3110171, p. 171; "delays," hearings, p. 205.
175 Landers, letters, Rettig, *Cancer Crusade,* pp. 175ff; Lasker forces, James Cavanaugh to Ken Cole 4/30/71, NPMP, WHCF, HE Box 9, folder HE 1-3 Cancer 1/1/71–5/31/71.
175 Woods, Ehrlichman memos NPMP, WHCF [EX] HE 1-3 Cancer 5/1/71–5/31/71.
175 "Our own agency," Haldeman diary 4/28/71; Ehrlichman orders, Kenneth Cole int 2/28/95.
175 Policy shift, *PPP Nixon 1971,* pp. 629ff.
175 EMK doubts, Schmidt int 12/14/93; $100 million, *BG* 5/20/71; legislation, Cole to Cavanaugh 5/29/71, NPMP, WHCF, Subject Files HE Box 8, folder [EX] HE 1-3 Cancer.
176 Nixon, Cole, Schmidt int 12/14/93, EMK int 9/10/92, Cole int 2/28/95.
176 "Peter, why don't," Rettig, *Cancer Crusade,* p. 194.
176 "Conquest," *CR* 92:I, p. 23768.
176 Rogers's caution, EMK int 4/12/96; doubts, Paul Rogers int 5/28/95; "necessary tools," *CR* 92:I, p. 45836.
176 Reluctance, Rogers int 5/18/95.
177 Ceremony, *WP* 12/14/71.
177 EMK call, Cole int 2/28/95.
177 Nixon on health, *NYT* 6/23/71; EMK, *NYT* 2/1/71, 7/15/71; "kicking," Cole to Ed Morgan 2/8/71, NPMP, Cole int 2/28/95.
177 EMK to Rogers, "sad and tragic," unpublished hearing of Judiciary Subcommittee on Refugees and Escapees, NA, RG 46, Box 62, 1972–1977; Indo-China Hearings, Vietnam folder, NA; "mad spell," *BH* 5/6/70; Agnew, *NYT* 6/20/70.
177 Diplomats, Walter Isaacson, *Kissinger: A Biography* (New York: Simon & Schuster, 1992), p. 372.
178 Vietnam comparison, EMK int 10/9/96; "slaughter," *BEG* 8/26/71; upstaging, Honan, *Ted Kennedy,* pp. 24ff.
178 April 9, Nixon tape 494-4, NPMP, NA; May 28, June 23, Haldeman diary; July 27, Nixon tape 557-1; September 8, Nixon tape 274-44; September 18, Nixon tape 576-6; October 14, Nixon tape 592-10.
178 Cartoon, Flanigan to Haldeman and Colson 10/28/71, NPMP; Nixon discussions 10/20/71, Nixon tapes 609-3, 609-13; 11/3/71, Nixon tape 301-25; 11/12/71, Nixon tape 14-83; 11/29/71, Nixon tape 306-10.
179 Chappaquiddick, Nixon tape 274-44.
179 Polls, Gwirtzman int 4/3/93; Wechsler, James Wechsler memo to himself 5/15/71, in Rauh papers, LOC, Box 37, Kennedy correspondence folder; EMK int with Apple, *NYT* 5/23/71.

CHAPTER 15 INVESTIGATIONS BY AND OF NIXON

page
180 London, EMK int 6/2/97; camps, King int 2/22/97.
180 "Ulster is becoming," *CR* 92:I, p. 36973.
181 Cromer, *BG* 10/21, MPs, Faulkner, *BEG* 10/22/71; conscience, *BH* 10/25/71; Heath, *NYT* 10/27/71.
181 1938, Jean Kennedy Smith int 5/4/95; television specials, EMK int 10/9/96, Theodore C.

Sorensen, "Ted Kennedy's Memories of JFK," *McCall's,* 11/73; crowds, visitors, students, Irish diplomat int 5/8/95.

182 Resolution, *CR* 92:I, p. 36970.

182 *Times* letter, *BEG* 12/8/71; attacks, *Boston Record* 12/10/71, *BEG* 12/10/71; My Lai, *BG* 2/29/72.

182 Lynch, *BG* 3/3/72; "world stage," *Irish Times* 10/23/71.

182 Irish diplomats and IRA, *Irish Times,* Sean Donlon, 1/25/93.

182 Meets Hume, John Hume int 3/17/95.

183 Mississippi, Charles Evers in Stein and Plimpton, *American Journey,* p. 279.

183 Percy, EMK in *CR* 91: II, p. 43447; EMK testimony, Senate Labor Subcommittee on Aging, hearing on elderly nutrition 6/2/71, pp. 138f.

184 Ella Reason, Hearings, p. 163.

184 Appease, Ken Cole to Leonard Garment 6/9/71, WHCF Vicki Keller, Box 13, NPMP; administration arguments, Hearings, p. 247; group feeding, Joe Carlin int 3/28/97.

184 "Failure," *CR* 92:I, p. 3948; Mondale, p. 3955.

185 Division over national board, *CR* 93:I, p. 35639; EMK, p. 35643; $390 million, *WP* 10/9/71; credited, Frank Ducheneaux int 8/11/97, Patricia Lock int 8/18/97.

185 EMK support, *CR* 93:II, p. 9372.

185 War politics, *BG* 6/10/71.

186 Attack on Nixon, *BG* 1/14/72; visit to Dacca, *Boston Record* 2/15/72, 2/16/72.

186 Nixon analysis, Haldeman diary 4/5/72.

186 Deadlock *BG* 3/27/72; Daley, *NYT* 4/28/72; Meany, *BG* (Rowland Evans and Robert Novak) 6/16/72, EMK int 10/9/95.

187 Martinis, Robert S. Strauss int 3/31/95; admired McGovern, *NYT* 1/13/72, EMK int 10/9/95; RFK's children, *NYT* 4/28/72; Lewis column, *NYT* 4/29/72; told him, Anthony Lewis int 4/8/97.

187 EMK–Mills, Stan Jones int 11/11/93, Lee Goldman int 12/13/93; statement, JFKL-EMK.

188 Unshakable, Ted Van Dyk int 3/27/96, Richard Dougherty, *Goodbye, Mr. Christian* (Garden City, N.Y.: Doubleday, 1973), p. 150.

188 No draft, *NYT* 6/14/72; bid from Troy, EMK response, Troy int 3/15/94; Apple interpretation, *NYT* 6/21/72.

188 McGovern–EMK conversations, McGovern int 4/14/94.

188 Muskie, Humphrey, and Askew, Dougherty, *Goodbye, Mr. Christian,* pp. 150, 155; Carter, Peter G. Bourne, *Jimmy Carter: A Comprehensive Biography from Plains to Postpresidency* (New York: Scribner's, 1997), p. 226; White, McGovern int 4/14/94.

189 EMK on Kevin White, McGovern int 4/14/94, Mankiewicz int 7/5/95; Pierre Salinger, "Four Blows That Crippled McGovern's Campaign," *Life* 12/29/72; visited White, Kevin White int 10/27/94; EMK denies, EMK int 10/9/95; White's photograph, Lupo, *Liberty's Chosen Home,* p. 207.

189 Convention, Theodore H. White, *The Making of the President 1972* (New York: Atheneum, 1973), p. 185.

189 Practicing, Senate aide int 6/8/92.

189 Unique tribute, Robert Dallek, *Flawed Giant: Lyndon Johnson and His Times, 1961–1973* (New York: Oxford University Press, 1998), p. 617.

190 EMK convention speech, JFKL-EMK; Daley in mind, *NYT* 7/14/72; end war, George McGovern, *An American Journey: The Presidential Campaign Speeches of George McGovern* (New York: Random House, 1974), p. 20.

190 Serve simultaneously, McGovern int 4/14/94.

190 EMK on Shriver, McGovern int 9/8/94, *Boston Record* 8/12/72.

191 Bomber pilot, *McGovern, An American Journey,* p. 38; troops, "Peace is," Isaacson, *Kissinger,* pp. 440, 459.

191 Minnesotans, *NYT* 9/12/72; Chicago, *NYT* 9/13/72; Halloween, *BG* 11/1/72.

191 Vermont, Tom Salmon int 6/21/96; West Virginia, *NYT* 10/8/72, John D. Rockefeller IV int 5/8/97.
191 Telegrams, NPMP, Nixon tape 22-32, 3/16/72.
191 Rose Kennedy worried, *WP,* 2/8/97; Tunney, Burden, *New Hampshire Sunday News* 9/3/72; pious warning, Joan Kennedy int 1/18/96.
192 Nixon news summary, Box 56 President's Office Files, WHSF, NPMP; Secret Service, Nixon tapes 772-6, 772-15, 9/7/72; Newbrand, WHSF Files, Haldeman, Box 30 NPMP, and Alexander Butterfield, quoted in Gerald S. and Deborah H. Strober, *Nixon: An Oral History of His Presidency* (New York: HarperCollins, 1994), p. 263; nothing juicy, *WP,* Lardner, 2/8/97.
193 "We wanted them to know," Flug int 4/3/92; Mardian, EMK to Leon Jaworski 7/30/74, NA.
193 Segretti reports, *WP* 10/10/72, 10/13/72; investigation, Flug int 4/3/92, Flug material, Boxes 115, 117, Lenzner files, Records of the Senate Select Committee on Presidential Campaign Activities, NA.
193 EMK letter to Eastland, *CR* 93:I, p. 3552.
194 Helena, *Great Falls Tribune* 10/30/72; Mansfield to Eastland and Ervin, *WP* 1/7/73; no Presidential ambitions, Mike Mansfield int 1/27/93; "very obvious," EMK int 4/12/96.

CHAPTER 16 A CRISIS AT HOME

page
195 Bombing, Stephen E. Ambrose, *Nixon: Ruin and Recovery 1973-1990* (New York: Simon & Schuster, 1991), pp. 30-40.
195 Saxbe, *Cleveland Plain Dealer* 12/29/72; Reston, *NYT* 12/17/72; "olive branch," *BEG* 12/12/72; op-ed, *NYT* 12/27/72.
195 Caucus, Minutes of the Democratic Conference, af.
196 Warnings, *NYT* 12/22/72; "priority," *NYT* 1/21/73.
196 O'Higgins, EMK int 3/26/99.
197 McDuffee, Senate Judiciary Subcommittee on Refugees and Escapees, hearing on Chile 9/28/73, p. 8; Garrett-Schesch, pp. 14f; Kubisch, p. 20.
197 "no hurry," EMK hearings, pp. 2f; resolution, Mark L. Schneider int 11/17/97.
197 EMK concerns, Gwirtzman notes 12/5/72, af.
197 Polls, *NYT* 4/30/73, 5/31/73; Apple, *NYT* 5/21/73.
198 Planeload, *BG* 7/4/73; EMK speech, *WP* 7/5/73.
198 Goldwater, *BG* 7/13/73; Chappaquiddick, Gwirtzman notes 7/18/73, af.
198 Market access, Stuart Altman int 9/17/93, S. Philip Caper int 5/13/97.
199 Weinberger approach, Caspar W. Weinberger int 1/8/97.
199 EMK leaks, complaints, *NYT* 9/4/73; in touch, Altman int 9/17/93.
199 Expectations, Knowles, Kennedy, *Medical Economics* 10/29/73; "no question," Altman, address at Columbia University, 2/8/93, af.
200 Ervin committee powers, Flug int 6/15/97.
200 Prosecutor, Eastland, James Doyle, *Not Above the Law* (New York: Morrow, 1977), pp. 38, 40.
200 Rewrite, EMK int 4/12/96, Flug ints 4/12/92, 5/15/97; access to Congress, funds, Archibald Cox's notes on Elliot Richardson to Cox 5/18/73, tab G, Archibald Cox papers, Special Collections, Harvard Law School Library, Box 21-8.
201 Charter, Flug int 4/12/92.
201 "reckless," *BG* 10/21/73; Eastland, int with government official; impeachment, EMK int 4/12/96.
202 "ring down," *BG* 5/5/73; "torrents," Senate Rules Subcommittee on Privileges and Elections, hearings on election reform 6/6/73, pp. 161, 168.
202 Scott's primacy, Kennedy's approach, Robert E. Mutch, *Campaigns, Congress and Courts: The Making of Federal Campaign Finance Law* (New York: Praeger, 1988), pp. 122-23.

203 Scott, EMK *CR* 93:I, p. 25318; Common Cause, Fred Wertheimer int 5/7/97.
203 Coalition, Wertheimer int 5/7/97, Mutch, *Campaigns,* p. 124.
204 "A different America," *CR* 93: I, pp. 38177, 38201; Allen, p. 38187; Bennett, p. 38191.
204 "better this way," *CQA 1973,* p. 753.
204 Lobbying, Wertheimer int 4/29/97, Mutch, *Campaigns,* p. 126.
204 Mansfield, Cannon, *CR* 93:I, pp. 39175f.
205 For the account of Teddy's cancer, I have generally relied, at the suggestion of Dr. S. Philip Caper, on Geraldo Rivera, *A Special Kind of Courage: Profiles of Young Americans* (New York: Simon & Schuster, 1976), pp. 191–214; Chesire, *BG* 9/9/73, Joan Kennedy int 10/1/98.
205 Boston City Hospital, S. Philip Caper int 5/13/97.
206 Drayne's calls, author's recollection.
207 Crying, reassurance, *BH,* Eleanor Roberts, 12/2/73.
207 Kathleen's day, signal, Melissa Ludtke int 5/17/97; EMK to Kathleen, Kathleen Kennedy Townsend int 11/3/97.
207 Wallace's, soldiers, *WP* 3/1/74; tired, spoilsport, Leamer, *Kennedy Women,* p. 676, Joan Kennedy int 10/1/98.
207 Bone cancer cells, Europe, Dr. Larry Horowitz int 5/30/97; fears, EMK int 4/12/96; plane, Stuart Auerbach int 5/18/97.
208 Meeting, treatment, *WP* 3/8/74, Lester David, "Teddy, Jr., Bravest of the Kennedys," *Good Housekeeping* 10/74; treatment begins, *NYT* 2/3/74.
208 Sun Valley, Joseph P. Kennedy II int 1/10/95.

CHAPTER 17 NO CAMPAIGN IN 1976

page
209 White, *BG* 5/3/74.
209 Everett, *BH* 1/16/74; Lydon, *NYT* 2/15/74; angry, *NYT* 2/16/74, R. W. Apple int 5/30/97; Hamburg, *BG* 4/9/74; New York, *BG* 4/25/74; Healy, *BG* 6/5/74.
209 Joan, Leamer, *Kennedy Women,* p. 677; sanitariums, *BG* 6/7/74, *BH* 6/22/74, *NYT* 9/14/74; driving, *BEG* 11/6/74.
210 Chappaquiddick, int with former government official, John Culver int 4/16/93; "deep down," *BG* 5/7/74.
210 EMK, *CR* 93:II, p. 8209.
210 Dole, *CR* 93:II, p. 10503.
211 "finest hours," *CR* 93:II, p. 10964.
211 Berlin speech, *CR* 92:II, p. 16628.
211 Hiring, Robert E. Hunter int 8/23/94.
212 Disarmament, Andrei Pavlov int 11/24/97; snowball, *BH* 4/21/74.
212 Presidential treatment, cable Walter Stoessel to SecState 4/25/74, af; Watergate, Jim King int 2/14/94.
212 Arms control, *CR* 92:II, p. 16623; luncheon, Stoessel cable 4/22/74.
212 University speech, *CR* 92:II, p. 16626; translation, Grace Warnecke int 12/15/94; questions, *NYT* 4/22/74.
213 Hunter, reprimand, Pavlov int 11/24/97.
213 Brezhnev meeting, EMK int 6/2/97.
213 Bernstein, Joan Kennedy int 1/18/96; Furtseva, already left, Robert E. Hunter to author 4/19/96.
214 Play to Russians, EMK int 6/2/97.
214 Old man, Warren Zimmerman int 9/8/94; monastery, Shevardnadze, Warnecke int 12/15/94; toast, Pavlov int 11/24/97.

214 Epelman, Stoessel cable 4/25/74, af.

214 KGB, driver, Warnecke int 12/15/94.

215 EMK questions, *Christian Science Monitor* 4/26/74; protests, "pocketful," *NYT* 4/26/74; "sense of serenity," EMK int 6/2/97.

215 Atlanta speech, JFKL-EMK.

215 Defeating EMK, Peter Bourne int 11/7/96; impressing EMK, Hamilton Jordan int 3/2/95.

216 EMK speech, "Watergate," JFKL-EMK.

216 Plane, *Rolling Stone* 6/3/76; Carter speech, JCL.

216 Impressed, EMK int 6/2/97.

216 Martin cable, *NYT* 4/3/74; Kissinger letter, *WP* 4/1/74.

217 Stennis, *CR* 93:II, p. 13246; EMK, *CR* 93:II, p. 13252.

217 "moral commitment," Senate Subcommittee on Refugees and Escapees, hearing on Indochina, 7/8/74, p. 2; "amazes," *CR* 93:II, p. 19180.

217 Rescue, *BEG* 7/24/74, *BH* 7/25/74, EMK int 4/12/96.

218 "serious," *NYT* 1/1/74.

218 Mills meetings, EMK int 4/12/96; "encouraging," *NYT* 4/3/74.

218 Fine, *BEG* 4/16/74; union halls, Stan Jones int 11/11/93; *Progressive,* 5/74; told leaders, EMK int 4/12/96.

218 EMK called, needed success, Caspar W. Weinberger int 1/8/97.

219 Church basement, Jones int 11/11/83, Stuart Altman int 9/17/93; Nixon, *NYT* 5/21/74; EMK response, 4/22/74 radio text, JFKL-EMK.

219 Governors, *Seattle Post-Intelligencer* 6/4/74, *NYT* 6/4/74, *WP* 6/4/74.

219 EMK called, Weinberger int 3/25/97.

220 "armed troops," *WP* 6/1/74; "No more," *WP* 8/15/74.

221 Conference, EMK int 6/2/97, Mutch, *Campaigns,* p. 130.

221 "Abuse," *CR* 93:II, p. 633.

222 Barnicle, *BG* 9/8/74.

222 Suit, J. Anthony Lukas, *Common Ground: A Decade in the Lives of Three American Families* (New York: Alfred A. Knopf, 1985), p. 236.

223 Hicks, TV tape 9/9/74

223 Asked to meet, Lupo, *Liberty's Chosen Home,* p. 202; black fears, Bob Bates int 9/21/94.

223 Called office, *BG* 9/10/74; told of delegation, Lupo, *Liberty's Chosen Home,* pp. 202f, Eddie Martin int 11/18/94.

223 Police, Irish, Joseph P. Kennedy II int 1/10/95; "wacko," King int 2/14/94.

224 Taunts, *BG, BEG* 9/10/94.

224 Told reporters, TV tape 9/9/74; "deplorable," Lupo, *Liberty's Chosen Home,* p. 205.

224 Staff, White, Lupo, *Liberty's Chosen Home,* pp. 205, 208; White, *NYT* 9/10/74.

225 Violence, Lukas, *Common Ground,* p. 241.

225 Band, *Baltimore Sun* 9/16/74; Raftery, pardon, *BH* 9/14/74; San Francisco, *NYT* 9/16/74.

225 Joan, *Newsweek* 10/7/74; EMK statement, *BEG* 9/23/74.

225 Tough on family, *Newsweek* 10/7/74, *BG* 9/24/74; "where it never served much purpose," *BG* 9/24/74.

226 "cold turkey," *Newsweek* 10/7/74; McGrory, *WES* 9/25/74; "on my mind," EMK int 4/12/96.

CHAPTER 18 UNCONVENTIONAL WISDOM

page

227 Breyer–EMK meeting, EMK int 6/2/97, Stephen G. Breyer int 7/26/97; 20 percent, Kathryn Harrigan and Daniel M. Kasper, "Senator Edward Kennedy and the Civil Aeronautics Board," Harvard Business School Case Study, 1977.

228 Miller, James C. Miller to Alan Greenspan 8/14/74, CEA Records, RG 459, Staff Files 138, File James Miller June 1974–Oct 1975 (9), GRFL; Areeda, Breyer int 7/16/97.
228 Ford, *PPP Ford 1974,* p. 202; Pan American, Harrigan and Kasper, "Senator Edward Kennedy and the Civil Aeronautics Board," p. 7.
228 Protests, *NYT* 10/7/74; Lippman, *Senator Ted Kennedy,* pp. 160–62.
228 Complaints, Senate Judiciary Subcommittee on Administrative Practice and Procedure, hearings on air charter fares 11/4/74, pp. 4–28, 79–187.
229 Laker, hearings, pp. 188–93.
229 Clearwaters, hearings, pp. 210–19.
229 Liebeler, hearings, pp. 333–36.
230 Timm, hearings, pp. 379, 404.
230 Staff Report to the Subcommittee on Administrative Practice and Procedure, "Procedures Relating to Minimum Charter Air Fares," pp. III, 8, 14, 16, 20.
230 EMK, *CR* 93:II, pp. 40003f.
231 Epelman, King int 2/14/94, EMK to Polina and Michael Epelman 12/6/74, EMK Foreign Policy Files; blessing, Robert Hunter int 8/23/94.
231 Urged not to go, Hunter int 8/23/94.
232 Kissinger meeting, Hunter int 8/23/94; news conference, *BEG* 11/20/74; Soares, EMK to Mario Soares 11/27/74, EMK Foreign Policy Files; exciting, EMK int 10/31/97.
232 EMK, *CR* 93:II, p. 38131; State Department surprised, *NYT* 12/14/74; aid, Kissinger cable to Amembassy Lisbon 12/13/74, af.
232 Broaden arms control approach, Hunter int 8/23/94.
232 EMK, *CR* 93:II, p. 39388; Mondale, p. 39389.
232 January version, *NYT* 1/18/75, 1/30/75.
233 *Pravda, NYT* 2/5/75.
233 Brezhnev, Joan Kennedy int 1/18/96; concert, party, *WP* 3/1/75; Teddy, EMK int 6/2/97.
233 Common position, Dudley Chapman to Phil Areeda 1/29/75, Schmults papers, Box 28, folder Airlines 1/75–6/75, GRFL; unrealistic, Snow to Areeda 1/29/75, same folder.
233 Barnum, Senate Judiciary Subcommittee on Administrative Practice and Procedure, hearings on Civil Aeronautics Board 2/6/75, p. 6.
234 Six weeks, hearings, p. 44.
234 Engman, hearings, p. 24; Kahn, p. 92.
234 "That's why," Breyer int 7/26/97; Boston heckling, *BG* 3/8/75; North Quincy, *BH* 3/11/75.
234 Jordan, hearings 2/14/75, pp. 454f; Murphy, p. 527.
235 O'Melia, hearings 2/18/75, p. 648; Kutzke, p. 672.
235 Muse, hearings 2/25/75, p. 1249; EMK, pp. 1327, 1355.
235 Yohe, hearings, p. 1375; press, pp. 1378–80.
235 "Informal," hearings, p. 1358.
235 Investigations, hearings 4/21/75, pp. 2311, 2325, 2327, 2331; O'Melia, p. 2359; Timm, p. 2380; Justice Department, EMK to Attorney General Edward Levi 5/12/75, pp. 2385f.
236 *Aviation Week & Space Technology* 4/7/75, 4/28/75.
236 "enthusiastic," EMK press release, af; delay, EMK to Ford 6/26/75, Schmults papers, GRFL.
236 Draft, *NYT* 6/30/75; staff study, *Aviation Week & Space Technology* 7/28/75.
236 Deregulators, Miller to Domestic Council Regulatory Reform Group 7/3/75, Miller to Rod Hills 7/22/75, Schmults papers, Box 28, GRFL. "Re-Regulation," *Aviation Week & Space Technology* 11/17/75.
237 Legitimacy, John E. Robson OH, GRFL; "crusading," *WSJ* 4/24/75; leadership, *Fortune* 8/75.

237 Indiscriminate, *CR* 93:I, p. 3979; Schlesinger, *NYT* 2/24/75.
237 Teheran, *NYT* 5/28/75; "incredible," Senate Foreign Relations Committee, hearings on foreign aid 6/18/75, pp. 106ff.
237 Modernization, *Foreign Affairs* 10/75, pp. 14ff.
238 Funeral, Lippman, *Senator Ted Kennedy,* p. 153; "unconscionable," *CR* 93:I, p. 7778.
238 Cronin, *BG* 5/4/75; embargo, *NYT* 3/5/75; flag, *NYT* 3/11/75; antagonize, James Stevenson, "A Reporter at Large," *New Yorker* 8/25/75.
239 Chemotherapy, *BH* 6/7/75; Apple, *NYT* 4/28/75; "Ready," *Newsweek* 6/2/75; Rose Kennedy, *BH* 7/28; O'Neill, *NYT* 7/25/75, 8/1/75.
239 Nolan, *BG* 6/5/75.

CHAPTER 19 CRIME AND TAXES

page
240 Feinberg-EMK interview, Kenneth Feinberg int 9/18/97.
240 Crime speech, af, *Chicago Tribune* 10/21/75.
240 Negotiations, ACLU letter, Feinberg int 9/18/97.
241 Elevator, EMK int 4/12/96, *WP* 2/10/76.
242 Recommendations, Report of the Subcommittee on Administrative Practice and Procedure, p. 19; Ford, Schmults to Ford, 2/18/76; Ford to EMK 2/19/76, Schmults papers, GRFL, *NYT* 2/22/76.
242 EMK-Cannon, Senate Commerce Subcommittee on Aviation, hearings on airline regulation 4/6/76, pp. 163ff; loss of service, EMK int 10/24/97; Cannon, Robson, *Aviation Week & Space Technology* 4/29/76.
243 "Outrageous," *CR* 97:II, p. 20706; "human rights," p. 33232.
243 Stopped, *National Enquirer* 1/20/76; Smithers, 4/13, 4/20/76.
243 Shriver, *WES* 4/11/76.
243 Drilling, Paul McDaniel int 9/29/97.
244 EMK analysis, *CR* 94:II, p. 30724.
244 Angry colleagues, Senate Finance Committee, hearings on tax bill HR 10612, 7/20/76, pp. 16–26; Curtis helped, *NYT* 7/19/76.
244 Success, Stanley S. Surrey, "Reflections on the Tax Reform Act of 1976," *Cleveland State University Law Review,* vol. 25, no. 3 (1976).
245 Cape, *NYT* 2/12/76.
246 Aides, Woodcock int 2/12/96, Max W. Fine int 11/15/97; Carter, UAW, Woodcock int.
246 Carter speech, Jimmy Carter, *The Presidential Campaign of 1968,* vol. 1, part 1, (Washington, D.C.: Government Printing Office, 1978), p. 133.
246 Comparisons, *Washington Star-News* 4/16/76, *WP, NYT, Atlanta Constitution* 4/17/76.
246 Doubts, James McGregor Burns to Michael Beschloss 7/14/76, af; "indefinite," *BG* 5/29/76.
246 Wieghart, *New York Daily News* 5/21/76.
247 Denial, *NYT* 5/21/76; Nordan, *Atlanta Journal* 5/29/76.
247 Apple, *NYT* 5/30/76; Nolan, *BG* 5/23/76.
247 Appreciated, Hamilton Jordan int 3/2/95; "friend," *WP* 5/29; "animosity," *Women's Wear Daily* 7/7/76.
247 Morally, Peter Bourne int 11/7/96; "woman-killer," Douglas Brinkley int 8/1/97.
247 Larger role, Patrick Caddell int 1/1/98, Burns to Beschloss, 7/14/76, *NYT* 7/15/76.
248 1968, *WP* 7/14/76.
248 O'Neill, *BG* 7/14/76; ads, *BG* 7/15/76; Berta, *NYT* 7/15/76.
248 Carter call, *BG* 7/15/76; delegation, *BH* 7/16/76; Kirk, *BG* 7/17/76; Healy, *BG* 7/23/76; confidence, help, *BH* 8/8/76; shoulder, *BG* 8/24/76.
248 EMK got Doherty, Gerard Doherty int 11/7/97.

249 Meeting, *BG* 9/1/76.
249 Boston, *Baltimore Sun* 10/1/76, *NYT* 12/13/78.
249 Warning, Joseph P. Kennedy II int 1/10/95.
249 Campaign, *BG* 9/7/76, 9/17/76.
250 Debate, *BG* 11/1/76.
250 Handguns, *BH* 10/22/76.
250 *Words Jack Loved*, Edward M. Kennedy, ed. (privately printed, 1977).

CHAPTER 20 STRAINS WITH CARTER, STRAINS ON JOAN

page
252 Guilty, Anne Taylor Fleming, "Kennedy: Time of Decision," *NYT Magazine* 6/24/79; divorce, Martin Nolan int 11/5/97.
252 Hart, *NYT* 3/5/77.
253 No need, Walter F. Mondale int 3/27/97.
253 Atlanta, *NYT, LAT* 1/16/77.
254 Helpful, Hamilton Jordan int 3/2/95.
254 "not fair," *CR* 95:I, p. 11389.
254 Northern Ireland, Tim Pat Coogan, *The Troubles: Ireland's Ordeal 1966–1996 and the Search for Peace* (Boulder, Colo: Roberts Rinehart, 1996), pp. 179ff, 191f; Carter button, *NYT* 8/31/77; Democrats, Irish alarm, Coogan, *The Troubles*, p. 346, *NYT* 3/20/77.
255 Carter statement, British, Sean Donlon int 5/8/95.
255 Hume's help, John Hume int 3/17/95; statement, af.
255 $50 million, British, Zbigniew Brzezinski to Carter 8/23/77, JCL.
255 Carter on settlement, *PPP Carter 1977,* Book 2, p. 1524.
255 Statement, *NYT* 8/31/77; note, EMK to Carter 9/13/77, Name File Edward M. Kennedy, JCL.
255 Carter assurance, EMK int 10/31/97.
256 May 2 meeting, Joseph A. Califano, *Governing America* (New York: Simon & Schuster, 1981), pp. 97f, Larry Horowitz int 6/23/95, Leonard Woodcock int 2/12/96.
256 "Wrong," Califano, *Governing America,* p. 100.
256 EMK speech, af; "one of us," *LAT* 5/18/77.
256 Deadline, *PPP Carter 1977,* Book 1, p. 892; EMK, *LAT* 5/18/77.
257 Summit, Sneeden, Dershowitz, Kenneth Feinberg int 9/18/97; Lewis, *NYT* 3/14/77.
257 Dinner, *NYT* 6/16/76.
258 Conservatives, Feinberg int 3/13/95.
258 Holbrooke, Oksenberg, Jerome Alan Cohen int 2/4/95; Huang Hua, Kissinger, Cohen, "China Sweet and Sour," *Nieman Reports* Autumn 1978.
258 Legal advisers, *WP* 8/16/77; speech, af.
259 Shown to State Department, Warren Christopher to Cyrus Vance 8/10/77, af.
260 David, "Joan Kennedy: Her Search for Herself," *Ladies' Home Journal* 5/79; Mudd, *CBS Reports* 11/4/79, transcript, p. 9, af.
261 Braden, "Joan Kennedy Tells Her Own Story," *McCall's* 8/78.
261 Gossip, David, "Joan Kennedy: Her Search for Herself," *Ladies' Home Journal* 5/79.

CHAPTER 21 BEIJING AND MOSCOW

page
261 Chinese reaction, Cohen, "China Sweet and Sour," *Nieman Reports* Autumn 1978.
261 Message, Zbigniew Brzezinski to EMK 12/21/77, Name File Edward M. Kennedy, JCL; exchanges, *BG* 1/5/78.
261 Bedroom, Leonard W. Woodcock int 2/12/96.
262 University, Cohen, "China Sweet and Sour."

262 Prison, Cohen int 2/4/95; opera, Cohen, "China Sweet and Sour."
262 Foo, *BG* 1/1/78; State Department, Jan Kalicki int 12/9/93.
262 Shanghai boss, Kalicki int 12/9/93.
263 Phone calls, Cohen int 2/4/95.
263 EMK–Hua transcript, David Dean to Cyrus Vance 1/4/78, af.
264 EMK–Teng transcript, Dean to Vance 1/5/78, af.
264 Ting to Kalicki, Dean to Vance 1/6/78, af.
265 "private too," *BH* 1/5/78.
265 Shaoshan, Cohen, "China Sweet and Sour"; feudal, *BH* 1/10/78; hamburgers, *BG* 1/3/78; souvenirs, *BG* 1/8/78; circus, EMK Jr. int 1/31/96; Polaroids, Sharon Woodcock int 2/12/96; tooth, *WP* 2/1/78.
265 Taiwan, exchanges, *BEG* 1/9/78; press, *LAT, WP* 1/10/78.
266 symphony, EMK statement 1/9/78, af.
266 "three thousand laws," *CR* 95:II, p. 9, Bayh, p. 10; controversy, p. 10.
266 Sentencing, rehabilitation, *CR* 95:II, pp. 294f.
266 Amendments, *CR* 95:II, p. 392; Helms, p. 1369.
267 "bipartisan," *NYT* 1/31/78.
267 Cannon, *CR* 95:II, p. 10647.
267 EMK, *CR* 95:II, p. 10658f.
267 McGovern, *CR* 95:II, p. 10658; EMK, p. 10666.
268 EMK, Bumpers, *CR* 95:II, p. 10686; Carter, *PPP Carter 1978,* Book 1, p. 762.
268 Running, *Christian Science Monitor* 4/13/78, *BH* 4/14/78; Gallup Poll, *NYT* 5/7/78.
268 Judges, Judith Lichtman int 3/30/95; mayors, *NYT* 6/8/78, 6/20/78.
269 Carter view, meetings, Califano, *Governing America,* pp. 104–18; EMK–labor, *WP* 4/7/78.
269 Schultze, Blumenthal, *NYT* 6/21/78; suspicions, Rashi Fein to EMK 5/30/78, af.
270 EMK news conference, *NYT* 7/29/78.
270 Califano, *NYT* 7/30/78.
270 Kennedy's commitment, former Carter administration official int 1997; EMK suggestion, *WP* 1/9/78.
270 Foo arrives, *BG* 8/5/78
271 Soviet view of Carter, Anatoly Dobrynin, *In Confidence: Moscow's Ambassador to America's Six Cold War Presidents* (New York: Random House, 1995), pp. 374–414; threat, *Pravda* 6/15/78.
271 Brezhnev view of EMK, talks with Kalicki, Andrei Pavlov int 11/24/97.
271 Meeting plans, Pavlov int 11/24/97.
272 Melons, bread, EMK int 6/2/97, Kalicki int 12/9/93, Pavlov int 11/24/97.
272 Brezhnev statement, JCL.
273 EMK remarks, *Science* 9/22/78.
273 Shulman to Vance 9/16/73, af.
273 Meeting, *WP* 9/13/78, UPI in *City News* (New York City) 9/13/78, Boris Katz int 2/1/96.
274 Roitburg, Kalicki int 12/9/93; Katz, Katz int 2/1/96.
274 Levich, Pavlov int 11/24/97.
274 News conference, *WP* 9/12/78.
275 Departure, Katz int 2/1/96.
275 Meeting, memo, EMK to Carter 9/12/78, 9/25/78, af.
275 Airline representatives, *CQA 1978,* p. 497.

CHAPTER 22 AGAINST THE WIND

page
276 "sail against the wind," videotape, af.
277 Studying, *Arkansas Democrat* 12/10/78; Jordan, Caddell int 1/1/98.

277 "demagoguery," *BG* 12/11/78.
277 Rationale, Mondale int 3/27/97.
278 minimized, Carter diary 12/10/78; tentative, James Earl Carter int 5/15/98.
278 "fear," Jordan to Carter 1/17/79 in *WP* 6/8/80.
278 Carter news conference, *NYT* 12/13/78.
278 Timing, Jaw-Ling Joanne Chang, *United States–China Normalization: An Evaluation of Foreign Policy Decision-Making* (Denver: University of Denver, 1986), pp. 139f; Vance told EMK, Michael Oksenberg to Zbigniew Brzezinski 3/3/78, JCL, National Security Affairs, Brzezinski Material Country File, Box 8, folder China PRC 2-5-78; notice, Byrd, dinner, Oksenberg int 3/3/98.
278 Goldwater, Democrats, *NYT* 12/16/78.
279 Resolution, *CR* 96:I, p. 1715.
279 Worked out language, Jan Kalicki int 2/17/98; American weakness, Senate Committee on Foreign Relations, hearings on relations with China 2/7/79, p. 372.
279 Teng reassurance, *CQWR* 2/3/79, p. 207.
279 Rosalynn Carter, Senate Labor Subcommittee on Health and Scientific Research, hearing on mental health policy 3/21/79, pp. 1ff.
279 "tentatively," *NYT* 6/14/79, Carter int 5/1/98.
280 "history," *CR* 96:I, p. 15475.
280 EMK to Carter 5/23/79, JCL, Name File Edward M. Kennedy.
280 "Intimidated," "baloney," *NYT* 5/1/79.
280 Under their skins, Caddell int 1/1/98; March promise, *NYT* 6/14/79; polls, *NYT* 6/10/79.
280 Dinner, Tom Downey int 10/8/95.
281 Accountant, Eleanor Randolph, "The 'Whip-His-Ass' Story, or the Gang That Couldn't Leak Straight," *Washington Monthly* 9/79.
281 "whip inflation," *NYT* 6/14/79; Brokaw, videotape, af.
281 Lipshutz to Carter 5/3/79, JCL, File Staff offices, counsel, Lipshutz, Box 11; Udall, EMK int 10/31/97; age only, Carter int 5/15/98.
281 Caddell, Jonathan Moore, ed., *The Campaign for President: 1980 in Retrospect* (Cambridge: Ballinger, 1981), p. 30.
282 Senators, Peter Hart int 1/9/98; Symons, Moore, *Campaign,* p. 28.
282 Draft movement, Mark Siegel int 7/18/94, Kirk int 7/6/94; "great job," *BH* 8/8/94.
282 Aesop, Walter Mears column, Associated Press 5/2/79; vision, *NYT* 5/1/79.
283 "confidence," *NYT* 7/16/79.
283 Califano, EMK int 10/31/97.
283 Telephone, EMK int 10/31/97; spirit, Moore, *Campaign,* pp. 21f.
283 Carter's message, EMK int 10/31/97.
284 Carter insiders, Eizenstat int 5/19/92, Jordan int 3/2/95.
284 Tunney, EMK Jr. int 1/31/96; protection, Larry Horowitz int 6/23/95.
284 Joan, Horowitz int 6/23/95.
284 Joan, Rose, *NYT, New York Daily News* 9/7/79; "certain," *WP* 6/8/80; coordinators, Carter int 5/15/80.
284 Glance, Melody Miller int 9/24/92.
284 "run sometime," EMK int 3/27/98.
285 Frost, David Burke int 3/22/98; "It's Hard to Keep from Being King," *The Poetry of Robert Frost,* ed. Edward Connery Lathem (New York: Holt, Rinehart & Winston, 1970), p. 462.
285 Moral issue, Thomas P. O'Neill int 11/23/92.
286 Surprise, Roger Mudd int 12/1/97; life on Cape, Dick Goodwin int 9/17/93.
286 Interview, "Teddy" videotape, af.
286 "panicked," *NYT* 9/26/79; "habit," *BH* 9/27/79.
287 Carter aides, Jack Watson int 12/9/97.
287 Inhibition, EMK int 10/31/97, Larry Horowitz int 6/23/95.

288 Hart, Peter D. Hart int 1/9/98; Dees, Morris Dees with Steve Fisher, *A Season for Justice* (New York: Scribner's, 1991), p. 197.
288 "voluntarily," Walter F. Mondale int 3/27/97.
288 Schedule, Hamilton Jordan int 2/2/95; governors, Watson int 12/9/97.
289 Dedication, videotape, JFKL.
290 Carter airport interview, *PPP Carter 1979,* Book 2, pp. 1982ff.
290 Philadelphia, *NYT* 10/23/79; calls, *BH* 10/25/79; corridors, Carl Wagner int 3/12/92.
291 Prodded, State Department official int 1995.
291 "screwing around," *BG* 10/28/79.
291 Stearns, *NYT* 9/23/79, *BH* 10/30/79.

CHAPTER 23 LOSING

page
292 EMK speech, *NYT* 11/8/79.
293 Planted question, Joan Kennedy int 5/12/98.
293 Volcker, *NYT* 10/18/79; EMK answer, *LAT* 11/8/79.
294 Fumble-mouthed, *BH* 11/12/79, *Des Moines Register* 12/7/79.
294 Chicago, *WP* 11/9/79; Grinnell, Macalester, John Walcott memo 11/15/79.
295 "protect the Americans," *BG* 11/9/79.
295 Shah comments, *NYT* 12/4/79; Hart poll, af.
296 Smith as manager, Bill Carrick int 5/27/92, Hart int 1/9/98, Harold Ickes int 4/21/98.
296 Lose Iowa, Mondale to Carter 12/12/79, JCL; Woodcock, Leonard Woodcock int 2/12/96.
296 Schram, *WP* 11/13/79.
296 Kelderman, Walcott memo 1/11/80; Keokuk, Patti Saris int 10/12/95.
296 Wicker, *NYT* 11/9/79.
297 Breslin, *New York Daily News* 11/6/79; editorial, *BG* 11/11/79.
297 Iowa poll, *Des Moines Register* 12/16/79; Hart poll, af.
298 Mitgang, Associated Press 9/20/79; womanizing, Suzannah Lessard, "Kennedy's Woman Problem/Women's Kennedy Problem," *Washington Monthly* 12/79.
298 Speeches, photographs, Susan Estrich int 3/22/95.
298 National Organization for Women, *NYT* 12/11/79.
298 Palm Beach, Moore, *Campaign for President: 1980,* pp. 61f.
299 Avoid debate, Al McDonald to Hamilton Jordan and Jody Powell 12/14/79, Greg Schneiders to Jody Powell and Jerry Rafshoon 12/18/79, Staff Offices Press Powell Box 8 File Debate/Iowa/Carter/Kennedy (cancelled); never campaigning, Powell to Carter 12/28/79; "my duties here," Carter to Powell 12/28/79, Staff Offices Press-Jenkins Box 2 Folder Debates 1979, all JCL.
299 Pope, *NYT* 1/19/80.
300 Poll, Mondale, *Des Moines Register* 1/11/80; Mondale, EMK, *NYT* 1/12/80; Mondale opposition, Mondale int 3/27/97; Waterloo, Walcott memo 1/14/80.
300 Culver, John Culver int 2/1/93.
300 Joan, *Des Moines Register* 1/19/80.
301 "No," Walcott memo 1/25/80; Spanish, Charles Cook int 1/15/97.
301 Georgetown speech, *NYT* 1/29/80.
302 Chappaquiddick, *BG* 1/29/80.
302 Staff, Peter Hart int 1/9/98; Safire, *NYT* 2/25/80.
302 Kara, Collier and Horowitz, *The Kennedys,* p. 25.
303 "rough," EMK Jr. int 1/31/96; Shriver, Maria Shriver int 2/22/98; Patrick, Walcott memo 12/28/79; calls, Patrick Kennedy int 11/1/95.
303 "awful," memories, Kathleen Kennedy Townsend int 11/3/97.
304 Ads, *Chicago Tribune* 3/26/80.

304 Parade, *Chicago Tribune, Chicago Sun-Times, BG* 3/18/80; assassination threats, Joan Kennedy int 5/12/98.
304 Delegates, *NYT* 3/20/80; aides, *Chicago Tribune* 3/19/80, 3/20/80; Stearns, Moore, *Campaign for President: 1980,* p. 72.
305 Carey, Moynihan encouragement, *BG* 6/8/80; eggs, tomatoes, *BG* 3/22/80.
305 Ad budget, Walcott memo 3/27/80; ads, *BG* 3/23/80.
305 UN vote, *NYT* 3/2/80, 3/4/80.
306 "betrayal," *NYT* 3/14/80; Strauss, *NYT* 3/20/80; Vance, *NYT* 3/21/80, 3/22/80.
306 Pavlov message, Pavlov int 11/24/97, John Tunney int 1/8/98.
306 Proposal, Jan Kalicki int 2/17/98, Marshall Shulman int 3/30/98.
306 Expectations, EMK int 3/27/98.
306 Carter polls, Caddell int 1/1/98, Washington Post, *Pursuit of the Presidency 1980* (New York: Berkley, 1980), p. 112.
307 meeting, Al Kamen int 7/25/97; Parker House, Paul Kirk int 7/6/94; withdrawal, Shrum int 4/8/94.
307 Withdrawal text, af.
307 Joan, *BG* 3/26/80; "good advice," *WP* 3/26/80.
307 Percentages, *NYT*/CBS News exit poll; Midwood, *NYT* 3/26/80.
307 "The poor," 3/5/80 New York City speech, af.
308 EMK to Oliphant, Thomas Oliphant int 5/28/98, *BG* 3/27/80.
308 Carter on hostages, *WP* 4/2/80, *PPP Carter 1980–81,* Book 1, p. 576.
308 "Good news," Hamilton Jordan, *Crisis: The Last Year of the Carter Presidency* (New York: Putnam, 1982), p. 245.
308 Reagan campaign, Richard B. Wirthlin int 7/7/93.
309 Caddell memo, Susan Clough Files, Box 34, JCL; fears, Caddell int 1/1/98, Moore, *Campaign for President: 1980,* p. 86.
309 Commercials, *Philadelphia Inquirer* 4/17/80, *WP* 4/22/80, videotape, af.
309 Edited, "nuclear," Caddell int 1/1/98; Rafshoon, *Philadelphia Inquirer* 4/17/80.
309 "abandon," *BG* 4/17/80; "pale carbon copy," *NYT* 4/11/80.
310 "Turning the tide," *PPP Carter, 1980–81,* Book 2, p. 993; unemployment, *WP* 6/7/80; cost, Moore, *Campaign for President: 1980,* p. 91.
310 TV ads, *NYT* 5/26/80, 5/28/80, *WP* 5/27/80; Ron Brown int 6/80.
310 Release, *NYT, WP* 5/30/80; withdraw, Moore, *Campaign for President: 1980,* pp. 78f, 88.
311 Promises, EMK speech to National ERA Evening 6/18/80, af.
311 Executive order, *Advocate* 2/21/80; District of Columbia, *San Francisco Sentinel* 5/16/80.
311 May 24 event, Dudley Clendinen and Adam Nagourney, *Out for Good: The Struggle to Build a Gay Rights Movement in America* (New York: Simon & Schuster, 1999), p. 417, *LAT, Los Angeles Herald-Examiner* 5/25/80, Thomas Oliphant int 6/25/98; San Francisco rally, *San Francisco Chronicle* 5/31/80.
311 Patience, Richard Moe to Hamilton Jordan 5/29/80, JCL.
312 Conciliate, *NYT* 6/1/80; plea, Oliphant int 5/28/98, Susan Spencer int 5/26/98.
312 "first night," 6/3/80 statement, af.
312 Kirk's stand, Richard G. Stearns int 12/31/92, Kirk int 6/3/98; shortchanged, EMK int 9/11/98.
312 "fumble," "obsessed," Jimmy Carter, *Keeping Faith: Memoirs of a President* (New York: Bantam, 1982), p. 593.
313 Shrum notes, af; "Planning," *WP* 6/6/80.
313 "reserve judgment," *PPP Carter 1980–81,* Book 2, p. 1039.
313 Strauss, Robert S. Strauss int 5/29/98; Carter views, Carter private diary.
313 June 6 meeting, Caddell int 1/1/98, Jerry Rafshoon int 4/8/83 at Miller Center, Carter Presidency Project, JCL; Mondale int 1/5/98.
314 Trucking deregulation, *NYT* 7/2/80.

315 Vice Presidency, Robert C. Byrd int 2/12/93.
315 Slippage, Thomas A. Donilon ints 11/4/97, 7/14/98; Kennedy sense, Richard G. Stearns int 7/8/97; Carter calls, Tad Devine int 7/14/98.
315 Poll, *NYT* 8/10/80.
316 Tuesday speech, *NYT* 8/13/80. For analysis of the speech, see William Safire's discussion in *Lend Me Your Ears,* pp. 853f.
318 Jordan, *NYT* 8/12/80.
318 Delay, Harold Ickes int 4/21/98; unanimous, *BH* 8/14/80; waving standard, *NYT* 8/17/80.
319 Former aides, Peter Bourne int 11/7/96; Jordan, Caddell int 1/1/98.
319 Issues, EMK int 9/11/98; Carter, Jimmy Carter int 11/29/82 at Miller Center, Carter Presidency Project, JCL; Jordan, Miller Center, 11/6/81, JCL.
319 Voter shifts, *NYT* 11/12/80.
320 Called Clinton, William J. Clinton int with Elsa Walsh 2/24/97, af.

CHAPTER 24 IN THE MINORITY, LEADING

page
321 Liberal, EMK speech to ADA 4/4/81, af.
321 "education," EMK int 9/11/98.
321 "intense," EMK int 10/9/96; "epiphanies," Nancy Korman int 3/8/95.
322 Upside, Clendinen and Nagourney, *Out for Good,* p. 417; platform, Susan Estrich int 3/22/95.
322 Campaigns, EMK int 3/27/98.
322 Campaign staff, Carl Wagner int 3/12/92, Estrich int 3/22/95.
322 "cutting edge," *BG* 12/3/80; "career-wise," EMK int with John Aloysius Farrell 2/92, af.
323 Dangers, Orrin Hatch int 2/27/92.
323 "drastic," *CR* 96:II, p. 32987.
324 Byrd, p. 32988.
324 Baker, *CR* 96:II, p. 32988.
324 Conditions, EMK int 10/31/97; stuffing, Eddie Martin int 6/2/93.
324 Work together, Stephen Breyer int 7/6/93.
325 Thurmond, *CR* 96:II, p. 31442.
325 "go it alone," Joan Kennedy int 5/12/98.
325 Statements, *BH* 1/22/81.
325 Block grants, *PPP Reagan 1981,* pp. 110, 111.
326 EMK criticism, *BG* 2/19/81; inefficiency, Senate Labor Committee, hearing on block grants 4/2/81, p. 2.
326 States, *CR* 97:I, p. 13323.
326 Community health centers, David A. V. Reynolds, "An Analysis of the Political and Economic Viability of Community Health Centers: Implications for Their Future" (dissertation, University of Michigan School of Public Health, 1998), pp. 17–29, 118.
326 Cut spending, *NYT* 11/20/80.
327 Ancestry, St. Patrick's Day, Irish diplomat int 5/8/95.
327 Reagan notes, PR007-01, RRL.
327 Ceremony, *WP* 6/6/81; Reagan and EMK, videotape, *PPP Reagan 1981,* pp. 488f.
327 Desk drawer, Frank Mankiewicz int 7/97; did not withhold, Jimmy Carter int 5/15/98.
328 Delivered, *WP* 6/6/81; integration, *Baltimore Sun* 11/21/75, 11/23/75.
328 EMK concern, Jack Leslie int 4/9/92.
328 NCPAC, Leslie int 4/9/92; Lefkowich, *BH* 8/21/81.
329 "The answer," *NYT* 4/13/81.
329 "As Democrats," EMK Communications Workers speech 4/1/81, af.
329 Reclaim, forums, Larry Horowitz int 5/29/92; forums, *WP, NYT* 5/22/81.

329 Effectiveness, *BG* 2/14/81.
330 "Punitive," *CR* 97:I, p. 25027.
330 "Murder is murder," *CR* 97:I, p. 25045.
330 "unmistakable message," *CR* 97:I, p. 25031.
330 Colloquy, *CR* 97:I, p. 25049.
331 Dismay, Pam Solo, *From Protest to Policy: Beyond the Freeze to Common Security* (Cambridge: Ballinger, 1988), p. 76; "voodoo," *BH* 5/11/82.
331 Europe, David Gergen int 9/22/98; "annoyance," James A. Baker III int 8/98.
331 Impact, EMK int 9/11/98.
331 Hasten arms control policy, Robert MacFarlane int 9/28/98, Jeffrey W. Knopf, "The Nuclear Freeze Movement's Effect on Policy," in Thomas R. Rochon and David S. Meyer, *Coalitions and Political Movements: The Lessons of the Nuclear Freeze* (Boulder, Colo.: Rienner, 1997), p. 136.
332 Learning, Larry Horowitz int 7/21/98; trips, Horowitz int 1/7/99, Jack Matlock int 1/6/99.
332 61 percent, *NYT* 11/8/84.
332 Lyndon Johnson, *CR* 97:I, p. 6696; cosponsors, "monumental task," EMK int 9/11/98.
333 Leadership Conference, hearings, Ralph Neas int 2/10/95, Michael Pertschuk, *Giant Killers* (New York: Norton, 1986), pp. 154–59.
333 Decision, Smith, *LAT* 11/6/81, 11/7/81.
334 Strategy, dumbfounded, Ralph Neas int 2/10/95; "magic number," *NYT* 12/17/81.
334 Tax-exempt, Thurmond, *NYT* 1/8/82, 1/9/82; moderates, Pertschuk, *Giant Killers,* p. 177.
334 Hatch, Senate Judiciary Subcommittee on the Constitution, hearing on the Voting Rights Act 1/27/82, p. 3; Smith, p. 69; EMK, p. 77.
335 Worried, Burt Wides int 6/15/98.
335 Pressure, Burt Wides ints 3/18/96, 6/15/98, Neas int 6/16/98.
336 Compromise, *NYT, WP* 5/4/82; "agreement," EMK statement, af; Rauh, *WP* 5/9/82.
336 Dole, Reagan, Neas int 2/10/95; East, *NYT* 5/5/82.
336 Helms, *CR* 97:II, p. 13111; EMK, p. 13171; Hatch, p. 13112.
337 "some months," *LAT* 6/30/82; NAACP, *WP* 6/29/82.
337 A comprehensive account of the Quayle-Kennedy effort on job training appears in Richard F. Fenno, Jr., *The Making of a Senator: Dan Quayle* (Washington, D.C.: Congressional Quarterly Press, 1989), pp. 35–118.
337 "Administration endorsed," *NYT* 10/13/82; effectiveness, Abt Associates, *Does Training for the Disadvantaged Work?* (Washington, D.C.: Urban Institute Press, 1996), pp. 215–18; comparison with CETA, Larry Orr int 5/24/99.
338 Running again, Horowitz int 5/29/92.
338 Waldorf, *NYT* 2/2/82.
339 Mistakes, *Philadelphia Inquirer* 6/28/82.
339 Mondale speech, applause, *NYT* 6/29/82.
339 Coin toss, Bob Shrum int 3/10/92; floor passes, Bill Sweeney int 8/10/92.
339 EMK conference speech, af.
340 Freeze, *WSJ* 10/5/82.
340 Kaye spots, af.
342 Nomination speech, af; not interested, Caddell int 1/1/98; Mondale doubts, Walter Mondale int 3/27/97; Healy, *BG* 11/13/82.

CHAPTER 25 LITTLE MUD AND CENTRAL AMERICA

page
345 Metzenbaum, EMK int 9/11/98, Thomas Eagleton int 9/29/98.
346 Stevens, *CR* 97:II, p. 31500.
346 EMK role, *BG* 1/23/83.

346 Senate pay, *CR* 97:II, p. 31500.
346 Spending, *NYT* 2/2/83, *BG* 2/2/83.
347 Factories closed, *NYT* 2/6/83.
347 Reassurance, House Post Office and Civil Service Subcommittee on Census and Statistics, hearing on Martin Luther King, Jr., holiday 6/7/83, p. 18; "Joggers Day," p. 23.
347 Cooperation, Charles McC. Mathias int 4/27/92; Marxist, *CR* 98:I, p. 26868.
348 Racism, *CR* 98:I, p. 26869.
348 "dead brother," *CR* 98:I, p. 14008.
348 "my brother Robert," *CR* 98:I, p. 14012; wiretaps, p. 14027.
348 Bradley, *CR* 98:I, p. 14119.
348 Reagan, *NYT* 10/20/93; "unworthy," EMK statement 10/20/83, af.
349 Visit to Falwell, *NYT, WP* 10/4/83.
349 Separation, EMK text, af.
350 Attacks, Leslie, Melody Miller int 8/10/92.
350 EMK purpose, *BG* 11/20/83.
350 Childhood, Ireland, EMK, *Parade,* 11/13/83.
350 St. Anthony's, EMK text, af.
351 Brothers, *BG* 11/20/83.
351 Detroit visit, *BG* 11/21/83.
352 Eulogy, *NYT* 11/23/83.
352 Davis, Reynolds, *Frankfort State-Journal* 11/23/83; Thanksgiving, *Louisville Courier-Journal* 11/24/83.
352 Hall, Deal, *Louisville Courier-Journal* 11/24/83; Damron, sign, *Lexington Herald-Leader* 11/24/83.
353 Fleming-Neon, *Louisville Courier-Journal* 11/24/83.
353 Arrogance, *Newsday* 11/27/83.
353 New missiles, *CR* 98:I, p. 30061.
353 Degradation, UPI 12/9/83; Sagan, EMK, Kapitsa, *Christian Science Monitor* 12/14/83.
354 Divorce, Joan Kennedy int 9/19/98; family home, *BH* 2/11/84.
354 David Kennedy, *NYT* 4/26/84, 4/27/84.
355 Dodd, difference, Greg Craig int 2/13/95.
355 EMK op-ed, *WP* 1/15/84.
356 Lugar, Craig ints 2/13/95, 9/27/98.
356 More guns, blindness, *CR* 98:II, p. 7354.
356 "freedom fighters," *NYT* 3/29/84.
357 "see no evil," *CR* 98:II, p. 7705; Reagan letter, p. 7713.
357 Biden, deletion, Jim Wright, *Worth It All: My Fight for Peace* (Washington, D.C.: Brassey's, 1993), p. 69.
357 Goldwater letter, *WP* 4/11/84.
357 Respect, Barry Goldwater to EMK 7/1/84, af; damages, deaths, *CR* 98:II, p. 8410.
357 Congress's duty, *CR* 98:II, p. 8536.
358 White House did not fight, *WP* 4/11/84; Casey, *NYT* 4/11/84.
358 "first step," *NYT* 4/11/84; "not binding," mining stopped, Wright, *Worth It All,* p. 71.
358 Accountable, *BG* 4/6/84.
358 Manual, aid votes, *CQA 1984,* pp. 86–93.
358 McFarlane, North, Cynthia J. Arnson, *Crossroads: Congress, the Reagan Administration and Central America* (New York: Pantheon, 1989), pp. 163–65.
359 "dragging," *NYT* 2/23/84.
359 Conference, former Kennedy aide int 5/11/98; EMK statement, af.
359 Mondale, *BG* 3/7/84, *NYT* 3/7/84, 6/8/84.
360 Neutrality, Willie Brown, *WP* 4/16/84; Hollings, *NYT* 3/26/84; Goldwater, *NYT* 3/31/84.
360 EMK to Hart, delegates, rules, *BG, BH, Minneapolis Star-Tribune* 6/26/84.

360 St. Paul, *NYT* 6/26/84; EMK convention speech, af.
360 Reagan and JFK, *NYT* 11/3/84.
361 Catholics, abortion, O'Connor, *NYT* 9/23/84, 9/24/84.
361 Eulogies, af.
362 With Church, LeRoy Ashby and Rod Gramer, *Fighting the Odds: The Life of Senator Frank Church* (Pullman: Washington State University Press, 1994), p. 615, F. Forester Church, *Father and Son* (New York: Harper & Row, 1985), pp. 171f.

CHAPTER 26 A PILGRIMAGE TO SOUTH AFRICA

page
363 October 5 meeting, Greg Craig int 2/13/95, Desmond Tutu int 5/2/95; RFK speech, Schlesinger, *Robert Kennedy and His Times,* p. 746.
364 EMK concerns, EMK int 9/10/92.
364 Famine, Edward M. Kennedy, "Land of Death and Desolation," *People* 1/26/85; dignity, EMK Jr. int 11/6/98.
364 Elahmadi, *BG* 1/1/85.
365 Safari, *BG* 1/11/85; Patrick, EMK int with Chris Black 7/19/97, af.
365 EMK comments, Reuters 1/5/85; Nickel, Craig int 2/18/95; police, Tutu int 2/2/95; resolution, cable Herman Nickel to State Department 12/18/84, af.
365 Police, welcome, *Sunday Star* (Johannesburg) 1/6/85.
366 Hostel, *BG* 1/7/85; Mathopestad, *BG* 1/8/85; Botha, Reuters 1/7/85.
366 Meeting, cable Nickel to State Department 1/8/85, af; tone, Craig int 2/23/95; letter, Botha to EMK 1/11/85, af.
366 Mandela, Craig int 2/23/95.
367 Business resolution, *Rand Daily Mail* 1/8/85; challenge, EMK speech, af.
367 Nickel text, af; clearance, Robert Healy int 10/21/98, *BG* 1/9/85; "impressed," cable Nickel to State Department 1/10/85, af; different views, Reuters 1/8/85.
367 Chappaquiddick, *Sunday Times* (Johannesburg) 1/13/85; Harvard, *Beeld* (Johannesburg) 1/14/85; Negro vote, *Die Vaderland* 1/4/85.
367 Onverwacht, Johannesburg, *Citizen* 1/10/85; Brandfort, Craig int 2/18/95, Reuters 1/10/85.
368 Welcome, *Cape Times* 1/11/85; Crossroads, *Citizen* 1/11/85, *BG* 1/13/85.
368 UDF, Molefe, Lekota, Greg Craig int 2/18/95; divisions, cable U.S. Embassy to State Department 1/11/85, af; "these walls," Reuters 1/13/85; Mandela knew, *BH* 6/24/90.
368 Rally, *Cape Times* 1/12/85.
369 Windhoek, cable Nickel to State Department 1/14/85.
369 Cathedral, *BG* 1/14/85, Tutu int 5/2/95; police, Leslie Dach int 11/20/98.
369 EMK departure statement 1/13/85, af.
370 Tambo, Reuters 1/14/85.
370 Impact, liberals, David Welsh int 10/19/98; press, Bloemfontein, *Die Vaderland* 1/4/85, *Die Volksblad* 1/10/85; attention, cable Nickel to State Department 1/23/85, Tutu int 2/2/95.
371 Unimportant, *Sunday Times* (Johannesburg) 1/13/86.
371 Now or never, EMK Jr. int 1/31/96.
371 "try the idea," EMK Jr. int 11/6/98; Farmer, *BH* 7/4/85; lukewarm, Caddell int 1/1/98, Nancy Korman to EMK 3/13/85; mother, Joan Kennedy int 12/12/96; "proud," EMK statement 12/4/85, af.
371 Long, Robert Mann, *Legacy to Power: Senator Russell Long of Louisiana* (New York: Paragon House, 1992), p. 392.
372 Mikulski, Barbara Mikulski int 7/12/95; visit Mathias, Charles McC. Mathias int 4/2/92.
372 Hofstra speech, af.
373 McLean, videotape recording, RRL.

373 "do nothing," *NYT* 10/10/85; furious, unions, Bill Carrick int 4/15/92; "programs," EMK statement 12/11/85, af.

374 Packwood, EMK, *NYT* 10/3/84.

374 Rewriting, former Kennedy aide int 5/11/98.

375 architect of retreat, Senate Judiciary Committee, hearings on nomination of William Bradford Reynolds 6/4/85, pp. 5f.

375 Biden, hearings 6/4/85, pp. 28–31; EMK, pp. 36–40; Specter, pp. 40–44; Mathias 6/5/85, pp. 127–30.

375 "Recollection," hearings 6/18/85, p. 845; Specter, p. 887.

376 EMK, p. 980; broadcast, *NYT* 6/16/85.

376 "burning, burning desire," EMK int 3/26/99.

376 EMK, Cranston, Craig int 2/27/95.

376 EMK statement on South Africa legislation 3/7/85, af.

377 Reforms, *NYT* 3/2/85; Shultz, *NYT* 3/22/85; McConnell, Senate Committee on Foreign Relations, hearing on policy toward South Africa 4/24/85, p. 3; Lugar, Richard G. Lugar, *Letters to the Next President* (New York: Simon & Schuster, 1988), pp. 214–19.

377 "one voice," *CR* 99:I, p. 7334.

377 EMK meetings, lobbying, Greg Craig int 3/12/99.

377 "some day," Senate Committee on Banking, Housing and Urban Affairs, hearing on anti-apartheid legislation 4/16/85, p. 6; Shultz, *NYT* 4/17/85.

378 EMK testimony, hearings, pp. 9f; Cranston, p. 29.

378 Amendments, *WP* 7/12/85.

379 Lugar, Goldwater, *NYT* 7/12/85; EMK, *CR* 99:I, p. 17866.

379 Delay, *WP* 7/27/85; arrests, *WP* 7/14/85; protocol, Dole, *CR* 99:I, pp. 20435f.

379 Conference, *NYT* 8/1/85.

379 Gingrich, *WP* 8/2/85.

379 Negotiations, Lugar, *Letters to the Next President,* pp. 218–20; segregation, *PPP Reagan 1985,* Book 2, p. 1011.

380 Persuaded, Lugar, *Letters to the Next President,* p. 221; "focus," *CR* 99:I, p. 23035.

380 EMK, *CR* 99:I, p. 23036; O'Neill, Botha, Tutu, *WP* 9/10/85; "Republican Party," "self absorption," *WP* 9/11/85; Weicker, Glenn, *CR* 99:I, p. 23295; Dole, p. 23299.

381 Stafford, EMK, *WP* 9/12/85; Weicker, *NYT* 9/12/85.

381 Conference report removed, debated, *CR* 99:I, p. 23582, Richard Lugar int 12/23/98; Krugerrand, Lugar int 12/23/98.

382 EMK and Stark, EMK int 11/11/98.

382 Pet projects, David Nexon, "The Politics of Congressional Health Care Policy," *Medical Care Review* Spring 1987; labor bill, Nexon int 11/19/98, memo, Nexon to EMK 9/30/85, af; Grassley, Report of the Budget Committee on S. 1730, 10/2/85, pp. 445–98.

383 Conference, David Nexon int 11/19/98.

383 "requisitioned," Patricia Schroeder int 12/4/98.

383 "family issues," *CR* 99:I, p. 12332.

384 Conference, *Air Force Times* 8/5/85, 8/12/85; *CR* 99:I, p. 28572.

384 Kennedy, O'Neill, Clark, Irish diplomat int 5/8/95; Clark role, EMK int 6/2/97.

384 Extremists, EMK statement 11/15/85, af.

385 Broadcast, *BG* 12/20/85.

385 "activist," *BG* 11/10/85.

385 Thanksgiving, EMK int 3/26/99.

385 Horowitz, *BG* 12/20/85; Kaye, Horowitz int 5/29/92; Hyannis, Robert Shrum int 3/10/92.

386 Cooper, Horowitz int 5/29/92.

386 Statement, *BG* 12/20/85.

386 "Presidency," *NYT* 12/21/85.

386 Taping, Horowitz int 5/29/92.
386 Shoop, *St. Louis Post-Dispatch* 12/22/85.

CHAPTER 27 AN ADVOCATE FOR ARMS CONTROL

page
387 Menem, Greg Craig int 2/27/95.
387 Visa, "unfriendly," *BG* 1/8/86; irregular, Harry Barnes int 12/4/98.
388 Ads, Dach int 11/20/98; CIA, Craig int 2/27/95; EMK comments, *BG, WP* 1/15/86.
388 Arrival, Craig, EMK comments, Associated Press 1/15/86; EMK told Craig, Craig int 2/27/95.
388 Helicopters, meetings, mothers, flag, Dach int 11/20/98; EMK to crowd, *LAT* 11/16/86.
388 Courage, EMK int 11/11/98.
389 EMK text, af.
389 Singing, Craig int 2/27/95, Dach int 11/20/98, Barnes int 12/4/98.
390 Kremlin on EMK, Andrei V. Pavlov int 11/24/97, EMK int 10/31/97; "God," Larry Horowitz int 5/29/92.
390 Shcharansky or Sakharov, George P. Shultz, *Turmoil and Triumph: My Years as Secretary of State* (New York: Scribner's, 1993), p. 703; Shcharansky, Horowitz int 1/12/99.
390 January meeting with Pavlov, Horowitz int 4/1/99 while reading from his trip diary.
391 Credit, Avital Shcharansky, Horowitz int 4/1/99.
391 EMK worries, Horowitz int 4/1/99.
391 Vague, Jack Matlock int 1/6/99; Hartman, Horowitz int 4/1/86.
391 Shevardnadze on Shcharansky, Horowitz int 4/1/99; families, Horowitz int 5/29/92.
391 EMK on SDI, Horowitz int 6/23/95.
392 Shevardnadze on linkage, Horowitz int 4/1/99.
392 Academy speech, af; Alexandrov, EMK, Horowitz int 4/1/99.
392 *Challenger,* Reagan, INF, Horowitz int 4/1/99.
392 Gorbachev and SDI, EMK statement 2/8/86, *BG* 2/9/86; human rights, Horowitz ints 6/23/95, 4/1/99.
393 Veins, Horowitz int 4/1/99; news conference, *BG* 2/9/86.
393 Shcharansky, professional, John-John, Horowitz int 4/1/99; other families, *BG* 2/9/86.
393 Lerner, EMK statement 2/8/86, af; dissidents, Horowitz int 4/1/99; EMK broadcast 2/7/86, af.
394 Did not brief embassy, Horowitz int 6/23/95; Shultz informed, Matlock int 1/6/99; Shultz account, Shultz, *Turmoil and Triumph,* p. 903.
394 EMK impact, Matlock int 1/6/99.
395 Gridiron text, af.
395 Food stamps, Tom Rollins int 4/21/98; EMK statement, af; Reagan, *BH* 5/22/86.
396 CIA, Casey, Botha, Shultz, *Turmoil and Triumph,* pp. 1115–20; "milestone," *NYT* 4/25/86.
396 Sanctions, *NYT* 7/21/86.
396 Lugar, *NYT* 6/16/86; Reagan, *NYT* 6/14/86; House, *NYT* 6/19/86.
396 EMK on Reagan, *NYT* 7/21/86.
397 Forum, Fraser, UPI 7/21/86.
397 Dinner, Greg Craig int 2/27/95.
398 "very helpful," Lugar int 12/23/98.
398 Lugar, others to White House, Lugar, *Letters to the Next President,* p. 227; Buchanan, desperation, Shultz, *Turmoil and Triumph,* p. 1122; "transition," *PPP Reagan 1986,* Book 2, p. 985; Botha, *NYT* 7/24/86.
398 Biden, Senate Committee on Foreign Relations, hearings on Africa 7/23/86, p. 94; EMK, p. 116.

399 Lugar, EMK, *NYT* 7/23/86; EMK broadcast, af.
399 Lugar attitude, Lugar, *Letters to the Next President,* p. 235; Dole, *BG* 8/3/86.
399 Friday, *NYT* 8/2/86, Lugar, *Letters to the Next President,* p. 231.
400 Lugar, *CR* 99:II, p. 21471; EMK, pp. 21375, 21823.
400 EMK amendments, *CQWR* 8/16/86, p. 1861.
400 Distrust, Lugar, *Letters to the Next President,* p. 234; Black Caucus, Greg Craig ints 2/27/95, 3/12/99; Dellums, EMK int 3/26/99; Gray, *CR* 99:II, p. 23142.
401 Veto, *PPP Reagan 1986,* Book 2, pp. 1278–80.
401 "backward," *NYT* 9/30/86.
401 Botha, Lugar, *NYT* 10/2/86; EMK, *CR* 99:II, p. 27655.
401 EMK on apartheid, *CR* 99:II, pp. 27854f.
402 Dole, *CR* 99:II, p. 27858.
402 Montero-Duque, Craig int 2/27/95.
403 "Framers," Senate Judiciary Committee, hearings on nomination of William H. Rehnquist 7/29/86, p. 14.
403 Challenging, *CR* 99:II, p. 22803.
404 Plessy, *CR* 99:II, p. 22802; vote, *NYT* 8/15/86.
404 EMK floor speech, *CR* 99:II, pp. 22802ff.
405 "discrimination," *CR* 99:II, p. 32377.
405 Safety regulations, Senate Labor and Human Resources Committee, hearings on mine safety 9/25/86, pp. 73–77.
405 "lax attitude," hearings, p. 2.

CHAPTER 28 DEFEATING BORK

page
406 Biden, Mark Gitenstein, *Matters of Principle: An Insider's Account of America's Rejection of Robert Bork's Nomination to the Supreme Court* (New York: Simon & Schuster, 1992), p. 23; "help people," Tony Podesta int 2/26/99; promised, Ralph Neas int 2/12/99.
407 Labor Committee, *NYT* 11/9/86.
407 "Dukakis," *BG* 12/7/86; "affirmation," Michael S. Dukakis int 12/13/94.
407 Aground, *BH* 11/30/86; Patrick, Associated Press 12/9/91.
407 RFK Award, text, af; trip barred, *NYT* 12/10/86.
408 Messages, Harrison Rainie with Ted Gest and Gillian Sandford, "The King of Capitol Hill," *U.S. News & World Report* 4/4/88; Israel, *BG* 12/12/86; Jordan, *BG* 12/14/86; Egypt, *BG* 12/15/86, 12/16/86.
408 Subcommittee, Tom Harkin int 6/8/95; Appropriations, Barbara Mikulski int 7/12/95.
408 Agenda, af.
409 Mine inspectors, Senate Labor Committee, hearings on mine safety 3/11/87, p. 185.
409 Trumka, hearings 3/12/87, p. 661.
410 "shameful and tragic," hearings, p. 492.
410 Crippled efforts, Senate Labor Committee, hearings on Civil Rights Restoration Act 3/19/87, p. 1; Greenberger, p. 256.
410 Disler, hearings 4/1/87, pp. 461ff; Kicklighter, hearings, pp. 487f.
411 Midnight meeting, *Atlanta Journal* 5/7/87, *BG* 5/10/87.
412 Neas, *NYT* 5/21/87.
412 Hart Building remarks, af.
412 Irish, EMK arrival statement, af.
413 Jaruzelski, Michnik, Michael T. Kaufman int 2/8/99; expulsion threat, Greg Craig int 2/27/95.
413 Grave, Reuters 5/23/87.
413 Presentation, Associated Press 5/23/87, *NYT* 5/24/87.
414 Economic help, *BG* 5/24/87.

414 Bus, Greg Craig int 2/27/95; Gdansk, Associated Press, *Chicago Tribune* 5/25/87.
414 EMK on RFK and JFK, af.
414 Michnik, Patrick, Kaufman int 2/8/99.
415 Black Madonna, Kerry Kennedy Cuomo int 1/22/97.
415 Departure, *BG* 5/26/87.
415 "not optimistic," *BG* 5/27/87; blessing, Craig int 2/18/95.
415 Hugging, public health, Orrin Hatch int 2/3/99; "good in terms of public health," EMK int 9/11/98.
415 Advertising, Mason, Senate Labor Committee, hearing on AIDS reasearch 5/15/87, p. 59.
416 Pierce, Senate Judiciary Subcommittee on Constitutional Rights, hearing on Fair Housing Act 3/31/87, p. 52; Simmons, p. 61.
417 Bork's worry, Ethan Bronner, *Battle for Justice: How the Bork Nomination Shook America* (New York: Norton, 1989), p. 28.
417 Speech preparation, former Senate aide int 9/25/98.
417 EMK–Cox conversation, former Kennedy aide int 9/25/98, EMK int 11/11/98.
417 EMK's Bork speech, C-SPAN videotape, af.
418 Intentions, EMK int 11/11/98.
418 EMK defends, *BG* 10/11/87, EMK int 11/11/98.
419 Bork objection, *BG* 2/9/88; analysts, *NYT* 7/5/87.
419 "incredulous," Robert H. Bork, *The Tempting of America* (New York: Free Press, 1990), p. 268.
420 "core values," EMK int 11/11/98.
420 Biden, *CR* 100:I, pp. 20909, 20915.
420 Vote for Bork, *Philadelphia Inquirer* 11/14/86; Biden and civil rights groups, Bronner, *Battle for Justice,* pp. 139–43; leaks, *NYT, WP* 8/9/87.
420 Letter, Bronner, *Battle for Justice,* p. 106; EMK message, effort, Podesta int 2/26/99.
421 Lowery, Bronner, *Battle for Justice,* p. 105, *New Orleans Times-Picayune* 8/15/87.
421 Hanrahan, moneymen, churches, Bronner, *Battle for Justice,* p. 107; Griggs, p. 145; Bork as moderate, pp. 195–98.
421 Unprepared, Bork, *Tempting of America,* pp. 283f.
421 Tribe, former Kennedy aide int 9/25/98, Gitenstein, *Matters of Principle,* p. 212.
422 EMK Georgetown speech, af.
422 Bork on overruling, Bronner, *Battle for Justice,* pp. 258f.
422 "profound," *NYT* 10/8/87.
422 "sparingly," Senate Judiciary Committee, hearings on nomination of Robert H. Bork 9/15/87, p. 104.
422 "Activist," hearings, pp. 33f.
423 "clock," p. 151; "second-class," p. 160; poll tax, p. 156.
423 Abortion, hearings, p. 149; hypotheticals, p. 151.
423 Notes, former Kennedy aide int 9/25/98; "my rights," hearings, p. 97; "indulged," p. 116.
424 "conversion," Gitenstein, *Matters of Principle,* p. 234; "rationale," hearings 9/16/87, pp. 287f.
424 Controversial, hearings, p. 428; EMK, pp. 336–45.
425 Writings, hearings 9/17/87, pp. 648–63; Canisius tape, p. 663.
425 American Cyanamid case, propaganda, Bronner, *Battle for Justice,* pp. 160–79; rationale, Bork, *Tempting of America,* p. 328.
426 Metzenbaum, hearings, p. 469; Bork, pp. 468, 470.
426 Groups, Bronner, *Battle for Justice,* p. 236; Riggs telegram, hearings, p. 678.
426 Specter–Bork, hearings 9/18/87, pp. 815–42; Hatch–Bork, pp. 848–50; Simpson–Bork, p. 854.
427 Conservative ads, Bronner, *Battle for Justice,* pp. 297–99.
427 Groups, former Senate aide int 9/25/98, Gitenstein, *Matters of Principle,* pp. 261f, Bronner, *Battle for Justice,* pp. 300f.

427 "fair," next nominee, *CR* 100:I, p. 29102.
428 Anthony Kennedy hearing, *NYT* 12/20/87.
428 Anthony Kennedy record, *NYT* 10/5/97.

CHAPTER 29 PICKING UP HIS BROTHERS' MANTLE

page
429 "self-righteousness," *BH* 1/26/87; "Bolshevism," *BH* 7/26/87; "character assassination," *BH* 7/2/87; "kow-towing," *BH* 11/12/86; "confetti," *BH* 1/20/87; "Worst Shapes," *BH* 1/23/87; "trims fat," *BH* 2/8/87; blondes, *BH* 9/13/87; EMK disliked, Jeff Smith int 2/25/99.
430 Assured Congress, *CR* 100:II, p. 122.
430 Fowler, *CR* 100:II, p. 115; comment, *BH* 1/6/88.
430 Koch, Press Club, *BG* 1/7/88; "defamed," *WP* 1/6/88.
431 "cornerstone," EMK statement 1/5/88, af.
431 Nyhan, *BG* 1/11/88; *PPP Reagan 1988,* p. 111.
431 Symms, *CR* 100:II, p. 110.
431 Hollings, *CR* 100:II, pp. 112, 114; D'Amato, Weicker, p. 220; EMK, p. 221.
432 Surprised, Smith int 2/25/99; Court of Appeals, *News America v. Federal Communications Commission* 844 F.2d 800.
432 Surprised again, Senate aide int 3/3/99.
432 "risk," *CR* 100:II, p. 97.
432 Danforth, *CR* 100:II, p. 243.
433 Packwood, p. 360.
433 Dilution, *WP* 1/29/88.
433 "expand the power," *PPP Reagan 1988,* p. 345.
434 Boschwitz, Fahrenkopf, *CQA 1988,* p. 67; Bush, *NYT* 3/25/88.
434 Religious right, *NYT* 3/22/88; Falwell, *BG* 3/22/88.
434 Phone calls, *CR* 100:II, p. 4666.
435 "260,000," "altar boy," *CR* 100:II, p. 2823.
435 Movie, Dr. J. Larry Brown and H. F. Pizer, *Living Hungry in America* (New York: Mentor, 1989), p. 231.
435 EMK, *CR* 100:II, p. 2875; Leahy, p. 18669.
436 1988 campaign, EMK Jr. int 1/31/96.
436 Surgery, *BG* 4/22/88.
436 Pollster, Skeffington, *BG* 6/8/88.
436 Patrick's idea, Patrick J. Kennedy int 11/1/95, EMK int 11/11/98.
436 Door-to-door, Joan Kennedy int 12/12/96; house parties, EMK int 11/11/98.
437 McCormack, bitterness, EMK int 11/11/98.
437 Primary Day, *BH, BG* 9/15/88.
437 Oppose amendments, Tom Rollins int 4/21/98.
437 EMK, *CR* 100:II, p. 9310; Helms, pp. 9311f.
438 Staff warning, Orrin Hatch int 2/3/99; "homosexuals," *CR* 100:II, pp. 9312ff.
438 Hatch influence, EMK int 3/26/99; "declared war," *NYT* 4/29/88.
439 "not only pain," *CR* 100:II, p. 30231.
439 Iowa, *BG, NYT* 1/22/88; grateful, Michael S. Dukakis int 12/13/94; Carter, *BH* 7/19/88.
439 JFK Jr., EMK, *Official Proceedings of the 1988 Democratic National Convention* (Washington, D.C.: DNC Services Corp., 1988), p. 313.
440 Rogers, *WP* 8/9/88.
440 Compromise, *NYT* 6/28/88.
440 EMK statement, af.
440 Acting chairmanship, two former Senate aides ints 3/11/99.

441 Persuaded Byrd, former Senate aide int 3/11/99.
441 Muhammad Ali, *WP* 8/13/88.
441 "beautiful moment," "separate," *CR* 100:II, p. 19900.
442 "shackle," *PPP Reagan 1988,* p. 540.
442 "fabricated," *CR* 100:II, p. 13797.
442 Veto expectation, impact, Jay Harvey int 4/11/99.
443 "Important step," *CR* 100:II, p. 32473.
443 Quayle not dumb, *BH* 10/21/88; cheering messages, Rich Galen int 3/15/99.
443 Teddy, Kara, EMK Jr. int 1/31/96; Malone, *BG* 10/30/88; "pork," *BG* 11/4/88.
444 Barnes appreciated message, Barnes to EMK and David Durenberger 11/9/88, Barnes papers, LOC.
444 Grove City impact, Marcia Greenberger int 2/8/99; housing impact, Robert Schwemm int 3/17/99.
444 From, *NYT* 11/11/88; Kennedy School speech, *BH* 11/15/88, *BG* 11/21/88.

CHAPTER 30 RIGHTS FOR THE DISABLED

page
445 Patrick, *BH* 1/4/89.
445 American Trash, *BH* 1/20/89.
445 "accused," *NYT* 3/5/89.
445 Agenda, 3/6/89, EMK Chubb Fellowship lecture, af.
446 JFK association, Paul Donovan int 11/14/98; EMK effort, Nick Littlefield int 1/31/99; reluctant unions, Democrats, William Greider, *Who Will Tell the People: The Betrayal of American Democracy* (New York: Simon & Schuster, 1992), pp. 195f, Jay Harvey int 4/11/99.
446 Elizabeth Dole, Senate Committee on Labor and Human Resources, hearing on minimum wage 3/3/89, p. 30; Labor Committee approval, *NYT* 3/9/89.
447 Chamber of Commerce, *CR* 101:I, pp. 5710f.
447 "CEOs," *CR* 101:I, pp. 5810f.
447 Irrigation, *CR* 101:I, p. 6148, Harvey int 2/19/99.
447 Computer experts, *CR* 101:I, pp. 6151f, Harvey int 4/11/99.
448 Goodling, *CQA 1989,* pp. 334f; Domenici, Harvey int 4/11/99, EMK plan, Jay Harvey to EMK 9/6/89, Nick Littlefield files.
448 Kirkland, Dole, *NYT* 11/1/89.
449 "complex," *CR* 101:I, p. 27859.
449 "anti-poverty," *CR* 101:I, p. 27878.
449 "big spending," EMK int 2/5/93.
449 Organizing, Pat Wright int 4/1/99; Bush, Robert Silverstein int 3/23/99, Boyden Gray int 4/7/99, UPI 8/10/88; Coelho, Kennedy, Tony Coelho int 4/20/99.
450 Strategy, label, former Kennedy aide int 4/8/99, Silverstein int 3/23/99; momentum, Coelho int 4/20/99.
450 Obstacles, Wright int 4/16/99, former Kennedy aide int 4/8/99.
450 Drafts, delay, Silverstein int 3/23/99.
450 Dart, Senate Committee on Labor and Human Resources and Subcommittee on the Handicapped, hearings on Americans with Disability Act 5/9/89, pp. 18f; inexpensive, p. 30.
451 Hartigan, EMK, hearings 5/10/89, pp. 81ff.
451 Dole, hearings 5/16/89, pp. 87ff; Hatch, Harkin, EMK, pp. 133ff.
451 Figures, hearings, pp. 172f; EMK, p. 149; Cook, p. 169.
452 Welcome, hearings 6/22/89, pp. 195f.
452 Thornburgh testimony, hearings, pp. 195–202.
452 Reelection, Tom Harkin int 6/8/95, Silverstein int 3/23/99; Osolinik, former Kennedy aide int 4/8/99; July 6 agreement, Silverstein int.

453 Silverstein to Sununu, Silverstein int 3/23/99.
453 Raging Sununu, hesitant Harkin, angry Kennedy, Harkin int 6/8/95, Silverstein int 3/23/99, *WP* 5/27/90.
453 Dole's eyes, Kennedy's optimism, Harkin int 6/8/95; Thornburgh to Sununu, Wright int 4/1/99.
453 Positive emphasis, Silverstein int 3/23/99.
453 Exchange, *CQWR* 8/5/89, *NYT* 9/17/89.
454 Compromise, White House, *CQWR* 8/5/89.
454 "great civil rights laws," *CR* 101:I, p. 19807.
454 Interpreter, *CR* 101:I, p. 19831.
455 "mom and pop," *CR* 101:I, pp. 19859f; hiring judgments, p. 19850; Fulco, *NYT* 8/14/89.
455 Hatch importance, Coelho int 4/20/99; Mormon faith, Orrin Hatch to author 5/5/99; "inadequately," *CQWR* 8/5/89; tax credit, *CR* 101:I, p. 19836; "fight," p. 19848.
455 "moral standards," *CR* 101:I, p. 19864; HIV, p. 19866; transvestites, pp. 19863f, 19875.
455 Armstrong's list, *CR* 101:I, p. 19853; "could not believe," EMK int 3/26/99; limited list, *CR* 101:I, p. 19884.
456 Reception Room, Paul Donovan int 11/14/98.
456 Contract, *BH* 12/2/88.
456 AFL-CIO, Jay Harvey int 1/15/99; another trade, Harvey to EMK 11/19/89, Littlefield files.
456 Pleaded, Littlefield int 12/10/98.
457 "fair exchange," Orrin Hatch int 2/3/99.
457 Helms, Hatch int 2/3/99; Bozzotto, *BG* 11/23/89.
457 Elizabeth Dole, *BH* 11/22/89; pressure, *BG* 4/24/90.
457 Brokaw report, af.
458 Arrival statement, af.
458 Volkspolizei, Nancy Soderberg int 3/31/97.
458 Berlin speech, af.
458 Corny, Soderberg int 3/31/97.
458 East German leaders, cable Harry J. Gilmore to Secretary of State 11/29/89, Kennedy Foreign Policy Files.
459 Crowd, *WP* 11/29/89.
459 "Emotionally," *BG* 11/29/89.
459 "basic right," Senate Labor Committee, hearings on health care crisis 12/11/89, p. 2.
460 Emergency rooms, hearings 12/12/89, pp. 15–18.
460 Harbor-UCLA, "The Health Care Crisis," Report of Committee on Labor and Human Resources, June 1990, p. 36.
460 Krause, hearings, pp. 96ff; Van Tassel, pp. 101ff.
460 Gates, hearings, pp. 104f.
460 McDaniel, hearings 12/13/89, pp. 123f.
461 Baretta, hearings, pp. 125ff.
461 Kerr, hearings, pp. 133f.
461 Holloway privately, Michael Iskowitz int 2/18/99, Mark Childress int 5/3/99; "several patients," hearings 12/14/89, p. 107.
461 Ayer, *CQA 1990,* p. 463.
462 *Harvard Crimson,* Ralph Neas int 5/10/99; "bums," *CR* 101:II, p. 17672.
462 "renege," Coleman, *WP* 7/18/90; "furthered," CR 101:II, p. 17673.
462 "Quotas, shmotas," *CR* 101:II, p. 18029.
463 Earthquake rationale, Nick Littlefield int 12/10/98; Hatch and rural areas, Michael Iskowitz int 2/18/99.
463 Money, Littlefield ints 12/12/98, 5/7/99.
463 "thirteen-camera," Littlefield int 12/12/98; "We need money," "America responded," *CR* 101:II, pp. 3538, 3530; "resources," *BH* 3/7/90.

463 Ryan White, *NYT* 4/9/90.
464 Press releases, af.
464 "By dedicating," "I believe Ryan," *CR* 101:II, p. 10254; Helms, Iskowitz-Beirn urging, Iskowitz int 5/9/99.
464 Filibuster, Iskowitz int 2/18/99; Hatch, *CR* 101:II, p. 10244.
464 Helms, Hatch exchanges, *CR* 101:II, pp. 10246–48.
465 "callously," *CR* 101:II, pp. 10248–51.
465 Epidemic, *CR* 101:II, p. 10713.
465 Emergency relief, *CR* 101:II, p. 10737.
466 "pain, suffering," *CR* 101:II, p. 23390.
466 Rare trip, Soderberg int 3/31/97; swearing-in, *NYT* 3/12/90, EMK int 3/26/99.
466 "Transition," EMK int 3/26/99; stadium, EMK int 11/11/98; "hateful and blind days," *CR* 101:I, p. 4675.
467 Bush–Gorbachev, Michael R. Beschloss and Strobe Talbott, *At the Highest Levels: The Inside Story of the End of the Cold War* (Boston: Little, Brown, 1993), p. 164.
467 Ultimatum, party headquarters, tanks, Beschloss and Talbott, *At the Highest Levels,* pp. 195–99; diplomats, Jack F. Matlock, Jr., *Autopsy of an Empire: The American Ambassador's Account of the Collapse of the Soviet Union* (New York: Random House, 1995), p. 339.
467 Arrival, Matlock, *Autopsy of an Empire,* p. 341.
467 Kennedy–Gorbachev meeting, cable Matlock to Secretary of State 3/27/90, af.
468 Five times, EMK "Notes from Soviet Trip," Kennedy Foreign Policy Files.
469 Press conference, *NYT* 3/27/90.
469 EMK on Kryuchkov, EMK to Leonard S. Spector 4/10/90, EMK Foreign Policy files.
469 Tactics, *BG* 3/31/90; American opinion, *WP* 3/30/90.
469 Bush on Gorbachev, Soderberg notes, EMK Foreign Policy Files; "Cut Lithuania off," *WP* 3/30/90.
470 Fitzwater, letter, Beschloss and Talbott, *At the Highest Levels,* p. 201; Bush comment, *NYT* 3/31/90.
470 Mandela introduction, "honorary Irishman," *BG* 6/24/90.
470 "frustrating," *BH* 6/24/90.
471 Restaurant industry, *CQA 1990,* p. 451; Osolinik, Michael Iskowitz to EMK 6/7/90, Littlefield files.
471 "gut," *CR* 101:II, p. 17036.
471 "witches," *CR* 101:II, p. 17046.
472 "walls," *CR* 101:II, p. 17376.
472 "brother Frank," *CR* 101:II, p. 17369.
472 "touched by others," *CR* 101:II, p. 17370.
472 Raymond Hansen, *CR* 101:II, p. 17375.
472 Weicker, Wright, "beautiful day," EMK int 3/26/99; Sununu, Lowell Weicker int 5/6/99.
473 Wheelchair upstairs, Leamer, *Kennedy Women,* p. 748; doubtful quotes, p. 747; tennis racket, *NYT* 7/13/90; "old wine," 7/16/90.
473 "Rosie O'Grady," RFK Jr. int 12/19/97; Irish ditties, Desmond Tutu int 5/2/95.
473 EMK–Steve Smith relationship, funeral, Melody Miller int 3/13/95.
473 Wedding, *BG* 9/9/90, 9/10/90, *NYT* 9/9/90.
474 Drinking, women, *WP* 4/29/90, former Kennedy aides ints.

CHAPTER 31 BATTLES IN THE SENATE—AND THE GULF

page
475 "tragedy," *WP* 10/11/90; argument, Nick Littlefield int 5/7/99; encouraged others, memos Michael Iskowitz to EMK 9/14/90, 9/21/90, Littlefield files.

476 EMK statement 10/10/90, af.
476 Little risk, Littlefield int 12/10/98.
476 "political cover," *CR* 101:I, p. 33463.
477 Government officials, *CR* 101:I; EMK, p. 33470.
477 EMK and City Year, Littlefield int 5/9/99.
477 "young citizens," *CR* 101:I, p. 29581.
478 "paid volunteers," *PPP Bush 1990,* Part 2, p. 1614.
478 Fifty-four bills, *CR* 102:I, p. 8569.
478 Brooks, *CQA* 1990, p. 477.
479 "Our goal," *CR* 101:II, p. 35609.
479 Enrolling clerk, former Kennedy aide int 5/10/99.
480 "maze," *PPP Bush 1990,* Part 2, p. 1438.
480 "good faith," *CR* 101:I, p. 33379.
480 "will not stand," *PPP Bush 1990,* Part 2, p. 1102; Iraq as Republican reason, *CQA 1990,* p. 473.
481 Discussions about Souter, call to Rudman, former Kennedy aide int 9/25/98.
481 "Judge Souter's," "vote our fears," *CR* 101:II, p. 26966.
481 Thomas-EMK exchange, Senate Judiciary Committee, hearings on nominations 2/6/90, Part 4, p. 36.
482 Coleman, Danforth, "radar screens," EMK int 11/11/98; black advantage, Howard M. Metzenbaum int 5/12/99; successor to Marshall, *NYT* 2/23/90; Metzenbaum, Pryor, *CQA 1990,* p. 518.
482 "blank check," *CR* 101:II, p. 26726.
483 Rayburn, *CR* 77:I, p. 7; Bush, *NYT* 1/4/91; "King George Bush," *CR* 102:I, p. 13.
483 "War is not," *CR* 102:I, p. 433.
484 "defending the throne," Senate Labor Committee, hearing on recession 1/7/91, p. 1.
485 Women flying in combat, *CR* 102:I, p. 20713.
485 Unemployment benefits, *NYT* 4/27/91.
485 New proposal, *NYT* 3/2/91.

CHAPTER 32 THE TURNING POINT

page
487 Cover-up, *NYT* 4/3/91, 4/5/91; $1,000, Elizabeth A. Murphy deposition 6/6/91, p. 93, af.
487 First rumors, Dennis E. Spear deposition 5/21/91, af.
487 Romp, *New York Post* 4/5/91; Cassone int in *BG* 4/6/91.
487 Leno, *Hotline* 4/18/91, 4/16/91; Carr, *BH* 4/5/91.
488 Staff, Littlefield int 1/30/99, railroad strike memos, af.
488 Skirmish, *CQWR* 4/20/91.
489 Mitchell, Simpson, Specter, *BG* 4/28/91.
489 Bullshot etc., Tatianna Lee Pritchard deposition 5/23/91, af.
489 "Mormon missionaries," Hatch int 2/27/92; "countess," Nancy Korman to EMK 4/16/91; "settle down," Korman to EMK 5/17/91, af.
490 Behavior, drinking, *BG* 6/9/91.
490 Hour and a quarter, EMK deposition 5/1/91, p. 43, af; snack, p. 54; nightshirt, p. 44; good night, p. 57; no screams, p. 58; "serious" offense, pp. 74, 76; sexual harassment, p. 85.
490 Nightshirt, Patrick J. Kennedy deposition 5/1/91, p. 109, af; Willy and his date, pp. 169ff.
490 "deliberately misled," *NYT* 5/17/91.
490 Joan, *BG* 5/31/91; Teddy, *BG* 7/12/91, 10/7/93.
491 No legislation needed, *NYT* 10/11/92.
491 "no American family," EMK statement, af; Chafee, *WP* 6/6/91.

491 Business Roundtable, *CQWR* 4/20/91, *NYT* 4/20/91.
491 Strategy, John C. Danforth int 7/7/95, former Kennedy aide int 9/25/98, Peter Leibold int 7/7/95.
492 Believed Bush, *NYT* 7/7/91.
492 June 17 hearing, statement, af, *BG* 6/18/91.
492 "Commander," *Today* show 9/8/92.
492 Unless noted otherwise, the account of Kennedy's courtship of Vicki Reggie comes from a 5/13/99 interview with her; vegetables, *BG* 9/8/92.
492 "known Vicki before," *Today* show 9/8/92.
492 Choice of Thomas, Jane Mayer and Jill Abramson, *Strange Justice: The Selling of Clarence Thomas* (New York: Houghton Mifflin, 1994), pp. 11–20.
493 EMK role, EMK int 11/11/98, former Kennedy aide int 9/25/98.
494 Floyd Brown ad, *NYT* 9/4/91, 9/5/91, 9/6/91, Mayer and Abramson, *Strange Justice,* pp. 199–201.
494 Natural law, abortion, Senate Judiciary Committee, hearings on nomination of Clarence Thomas 9/10/91, pp. 146–48.
495 Thomas on *Roe,* hearings 9/11/91, p. 222, "lying," EMK int 11/11/98.
495 "vanishing views," hearings 9/16/91, p. 452.
495 Griswold, hearings 9/17/91, p. 233.
496 Seidman's calls, instructions, Ricki Seidman int 5/19/99.
496 Accusers, *NYT* 7/23/91; polls, *WP* 8/2/91; grand jury, *NYT* 8/31/91.
496 Unless noted otherwise, the account of the behind-the-scenes aspects of the Thomas–Hill issue relies on Mayer and Abramson, *Strange Justice,* the thorough, authoritative account of Thomas's confirmation.
497 "rope," hearings 10/11/91 part 4, p. 10; fantasy, p. 83; spurned woman, p. 60; conspiracy, p. 133; "high-tech," p. 157; "perjury," hearings 10/12/91, p. 228.
498 Negotiating, Danforth int 7/7/95; words in mouth, hearings, p. 118.
498 "unworthy, unsubstantiated," hearings 10/13/91, p. 307.
498 Psychiatrist, *NYT* 10/14/91.
499 Specter, *CR* 102:I, p. 26292; Hatch, p. 26306.
499 Barry, *Miami Herald* 10/15/91; Quindlen, *NYT* 10/19/91; "womanizer," *BG* 10/20/61.
500 "grown-ups," *WP* 9/20/91; too-tolerant liberals, James Carroll, "The End of the Dream," *New Republic* 6/24/91.
500 Harassment, *NYT* 10/21/91; Danforth, Duke, *NYT* 10/25/91; decent terms, Danforth int 7/7/95.
501 "no quota bill," *NYT* 10/25/91.
501 "as good a deal," Neas int 2/10/95.
502 "special obligation," EMK speech, C-SPAN tape, af.
503 "profile," "satisfying," *CR* 101:I, p. 29058; signing, *NYT* 11/22/91.
504 Jurors, *WP* 11/1/91.
504 Bars testimony, *NYT* 12/3/91, *BG* 12/3/91.
504 Bowman testimony, *NYT* 12/5/91.
504 EMK testimony, Court TV tape, af.

CHAPTER 33 A PRESIDENT AND AN ALLY

page
506 Variety of approaches, *CR* 102:II, p. 90; Bush, Hatch, *BG* 1/22/92; "Republican Presidents," EMK statement 1/28/92, af.
507 "marketplace forces," *NYT* 1/24/92; "taxpayers' money," *CR* 102:II, p. 375.
508 Department of Education, Senate Labor Committee, hearing on direct student loans 2/25/92, p. 14.

508 $3,700 grants, Kennedy, *CR* 101:II, p. 16993; Pell, p. 16995; Durenberger, p. 17001; Cochran, p. 17004; Dodd, p. 17005.
508 "guarantee," EMK statement 1/22/92, af.
508 Sullivan, Altman, Senate Labor Committee, hearing on comprehensive health reform, 3/4/92, pp. 2ff.
508 Health insurers, *BG* 4/28/92; EMK text, af; small groups, EMK int 2/5/93.
509 Thurmond's daughter, *CQA 1992,* p. 394.
509 Waxman, *CQA 1992,* p. 395.
510 "craven," "specious," "murder," *CR* 101:II, pp. 13440f.
510 Smith's father, *CR* 101:II, p. 13452.
510 Fitzwater, *CQA 1992,* p. 340.
510 Dinner, Littlefield int 1/30/99; "decade of neglect," EMK statement 5/14/92, af; "idly by," *BG* 5/15/92.
510 "small step," *CR* 101:II, pp. 15315f.
511 Engagement story, former Kennedy aide int, former Washington reporter int.
511 Proposal, Vicki Kennedy int 5/13/99.
512 Gorbachev luncheon, *NYT* 5/16/92; menu, af; "defining moment," EMK text, af.
513 Reporters never asked, Donovan int 5/25/99; Tsongas, Paul E. Tsongas int 12/12/94; endorsement, *BG* 3/13/92.
513 First call, *BG* 10/28/93; wariness toward Clinton, Littlefield int 1/17/99.
514 Endorsement of Clinton, *BG* 4/24/92; Clinton would win, Soderberg int 3/31/97.
514 EMK convention text, af.
514 Clinton promise, *CQA 1992,* p. 55–A; socialized medicine, *PPP Bush 1992,* Part 2, p. 1299; "tired old," EMK statement 8/4/92, af.
515 Veto, *CR* 101:II, p. 27494; Dodd, p. 27495.
515 "medical crisis," *CR* 101:II, p. 27502; Bill Clinton, p. 27503.
515 "Governor Clinton," "political football," *CR* 101:II, p. 31491.
516 Community health centers, *NYT* 10/18/92, *CQA 1992*, p. 435, Littlefield int 1/30/99.
516 $45 million, *CR* 101:II, p. 34312.
516 Hawkins, *NYT* 10/18/92.
517 "tax on innovation," Senate Labor Committee, hearing on user fees for FDA 9/22/92, p. 26
517 Negotiations, Mark Childress int 5/3/99, *Business Week* 8/24/92, *Time* 8/24/92.
517 "mismatch," hearing, p. 2; Mossinghoff, pp. 26ff.
517 "performance goals," *CR* 101:II, p. 33610.
518 1998 rate, FDA report on new health care products approved in 1998, 1/15/99, af.
518 "still-frisky," *WP* 3/15/92; "aging rogue," *BG* 3/16/92; "marriage of convenience," *NYT* 10/1/92.
518 *Today* transcript, 9/8/72, af.
519 Perrier, *Springfield Republican* 9/6/92; "elegant portrait," *Quincy Patriot-Ledger* 9/8/92.
519 "wry," *BG* 9/8/92.
520 "one-stop shopping," Downey int 5/28/99.
520 NORAID, Conor O'Clery, *The Greening of the White House* (Dublin: Gill & Macmillan, 1996), pp. 7–11.
520 "visionary," visit, Jean Kennedy Smith int 5/4/95; visit, John Hume int 3/17/95.
521 Hillary Rodham Clinton, Tom Oliphant int 5/6/99.
521 Blundered, Littlefield int 5/26/99.
522 Dole, *NYT* 2/5/93; Kassebaum, *CR* 102:I, p. 1694.
522 "sound family policy," *CR* 102:I, p. 651.
522 First legislation, *CR* 102:I, p. 1697.
522 Murray, Feinstein, *NYT* 2/11/93; Murray on EMK, Patty Murray int 5/26/99.
523 Dodd on EMK, *CR* 102:I, p. 2264.
523 Doubts, Littlefield int 1/17/99.

523 "right time," EMK int 2/5/93.
525 Nexon memo, Littlefield files.
525 EMK, Clinton on reconciliation, Hillary Rodham Clinton int 5/21/98, Littlefield int 1/17/99.
525 Byrd rebuff, Haynes Johnson and David Broder, *The System: The American Way of Politics at the Breaking Point* (Boston: Little, Brown, 1996), p. 126; Byrd and Rockefeller, EMK, and Mitchell, Littlefield int 4/5/93.
526 "shocking lack," *CR* 103:I, p. S6882.
526 Strategy in conference, Littlefield int 7/29/93.
527 T-bone steaks, EMK int 10/9/95.
527 Front-runner, *NYT* 6/11/93; Breyer press release, af.
527 House cleaner, George Stephanopoulos, *All Too Human: A Political Education* (Boston: Little, Brown, 1999), pp. 169f; Breyer did not impress, *NYT* 5/15/94, Stephanopoulos int 5/29/99.
527 Fear of offending, Stephanopoulos int 5/29/99; McLarty, *BG* 11/1/93; $500,000 fund-raiser, *BG* 6/20/93.
528 Third priority, *PPP Clinton 1993,* Part 1, p. 1147.
528 Elizabeth Dole, *NYT* 7/27/93.
528 Telling reporter, Dole glared, author's recollection; not filibustering, *NYT* 7/30/93.
528 "bipartisan," *CR* 103:I, pp. S10113f.
529 "wonderful moment," news conference transcript, af; tears, author's recollection.
529 Greet President, EMK eulogy to Mrs. Onassis, *WP* 5/24/94.
529 EMK and Clintons, Kennedy aide int 9/13/93.
529 "relaxed," Clinton int with Elsa Walsh 2/14/97, af.
530 Prepped on jurisdiction, not right occasion, Littlefield int 5/28/99; Southbridge, Steve Wolfe int 5/31/99.
530 Motley, Johnson and Broder, *The System,* p. 159.
530 Clinton, *NYT* 9/23/93; shook hands, *BG* 9/24/93; "days of FDR," EMK statement 9/22/93, af.
531 Confrontation, *LAT* 6/19/94, *BG* 11/23/93, *NYT* 11/25/93, Lawrence O'Donnell int 6/21/94, Littlefield int 6/21/94, Daniel P. Moynihan int 7/26/95.

CHAPTER 34 STILL CAMPAIGNING

page
533 Romney letter, *BG* 11/9/93.
533 Luncheon remarks, Doak Shrum videotape, af; rosary, Vicki Kennedy int 6/28/95.
535 "tactics of violence," *CR* 103:I, p. S15658.
535 "pro-life expression," *CR* 103:I, p. S15664.
535 Boxer, *NYT* 11/17/93.
535 EMK asked Dole, Bob Dole int 11/21/93.
536 Simple family visit, EMK int 6/5/99; visa for Adams, "terrorist," Jean Kennedy Smith int 5/4/95.
536 Hume, Reynolds, Smith int 5/4/95.
536 Coogan argument, Tim Pat Coogan, *The Troubles: Ireland's Ordeal 1966–1996 and the Search for Peace* (Boulder, Colo.: Roberts Rinehart, 1996), p. 373.
536 Reynolds on visa, Smith int 5/4/95, Albert Reynolds int 5/8/95.
536 Embassy, O'Clery, *Greening of the White House,* p. 94.
537 Foley, O'Clery, *Greening of the White House,* pp. 95ff.
537 Pandering, *NYT* 2/2/94; Moynihan's support, Stephanopoulos int 5/29/99; "tangible evidence," O'Clery, *Greening of the White House,* p. 104; Clinton's anger, EMK int 10/17/94.
537 Mitchell, Johnson and Broder, *The System,* p. 348, Littlefield int 5/31/99.

538 Worry aloud, *BG* 1/23/94; no crisis, *NYT* 1/26/94.
538 Reischauer, Johnson and Broder, *The System,* pp. 84f.
538 Readiness to compromise, Littlefield int 5/31/99.
539 Hatch to Clinton, EMK, *NYT* 5/15/94.
540 Reporter, "great bill," author's notes; conversation, *NYT* 5/15/94.
540 Cutler, IRS, former Kennedy aide int 5/28/99; videotape, Stearns int 5/28/99.
540 "political skills," *NYT* 5/14/94.
541 "Joseph Story," *CR* 103:II, p. 18643.
541 "belly up," *NYT* 4/23/94.
541 Hatch, Kassebaum, *NYT* 5/19/94.
542 Wanted to be there, *BG* 5/19/94.
542 Eulogy, *WP* 5/24/94.
543 "heart and soul," author's notes.
544 Dole, *NYT* 6/12/94, 6/29/94, 7/1/94; Gingrich, *NYT* 6/17/94.
544 EMK on Republicans, Johnson and Broder, *The System,* p. 447.
544 Mitchell, *NYT* 8/3/94.
545 EMK, Clinton, McDermott, *NYT* 8/3/94.
545 August 9 debate, *NYT* 8/10/94.
545 Put off "tough stuff," *BG* 1/23/94.
545 "special moment," *CR* 103:II, pp. 20414ff.
546 Kerrey, EMK, Senate aide's notes.
547 Littlefield to Johnson, Johnson and Broder, *The System,* p. 519; last to give up, Littlefield int 5/31/99.
547 "never give up," *CR* 103:II, p. 25597.
548 Businessmen, O'Clery, *Greening of the White House,* pp. 149ff.
548 Cahill visa, Reynolds int 5/8/95.
548 Smith, O'Clery, *Greening of the White House,* pp. 150ff; saints, Reynolds int 5/8/95.
548 Efforts vindicated, Reynolds int 5/8/95; joyous day, EMK statement 8/31/94, af.
549 Reynolds on EMK, Reynolds int 5/8/95.
549 Nastiness, Michael Kennedy int 12/13/94.
549 Dieting, EMK int 10/15/94; Romney ad, *NYT* 10/19/94; "whale," *WP* 10/7/94.
550 Negative ads, Hillary Rodham Clinton int 5/21/98; surge, Tom Kiley int 2/17/95; debt, Michael Kennedy int 12/13/94.
550 Venture capitalists, Michael Kennedy int 12/13/94, Kiley int 2/17/95, Vicki Kennedy int 6/28/95.
550 Hired Lenzner, *BG* 8/19/94; found Ampad, Terry Lenzner int 6/2/99; management eliminated, *NYT* 10/10/94.
551 Romney complaint, *BH* 9/19/94; "mistake," *BH* 9/22/94.
551 EMK on Romney's faith, *BH* 9/27/94; Romney on JFK, *BH* 9/28/94; religion should not be issue, *BG* 9/28/84.
551 Back and forth, EMK int 10/15/94; might lose, former Kennedy aide int 5/22/95.
552 Poll, Kiley ints 2/17/95, 6/1/99, Shrum int 5/21/99.
553 Romney victory speech, author's notes.
553 Foot, Vicki Kennedy int 6/28/95; working families, *BG* 8/21/94.
554 Shrum opposition, Shrum int 5/21/99; fear, Burke, Michael Kennedy int 12/13/94.
555 Worcester, Taunton, my notes; road race, Burke int 12/4/95; Goodman, *BG* 10/2/94.
555 Reporters, Dartmouth, author's notes.
555 Faneuil Hall speech, author's notes.
556 "danger," *BG* 10/17/94; "incoherent," *BH* 10/14/94; debate, videotape, af.
557 Brandeis, Abrams int 3/15/95, Boston *Jewish Advocate* 11/4/94.
557 "lonely," EMK–Vicki Kennedy int with Natalie Jacobson 10/29/95, transcript, af.
558 Truman rally, author's notes.

CHAPTER 35 THE COUNTERREVOLUTION

page

560 Press Club speech, 1/11/95, af.
561 "live with this," William J. Clinton int 1/29/99.
561 EMK talking points 12/13/94, af.
562 "never to despair," Clinton int with Walsh 2/14/97, af; "back and forth," Stephanopoulos int 6/5/99; "left an impression," Leon Panetta int 6/4/99.
562 Middle class, *PPP Clinton 1994,* pp. 2182ff.
563 "structure his day," Douglas Kennedy int 10/22/97.
563 "joys and sorrows," *BG* 1/25/96; St. Stephen's, *NYT* 1/29/95.
563 Quotations from funeral, videotape, af.
564 Annulment, Joan Kennedy int 10/12/98.
565 Promised Kennedy, Clinton int with Walsh 2/14/97, af; make a living, *CQA 1995,* p. D-9.
565 "don't deserve," Senate aide int 8/9/96.
565 Daschle would back increase, Tom Daschle int 8/8/96.
566 "working families," *CR* 104:I, p. 7867.
566 Restore spending, Littlefield int 6/6/99.
567 Protect children, *CR* 104:I, pp. S4966ff.
567 "dozen very wealthy individuals," *CR* 104:I, p. S4983.
567 Republicans winning, Dick Morris, *Behind the Oval Office: Winning the Presidency in the Nineties* (New York: Random House, 1997), pp. 160ff.
568 EMK–Clinton meeting, Clinton int 1/29/99.
568 "triangulation," Morris, *Behind the Oval Office,* p. xii.
569 Medicare, Littlefield int 8/14/95.
569 "far better," EMK statement 6/14/95, af.
569 Breaux, Obey, Schroeder, DeLauro, *NYT* 6/16/95.
569 Aggressive, Clinton int 1/29/99; kept complaints private, Clinton int with Walsh 2/14/97.
570 "will not cure," *WP* 7/15/95.
570 "life better," *NYT* 7/14/95; eating crow, author's notes.
571 Needy, greedy, *CR* 104:I, p. S15617; "wither," *NYT* 10/27/95.
571 "petty," *WP* 11/16/95.
572 "tougher," *NYT* 1/22/96.
572 "absolutely committed," *CR* 104:II, p. S372.
572 "only opposition," *CR* 104:II, p. S374.
573 Kassebaum, Senate Labor Committee, hearing on minimum wage 12/15/94, p. 1; Reich, p. 4; Kennedy, pp. 42f.
573 40-year low, *CQA 1996,* p. D-6.
574 Parliamentary slip, *NYT* 3/27/96.
574 Majority for increase, *NYT* 3/29/96.
574 "Doledrums," *CR* 104:II, p. S4953.
575 Wealthy, young, *CR* 104:II, p. S3561.
575 Vote on MSAs, *NYT* 4/19/96.
576 Appointing conferees, *NYT* 4/26/96.
576 "rule the world," *NYT* 6/8/96.
576 Historical parks, *BG* 5/2/96.
577 Delighted, my recollection; assault weapons, *NYT* 3/23/96.
577 Lunch, EMK int with Chris Black 7/19/97.
578 Equal footing, Patrick J. Kennedy int 11/1/95.
578 Shannon, David Obey int 12/11/95.
578 Joan's settlement, *BG* 6/26/96.

578 Morris-Lott, Morris, *Behind the Oval Office,* p. 294.
579 *Times to Remember,* Littlefield int 8/6/96; Archer touched, Republican mother, Bill Archer int 8/7/96; impressed with Kennedy, Charles Kahn int 12/9/98.
579 Compromise numbers, *NYT* 7/26/96; "tough bargainer," *NYT* 8/11/96.
580 Ethical breach, *Congress Daily* 8/1/96.
580 Many Senators, Vicki Hart, *Congress Daily* 8/2/96; Specter, Santorum, *NYT* 8/5/96.
581 EMK–Lott discussion, Littlefield int 8/6/96.
581 "enthusiastic freshman," Daschle int 8/8/96.
582 "reform," EMK statement 8/1/96, af; "enemies," *BG* 8/30/96.
582 "united voice," EMK statement 6/19/96, af; "despicable," *CQA 1996,* pp. 5-32; other staffers, Tom Perez int 6/9/99.
583 "mean-spirited," *CR* 104:II, p. S10101.
583 "irrational animus," *CR* 104:II, p. 10055.
583 "shameful," *CR* 104:II, p. 12182; "since 1948," "extraneous," p. 12221; "spewed," p. 12100.
584 "conspired," *CR* 104:II, p. 12219.
584 "technicalities," *CR* 104:II, p. 11885.

CHAPTER 36 HEALTH CARE, BIT BY BIT

page
585 Inauguration Day, *BG* 1/21/97, Tim Shriver int 10/3/97.
586 Republicans, Nick Littlefield draft on children's health 1/14/99, af.
587 "odd couple," *BG* 3/14/97.
587 Withdrawing support, Lott, Nickles, *NYT* 4/10/97, 4/12/97; Johnson, *NYT* 4/15/97.
587 Radio ads, Associated Press 4/30/97.
589 Wasserman, Littlefield draft.
589 State revenues, Lott, Nickles, *NYT* 5/21/97.
589 "Big Tobacco," *CR* 105:I, p. S4783; part of budget, p. S4785.
590 Not needed, *CR* 105:I, p. S4787; "excessive," pp. S4808f.
590 President, Vice President, *CR* 105:I, p. S4809.
590 Calls, Littlefield draft.
590 Flurry of debate, *CR* 105:I, p. S4827.
590 Angry, "mistaken," ducks question, *NYT* 5/22/97.
591 Misunderstanding, *BG* 5/23/97; annoyed, EMK int 6/2/97.
591 20-cent increase, Littlefield draft, *CQA 1997,* p. 2-34, p. 6-9, *BG* 6/25/97, 6/28/97; "seized the moment," *WP* 6/30/97.
591 "every child," "indispensable," EMK statements 7/31/97, af.
592 $6 million, *BG* 2/6/97; "Lincoln Bedroom," Hillary Rodham Clinton int 5/21/98.
592 Eddie McCormack, EMK statement 3/4/97, af; better Senator, McCormack int 5/15/92.
594 FDA passage, EMK statement 11/9/97, af.
594 Irish speech 1/9/98, af.
594 Trimble, *BG* 1/11/98.
595 Patients' rights, *BG* 2/26/97.
596 Jeffords, *BG* 11/6/97; "wallets," *NYT* 11/4/97.
596 Minimum wage, *WP* 7/12/97.
596 $1.50 per pack, *NYT* 1/10/97.
597 Press Club speech 12/11/97, af.
597 Troubled child, *CR* 105:II, p. S3386.
598 "local communities," *CR* 105:I, p. S3412.
598 Meetings, Danica Petroshius int 6/5/99.

CHAPTER 37 THE JOB YOU ASKED THE PEOPLE FOR

page

601 Early vote, former White House aide int 4/30/99.

601 Byrd, Dodd, caucus generally, *NYT* 1/9/99.

601 Sense of failure, Richard Durbin int 1/13/99.

603 "friend," William J. Clinton int 1/29/99.

603 Senate, wedding, reunion, EMK int with Robin Toner, 7/16/99, af.

604 EMK and JFK Jr. close, *NYT* 7/22/99.

604 Mass, *NYT* 7/24/99.

604 eulogy, *NYT* 7/24/99.

604 Patriarch, Maria Shriver int 2/22/98.

605 Cape summers, Kerry Kennedy Cuomo int 10/22/97; available, RFK Jr. int 12/10/97.

605 "going badly," Joseph P. Kennedy II int 1/10/95.

608 Byrd on EMK, *CR* 105:I, p. S12449.

Bibliography

Abt Associates. *Does Training for the Disadvantaged Work?* Washington: Urban Institute Press, 1996.

Ambrose, Stephen E. *Nixon: The Triumph of a Politician.* New York, Simon & Schuster, 1989.

——. *Nixon: Ruin and Recovery 1973–1990.* New York: Simon & Schuster, 1991.

America's Health Centers. Washington, D.C.: National Association of Community Health Centers, 1995.

Arnson, Cynthia J. *Crossroads: Congress, the Reagan Administration and Central America.* New York: Pantheon, 1989.

Ashby, LeRoy, and Rod Gramer. *Fighting the Odds: The Life of Senator Frank Church.* Pullman: Washington State University Press, 1994.

Berman, Edgar. *Hubert: The Triumph and Tragedy of the Humphrey I Knew.* New York: Putnam, 1979.

Beschloss, Michael R. *Kennedy and Roosevelt: The Uneasy Alliance.* New York: Norton, 1980.

Beschloss, Michael R., and Strobe Talbott. *At the Highest Levels: The Inside Story of the End of the Cold War.* Boston: Little, Brown, 1993.

Bork, Robert H. *The Tempting of America.* New York: Free Press, 1990.

Bourne, Peter G. *Jimmy Carter: A Comprehensive Biography from Plains to Postpresidency.* New York: Scribner's, 1997.

Bradlee, Benjamin C. *Conversations with Kennedy.* New York: Norton, 1975.

Bronner, Ethan. *Battle for Justice: How the Bork Nomination Shook America.* New York: Norton, 1989.

Brown, Dr. J. Larry, and H. F. Pizer. *Living Hungry in America.* New York: Mentor, 1989.

Bunker, Ellsworth. *The Bunker Papers: Reports to the President from Vietnam, 1967–1973.* Ed. Douglas Pike. 3 vols. Berkeley, Calif.: Institute of East Asian Studies of the University of California, 1990.

Burns, James MacGregor. *Edward Kennedy and the Camelot Legacy.* New York: Norton, 1976.

Byrd, Robert C. *The Senate 1789–1989.* Washington, D.C.: Government Printing Office, 1994.

Califano, Joseph A. *Governing America.* New York: Simon & Schuster, 1981.

——. *The Triumph and Tragedy of Lyndon Johnson.* New York: Simon & Schuster, 1991.

Bibliography

Carter, Jimmy. *The Presidential Campaign of 1968,* Vol. I, Part 1. Washington, D.C.: U.S. Government Printing Office, 1978.

——. *Keeping Faith: Memoirs of a President.* New York: Bantam, 1982.

Chang, Jaw-Ling Joanne. *United States–China Normalization: An Evaluation of Foreign Policy Decision-Making.* Denver: University of Denver, 1986.

Chellis, Marcia. *Living with the Kennedys: The Joan Kennedy Story.* New York: Simon & Schuster, 1985.

Chester, Lewis, Geoffrey Hodgson, and Bruce Page. *An American Melodrama: The Presidential Campaign of 1968.* New York: Viking, 1969.

Church, F. Forester. *Father and Son.* New York: Harper & Row, 1985.

Clendinen, Dudley, and Adam Nagourney. *Out for Good: The Struggle to Build a Gay Rights Movement in America.* New York: Simon & Schuster, 1999.

Clifford, Clark, with Richard Holbrooke. *Counsel to the President: A Memoir.* New York: Random House, 1991.

Collier, Peter, and David Horowitz. *The Kennedys: An American Drama.* New York: Summit Books, 1984.

Coogan, Tim Pat. *The Troubles: Ireland's Ordeal 1966–1996 and the Search for Peace.* Boulder, Colo.: Roberts Rinehart, 1996.

Cortright, David. *Peace Works: The Citizen's Role in Ending the Cold War.* Boulder, Colo.: Westview, 1993.

Cutler, John Henry. *"Honey Fitz": Three Steps to the White House.* Indianapolis: Bobbs-Merrill, 1962.

Dallas, Rita, with Jeanira Ratcliffe. *The Kennedy Case.* New York: Putnam, 1973.

Dallek, Robert. *Flawed Giant: Lyndon Johnson and His Times, 1961–1973.* New York: Oxford University Press, 1998.

Damore, Leo. *Senatorial Privilege: The Chappaquiddick Cover-Up.* Washington, D.C.: Regnery Gateway, 1988.

David, Lester. *Ted Kennedy, Triumphs and Tragedies.* New York: Grosset & Dunlap, 1971.

——. *Joan: The Reluctant Kennedy.* New York: Funk & Wagnalls, 1974.

Dees, Morris, with Steve Fisher. *A Season for Justice.* New York: Scribner's, 1991.

Dobrynin, Anatoly. *In Confidence: Moscow's Ambassador to America's Six Cold War Presidents.* New York: Random House, 1995.

Dougherty, Richard. *Goodbye, Mr. Christian.* Garden City, N.Y.: Doubleday, 1973.

Douglas, Paul H. *In the Fullness of Time.* New York: Harcourt Brace Jovanovich, 1972.

Doyle, James. *Not Above the Law.* New York: Morrow, 1977.

Exner, Judith. *My Story.* As told to Ovid Demaris. New York: Grove Press, 1977.

Fay, Paul B., Jr. *The Pleasure of His Company.* New York: Harper & Row, 1966.

Fenno, Richard F., Jr. *The Making of a Senator: Dan Quayle.* Washington, D.C.: Congressional Quarterly Press, 1989.

Frank, John P. *Clement Haynsworth, the Senate and the Supreme Court.* Charlottesville: University of Virginia Press, 1991.

Frischauer, Willi. *Jackie.* London: Michael Joseph, 1976.

Germond, Jack W., and Jules Witcover. *Blue Smoke and Mirrors: How Reagan Won and Why Carter Lost the Election of 1980.* New York: Viking, 1981.

Gillon, Steven M. *The Democrats' Dilemma: Walter F. Mondale and the Liberal Legacy.* New York: Columbia University Press, 1992.

Gitenstein, Mark. *Matters of Principle: An Insider's Account of America's Rejection of Robert Bork's Nomination to the Supreme Court.* New York: Simon & Schuster, 1992.

Goldman, Peter, and Tony Fuller. *The Quest for the Presidency 1984.* New York: Bantam, 1987.

Goodwin, Doris Kearns. *The Fitzgeralds and the Kennedys*. New York: Simon & Schuster, 1987.

Goodwin, Richard N. *Remembering America*. Boston: Little, Brown, 1988.

Gormley, Ken. *Archibald Cox: Conscience of a Nation*. Reading, Mass.: Addison-Wesley, 1997.

Greider, William. *Who Will Tell the People: The Betrayal of American Democracy*. New York: Simon & Schuster, 1992.

Guthman, Edwin O., and C. Richard Allen, eds. *RFK: Collected Speeches*. New York: Viking, 1993.

Halberstam, David. *The Unfinished Odyssey of Robert Kennedy*. New York: Random House, 1968.

Harris, Richard. *Decision*. New York: Dutton, 1970.

Hersh, Burton. *The Education of Edward Kennedy*. New York: Morrow, 1972.

Hollister, Robert M., Bernard M. Kramer, and Seymour C. Bellin, eds. *Neighborhood Health Centers*. Lexington, Mass.: D. C. Heath, 1974.

Honan, William H. *Ted Kennedy, Profile of a Survivor*. New York: Quadrangle, 1972.

Humphrey, Hubert H. *The Education of a Public Man: My Life in Politics*. Minneapolis: University of Minnesota Press, 1991.

Huthmacher, J. Joseph. *Senator Robert F. Wagner and the Rise of Urban Liberalism*. New York: Atheneum, 1968.

Isaacson, Walter. *Kissinger: A Biography*. New York: Simon & Schuster, 1992.

Jacobs, John. *A Rage for Justice: The Passion and Politics of Philip Burton*. Berkeley: University of California Press, 1995.

Johnson, Haynes, and David Broder. *The System: The American Way of Politics at the Breaking Point*. Boston: Little, Brown, 1996.

Jordan, Hamilton. *Crisis: The Last Year of the Carter Presidency*. New York: Putnam, 1982.

Kampelman, Max. *Entering New Worlds: The Memoirs of a Private Man in Public Life*. New York: HarperCollins, 1991.

Karnow, Stanley. *Vietnam*. New York: Penguin, 1991.

Kearns, Doris. *Lyndon Johnson and the American Dream*. New York: Harper, 1976.

Kennedy, Edward M. *Decisions for a Decade*. Garden City, N.Y.: Doubleday, 1968.

———. *In Critical Condition: The Crisis in America's Health Care*. New York: Simon & Schuster, 1972.

———, ed. *The Fruitful Bough: A Tribute to Joseph P. Kennedy*. Privately printed, 1965.

———, ed. *Words Jack Loved*. Privately printed, 1977.

Kennedy, John F. *A Nation of Immigrants*. New York: B'nai B'rith, 1959. Reprint, New York: Harper, 1964.

———, ed. *As We Remember Joe*. Privately printed, 1945.

Kennedy, Robert F. *Robert Kennedy in His Own Words: The Unpublished Recollections of the Kennedy Years*. Ed. Edwin O. Guthman and Jeffrey Shulman. New York: Bantam, 1988.

Kennedy, Rose Fitzgerald. *Times to Remember*. Garden City, N.Y.: Doubleday, 1974.

Koskoff, David E. *Joseph P. Kennedy: A Life and Times*. Englewood Cliffs, N.J.: Prentice-Hall, 1974.

Kutler, Stanley I. *The Wars of Watergate: The Last Crisis of Richard Nixon*. New York: Knopf, 1990.

Lange, James E. T., and Katherine DeWitt, Jr. *Chappaquiddick: The Real Story*. New York: St. Martin's, 1992.

Lawford, Patricia Kennedy, ed. *That Shining Hour*. Privately printed, 1969.

Leamer, Laurence. *The Kennedy Women: The Saga of an American Family*. New York: Villard, 1994.

Lerner, Max. *Ted and the Kennedy Legend: A Study in Character and Destiny*. New York: St. Martin's, 1980.

Levin, Murray B. *Kennedy Campaigning*. Boston: Beacon Press, 1966.

Lewis, Finlay. *Mondale: Portrait of an American Politician*. New York: Harper, 1980.
Lincoln, Evelyn. *My Twelve Years with John F. Kennedy*. New York: David McKay, 1965.
Lippman, Theo. *Senator Ted Kennedy: The Career Behind the Image*. New York: Norton, 1976.
Lugar, Richard G. *Letters to the Next President*. New York: Simon & Schuster, 1988.
Lukas, J. Anthony. *Nightmare: The Underside of the Nixon Years*. New York: Viking, 1976.
Lupo, Alan. *Liberty's Chosen Home: The Politics of Violence in Boston*. Boston: Little, Brown, 1977.
MacNeil, Neil. *Dirksen: Portrait of a Public Man*. Cleveland: World, 1970.
Manchester, William. *Death of a President*. New York: Harper & Row, 1967.
Mann, Robert. *Legacy to Power: Senator Russell Long of Louisiana*. New York: Paragon House, 1992.
——. *The Walls of Jericho*. New York: Harcourt Brace, 1996.
Massaro, John L. "Advice and Dissent: Factors in the Senate's Refusal to Confirm Supreme Court Nominees, with Special Emphasis on the Cases of Abe Fortas, Clement F. Haynsworth, Jr., and G. Harrold Carswell." Ph.D. dissertation, Southern Illinois University at Carbondale, 1973.
Matlock, Jack F., Jr. *Autopsy of an Empire: The American Ambassador's Account of the Collapse of the Soviet Union*. New York: Random House, 1995.
Mayer, Jane, and Jill Abramson. *Strange Justice: The Selling of Clarence Thomas*. New York: Houghton Mifflin, 1994.
McGovern, George. *An American Journey: The Presidential Campaign Speeches of George McGovern*. New York: Random House, 1974.
McNamara, Robert S., with Brian VanDeMark. *In Retrospect: The Tragedy and Lessons of Vietnam*. New York: Times Books, 1995.
Mizell, Winston Broderick. "The United States Senate Committee on the Judiciary and Presidential Nominations to the Supreme Court, 1965–1971: A Study of Role and Function of a Legislative Subsystem." Ph.D. dissertation, University of Oklahoma, 1974.
Moore, Jonathan, ed. *The Campaign for President: 1980 in Retrospect*. Cambridge: Ballinger, 1981.
Morris, Dick. *Behind the Oval Office: Winning the Presidency in the Nineties*. New York: Random House, 1997.
Mutch, Robert E. *Campaigns, Congress and Courts: The Making of Federal Campaign Finance Law*. New York: Praeger, 1988.
Newfield, Jack. *Robert Kennedy: A Memoir*. New York: E. P. Dutton, 1969.
O'Brien, Lawrence F. *No Final Victories: A Life in Politics from John F. Kennedy to Watergate*. Garden City, N.Y.: Doubleday, 1974.
O'Clery, Conor. *The Greening of the White House*. Dublin: Gill & Macmillan, 1996.
O'Donnell, Kenneth P., and David F. Powers with Joe McCarthy. *Johnny, We Hardly Knew Ye*. Boston: Little, Brown, 1972.
Olsen, Jack. *The Bridge at Chappaquiddick*. Boston: Little, Brown, 1970.
O'Neill, Thomas P., Jr., with William Novak. *Man of the House*. New York: Random House, 1987.
Palmer, Bruce. *The 25-Year War: America's Military Role in Vietnam*. Lexington: University Press of Kentucky, 1984.
Patterson, James T. *Mr. Republican: A Biography of Robert A. Taft*. Boston: Houghton Mifflin, 1972.
Pertschuk, Michael. *Giant Killers*. New York: Norton, 1986.
Rainie, Harrison, and John Quinn. *Growing Up Kennedy: The Third Wave Comes of Age*. New York: Putnam, 1983.
Reeves, Richard. *President Kennedy: Profile of Power*. New York: Simon & Schuster, 1993.
Remini, Robert V. *Henry Clay: Statesman for the Union*. New York: Norton, 1991.

Rettig, Richard. *Cancer Crusade: The Story of the National Cancer Act of 1971*. Princeton, N.J.: Princeton University Press, 1977.

Riedel, Richard L. *Halls of the Mighty: My 47 Years at the Senate*. Washington, D.C.: Robert B. Luce, 1969.

Rivera, Geraldo. *A Special Kind of Courage: Profiles of Young Americans*. New York: Simon & Schuster, 1976.

Rochon, Thomas R., and David S. Meyer. *Coalitions and Political Movements: The Lessons of the Nuclear Freeze*. Boulder, Colo.: Rienner, 1997.

Safire, William. *Before the Fall: An Inside View of the Pre-Watergate White House*. Garden City, N.Y.: Doubleday, 1975.

——, ed. *Lend Me Your Ears: Great Speeches in History*. New York: Norton, 1992.

Schlesinger, Arthur M., Jr. *A Thousand Days: John F. Kennedy in the White House*. Boston: Houghton Mifflin, 1965.

——. *Robert Kennedy and His Times*. Boston: Houghton Mifflin, 1978.

Shannon, William V. *The Heir Apparent: Robert Kennedy and the Struggle for Power*. New York: Macmillan, 1967.

Sherrill, Robert. *The Last Kennedy*. New York: Dial, 1976.

Shultz, George P. *Turmoil and Triumph: My Years as Secretary of State*. New York: Scribner's, 1993.

Sirkis, Nancy. *Boston*. Introduction by Edward M. Kennedy. New York: Viking, 1965.

Solberg, Carl. *Hubert Humphrey: A Biography*. New York: Norton, 1984.

Solo, Pam. *From Protest to Policy: Beyond the Freeze to Common Security*. Cambridge: Ballinger, 1988.

Sorensen, Theodore C. *The Kennedy Legacy*. New York: Macmillan, 1969.

Stacks, John F. *Watershed*. New York: Times Books, 1981.

Stein, Jean, interviews, and George Plimpton, ed. *American Journey: The Times of Robert Kennedy*. New York: Harcourt Brace Jovanovich, 1970.

Stephanopoulos, George. *All Too Human: A Political Education*. Boston: Little, Brown, 1999.

Stockdale, Jim and Sybil. *In Love and War*. Rev. ed. Annapolis: Naval Institute Press, 1990.

Strober, Gerald S. and Deborah H. *Nixon: An Oral History of His Presidency*. New York: HarperCollins, 1994.

vanden Heuvel, William, and Milton Gwirtzman. *On His Own: Robert F. Kennedy, 1964–1968*. Garden City, N.Y.: Doubleday, 1970.

Washington Post. *Pursuit of the Presidency 1980*. New York: Berkley, 1980.

Watson, Denton L. *Lion in the Lobby: Clarence Mitchell, Jr.'s Struggle for the Passage of Civil Rights Law*. New York: Morrow, 1990.

White, Theodore H. *The Making of the President 1960*. New York: Atheneum, 1961.

——. *The Making of the President 1972*. New York: Atheneum, 1973.

White, William S. *Citadel: The Story of the U.S. Senate*. New York: Harper, 1957.

Wiesner, Jerome, and Abram Chayes, eds. *ABM: An Evaluation of the Decision to Deploy an Antiballistic Missile System*. New York: Harper & Row, 1969.

Witcover, Jules. *85 Days: The Last Campaign of Robert Kennedy*. New York: Putnam, 1969.

Wofford, Harris. *Of Kennedys and Kings*. New York: Farrar, Straus & Giroux, 1980.

Wright, Jim. *Worth It All: My Fight for Peace*. Washington, D.C.: Brassey's, 1993.

Index

Index

Index

Index

Index